M. KRITZ

CAMBRIDGE, '66

ORDINARY DIFFERENTIAL EQUATIONS

This book is in the

ADDISON-WESLEY SERIES IN MATHEMATICS

ORDINARY DIFFERENTIAL EQUATIONS

by

WILFRED KAPLAN

Department of Mathematics
University of Michigan

ADDISON-WESLEY PUBLISHING COMPANY, INC.

READING, MASSACHUSETTS, U.S.A.

LONDON, ENGLAND

Library of Congress Catalog Card No. 58–9173

Third printing, January, 1961

Since the discovery of the calculus, the problem of integration of differential equations has been of major interest for mathematicians. For almost all problems of applied mathematics, and many of pure mathematics, lead to differential equations satisfied by a function to be determined. But the history of mathematics has shown that the formulation of the problem of solution of a differential equation had to go through a long series of stages of development, until it became fully clear to what extent a function was known when one had set up a differential equation which the function was known to satisfy.

LUDWIG SCHLESINGER

PREFACE

This book is intended to serve as an introduction to the theory of ordinary differential equations. A knowledge of calculus is assumed. Enough material is provided for a year's course in the subject.

Throughout it has been the author's aim to treat the subject from the point of view of "functional analysis." A differential equation is equivalent to a specification of certain functions; we try to obtain properties of the functions from the differential equation which determines them. Certain differential equations describe a relationship between two functions; if one (the *input*) is known and the differential equation is *stable*, the other (the *output*) is known approximately, the error being a *transient*. A study of the dependence of output function on input function provides much insight into differential equations.

The branch of engineering known as *systems analysis* or *instrumentation* is concerned to a great extent with the point of view described. It has now become clear that among the differential equations occurring in the physical sciences, those for which all questions are answered by a simple formula are very rare, and that a deeper understanding of the functional relationships is required. The concepts of input and output, which arose in engineering, are an important means to this end. These concepts are beginning to find wider applications: for example, in biology and economics.

Although this book has been written with such applications in mind, and many illustrations thereof are provided throughout, it is to be considered a textbook in mathematics. One of the by-products of the manifold applications of the subject has been a multiplication of terminology, one concept often being described by a different name in each field of application. An important role of mathematics is to counter this tendency by providing a *common language*. It is to be hoped that the study of "pure" mathematics by workers in physical and social sciences will become more widespread, so that the common problems can be recognized in the language of mathematics. For the same reason, it is desirable that pure mathematicians become more acquainted with other sciences and learn to recognize the mathematical ideas they employ.

As a course in mathematics, this book is required to meet standards of rigor. They have been met by careful formulation of definitions and theorems. The proofs of the principal theorems are rather difficult and, accordingly, are placed in the final chapter.

SUMMARY. Chapter 1 presents the important concepts and the main problems. By a study of simple numerical methods, an understanding of

vii

the existence theorem is gained. Chapter 2, devoted to equations of first
order and first degree, gives some special devices for solving problems in
explicit form, but places more emphasis on understanding the processes.
Chapter 3 is a detailed study of the first order linear equation from the
point of view of input and output. Chapter 4 considers linear equations
of arbitrary order, with emphasis on those with constant coefficients; the
methods of variation of parameters, operators, and Laplace transforms
are presented; complex solutions are studied and methods are given for
finding complex roots of an algebraic equation. Chapter 5 extends the
input-output analysis to linear equations of higher order. Chapter 6 is
devoted to simultaneous linear equations and gives methods of solution
and applications; an appendix explains the matrix approach to these
equations, with emphasis on the Jordan normal form. Chapter 7 considers integrals of differential equations of arbitrary order. Chapter 8
discusses equations not of first degree and introduces the concept of
singular solution. Chapter 9 is a thorough treatment of power series
solutions; estimates for errors are given; solutions near singular points
are obtained both for linear and for nonlinear equations. Chapter 10
presents several numerical methods for obtaining solutions. Chapter 11
is devoted to phase plane analysis, of much value for nonlinear equations
of first order and also a source of insight into the problems of higher
order. Singular points, index theorems, stability of equilibrium solutions
and periodic solutions are considered in detail. Chapter 12 supplies the
proofs of the main theorems, especially the general existence theorem;
solutions are considered both locally and in the large.

SUGGESTED COURSE SCHEDULE. The following is a sample outline for
a one-semester course meeting three hours a week. It has been used at the
University of Michigan in a course for engineers: 1–1 to 1–3, 1–4, 1–5 to
1–6, 1–7, 1–8 to 1–11, 2–1 to 2–2, 2–3 to 2–4, 2–5 to 2–6, 2–7 to 2–8, 2–10
to 2–11, 3–1 to 3–3, 3–4 to 3–8, 3–9 to 3–10, 3–11, 3–12, 3–13 to 3–14, 3–15,
4–1 to 4–3, 4–4 to 4–5, 4–7 to 4–8, 4–9, 4–10, 4–11 to 4–12, 4–13 to 4–14,
5–1 to 5–3, 5–4, 5–5, 5–6, 5–7, 5–8 to 5–9, 5–10, 5–11 to 5–13, 6–1, 6–2 to
6–3, 6–5 to 6–6, 9–1, 9–2 to 9–3, 10–1 to 10–3, 10–5 to 10–8, 11–1 to 11–3,
11–4 to 11–6. Two class hours were devoted to demonstrations of solutions by analog computer.

Sections marked with an asterisk can be omitted in a first course.

ACKNOWLEDGMENTS. The author must first acknowledge his great
debt to the late George D. Birkhoff, for his inspiring lectures on differential equations and for his wise counsel as teacher and advisor. It is
impossible to list all the mathematicians and engineers whose advice has
been helpful, but the author wishes to express special appreciation to
Professor Arnold E. Ross of the University of Notre Dame and Professor
Ernst Weber of the Polytechnic Institute of Brooklyn for inviting the

author to participate in symposia on differential equations. Among the
author's colleagues at the University of Michigan the following have been
especially helpful: Professors R. C. F. Bartels, J. W. Carr, C. L. Dolph,
G. E. Hay, R. M. Howe, G. Y. Rainich, L. M. Rauch, R. K. Ritt, E. H.
Rothe, H. Samelson, C. J. Titus. The author is especially grateful to
Mr. Ross Finney, who assisted greatly in reading proofs. He also offers
thanks to Mr. Ralph E. Drischell for his splendid assistance in typing the
manuscript. To his wife the author remains eternally grateful for her
steady encouragement throughout the lengthy task of completing the book.
And to Addison-Wesley the author expresses his appreciation for their
fine cooperation and for the high quality of their work.

WILFRED KAPLAN

September 1958

CONTENTS

CHAPTER 1

BASIC IDEAS

1–1 Significance of differential equations. By an *ordinary differential equation* is meant a relation between x, an unspecified function y of x, and certain of the derivatives y', y'', ..., $y^{(n)}$ of y with respect to x. For example,

$$y' = 1 + y, \tag{1-1}$$

$$y'' + 9y = 8 \sin x, \tag{1-2}$$

$$y'^2 - 2y' + y - 3x = 0 \tag{1-3}$$

are ordinary differential equations.

In the calculus we learn how to find the successive derivatives of a given function y of x. For example, if $y = e^x - 1$, then

$$y' = e^x = y + 1; \tag{1-4}$$

thus for this function the relation (1–1) is satisfied. We say that $y = e^x - 1$ is a *solution* of Eq. (1–1). However, this is not the only solution, for $y = 2e^x - 1$ also satisfies Eq. (1–1):

$$y' = 2e^x = y + 1.$$

One fundamental problem to be studied is that of *determining all solutions of a given differential equation*. This problem can be considered as a generalization of that of solving an equation such as

$$x^2 - x + 2 = 0 \tag{1-5}$$

in algebra. The solutions of Eq. (1–5) are *numbers;* the solutions of a differential equation are *functions*.

There is a second problem, more fundamental than the first. It will be seen that for many differential equations it is very difficult to obtain explicit formulas for all the solutions. However, a general "existence theorem" guarantees that there are solutions; in fact, infinitely many. The problem is *to determine properties of the solutions*, or of some of the solutions, from the differential equation itself. It will be seen that many properties can be found *without explicit formulas for the solutions;* we can,

1

in fact, obtain numerical values for solutions to any accuracy desired. Accordingly, we are led to regard a differential equation itself as a sort of explicit formula describing a certain collection of functions. For example, we can show that all solutions of Eq. (1–2) are given by

$$y = \sin x + c_1 \cos 3x + c_2 \sin 3x, \tag{1–6}$$

where c_1 and c_2 are "arbitrary constants." Equation (1–2) itself, $y'' + 9y = 8 \sin x$, is another way of describing all these functions.

The fact that a single differential equation can describe many functions makes it a remarkably concise form of expression. It is not surprising, therefore, that most laws of physics have been stated in the form of differential equations. An outstanding example is Newton's second law:

$$\text{Force} = \text{mass} \times \text{acceleration.}$$

For a particle of mass m moving on a line, this equation corresponds to the differential equation

$$m \frac{d^2 x}{dt^2} = F\left(t, x, \frac{dx}{dt}\right) \tag{1–7}$$

describing the position x at time t in terms of the force F, which may depend on time t, position x, and velocity dx/dt.

Throughout this book we shall be concerned with both of the problems referred to above. We shall try to obtain explicit expressions for the solutions where feasible; it will be found that the expressions may take a variety of forms, including infinite series. We shall also try to deduce properties of the solutions, such as numerical values, graphs, asymptotes, etc., directly from the differential equations. Experience has shown that, except for a few special types of equations, the problems are difficult. It is fortunate indeed that among the special types are the *linear equations*, for which most questions can be answered satisfactorily, and that these equations are sufficient for the majority of applications in the sciences. Linear equations will receive a large share of attention in this book.

1–2 Basic definitions. An *ordinary differential equation* of *order n* is an equation of form

$$F(x, y, y', \ldots, y^{(n)}) = 0, \tag{1–8}$$

which expresses a relation between x, an unspecified function $y(x)$, and its derivatives through the nth order. Thus

$$xy'' + 2y' + 3y - 6e^x = 0, \qquad (1\text{-}9)$$

$$(y''')^2 - 2y'y''' + (y'')^3 = 0 \qquad (1\text{-}10)$$

are ordinary differential equations of orders 2 and 3, respectively.

Equation (1-10) is a quadratic equation in the highest derivative y'''; we say that the equation is of *degree* 2, whereas Eq. (1-9) is of degree 1. In general, if a differential equation has the form of an algebraic equation of degree k in the highest derivative, then we say that the differential equation is of *degree* k. Most equations to be studied in this book are of *first degree* and are expressible in the form

$$y^{(n)} = G(x, y, \ldots, y^{(n-1)}). \qquad (1\text{-}11)$$

An ordinary differential equation is said to be *linear* if it has the form

$$a_0(x)y^{(n)} + a_1(x)y^{(n-1)} + \cdots + a_{n-1}(x)y' + a_n(x)y = Q(x). \qquad (1\text{-}12)$$

Thus the equation is linear in y and its derivatives. An example is provided by Eq. (1-9) above, in which

$$a_0(x) \equiv x, \qquad a_1(x) \equiv 2, \qquad a_2(x) \equiv 3, \qquad Q(x) \equiv 6e^x. \qquad (1\text{-}13)$$

A linear differential equation is always of first degree, but not conversely; thus

$$y' = 1 + xy^2 \qquad (1\text{-}14)$$

is of first degree but is nonlinear.

The word "ordinary" is used to emphasize that no partial derivatives appear, since there is just one independent variable. An equation such as

$$\frac{\partial^2 z}{\partial x^2} + \frac{\partial^2 z}{\partial y^2} = 0 \qquad (1\text{-}15)$$

is called a *partial differential equation*. Throughout this book, with very few exceptions, only ordinary differential equations are considered and, this being understood, the word "ordinary" will generally be omitted.

1–3 Solutions. By a *solution* (or *particular solution*) of a differential equation

$$F(x, y, y', \ldots, y^{(n)}) = 0 \qquad (1\text{-}16)$$

is meant a function $y = f(x)$, defined in some interval $a < x < b$ (perhaps infinite), having derivatives up to the nth order throughout the interval, and such that Eq. (1–16) becomes an identity when y and its derivatives are replaced by $f(x)$ and its derivatives. For example, $y = e^{2x}$ is a solution of the equation

$$y'' - 4y = 0, \qquad (1\text{–}17)$$

since $y'' = 4e^{2x}$, so that

$$y'' - 4y = 4e^{2x} - 4e^{2x} \equiv 0.$$

For many of the differential equations to be considered, it will be found that all solutions can be included in one formula:

$$y = f(x, c_1, \ldots, c_n), \qquad (1\text{–}18)$$

where $c_1, \ldots, c_n$ are arbitrary constants. Thus for each special assignment of values to the c's, (1–18) gives a solution of Eq. (1–16) and all solutions can be so obtained. (The range of the c's and of x may have to be restricted in some cases to avoid imaginary expressions or other degeneracies.) For example, all solutions of Eq. (1–17) are given by

$$y = c_1 e^{2x} + c_2 e^{-2x}; \qquad (1\text{–}19)$$

the solution $y = e^{2x}$ is obtained when $c_1 = 1$, $c_2 = 0$. When a function (1–18) is obtained, providing *all* solutions, it is called the *general solution*. In general, the number of arbitrary constants will equal the order n, as will be explained in Section 1–8.

The definitions can be illustrated by the simplest type of differential equation of first order:

$$y' = F(x), \qquad (1\text{–}20)$$

where $F(x)$ is defined and continuous for $a < x < b$. All solutions of Eq. (1–20) are obtained by integrating:

$$y = \int F(x)\, dx + C, \qquad a < x < b. \qquad (1\text{–}21)$$

Here the arbitrary constant appears as the familiar one associated with the indefinite integral. This can be generalized to equations of higher order, as the following example shows:

$$y'' = 20x^3. \qquad (1\text{–}22)$$

Since y'' is the derivative of y', we find by integrating twice in succession that

$$y' = 5x^4 + c_1, \qquad y = x^5 + c_1 x + c_2. \qquad (1\text{–}23)$$

Note that there are *two* arbitrary constants, c_1, c_2, and that Eq. (1–22) has order *two*.

PROBLEMS

1. For each of the following differential equations state the order and degree and whether or not the equation is linear:

(a) $y' = x^2 - y$,

(b) $y'' - (y')^2 + xy = 0$,

(c) $(y')^2 + xy' - y^2 = 0$,

(d) $x^3 y'' - xy' + 5y = 2x$,

(e) $y^{(vi)} - y'' = 0$,

(f) $\sin(y'') + e^{y'} = 1$.

2. Find the general solution of each of the following differential equations:

(a) $y' = e^{2x} - x$,

(b) $y'' = 0$,

(c) $y''' = x$,

(d) $y^{(n)} = 0$,

(e) $y^{(n)} = 1$,

(f) $y' = 1/x$.

3. Verify that the following are solutions of the differential equations given:

(a) $y = \sin x$, for $y'' + y = 0$;

(b) $y = e^{2x}$, for $y''' - 4y' = 0$;

(c) $y = c_1 \cos x + c_2 \sin x$ (c_1 and c_2 any constants), for $y'' + y = 0$;

(d) $y = c_1 e^{2x} + c_2 e^{-2x}$, for $y'' - 4y = 0$.

4. Given the differential equation $y' = 2x$,

(a) show that $y = x^2 + c$ is the general solution;

(b) choose c so that the solution passes through the point $(1, 4)$;

(c) choose c so that the solution is tangent to the line $y = 2x + 3$;

(d) choose c so that the solution satisfies the condition $\int_0^1 y \, dx = 2$.

5. Given the differential equation $y'' = x^2 - 1$,

(a) find the general solution;

(b) find a solution $y(x)$ such that $y(0) = 1$, $y'(0) = 2$;

(c) find a solution passing through the points $(1, 2)$ and $(3, 5)$;

(d) find a solution $y(x)$ such that $y(1) = 2$, and $y'(2) = 1$.

6. Find the general solution of the following equations:

(a) $xy' + y = \sin x$ [*Hint*: $xy' + y = (d/dx)(xy)$],

(b) $x^2 y' + 2xy = e^x$,

(c) $2xyy' + y^2 = 3x^2$,

(d) $x^2 y'' + 4xy' + 2y = e^x$.

ANSWERS

1. Orders: (a) 1, (b) 2, (c) 1, (d) 2, (e) 6, (f) 2; degrees: (a) 1, (b) 1, (c) 2, (d) 1, (e) 1, (f) undefined; (a), (d), (e) are linear.

2. (a) $\frac{1}{2}(e^{2x} - x^2) + c$, (b) $c_1 x + c_2$, (c) $\frac{1}{24}x^4 + c_1 x^2 + c_2 x + c_3$, (d) $c_1 x^{n-1} + c_2 x^{n-2} + \cdots + c_{n-1} x + c_n$, (e) $(x^n/n!) + c_1 x^{n-1} + \cdots + c_n$, (f) $\log |x| + c \ (x \neq 0)$.

4. (b) $c = 3$, (c) $c = 4$, (d) $c = \frac{5}{3}$.

5. (a) $y = \frac{1}{12}(x^4 - 6x^2 + 12c_1x + 12c_2)$, (b) $y = \frac{1}{12}(x^4 - 6x^2 + 24x + 12)$, (c) $y = \frac{1}{12}(x^4 - 6x^2 + 2x + 27)$, (d) $y = \frac{1}{12}(x^4 - 6x^2 + 4x + 25)$.

6. (a) $y = (c - \cos x)/x$, $x \neq 0$; (b) $y = (e^x + c)x^{-2}$, $x \neq 0$; (c) $y = \pm[(x^3 + c)/x]^{1/2}$, $x \neq 0$; (d) $y = (e^x + c_1x + c_2)x^{-2}$, $x \neq 0$.

1–4 Geometric interpretation of the first order differential equation. Graphical solution. An equation of first order and first degree can be written in the form

$$y' = F(x, y). \qquad (1\text{--}24)$$

The solutions sought are functions of form $y = f(x)$. Equation (1–24) prescribes the slope of the *tangent* to the solution at each point (x, y). Accordingly, it is possible to construct the tangent to a solution $y = f(x)$ through a specified point, even though the function itself is not known.

We can systematically plot many short tangent line segments through scattered points in the xy-plane, and thereby obtain a "field of line elements." This is shown in Fig. 1–1 for the differential equation

$$y' = -\frac{x}{y}. \qquad (1\text{--}25)$$

For example, at $(1, 1)$ the slope is -1; we draw a short line segment of slope -1 and midpoint $(1, 1)$. At $(3, 2)$ the slope is $-3/2$ and we draw

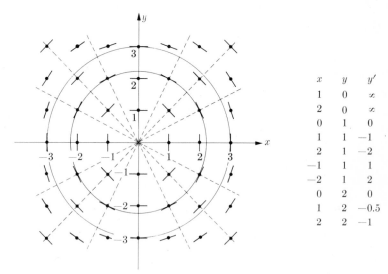

x	y	y'
1	0	∞
2	0	∞
0	1	0
1	1	-1
2	1	-2
-1	1	1
-2	1	2
0	2	0
1	2	-0.5
2	2	-1

FIG. 1–1. Line elements for the equation $y' = -x/y$.

the corresponding segment. The numerical work can be summarized in tabular form, as shown in Fig. 1–1.

If many judiciously spaced line elements are drawn, the figure begins to suggest a family of curves whose tangents are the line elements; in Fig. 1–1 the curves suggested are the circles

$$x^2 + y^2 = c^2, \tag{1-26}$$

for different values of the constant c. For each circle (1–26) the slope of the tangent at the point (x, y) is precisely $-(x/y)$; that is, each circle in (1–26) represents a solution of Eq. (1–25). It is plausible that these are the only solutions; a proof is easily given (Prob. 3 below).

The principles just discussed and illustrated suggest some basic conclusions about the general equation (1–24): *the solutions of Eq. (1–24) form a family of curves in the xy-plane; through each point (x, y) there is precisely one solution curve.* These conclusions can be fully justified, provided reasonable assumptions are made about the continuity of $F(x, y)$. (A full discussion is given in Section 1–8.) That some condition on $F(x, y)$ is needed is suggested by the example (1–25), for which $F(x, y)$ is discontinuous when $y = 0$. Actually, this causes no serious trouble for $x \neq 0$, since the slope is infinite; however, at $(0, 0)$ there is no solution in any sense.

We can apply the graphical procedure of the above example to any given first order equation. We simply plot line elements until the solutions begin to take shape; at the proper stage, the drawing of smooth curves with the line elements as tangents completes the analysis.

The above procedure is open to several criticisms. It gives us solution curves but no formulas. The curves, being graphically obtained, are only approximations to the solutions and we know nothing of their accuracy. The method is also time-consuming. In answer to these criticisms it should be noted that in many cases no explicit formulas (in terms of elementary functions) are available for the solutions. As remarked in Section 1–1, the differential equation itself can be considered as a sort of formula describing the solutions, and there may be no other simple way of describing them. Solutions by the graphical method can be obtained to any desired accuracy, as will be shown in the following section.

The speed with which solutions are obtained depends considerably on the skill of the investigator. It is possible to significantly accelerate the process by means of curves called *isoclines*. These are the loci along which the slope prescribed has a constant value m; that is, for Eq. (1–24) they are the curves

$$F(x, y) = m, \tag{1-27}$$

for different choices of the constant m. Once the curves (1–27) have been sketched, it is usually a simple matter to draw, for each m, a series of parallel line segments of slope m, all having their midpoints on the curve $F(x, y) = m$. For Eq. (1–25) the isoclines are the curves

$$-\frac{x}{y} = m; \qquad (1\text{–}28)$$

they are straight lines through $(0, 0)$ with slope $-1/m$. Thus, in this case, the line elements are perpendicular to the isoclines; the isoclines are shown as broken lines in Fig. 1–1.

It is to be stressed that *the isoclines are not the solutions sought*, although it will occasionally happen that an isocline is a solution curve; in such a case, the isocline must be a straight line of slope m.

EXAMPLES. For the differential equation

$$y' = 1 - \frac{1}{x + y}, \qquad (1\text{–}29)$$

the isoclines are the lines

$$x + y = \frac{1}{1 - m},$$

as shown in Fig. 1–2. The values of m are indicated on the corresponding lines.

For the differential equations

$$y' = \cos x - y, \qquad (1\text{–}30)$$

$$y' = x^2 + y^2, \qquad (1\text{–}31)$$

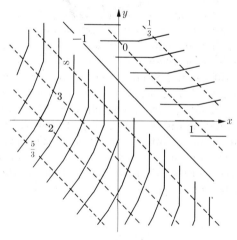

FIG. 1–2. Isoclines for the equation $y' = 1 - 1/(x + y)$.

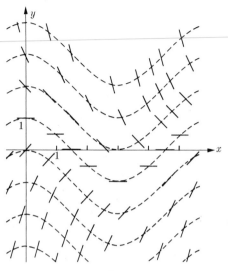

Fig. 1–3. $y' = \cos x - y$.

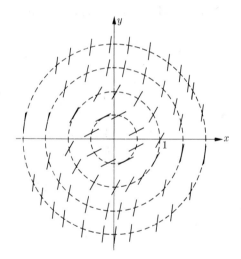

Fig. 1–4. $y' = x^2 + y^2$.

$$y' = \frac{x + y}{x - y},\qquad\qquad (1\text{--}32)$$

$$y' = \frac{(1 - x^2)y - x}{y},\qquad\qquad (1\text{--}33)$$

the solutions and isoclines are shown in Figs. 1–3, 1–4, 1–5, and 1–6. The

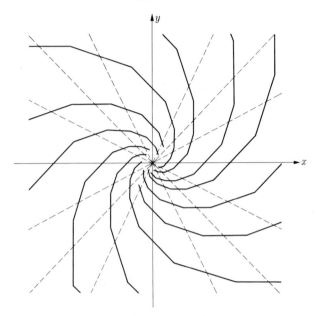

FIG. 1–5. $y' = (x + y)/(x - y)$.

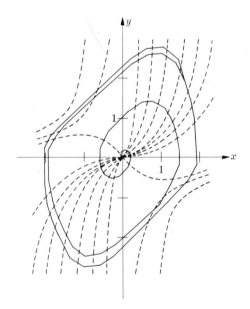

FIG. 1–6. $y' = [(1 - x^2)y - x]/y$.

verification of the graphs is left as an exercise (Prob. 4 below). The figures illustrate the great variety of configurations which can arise.

Equations of higher degree. An equation of first order and *second* degree has the form

$$Py'^2 + Qy' + R = 0, \tag{1–34}$$

where P, Q, and R depend on x and y. We can solve Eq. (1–34) for y' by the quadratic formula

$$y' = \frac{-Q \pm \sqrt{Q^2 - 4PR}}{2P}. \tag{1–35}$$

Thus Eq. (1–34) is really equivalent to *two* first degree equations:

$$y' = \frac{-Q + \sqrt{Q^2 - 4PR}}{2P} = F_1(x, y),$$

$$y' = \frac{-Q - \sqrt{Q^2 - 4PR}}{2P} = F_2(x, y), \tag{1–36}$$

both of which can be solved graphically. The solutions form two families of curves which are, in general, unrelated, although they may fit together smoothly along a borderline curve. If the two families are analyzed on the same graph, we will find two line elements through each point at which $Q^2 - 4PR > 0$, one where $Q^2 - 4PR = 0$, and none where $Q^2 - 4PR < 0$. The locus $Q^2 - 4PR = 0$ may happen to contain a solution curve, which then serves as a borderline curve.

EXAMPLE. For the differential equation

$$y'^2 + xy' - y = 0, \tag{1–37}$$

we find

$$y' = \frac{-x \pm \sqrt{x^2 + 4y}}{2}.$$

Hence $Q^2 - 4PR = x^2 + 4y$ and we have

for $x^2 + 4y > 0$,   2 elements;

for $x^2 + 4y = 0$,   1 element;

for $x^2 + 4y < 0$,   no element.

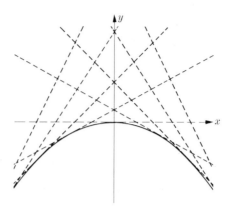

FIG. 1–7. $y'^2 + xy' - y = 0$.

The locus $x^2 + 4y = 0$ serves as a borderline curve common to the two families, for it is a solution curve (Prob. 5 below). The graphical analysis is shown in Fig. 1–7.

PROBLEMS

1. Plot line elements for each of the following differential equations and then plot the exact solution curves:

(a) $y' = 2x$,   (b) $y' = 1/x$,

(c) $y' = \cos x$,   (d) $y' = e^x$.

2. Obtain solutions graphically, with the aid of isoclines, for each of the following differential equations and compare the results with the general solutions given:

(a) $y' = 2y$, general solution $y = ce^{2x}$;

(b) $y' = x + y$, general solution $y = ce^x - x - 1$;

(c) $y' = -y/x$, general solution $xy = c$.

3. Write Eq. (1–25) in the form $(d/dx)(y^2) = -2x$, to show that all solutions are given by (1–26). [For each c, Eq. (1–26) defines two functions, $y = \pm\sqrt{c^2 - x^2}$, $-c < x < c$, each of which is a solution.]

4. Obtain graphical solutions for the following, with the aid of isoclines, and compare with the corresponding figures in the text:

(a) Eq. (1–30), Fig. 1–3,   (b) Eq. (1–31), Fig. 1–4,

(c) Eq. (1–32), Fig. 1–5,   (d) Eq. (1–33), Fig. 1–6.

5. (a) Obtain solutions of Eq. (1–37) graphically and compare with Fig. 1–7.

(b) Show that $y = -\frac{1}{4}x^2$ is a solution of Eq. (1–37).

6. For a particle moving on the x-axis with velocity $v = dx/dt$, the acceleration is

$$\frac{d^2x}{dt^2} = \frac{dv}{dt} = \frac{dv}{dx}\frac{dx}{dt} = v\frac{dv}{dx}.$$

Hence if the particle is subject to a force F which depends only on x and v, Newton's law, Eq. (1–7), gives

$$mv\frac{dv}{dx} = F(x, v).$$

This is a *first* order equation relating x and v. Its solutions can be studied graphically in the xv-plane, often called the *phase plane* (see Chapter 11). Reduce the following equations to the x, v form and analyze the solutions graphically in the xv-plane:

(a) $d^2x/dt^2 = 32$ (falling body),

(b) $d^2x/dt^2 = -x$ (undamped oscillations of a spring),

(c) $d^2x/dt^2 = -x - v$ (damped oscillations of a spring).

7. Show that the van der Pol equation

$$\frac{d^2x}{dt^2} + (x^2 - 1)\frac{dx}{dt} + x = 0,$$

in x, v form (Prob. 6), is the same as Eq. (1–33) (Fig. 1–6), with y replaced by v. This differential equation occurs in the theory of electric circuits which contain vacuum tubes. The solution by isoclines still remains one of the best methods for obtaining the solutions, for which no simple explicit formulas are known. [See Section 11–14 and pp. 250–252 of Reference, 1 at the end of this chapter.]

ANSWERS

6. (a) $v(dv/dx) = 32$, curves are parabolas $v^2 = 64x + c$; (b) $v(dv/dx) = -x$, curves are circles $x^2 + v^2 = c$; (c) $v(dv/dx) = -x - v$, curves are spirals winding about (0, 0).

1–5 Method of step-by-step integration. Given the first order differential equation

$$y' = F(x, y), \tag{1–38}$$

we may require one particular solution through a given point (x_0, y_0), rather than the family of all solutions, as in the preceding section. Finding this solution is the *initial value problem*. (The term "initial value" arose in mechanics, where it referred to specification of the values of positions and velocities at an initial time t_0.)

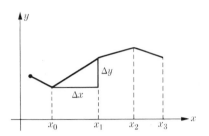

Fig. 1–8. Solution obtained by step-by-step integration.

We can obtain such a solution graphically with the aid of line elements, but we need not draw a whole field of such elements. We can simply draw a short segment through (x_0, y_0), with slope $F(x_0, y_0)$, and follow this to a nearby point (x_1, y_1), with $x_1 > x_0$. At this point we evaluate the slope $F(x_1, y_1)$ and draw a line with that slope from (x_1, y_1) to a nearby point (x_2, y_2), with $x_2 > x_1$. By repeating the process many times, we obtain a broken line which is an approximation to the solution sought; see Fig. 1–8. It is clear that the shorter the segments used, the more accurate the solution will be. By proceeding to the left from x_0, rather than to the right, we can extend the graphical solution to values of x less than x_0.

The procedure described is known as *step-by-step integration*. Actually, we need not use the graph at all, for the steps are purely numerical. The work can be recorded in tabular form in four columns, listing the values of x, y, $F(x, y)$, and Δy, as in Table 1–1. The increment Δx is chosen at discretion, while Δy is computed by the formula

$$\Delta y = F(x, y)\,\Delta x. \tag{1–39}$$

Thus Δy is really being computed by the use of the approximation $dy = \Delta y$. The increment Δx can be varied from step to step, although it is simpler to keep it constant, as is done in Table 1–1.

TABLE 1–1

x	y	$F(x, y)$	Δy
x_0	y_0	$F(x_0, y_0)$	$F(x_0, y_0)\,\Delta x$
$x_0 + \Delta x$	$y_0 + \Delta y$	$F(x_0 + \Delta x, y_0 + \Delta y)$	...
$x_0 + 2\,\Delta x$	...	...	...

The example

$$y' = x^2 - y^2, \tag{1–40}$$

with $x_0 = 1$, $y_0 = 1$, $\Delta x = 0.1$, is worked out in Table 1–2. The third and fourth columns are introduced to simplify the work.

TABLE 1–2

x	y	x^2	y^2	$x^2 - y^2$	$\Delta y = (x^2 - y^2)\,\Delta x$
1	1	1	1	0	0
1.1	1	1.21	1	0.21	0.021
1.2	1.021	1.44	1.042	0.398	0.040
1.3	1.061	...	...	...	...

1–6 The equation $y' = F(x)$. Let $F(x)$ be continuous for $a < x < b$. Then the general solution of the differential equation

$$y' = F(x) \tag{1–41}$$

is obtained by integration:

$$y = \int F(x)\,dx + C, \qquad a < x < b, \tag{1–42}$$

where $\int F(x)\,dx$ is an indefinite integral of $F(x)$. This can also be written as follows:

$$y = \int_{x_0}^{x} F(u)\,du + C, \qquad a < x < b, \tag{1–43}$$

where $a < x_0 < b$. Thus the indefinite integral has been **replaced by a** definite integral from x_0 to x, with u as a "dummy variable" of integration. It is a standard theorem of calculus that

$$\frac{d}{dx} \int_{x_0}^{x} F(u)\,du = F(x), \tag{1–44}$$

so that the integral on the right of Eq. (1–43) is indeed an indefinite integral of $F(x)$ (Prob. 5 below).

We now seek the solution of Eq. (1–41) such that $y = y_0$ when $x = x_0$. This forces us to choose C appropriately in Eq. (1–43):

$$y_0 = \int_{x_0}^{x_0} F(u)\,du + C = 0 + C.$$

Hence $y_0 = C$ and the solution sought is

$$y = \int_{x_0}^{x} F(u)\,du + y_0. \tag{1–45}$$

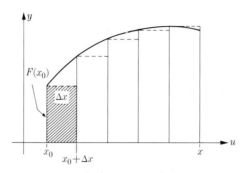

FIG. 1–9.   Integration and step-by-step integration.

Explicit evaluation of the integral would provide a satisfactory solution of the problem; however, it often happens that explicit evaluation in terms of elementary functions is not possible.   Equation (1–45) then still provides a significant solution and means for computing the value of y for each x.   For when x is fixed the definite integral can be evaluated approximately by any one of the standard procedures: the trapezoidal rule, Simpson's rule, etc. In particular, it can be evaluated by "rectangular sums" as follows. Fix x and divide the interval from x_0 to x into n equal parts Δx, so that $x = x_0 + n\,\Delta x$. Then, approximately,

$$y = y_0 + \int_{x_0}^{x} F(u)\,du = y_0 + F(x_0)\,\Delta x + F(x_0 + \Delta x)\,\Delta x + \cdots$$

$$+ F[x_0 + (n-1)\,\Delta x]\,\Delta x, \qquad (1\text{–}46)$$

as suggested in Fig. 1–9.

Equation (1–46) provides the same value of y at x as that given by step-by-step integration. The successive values of y for $x = x_0,\, x_0 + \Delta x,\, \ldots,\, x_0 + n\,\Delta x$ are

$$y_0,$$

$$y_0 + \Delta y = y_0 + F(x_0)\,\Delta x;$$

$$y_0 + F(x_0)\,\Delta x + F(x_0 + \Delta x)\,\Delta x,$$

$$\vdots$$

$$y_0 + F(x_0)\,\Delta x + \cdots + F[x_0 + (n-1)\,\Delta x]\,\Delta x.$$

In other words, step-by-step integration is in this case simply one method for approximate evaluation of a definite integral.

This example suggests the reason for the word "integration" in connection with the step-by-step procedure. In general, solution of a differential equation should be thought of as a kind of generalized process of integration. Certain mechanical and electrical devices exist for solving differential equations which are composed of "integrators"; these devices carry out the step-by-step process continuously; that is, effectively they proceed to the limit corresponding to $\Delta x \to 0$.

PROBLEMS

1. Using step-by-step integration, with $\Delta x = 0.1$, find the value of y for $x = 1.5$ on the solution of $y' = x - y^2$ such that $y = 1$ when $x = 1$. Plot the solution as a broken line.

2. (a) Using step-by-step integration, with $\Delta x = 0.1$, find the value of y at $x = 0.5$ on the solution of $y' = \sqrt{1 - y^2}$ such that $y = 0$ for $x = 0$. Compare the result with the exact solution, which is $y = \sin x$.
   (b) Proceed as in part (a), with $\Delta x = 0.05$, and compare again with the exact solution.

3. Show that the value of y when $x = 1$ on the solution of $y' = xe^x$ through $(0, 1)$ is given by $1 + \int_0^1 xe^x \, dx$. Obtain the value numerically by step-by-step integration with $\Delta x = 0.2$. Compare with the exact value. Also show how the steps correspond to rectangles whose total area approximates the integral.

4. Find the value of y for $x = 0.6$ on the solution of $y' = e^{-x^2}$ through $(0, 2)$
   (a) by step-by-step integration with $\Delta x = 0.2$;
   (b) by the trapezoidal rule, using subdivision points at $x = 0, 0.2, 0.4, 0.6$;
   (c) by Simpson's rule, using subdivision points at $x = 0, 0.1, 0.2, 0.3, 0.4, 0.5, 0.6$;
   (d) by employing tables of the error function.

5. Carry out the indicated steps to prove the rule (1–44). Let $G(x) = \int_{x_0}^x F(u) \, du$.
   (a) Show that $G(x + \Delta x) - G(x) = \int_x^{x+\Delta x} F(u) \, du$.
   (b) Apply the law of the mean to show that

$$G(x + \Delta x) - G(x) = F(x_1) \int_x^{x+\Delta x} du = F(x_1) \, \Delta x,$$

where x_1 is between x and $x + \Delta x$.

   (c) Conclude that $G'(x) = \lim_{\Delta x \to 0} \dfrac{G(x + \Delta x) - G(x)}{\Delta x} = F(x)$.

ANSWERS

1. 1.082. 2. (a) 0.4850, exact value 0.47943; (b) 0.4822. 3. Step-by-step: 1.7429, exact value 2. 4. (a) 2.56259, (b) 2.53235, (c) 2.53516, (d) 2.53515.

1–7 Higher order differential equations and systems of differential equations. Step-by-step integration. The second order differential equation

$$y'' = F(x, y, y') \tag{1–47}$$

can be written as a *system* of differential equations:

$$\frac{dy}{dx} = z, \qquad \frac{dz}{dx} = F(x, y, z). \tag{1–48}$$

The introduction of the variable z simply expresses the fact that the second derivative of y is the derivative of the first derivative. Therefore, for every solution $y = f(x)$ of (1–47), the functions $y = f(x)$ and $z = f'(x)$ satisfy the system (1–48). Conversely, if $y = f(x)$ and $z = g(x)$ satisfy (1–48) then, necessarily,

$$f'(x) = g(x), \qquad g'(x) = F[x, f(x), g(x)];$$

that is, $f''(x) = F[x, f(x), f'(x)]$, and $y = f(x)$ satisfies Eq. (1–47).

In general, a pair of equations

$$\frac{dy}{dx} = G(x, y, z), \qquad \frac{dz}{dx} = F(x, y, z) \tag{1–49}$$

is called a *system of two first order differential equations*. A solution of such a system is a pair of functions $y = f(x)$, $z = g(x)$, defined in the same interval of x, which satisfy (1–49) identically. For example,

$$y = e^{-x} + e^{3x}, \qquad z = e^{-x} - e^{3x}$$

is a solution of the system

$$\frac{dy}{dx} = y - 2z, \qquad \frac{dz}{dx} = z - 2y,$$

since

$$-e^{-x} + 3e^{3x} \equiv e^{-x} + e^{3x} - 2(e^{-x} - e^{3x}),$$

$$-e^{-x} - 3e^{3x} \equiv e^{-x} - e^{3x} - 2(e^{-x} + e^{3x}).$$

Approximate solutions of the general system (1–49) can be obtained by step-by-step integration. We simply replace Eqs. (1–49) by the system of "difference equations"

$$\Delta y = G(x, y, z)\, \Delta x, \qquad \Delta z = F(x, y, z)\, \Delta x. \tag{1–50}$$

Given initial values x_0, y_0, z_0 of all three variables, we choose Δx, compute

$$\Delta y = G(x_0, y_0, z_0)\, \Delta x,$$
$$\Delta z = F(x_0, y_0, z_0)\, \Delta x, \qquad (1\text{--}51)$$

and obtain the next set of values:

$$x_1 = x_0 + \Delta x, \qquad y_1 = y_0 + \Delta y,$$
$$z_1 = z_0 + \Delta z. \qquad (1\text{--}52)$$

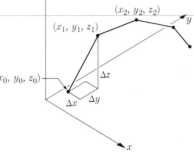

Fig. 1–10. Step-by-step integration for system.

Repetition of the process with x_1, y_1, z_1 as new initial values yields a new set x_2, y_2, z_2, ..., and so on. We can graph y as a function of x, as in Fig. 1–8, obtaining a broken line; similarly, we can graph z as a function of x. Also, the functions $y(x)$ and $z(x)$ can be represented together as one broken line in xyz-space, joining the successive points (x_0, y_0, z_0), (x_1, y_1, z_1), (x_2, y_2, z_2), ... as in Fig. 1–10. The numerical work can again be recorded in tabular form, as in Table 1–3.

TABLE 1–3

x	y	z	$G(x, y, z)$	Δy	$F(x, y, z)$	Δz
x_0	y_0	z_0	$G(x_0, y_0, z_0)$	$G(x_0, y_0, z_0)\, \Delta x$	$F(x_0, y_0, z_0)$	$F(x_0, y_0, z_0)\, \Delta x$
$x_0 + \Delta x$	$y_0 + \Delta y$	$z_0 + \Delta z$	...	...	...	...

The second order equation (1–47) can be replaced by the equivalent system (1–48) and particular approximate solutions can be obtained by step-by-step integration. For example, for the equation

$$y'' = xy' - y,$$

with initial values $x = 0$, $y = 1$, $y' = 0$, we write

$$\frac{dy}{dx} = z, \quad \frac{dz}{dx} = xz - y; \quad \Delta y = z\, \Delta x, \quad \Delta z = (xz - y)\, \Delta x.$$

With $\Delta x = 0.1$, the solution is then computed as in Table 1–4.

Unfortunately, there is no useful generalization of the graphical procedure which uses isoclines (Section 1–4). We can consider the family of

TABLE 1–4

x	y	z	$z' = xz - y$	$\Delta y = z\,\Delta x$	$\Delta z = z'\,\Delta x$
0	1	0	−1	0	−0.1
0.1	1	−0.1	−1.01	−0.01	−0.101
0.2	0.99	−0.201	. . .	. . .	. . .

solutions of (1–49) as a collection of curves $y = f(x)$, $z = g(x)$ in xyz-space. Equations (1–49) specify the direction of the tangent line at (x, y, z). Hence we have again a field of line elements. However, such a collection of line elements in space is exceedingly difficult to visualize even with graphical aids.

All the above remarks can be extended to the case of a *system of n first order differential equations:*

$$\frac{dy_1}{dx} = F_1(x, y_1, \ldots, y_n),$$

$$\frac{dy_2}{dx} = F_2(x, y_1, \ldots, y_n),$$

$$\vdots \tag{1–53}$$

$$\frac{dy_n}{dx} = F_n(x, y_1, \ldots, y_n).$$

Each solution of (1–53) is a set of n functions

$$y_1 = f_1(x), \qquad y_2 = f_2(x), \qquad \ldots, \qquad y_n = f_n(x) \tag{1–54}$$

which satisfy (1–53) identically. Approximate solutions are obtained by step-by-step integration:

$$\Delta y_1 = F_1(x, y_1, y_2, \ldots)\,\Delta x,$$

$$\Delta y_2 = \ldots, \tag{1–55}$$

$$\vdots$$

$$\Delta y_n = F_n(x, y_1, y_2, \ldots)\,\Delta x.$$

The nth order equation

$$\frac{d^n y}{dx^n} = F(x, y, y', \ldots, y^{(n-1)}) \tag{1–56}$$

can be replaced by such a system:

$$\frac{dy_1}{dx} = y_2, \qquad \frac{dy_2}{dx} = y_3, \qquad \ldots, \qquad \frac{dy_{n-1}}{dx} = y_n,$$

$$\frac{dy_n}{dx} = F(x, y_1, y_2, \ldots, y_n), \tag{1–57}$$

where $y_1 = y$, $y_2 = y'$, $y_3 = y''$, ..., $y_n = y^{(n-1)}$.

The equations (1–53) are, in a sense, the general form. We have just seen that a single nth order equation (1–56) is equivalent to such a system. The same is true of more complicated systems of differential equations. For example, the system

$$\frac{d^2y}{dx^2} = F(x, y, y', z, z'), \qquad \frac{d^2z}{dx^2} = G(x, y, y', z, z') \tag{1–58}$$

is equivalent to a system of four first order equations:

$$\frac{dy_1}{dx} = y_2, \qquad \frac{dy_2}{dx} = F(x, y_1, y_2, y_3, y_4),$$

$$\frac{dy_3}{dx} = y_4, \qquad \frac{dy_4}{dx} = G(x, y_1, y_2, y_3, y_4). \tag{1–59}$$

Here $y_1 = y$, $y_2 = y'$, $y_3 = z$, $y_4 = z'$.

An equation

$$F(x, y, y', \ldots, y^{(n)}) = 0,$$

not of first degree, can in general be solved for the highest derivative $y^{(n)}$; the equation is then replaced by one or more equations of form (1–56), each of which can be replaced by a system of first order equations. Similar remarks apply to simultaneous equations in several unknown functions.

The order of a system (1–53) is defined to be n; for more general systems the order can be defined to be the order of an equivalent system of form (1–53).

PROBLEMS

1. Replace each of the following differential equations by an equivalent system of first order equations:

(a) $y'' - xy' - x^2y = 0$,   (b) $y'' = xy'^2 - y'$,

(c) $y''' - 3y'' + 5y' - 6y = \sin x$,   (d) $d^5y/dx^5 = x$.

2. Determine by step-by-step integration the solution of each of the following equations with given initial values, choice of Δx, and number of steps. Graph each solution in the xy-plane.

(a) $y'' = yy' + x; y = 1, y' = 0$ for $x = 0; \Delta x = 0.5;$ 4 steps.

(b) $y'' = -y; y = 0, y' = 1$ for $x = 0; \Delta x = 0.1;$ 5 steps.

(c) $y''' - y'' - y' + y = x; y = 0, y' = 0, y'' = 1$ for $x = 0; \Delta x = 0.1;$ 5 steps.

(d) $y^{(iv)} + 2y'' + y = x^2; y = -1, y' = 0, y'' = 0, y''' = 0$ for $x = 0; \Delta x = 1;$ 5 steps.

3. Given the system

$$\frac{dx}{dt} = xy - z, \qquad \frac{dy}{dt} = x + z, \qquad \frac{dz}{dt} = xz,$$

find, by step-by-step integration, the solution such that $x = 1, y = 1, z = 1$ for $t = 0$. Use $\Delta t = 1$ and compute the solution up to $t = 4$. Graph the result in xyz-space, with t treated as a parameter.

4. Replace each of the following systems by a system of first order differential equations:

(a) $\dfrac{d^2 y}{dx^2} + y = z, \qquad \dfrac{d^2 z}{dx^2} + 4z = 2y;$

(b) $\dfrac{d^3 x}{dt^3} - \dfrac{dx}{dt} + \dfrac{dy}{dt} = t^2, \qquad \dfrac{d^2 x}{dt^2} + \dfrac{dx}{dt} - \dfrac{d^2 y}{dt^2} = e^t;$

(c) $2\dfrac{dx}{dt} - 3\dfrac{dy}{dt} + x - y = \sin t, \qquad 3\dfrac{dx}{dt} + 2\dfrac{dy}{dt} - x = \cos t.$

[*Hint:* For (c) solve for dx/dt and dy/dt.]

ANSWERS

1. (a) $y' = z, z' = xz + x^2 y;$ (b) $y' = z, z' = xz^2 - z;$ (c) $y' = z, z' = w, w' = 3w - 5z + 6y + \sin x;$ (d) $y' = z, z' = w, w' = u, u' = v, v' = x.$

2. (a) for $x = 0, 0.5, \ldots, y = 1, 1, 1, 1.125, 1.563;$ (b) $y = 0, 0.1, 0.2, 0.299, 0.396, 0.490;$ (c) $y = 0, 0, 0.01, 0.031, 0.0643, 0.1115;$ (d) $y = -1, -1, -1, -1, 0, 5.$

3. $x = 1, 1, 2, 10, 118; y = 1, 3, 6, 12, 34; z = 1, 2, 4, 12, 132.$

4. (a) $y' = w, w' = z - y, z' = u, u' = 2y - 4z;$ (b) $x' = z, z' = u, y' = v, u' = z - v + t^2, v' = u + z - e^t;$ (c) $13(dx/dt) = x + 2y + 2\sin t + 3\cos t, 13(dy/dt) = 5x - 3y + 2\cos t - 3\sin t.$

1–8 Existence theorem. Initial value problem and boundary value problem. The analysis in the preceding sections has strongly indicated that for the differential equation

$$\frac{dy}{dx} = F(x, y) \tag{1-60}$$

there is a unique solution $y = f(x)$ which passes through a *given initial point* (x_0, y_0). Under appropriate assumptions concerning the function $F(x, y)$, existence of such a solution can indeed be guaranteed. For the equation of nth order,

$$y^{(n)} = F(x, y, y', \ldots, y^{(n-1)}), \tag{1-61}$$

we expect that there will be a unique solution satisfying *initial conditions:*

when $x = x_0$,   then $y = y_0$, $y' = y_0'$, $\ldots$,

and $y^{(n-1)} = y_0^{(n-1)}$,   (1-62)

where $x_0, y_0, \ldots, y_0^{(n-1)}$ are given numbers. The following fundamental theorem justifies these expectations.

EXISTENCE THEOREM. *Let* $F(x, y, y', \ldots, y^{(n-1)})$ *be a function of the variables* $x, y, y', \ldots, y^{(n-1)}$, *defined and continuous when*

$$|x - x_0| < h, \qquad |y - y_0| < h, \qquad \ldots, \qquad |y^{(n-1)} - y_0^{(n-1)}| < h,$$

and having continuous first partial derivatives with respect to $y, y', \ldots,$ $y^{(n-1)}$. *Then there exists a solution* $y = f(x)$ *of the differential equation* (1-61), *defined in some interval* $|x - x_0| < h_1$, *and satisfying the initial conditions* (1-62). *Furthermore, the solution is unique; that is, if* $y = g(x)$ *is a second solution satisfying* (1-62), *then* $f(x) \equiv g(x)$ *wherever both functions are defined.*

A proof of this theorem is given in Chapter 12. One method of proof is to show that, as the increment Δx tends to zero, the broken-line solution obtained by step-by-step integration approaches as limit precisely the solution $y = f(x)$ sought. A second method, described in Chapter 12, is based on the Picard procedure of "successive approximation."

EXAMPLE. For the differential equation

$$y'' = \sin x, \tag{1-63}$$

the general solution is found by integrating twice:

$$y' = -\cos x + c_1, \qquad y = -\sin x + c_1 x + c_2. \tag{1-64}$$

If the given initial conditions are

$$\text{for } x = 0: \quad y = y_0, \quad y' = y_0', \tag{1-65}$$

then we obtain equations for c_1 and c_2:

$$y_0 = c_2, \quad y_0' = -1 + c_1. \tag{1-66}$$

Accordingly, $c_1 = y_0' + 1$, $c_2 = y_0$, and the solution sought is

$$y = -\sin x + (y_0' + 1)x + y_0. \tag{1-67}$$

This example shows why it is natural that the general solution (when obtainable) of an nth order equation should depend on n arbitrary constants. For the initial conditions (1–62) lead to n equations in the arbitrary constants, and n equations in n unknowns have "in general" one solution.

With this in mind, we can consider ways of determining the values of the arbitrary constants other than by initial conditions (1–62). The most common alternative is to require that certain *boundary conditions* be satisfied at the ends of an interval $a \leq x \leq b$ in which the solution is sought. For example, for Eq. (1–63) we could impose the conditions

$$y = y_1 \text{ for } x = 0, \quad y = y_2 \text{ for } x = \pi. \tag{1-68}$$

From the second part of (1–64), we then obtain the equations

$$y_1 = c_2, \quad y_2 = c_1 \pi + c_2; \tag{1-69}$$

accordingly, $c_1 = (y_2 - y_1)/\pi$, $c_2 = y_1$, and the solution sought is

$$y = -\sin x + \frac{y_2 - y_1}{\pi} x + y_1. \tag{1-70}$$

This particular example gave no difficulty, but it is not true that we can always find a solution which satisfies such boundary conditions or, if one does exist, that there is only one (see Prob. 4 in Section 1–11). There are existence theorems for such boundary value problems, but they involve complicated hypotheses and we attempt no formulation here.

An existence theorem analogous to the one stated above holds for systems of n first order equations:

$$\frac{dy}{dx} = F(x, y, z, \ldots), \quad \frac{dz}{dx} = G(x, y, z, \ldots), \quad \ldots \tag{1-71}$$

in the n unknown functions $y, z, \ldots$ of x. For simplicity we state the theorem for the case of three unknowns, y, z, w:

$$\frac{dy}{dx} = F(x, y, z, w), \qquad \frac{dz}{dx} = G(x, y, z, w), \qquad \frac{dw}{dx} = H(x, y, z, w). \quad (1\text{–}72)$$

If the functions F, G, H are continuous and have continuous first partial derivatives with respect to y, z, w for $|x - x_0| < h$, $|y - y_0| < h$, $|z - z_0| < h$, $|w - w_0| < h$, then there is a unique solution

$$y = f(x), \qquad z = g(x), \qquad w = p(x), \qquad |x - x_0| < h_1, \quad (1\text{–}73)$$

which passes through the initial point (x_0, y_0, z_0, w_0). A proof is given in Chapter 12.

This theorem includes the previous one as a special case, since Eq. (1–61) can always be replaced by an equivalent system (1–71). For the same reason, it also includes the case of a system such as the following:

$$\frac{d^2 x}{dt^2} = F\left(t, x, y, \frac{dx}{dt}, \frac{dy}{dt}\right), \qquad \frac{d^2 y}{dt^2} = G\left(t, x, y, \frac{dx}{dt}, \frac{dy}{dt}\right). \quad (1\text{–}74)$$

Newton's law (force = mass $\times$ acceleration), when applied to systems of particles, leads to systems of second order equations such as (1–74). The existence theorem ensures that there is a unique solution for (1–74) which satisfies given initial conditions:

$$\text{for } t = t_0: \qquad x = x_0, \qquad y = y_0, \qquad \frac{dx}{dt} = \left(\frac{dx}{dt}\right)_0, \qquad \frac{dy}{dt} = \left(\frac{dy}{dt}\right)_0;$$
$$(1\text{–}75)$$

that is, in the mechanical case the motion of the system is completely determined by the positions and velocities of all the particles at one instant of time. Similar laws apply to electric circuits.

1–9 Verification of a general solution. One important application of the existence theorem is to establish that a formula gives the general solution (*all solutions*) of a differential equation.

EXAMPLE. For the differential equation

$$y'' - 4y = 0, \qquad (1\text{–}76)$$

the function

$$y = c_1 e^{2x} + c_2 e^{-2x}, \qquad (1\text{–}77)$$

which depends on two arbitrary constants c_1, c_2, gives the general solution.

For

$$y'' = 4c_1 e^{2x} + 4c_2 e^{-2x} = 4y,$$

so that, no matter what values are assigned to c_1, c_2, (1–77) defines a solution of (1–76). Furthermore, for each set of initial conditions:

$$\text{for } x = x_0: \quad y = y_0, \quad y' = y_0', \tag{1–78}$$

there is a choice of c_1, c_2 such that (1–77) satisfies these conditions:

$$y_0 = c_1 e^{2x_0} + c_2 e^{-2x_0}, \qquad y_0' = 2c_1 e^{2x_0} - 2c_2 e^{-2x_0},$$

so that

$$c_1 = \tfrac{1}{4} e^{-2x_0}(2y_0 + y_0'), \qquad c_2 = \tfrac{1}{4} e^{2x_0}(2y_0 - y_0').$$

Hence (1–77) gives *all* solutions and must be the general solution of (1–76).

In general, for an equation of second order

$$y'' = F(x, y, y'),$$

verification of a general solution, $y = f(x, c_1, c_2)$, consists of two steps: first, substitution of $y = f(x, c_1, c_2)$ in the differential equation and determining that, for all allowed choices of c_1, c_2, the equation is an identity in x; second, showing that the equations

$$y_0 = f(x_0, c_1, c_2), \qquad y_0' = f_x(x_0, c_1, c_2)$$

can be solved for c_1, c_2 for every choice of x_0, y_0, y_0' for which F has continuous first partial derivatives. Similar remarks apply to equations of higher order and to systems of equations.

It should be stressed that the existence theorem is applicable only when the continuity conditions are satisfied. For the equation

$$y' = F(x, y) \equiv 3y^{2/3}, \tag{1–79}$$

the function $F(x, y)$ has a partial derivative $F_y = 2y^{-1/3}$ which is discontinuous when $y = 0$. The points for which $y = 0$ must be considered as *singular points*, and solutions passing through these points (if there are any) require special investigation. Methods given in Chapter 2 lead us to the expression

$$y = (x - c)^3 \tag{1–80}$$

for the general solution of (1–79). We can verify at once that Eq. (1–80) does provide one and only one solution through every point (x_0, y_0) for which $y_0 \neq 0$. In fact, Eq. (1–80) also provides a unique solution through each singular point $(x_0, 0)$. There is, however, more than one solution through each singular point, for the function $y \equiv 0$ is a solution of Eq. (1–79); it passes through all the singular points, and is not included in Eq. (1–80).

Other examples can be given in which there are no solutions through singular points or in which there is just one solution through each singular point. A full discussion is given in Chapter 8.

1–10 Finding differential equation from primitive. At times a function $f(x, c_1, c_2, \ldots)$ which depends on x and several arbitrary constants is given, and we are asked to determine a differential equation of which the function f is the general solution. We call f a *primitive* of the differential equation.

For example, let the primitive be

$$y = c_1 x^2 + c_2 x^3. \tag{1–81}$$

To find the differential equation, we consider the equations for y, y', y'':

$$y = c_1 x^2 + c_2 x^3,$$

$$y' = 2c_1 x + 3c_2 x^2, \tag{1–82}$$

$$y'' = 2c_1 + 6c_2 x.$$

Here we can eliminate c_1, c_2 by solving the first two equations for c_1, c_2 and substituting in the third equation:

$$c_1 = \frac{3y - xy'}{x^2}, \qquad c_2 = \frac{xy' - 2y}{x^3},$$

$$y'' = 2c_1 + 6c_2 x = \frac{4xy' - 6y}{x^2}.$$

This last equation provides the differential equation sought, of form $y'' = F(x, y, y')$; note that F is discontinuous when $x = 0$ (locus of singular points). To complete the solution of the problem, we should then verify that $y = c_1 x^2 + c_2 x^3$ is the general solution of

$$y'' = \frac{4xy' - 6y}{x^2} \qquad (x \neq 0). \tag{1–83}$$

Because of the special form of equations (1–82), c_1 and c_2 can also be eliminated by employing a theorem of algebra: $n + 1$ linear equations in n unknowns are consistent only if the determinant formed of the coefficients of the unknowns and the "constant" terms is 0. In (1–82), c_1, c_2 are regarded as unknowns and y, y', y'' as "constants." Hence

$$\begin{vmatrix} y & x^2 & x^3 \\ y' & 2x & 3x^2 \\ y'' & 2 & 6x \end{vmatrix} = 0,$$

$$x^4 y'' - 4x^3 y' + 6x^2 y = 0. \tag{1–84}$$

For $x \neq 0$, this is the same as (1–83); for $x = 0$, (1–81) is found to satisfy (1–84), even though the existence theorem gives no information for $x = 0$.

In general, if the primitive is $y = f(x; c_1, c_2, \ldots, c_n)$, then the differential equation is obtained by differentiating n times with respect to x and eliminating $c_1, \ldots, c_n$ from these n equations and the equation $y = f(x; c_1, \ldots)$. There are various pitfalls here. For instance, the function $f(x; c_1, \ldots, c_n)$ may be expressible in terms of fewer than n constants. For example,

$$y = x^2 + c_1^2 - c_2^2$$

depends effectively on only one constant, $c = c_1^2 - c_2^2$. In such cases, fewer equations are needed to eliminate the c's and a differential equation of order less than n should be obtained. In general, elimination of the c's may be difficult and may yield only a complicated implicit relation between $x, y, y', \ldots, y^{(n-1)}$.

The primitive itself may be given as an implicit equation

$$F(x, y; c_1, \ldots, c_n) = 0. \tag{1–85}$$

A similar procedure can be followed here. For example,

$$(x - c_1)^2 + (y - c_2)^2 = 1, \tag{1–86}$$

differentiated twice, yields

$$2(x - c_1) + 2(y - c_2)y' = 0,$$

$$2 + 2(y - c_2)y'' + 2y'^2 = 0.$$

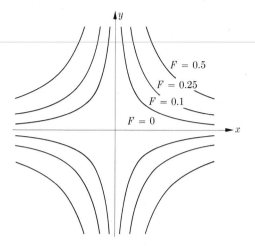

FIG. 1–11.   Level curves of the function xy.

Elimination of c_1, c_2 gives

$$y''^2 = (1 + y'^2)^3. \tag{1-87}$$

This second degree equation is equivalent to two first degree equations:

$$y'' = \pm(1 + y'^2)^{3/2}. \tag{1-88}$$

It can be verified that Eq. (1–86) provides the general solution of (1–88) (Prob. 6 below).

An important special case is the equation

$$F(x, y) = c, \tag{1-89}$$

which describes the *level curves* of the function F.   For example,

$$xy = c \tag{1-90}$$

describes the level curves of $F \equiv xy$ (Fig. 1–11).   Differentiation gives

$$xy' + y = 0, \qquad y' = -\frac{y}{x} \qquad (x \neq 0).$$

On each level curve, y can be considered as a function of x: $y = c/x$, ($x \neq 0$), and the functions so defined do give the general solution of the differential equation $y' = -y/x$, $(x \neq 0)$. We can treat the variables symmetrically by taking *differentials* rather than derivatives; from (1–90),

$$y \, dx + x \, dy = 0.$$

Here we think of y as a function of x or of x as a function of y, whichever is convenient. When $x = 0$, we think of x as a function of y, and the equation reads

$$y \frac{dx}{dy} + x = 0;$$

this is satisfied if $x \equiv 0$. Similarly, $y \equiv 0$ is a solution when we consider y as a function of x; hence there are two solution curves through the origin. In general, we shall agree to consider an equation

$$P(x, y) \, dx + Q(x, y) \, dy = 0 \tag{1–91}$$

as a differential equation of first order. A solution is either a function $y = f(x)$ or a function $x = g(y)$ which satisfies the corresponding differential equation of standard type:

$$Q(x, y) \frac{dy}{dx} + P(x, y) = 0 \quad \text{or} \quad P(x, y) \frac{dx}{dy} + Q(x, y) = 0. \tag{1–92}$$

Thus (1–91) really describes two differential equations; however, except where P or $Q = 0$, the solution curves are the same. A point at which $Q(x, y) = 0$ appears as a singular point in the first form of Eq. (1–92), but if $P(x, y) \neq 0$ at that point, the point will generally not appear as a singular point in the second form of the equation. At such a point we have $dy/dx = \infty$ and $dx/dy = 0$. A point at which both $P(x, y)$ and $Q(x, y)$ are 0 is a singular point for both forms of Eq. (1–92).

1–11 General remarks. This chapter is in a sense introductory, but the importance of the ideas presented must not be underestimated. The existence theorem answers the crucial question of the conditions under which solutions can be found. The method of step-by-step integration tells how to find these solutions; the accuracy obtained can be made as great as desired by employing sufficiently short steps. Accordingly, one could very well stop at this point and consider himself adequately educated on the subject of differential equations.

There are, however, several justifications for going further. We shall find that for certain important differential equations which are widely used there are especially simple methods for obtaining the solutions in explicit form. For a larger class, the solutions may be more difficult to obtain, but from the form of the equation important conclusions can be reached as to the nature of the solutions. Also we shall discover that for the general equation there are several worthy alternatives to the method of step-by-step integration.

PROBLEMS

1. For each of the following differential equations find a solution satisfying the given initial conditions:

(a) $y' = \sin x$; $y = 1$ for $x = 0$;
(b) $y'' = e^x$; $y = 1$ and $y' = 0$ for $x = 1$;
(c) $y' = y$; $y = 1$ for $x = 0$.

2. Find a solution of the differential equation satisfying the given boundary conditions:

(a) $y'' = 1$; $y = 1$ for $x = 0$; $y = 2$ for $x = 1$;
(b) $y^{(iv)} = 0$; $y = 1$ for $x = -1$ and $x = 1$; $y' = 0$ for $x = -1$ and $x = 1$.

3. Prove that $y = c_1 \cos x + c_2 \sin x$ gives the general solution of the differential equation $y'' + y = 0$.

4. Find (if possible) a solution of the differential equation of Prob. 3 which satisfies the following boundary conditions:

(a) $y = 1$ for $x = 0$; $y = -1$ for $x = \frac{1}{2}\pi$;
(b) $y = 1$ for $x = 0$; $y = 1$ for $x = \pi$;
(c) $y = 0$ for $x = 0$; $y = 0$ for $x = \pi$;
(d) $y = 0$ for $x = 0$; $y = 1$ for $x = \pi$.

5. Determine, on the basis of the existence theorem, the points (x_0, y_0) through which there is a unique solution of each of the following differential equations:

(a) $y' = \dfrac{x}{x^2 + y^2}$,   (b) $y' = \dfrac{y}{x}$,

(c) $y' = x \log x$,   (d) $y' = \dfrac{x + y - 1}{x + 2y}$.

6. (a) Verify that for each choice of c_1, c_2, Eq. (1–86) defines two functions y of x, both having derivatives for $c_1 - 1 < x < c_1 + 1$. Show that both of these functions satisfy (1–87) or (1–88). Show, on the basis of the existence theorem, that the functions (1–86) provide all solutions of (1–87).

(b) Show that every solution of (1–88) has constant curvature, equal to 1, and hence has form (1–86).

7. Obtain the differential equation having the given primitive:

(a) $y = c_1 x + c_2$,   (b) $y = cx + c^2$,
(c) $y = c_1 x + c_2 e^x$,   (d) $y = c_1 e^x + c_2 e^{2x} + c_3 e^{3x}$,
(e) $x^2 - xy = c$,   (f) $(x - c_1)^2 + y^2 = c_2^2$.

ANSWERS

1. (a) $2 - \cos x$,   (b) $e^x - ex + 1$,   (c) e^x.   2. (a) $\frac{1}{2}(x^2 + x) + 1$,   (b) 1. 4. (a) $\cos x - \sin x$,   (b) impossible,   (c) $c_2 \sin x$,   (d) impossible.   5. (a) all except $(0, 0)$,   (b) $x_0 \neq 0$,   (c) $x_0 > 0$,   (d) all except the points of the line $x + 2y = 0$.   7. (a) $y'' = 0$,   (b) $y'^2 + xy' = y$,   (c) $(x - 1)y'' - xy' + y = 0$,   (d) $y''' - 6y'' + 11y' - 6y = 0$,   (e) $(2x - y)\,dx - x\,dy = 0$, (f) $yy'' + y'^2 + 1 = 0$.

SUGGESTED REFERENCES

1. ANDRONOW, A., and CHAIKIN, C. E., *Theory of Oscillations.* Princeton: Princeton University Press, 1949.

2. MARTIN, W. T., and REISSNER, E., *Elementary Differential Equations.* Reading, Mass.: Addison-Wesley, 1956.

3. MILNE, WILLIAM E., *Numerical Solution of Differential Equations.* New York: John Wiley and Sons, Inc., 1953.

4. VON KÁRMÁN, T., and BIOT, M. A., *Mathematical Methods in Engineering,* Chap. 1. New York: McGraw-Hill, 1940.

CHAPTER 2

EQUATIONS OF FIRST ORDER
AND FIRST DEGREE

2–1 Different forms of the equation. By definition, a differential equation of first order and first degree has the form

$$Q(x, y)y' + P(x, y) = 0. \tag{2–1}$$

Where $Q(x, y) \neq 0$, this is equivalent to

$$y' = F(x, y), \qquad F = -\frac{P}{Q}. \tag{2–2}$$

In every portion of the xy-plane in which F and $\partial F/\partial y$ are continuous, the existence theorem (Section 1–8) guarantees existence of a unique solution of Eq. (2–2) through each point. The points at which $Q(x, y) = 0$ are singular points and must be investigated separately.

As remarked in Section 1–10, there are advantages in multiplying Eq. (2–1) by dx to obtain the equation

$$P(x, y)\, dx + Q(x, y)\, dy = 0, \tag{2–3}$$

which treats both variables on the same basis. Where $Q(x, y) \neq 0$, Eq. (2–3) can be divided by $Q(x, y)\, dx$, to yield Eq. (2–2). Where $P(x, y) \neq 0$, it can be divided by $P(x, y)\, dy$ to yield the equation

$$\frac{dx}{dy} + \frac{Q(x, y)}{P(x, y)} = 0, \tag{2–4}$$

for x as a function of y. We shall allow as a solution of Eq. (2–3) either a function $y(x)$ or a function $x(y)$ which satisfies (2–3) identically. At points (x, y) for which both P and Q are 0, both Eqs. (2–2) and (2–4) lose meaning. These are *singular points* for Eq. (2–3).

Concept of domain. By a *domain* or *open region* in the xy-plane is meant a portion of the xy-plane such that (a) each two points in the portion can be joined by a broken line within the portion, (b) for each point A in the portion there is a circle with center at A whose interior lies wholly in the portion. The following are examples of domains: all points inside a square; all points for which $x > 0$ (half-plane); all points except $(1, 0)$ and $(0, 1)$.

Throughout this chapter the emphasis will be on formal methods for finding explicit solutions. The functions that occur will be assumed continuous except for special points. Difficulties arising from discontinuities (usually caused by division by 0) will be illustrated by examples.

2–2 Exact differential equations. A differential equation of first order,

$$P(x, y) \, dx + Q(x, y) \, dy = 0, \tag{2-5}$$

is said to be *exact* if for some function $u(x, y)$

$$\frac{\partial u}{\partial x} = P, \qquad \frac{\partial u}{\partial y} = Q, \tag{2-6}$$

so that $du = P \, dx + Q \, dy$. More precisely, the equation (2–5) is exact in an open region D if (2–6) holds in D.

The solutions of an exact equation (2–5) are *level curves* of the function $u(x, y)$, that is, curves on which $u(x, y) \equiv$ const (Fig. 2–1). For example, if $y = f(x)$ is a solution of (2–5), then along the curve $y = f(x)$

$$\frac{du}{dx} = P + Q \frac{dy}{dx} = 0,$$

so that u is constant. In general, if a level curve of u is expressible in the form $y = f(x)$ [or $x = g(y)$], then $y = f(x)$ is a solution of (2–5). For if $y = f(x)$ satisfies

$$u(x, y) = c,$$

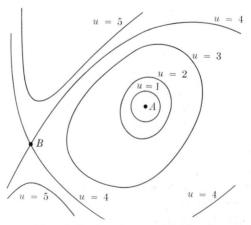

FIG. 2–1.   Level curves of $u(x, y)$.

then from calculus

$$\frac{\partial u}{\partial x} \cdot 1 + \frac{\partial u}{\partial y}\frac{dy}{dx} = 0, \qquad P + Q\frac{dy}{dx} = 0.$$

We summarize the situation by saying: *the solutions of an exact differential equation are the level curves of the function u.* It can of course happen that a level curve is not expressible as a function $y = f(x)$ or $x = g(y)$, and exceptional points may arise on particular level curves. These are suggested in Fig. 2–1 (points A and B).

EXAMPLE 1. $x\,dx + y\,dy = 0$. Here clearly $u = \frac{1}{2}(x^2 + y^2)$,

$$du = \tfrac{1}{2}d(x^2 + y^2) = x\,dx + y\,dy.$$

Hence the solutions are the circles

$$x^2 + y^2 = c.$$

For $c = 0$, the level curve reduces to a point which is a singular point of the differential equation. For $c < 0$ the locus is imaginary.

EXAMPLE 2. $2xy\,dx + x^2\,dy = 0$. Here $u = x^2y$. The solutions are the curves $x^2y = c$ shown in Fig. 2–2. For $c = 0$ the level curve is formed

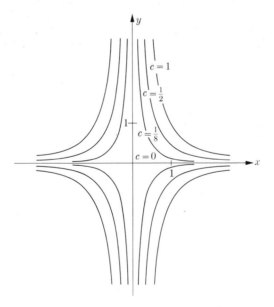

FIG. 2–2.   Level curves of $u = x^2y$.

of the lines $x = 0$, $y = 0$. The point $(0, 0)$ is a singular point, through which *two* solutions pass.

PROBLEMS

1. Plot the level curves of each of the following functions $u(x, y)$:
(a) $u = 2x^2 + y^2$,   (b) $u = 2x - 3y$,
(c) $u = y - 2x^2$,   (d) $u = y^2 e^{-x}$.

2. Find the differential equation for each of the families of level curves of Prob. 1. Indicate exceptional points for each.

3. Show that the following differential equations are exact, and find the general solution.

(a) $x \, dx + 3y^2 \, dy = 0$,   (b) $e^x \, dx - 2 \, dy = 0$,

(c) $y^2 \, dx + 2xy \, dy = 0$,   (d) $\dfrac{x \, dy - y \, dx}{x^2} = 0$.

4. (a) A first order equation $P(x, y) \, dx + Q(x, y) \, dy = 0$ can be replaced by a system of *two* first order equations:

$$\frac{dx}{dt} = Q(x, y), \qquad \frac{dy}{dt} = -P(x, y) \tag{*}$$

in terms of a parameter t. For elimination of t from equations (*) gives $P \, dx + Q \, dy = 0$. Show in more detail that, if $x = \phi(t)$, $y = \psi(t)$ is a solution of (*), then solving $x = \phi(t)$ for t in terms of x (if possible) and substitution in $y = \psi(t)$ yields a function $y = f(x)$ which satisfies $P \, dx + Q \, dy = 0$.
(b) Carry out the process of part (a) for the equation $y \, dx + x \, dy = 0$. Verify that the general solution of the equations (*) is $x = c_1 e^t$, $y = c_2 e^{-t}$. Eliminate t to find the general solution of the equation in x and y.

ANSWERS

2. (a) $2x \, dx + y \, dy = 0$, $(0, 0)$ exceptional;   (b) $2 \, dx - 3 \, dy = 0$;   (c) $-4x \, dx + dy = 0$;   (d) $-y^2 e^{-x} \, dx + 2y e^{-x} \, dy = 0$, $y = 0$ exceptional.

3. (a) $x^2 + 2y^3 = c$, (x, y) not $(0, 0)$;   (b) $e^x - 2y = c$;   (c) $xy^2 = c$; (d) $y/x = c \ (x \neq 0)$.

4. (b) $dx/dt = x$, $dy/dt = -y$, $xy = c$.

2–3 Equations with separable variables. If a differential equation of first order, after multiplication by a suitable factor, takes the form

$$X(x) \, dx + Y(y) \, dy = 0, \tag{2–7}$$

then the equation is said to have *separable variables*. For example, the equation

$$\frac{dy}{dx} = -\frac{x}{y}, \tag{2–8}$$

after multiplication by $y \, dx$, becomes

$$x \, dx + y \, dy = 0. \tag{2–9}$$

The equation is now exact, and we obtain the general solution as

$$x^2 + y^2 = c \qquad (c > 0). \tag{2–10}$$

An equation of form (2–7) is exact (except where X or Y is discontinuous), for we can choose

$$u = \int X \, dx + \int Y \, dy, \tag{2–11}$$

where the integrals are any indefinite integrals of the functions. Then

$$du = \frac{\partial u}{\partial x} \, dx + \frac{\partial u}{\partial y} \, dy = X \, dx + Y \, dy. \tag{2–12}$$

Hence the solutions of (2–7) are given by the level curves

$$\int X(x) \, dx + \int Y(y) \, dy = c. \tag{2–13}$$

An equation $y' = F(x, y)$ will have separable variables if F is expressible as a product of a function of x by a function of y:

$$y' = G(x)H(y). \tag{2–14}$$

For Eq. (2–14) can be written

$$G(x) \, dx - \frac{dy}{H(y)} = 0. \tag{2–15}$$

Equation (2–8) is of this type, with $G = x$, $H = -1/y$.

In converting an equation to form (2–7), we may introduce discontinuities (usually through division by 0) or extraneous solutions. Hence the general solution should be checked carefully against the original differential equation.

EXAMPLE 1. $y \, dx - x \, dy = 0.$

Solution. $\dfrac{dy}{y} - \dfrac{dx}{x} = 0 \qquad (y \neq 0, x \neq 0),$

$$\log y - \log x = c \qquad (x > 0, y > 0),$$

$$\log \frac{y}{x} = c,$$

$$\frac{y}{x} = c', \qquad y = c'x.$$

Hence $c' = e^c$ is a new arbitrary constant and $c' > 0$. However, the restrictions $x > 0$, $y > 0$, $c' > 0$ can be removed, since the solutions $y = c'x$ are valid for all x, y, c'. This is not quite the general solution, since the line $x = 0$ can also be considered as a solution. The restrictions to positive x, y could also have been avoided by noting that for $x \neq 0$, $y \neq 0$,

$$\int \frac{dx}{x} = \log |x| + \text{const}, \qquad \int \frac{dy}{y} = \log |y| + \text{const} \qquad (2\text{--}16)$$

(all logarithms to base e).

EXAMPLE 2. $$\frac{dx}{y} + \frac{dy}{x} = 0.$$

Solution. $$x \, dx + y \, dy = 0,$$

$$x^2 + y^2 = c \qquad (c > 0).$$

(In the differential equation itself x and y are not allowed to be 0. Hence the points $x = 0$, $y = 0$ should be deleted from the solution obtained.)

EXAMPLE 3. $$y' = xe^{y^2}.$$

Solution. $$e^{-y^2} \, dy = x \, dx,$$

$$\int e^{-y^2} \, dy = \int x \, dx = \tfrac{1}{2}x^2 + c.$$

The integral on the left can remain as indicated. It cannot be expressed in terms of elementary functions, but we can express it in terms of an infinite series:

$$\int e^{-y^2} \, dy = \int \left(1 - y^2 + \frac{y^4}{2!} - \frac{y^6}{3!} + \cdots \right) dy$$

$$= y - \frac{y^3}{3} + \frac{y^5}{5 \cdot 2!} - \frac{y^7}{7 \cdot 3!} + \cdots + \text{const}.$$

We can also write

$$\int e^{-y^2}\, dy = \int_0^y e^{-t^2}\, dt + \text{const.}$$

For each fixed y the integral on the right is a definite integral which can be computed to the desired accuracy (see Section 1–6 above).

2–4 Homogeneous equations. If in the differential equation

$$y' = F(x, y) \qquad\qquad (2\text{–}17)$$

the function $F(x, y)$ can be expressed in terms of the one variable

$$v = \frac{y}{x}, \qquad\qquad (2\text{–}18)$$

then the equation is said to be a *homogeneous* first order equation. The following equations are of this type:

$$y' = \frac{y}{x},$$

$$y' = \frac{x^2 - y^2}{xy} = \frac{1 - v^2}{v},$$

$$y' = \sin\frac{y}{x} = \sin v.$$

A homogeneous equation can be reduced to exact form as follows: $y = xv$, so that $dy = x\, dv + v\, dx$ and

$$x\, dv + v\, dx = F(x, y)\, dx = G(v)\, dx.$$

Accordingly, the differential equation becomes

$$\frac{dv}{v - G(v)} + \frac{dx}{x} = 0 \qquad \left(v = \frac{y}{x}\right). \qquad\qquad (2\text{–}19)$$

In other words, the substitution $v = y/x$ and the elimination of y lead to a separation of variables.

EXAMPLE.
$$y' = \frac{x^2 + y^2}{xy}.$$

Solution.

$$y' = \frac{x}{y} + \frac{y}{x} = \frac{1}{v} + v,$$

$$x\, dv + v\, dx = dy = \left(\frac{1}{v} + v\right) dx,$$

$$v\, dv - \frac{dx}{x} = 0,$$

$$\tfrac{1}{2}v^2 - \log|x| = \tfrac{1}{2}c,$$

$$y^2 = x^2 \log x^2 + cx^2.$$

No

(The points $x = 0, y = 0$ are ruled out by the differential equation itself.)

Remarks. The most common case of a homogeneous equation is that in which $F(x, y)$ is the ratio of two homogeneous polynomials in x, y, of the same degree. When the equation is written in the form $P\, dx + Q\, dy = 0$, it appears as

$$(a_0 x^n + a_1 x^{n-1}y + \cdots + a_{n-1}xy^{n-1} + a_n y^n)\, dx$$

$$+ (b_0 x^n + b_1 x^{n-1}y + \cdots + b_n y^n)\, dy = 0. \quad (2\text{–}20)$$

In general, a function $P(x, y)$ is said to be homogeneous of degree n if

$$P(tx, ty) \equiv t^n P(x, y). \quad (2\text{–}21)$$

Since

$$(tx)^k (ty)^{n-k} = t^n (x^k y^{n-k}),$$

we see that the coefficients P, Q of Eq. (2–20) are homogeneous of degree n. Also

$$\sqrt{x^2 + y^2}, \quad \sqrt{x^3 + x^2 y + y^3}, \quad xy \sin \frac{x+y}{x-y}$$

are homogeneous of degrees $1, 3/2$, and 2, respectively. Whenever $P(x, y)$, $Q(x, y)$ are homogeneous of the same degree n, the differential equation $P\, dx + Q\, dy = 0$ is homogeneous. For then

$$\frac{dy}{dx} = F(x, y) = \frac{-P(x, y)}{Q(x, y)} = \frac{-P(x, vx)}{Q(x, vx)}$$

$$= \frac{-x^n P(1, v)}{x^n Q(1, v)} = \frac{-P(1, v)}{Q(1, v)} = G(v).$$

PROBLEMS

1. Find all solutions by separation of variables:

(a) $y' = e^{x+y}$,   (b) $y' = 3y$,

(c) $y' = (y - 1)(y - 2)$,   (d) $y' = x^3 y^{-2}$,

(e) $\sin x \cos y \, dx + \tan y \cos x \, dy = 0$,   (f) $y' = xy^2 + y^2 + xy + y$.

2. Verify that the following equations are homogeneous and find all solutions:

(a) $y' = \dfrac{x - y}{x + y}$,   (b) $xy' - y = xe^{y/x}$,

(c) $(3x^2 y + y^3) \, dx + (x^3 + 3xy^2) \, dy = 0$,   (d) $y' = \dfrac{y}{x} + \sin \dfrac{y - x}{x}$.

3. (a) Show that the isoclines of a homogeneous equation of first order are straight lines through the origin.

(b) Show that the solutions of a homogeneous equation are *similar curves;* that is, if $y = f(x)$ is a solution, then so is $ky = f(kx)$ for each choice of the constant k.

(c) Solve Prob. 2(a) graphically, with the aid of isoclines.

4. Find the solution which satisfies the given initial conditions:

(a) $y^2 \, dx + (x + 1) \, dy = 0$; $x = 0, y = 1$;

(b) $\dfrac{dx}{dt} = \dfrac{xt}{x^2 + t^2}$; $t = 0$, $x = 1$;

(c) $y' = \dfrac{x(y^2 - 1)}{(x - 1)y^3}$; $x = 2$, $y = 1$;

(d) $y' = \dfrac{\sqrt{x^2 + y^2}}{x + y}$; $x = 1$, $y = 3$.

ANSWERS

1. (a) $e^x + e^{-y} = c$; (b) $y = ce^{3x}$; (c) $y - 2 = ce^x(y - 1)$, $y = 1$; (d) $4y^3 = 3x^4 + c$, $y \neq 0$; (e) $\cos xe^{-\sec y} = c$, $y \neq (2n + 1)\pi/2$ $(n = 0, \pm1, \ldots)$; (f) $\log |y/(y + 1)| = \frac{1}{2}x^2 + x + c$, $y = 0$, $y = -1$.

2. (a) $x^2 - 2xy - y^2 = c$, $x + y \neq 0$; (b) $e^{-y/x} + \log |x| = c$, $x \neq 0$; (c) $x^3 y + xy^3 = c$; (d) $\tan[(y - x)/(2x)] = cx$, $x \neq 0$.

4. (a) $y = 1/[1 + \log(x + 1)]$; (b) $x = e^{t^2/(2x^2)}$; (c) $y = 1$;

(d) $\displaystyle \int_3^v \dfrac{(1 + u) \, du}{\sqrt{1 + u^2} - u - u^2} = \log x$, $v = y/x$.

2–5 The general exact equation. If the equation

$$P(x, y) \, dx + Q(x, y) \, dy = 0 \qquad\qquad (2\text{–}22)$$

is exact, then for some function $u(x, y)$ we have

$$P(x, y)\, dx + Q(x, y)\, dy = du, \tag{2-23}$$

$$\frac{\partial u}{\partial x} = P(x, y), \qquad \frac{\partial u}{\partial y} = Q(x, y). \tag{2-24}$$

Therefore,

$$\frac{\partial^2 u}{\partial y\, \partial x} = \frac{\partial P}{\partial y}, \qquad \frac{\partial^2 u}{\partial x\, \partial y} = \frac{\partial Q}{\partial x}. \tag{2-25}$$

But from calculus

$$\frac{\partial^2 u}{\partial y\, \partial x} \equiv \frac{\partial^2 u}{\partial x\, \partial y}. \tag{2-26}$$

Therefore,

$$\frac{\partial P}{\partial y} \equiv \frac{\partial Q}{\partial x}. \tag{2-27}$$

Conversely, as will be shown below, if Eq. (2–27) holds, then the differential equation (2–22) is exact. *Hence* Eq. (2–27) *is a perfect test for exactness.* (We have tacitly assumed that P and Q have continuous derivatives in an open region of the xy-plane. Some further discussion of the regions concerned is required; this is given in the next section.)

EXAMPLE 1. $(3x^2y + 2xy)\, dx + (x^3 + x^2 + 2y)\, dy = 0.$

Solution. Here $P = 3x^2y + 2xy, \qquad Q = x^3 + x^2 + 2y,$ and

$$\frac{\partial P}{\partial y} \equiv 3x^2 + 2x \equiv \frac{\partial Q}{\partial x}.$$

Therefore the differential equation is exact. We may be able to recognize by inspection the function u whose differential is $P\, dx + Q\, dy$. If this fails, we can proceed as follows. We write

$$\frac{\partial u}{\partial x} = 3x^2y + 2xy, \qquad \frac{\partial u}{\partial y} = x^3 + x^2 + 2y. \tag{2-28}$$

We now integrate one (but not both) of the equations (2–28). From the first equation we obtain, on integration with respect to x,

$$u = x^3y + x^2y + \text{const.} \tag{2-29}$$

But the constant could very well depend on y; it would still drop out when (2–29) is differentiated with respect to x. Hence we write

$$u = x^3 y + x^2 y + \phi(y), \qquad (2\text{–}29\text{a})$$

where $\phi(y)$ is a function to be determined. From (2–29a) we find

$$\frac{\partial u}{\partial y} = x^3 + x^2 + \phi'(y). \qquad (2\text{–}30)$$

From this equation and the second of equations (2–28) we conclude that

$$x^3 + x^2 + \phi'(y) = x^3 + x^2 + 2y,$$

$$\phi'(y) = 2y,$$

$$\phi(y) = y^2 + c.$$

Accordingly,

$$u(x, y) = x^3 y + x^2 y + y^2 + c$$

is a function whose differential is $P\,dx + Q\,dy$. We can omit the c here, since we recover it in the general solution of the differential equation:

$$x^3 y + x^2 y + y^2 = c. \qquad (2\text{–}31)$$

[It can be verified that the level curves in (2–31) are actual solution curves; the points $(0, 0)$ and $(-1, 0)$ are singular points through which two solutions pass: $y = 0$ and $y = -x^2 - x^3$.]

EXAMPLE 2. $y^2\,dx + 2xy\,dy = 0$. The equation is exact:

$$\frac{\partial P}{\partial y} \equiv 2y \equiv \frac{\partial Q}{\partial x}.$$

By inspection, $y^2\,dx + 2xy\,dy = d(xy^2)$. The solutions are the level curves

$$xy^2 = c.$$

EXAMPLE 3. $(xy \cos xy + \sin xy)\,dx + (x^2 \cos xy + e^y)\,dy = 0$.

Solution. $P = xy \cos xy + \sin xy$, $Q = x^2 \cos xy + e^y$,

$$\frac{\partial P}{\partial y} \equiv 2x \cos xy - x^2 y \sin xy \equiv \frac{\partial Q}{\partial x}.$$

The equation is exact and we seek $u(x, y)$ such that

$$\frac{\partial u}{\partial x} = xy \cos xy + \sin xy, \qquad \frac{\partial u}{\partial y} = x^2 \cos xy + e^y.$$

We integrate the *second* equation with respect to y:

$$u = x \sin xy + e^y + \phi(x),$$

$$\frac{\partial u}{\partial x} = xy \cos xy + \sin xy + \phi'(x).$$

Hence

$$xy \cos xy + \sin xy + \phi'(x) = xy \cos xy + \sin xy,$$

$$\phi'(x) = 0, \quad \phi(x) = \text{const},$$

$$u = x \sin xy + e^y + \text{const},$$

$$x \sin xy + e^y = c \quad \text{(gen. sol.)}.$$

* 2–6 Justification of the test for exactness. Solution as a line integral.

Let the equation

$$P(x, y) \, dx + Q(x, y) \, dy = 0 \tag{2–32}$$

be given. We assume that $P(x, y)$ and $Q(x, y)$ have continuous first partial derivatives for all (x, y) and that

$$\frac{\partial P}{\partial y} = \frac{\partial Q}{\partial x} \tag{2–33}$$

for all (x, y). The effect of discontinuities will be discussed later.

We now seek a function $u(x, y)$ such that

$$\frac{\partial u}{\partial x} = P(x, y), \qquad \frac{\partial u}{\partial y} = Q(x, y). \tag{2–34}$$

Since addition of a constant to u does not affect (2–34), we can impose the additional condition

$$u(0, 0) = 0. \tag{2–35}$$

For by adding a constant to a solution of Eqs. (2–34), we can always ensure that Eq. (2–35) is satisfied.

To find the function u, we first seek the values of u along the x-axis. Here $y = 0$ and $\partial u/\partial x = P(x, 0)$. Since $u(x, 0)$ depends on x alone and

(2–35) holds,

$$u(x_1, 0) = \int_0^{x_1} \frac{\partial u}{\partial x}\, dx + u(0, 0) = \int_0^{x_1} \frac{\partial u}{\partial x}\, dx$$

$$= \int_0^{x_1} P(x, 0)\, dx. \tag{2-36}$$

Similarly, we can find the values of u along the y-axis:

$$u(0, y_1) = \int_0^{y_1} \frac{\partial u}{\partial y}\, dy + u(0, 0) = \int_0^{y_1} Q(0, y)\, dy. \tag{2-37}$$

From Eq. (2–36) we can find the value of u at (x_1, y_1):

$$u(x_1, y_1) - u(x_1, 0) = \int_0^{y_1} \frac{\partial u}{\partial y}(x_1, y)\, dy = \int_0^{y_1} Q(x_1, y)\, dy,$$

$$u(x_1, y_1) = u(x_1, 0) + \int_0^{y_1} Q(x_1, y)\, dy$$

$$= \int_0^{x_1} P(x, 0)\, dx + \int_0^{y_1} Q(x_1, y)\, dy. \tag{2-38}$$

Similarly, from (2–37)

$$u(x_1, y_1) = u(0, y_1) + \int_0^{x_1} P(x, y_1)\, dx$$

$$= \int_0^{x_1} P(x, y_1)\, dx + \int_0^{y_1} Q(0, y)\, dy. \tag{2-39}$$

We have now obtained two separate expressions (2–38) and (2–39) for the value of u at (x_1, y_1); however, the two are equal, for their difference is

$$\int_0^{x_1} P(x, 0)\, dx + \int_0^{y_1} Q(x_1, y)\, dy - \int_0^{x_1} P(x, y_1)\, dx - \int_0^{y_1} Q(0, y)\, dy$$

$$= \int_0^{y_1} [Q(x_1, y) - Q(0, y)]\, dy - \int_0^{x_1} [P(x, y_1) - P(x, 0)]\, dx$$

$$= \int_0^{y_1} \int_0^{x_1} \frac{\partial Q}{\partial x}(x, y)\, dx\, dy - \int_0^{x_1} \int_0^{y_1} \frac{\partial P}{\partial y}(x, y)\, dy\, dx.$$

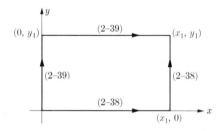

FIG. 2–3. Solution of exact equation by integration.

Both terms represent double integrals over the rectangle: $0 \leq x \leq x_1$, $0 \leq y \leq y_1$. Since $\partial Q/\partial x \equiv \partial P/\partial y$, the double integrals are equal and the difference is 0. Therefore (2–38) and (2–39) agree.

We now show that (2–34) holds at an arbitrary point (x_1, y_1). From Eq. (2–39), with y_1 held constant [see Eq. (1–44)], we find

$$\frac{\partial u}{\partial x_1} = \frac{\partial}{\partial x_1} \int_0^{x_1} P(x, y_1)\, dx = P(x_1, y_1).$$

Similarly, from Eq. (2–38), with x_1 held constant,

$$\frac{\partial u}{\partial y_1} = \frac{\partial}{\partial y_1} \int_0^{y_1} Q(x_1, y)\, dy = Q(x_1, y_1).$$

Thus (2–34) holds and Eq. (2–32) is exact.

The two expressions (2–38), (2–39) were obtained by examining the variation of u along the paths shown in Fig. 2–3. Thus in (2–38) the term $\int_0^{x_1} P(x, 0)\, dx$ represents the amount by which u increases in going from $(0, 0)$ to $(x_1, 0)$, and the second term gives the change in u in going from $(x_1, 0)$ to (x_1, y_1). We can, in fact, use a quite general path C from $(0, 0)$ to (x_1, y_1), as in Fig. 2–4. Let

$$x = x(t), \qquad y = y(t), \qquad 0 \leq t \leq t_1, \tag{2–40}$$

be the equations of the path in terms of a parameter t, $t = 0$ corresponding to $(0, 0)$, $t = t_1$ corresponding to (x_1, y_1). Then, along the path,

$$\frac{du}{dt} = \frac{\partial u}{\partial x} \frac{dx}{dt} + \frac{\partial u}{\partial y} \frac{dy}{dt} = P \frac{dx}{dt} + Q \frac{dy}{dt}. \tag{2–41}$$

Therefore

$$u(x_1, y_1) = \int_0^{t_1} \frac{du}{dt}\, dt + u(0, 0) = \int_0^{t_1} \frac{du}{dt}\, dt,$$

$$u(x_1, y_1) = \int_0^{t_1} \left\{ P[x(t), y(t)] \frac{dx}{dt} + Q[x(t), y(t)] \frac{dy}{dt} \right\} dt. \tag{2–42}$$

The last expression can be interpreted as a *line integral* along the path C; it is usually written in the form

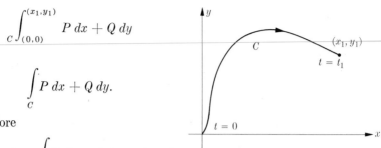

$$\int_{C}_{(0,0)}^{(x_1,y_1)} P\,dx + Q\,dy$$

or

$$\int_{C} P\,dx + Q\,dy.$$

Therefore

$$u(x_1, y_1) = \int_{C} P\,dx + Q\,dy, \qquad (2\text{–}43)$$

FIG. 2–4.   Path for line integral.

where C is an arbitrary path (2–40) from $(0, 0)$ to (x_1, y_1). It is assumed that C has a continuously turning tangent or, at the worst, has a finite number of corners, as in the case of the two paths of Fig. 2–3. We can, in particular, choose C as the line:

$$x = x_1 t, \qquad y = y_1 t, \qquad 0 \leqq t \leqq 1.$$

Then

$$u(x_1, y_1) = \int_{0}^{1} [x_1 P(x_1 t, y_1 t) + y_1 Q(x_1 t, y_1 t)]\,dt$$

or, more simply,

$$u(x, y) = \int_{0}^{1} [x P(xt, yt) + y Q(xt, yt)]\,dt. \qquad (2\text{–}44)$$

EXAMPLE. $(3x^2 y + 2xy)\,dx + (x^3 + x^2 + 2y)\,dy = 0$. This equation is that of Example 1 in the preceding section. We apply (2–44):

$$u(x, y) = \int_{0}^{1} [x(3t^3 x^2 y + 2t^2 xy) + y(x^3 t^3 + x^2 t^2 + 2ty)]\,dt$$

$$= 3x^3 y \int_{0}^{1} t^3\,dt + 2x^2 y \int_{0}^{1} t^2\,dt + x^3 y \int_{0}^{1} t^3\,dt$$

$$\qquad + x^2 y \int_{0}^{1} t^2\,dt + 2y^2 \int_{0}^{1} t\,dt$$

$$= \tfrac{3}{4}x^3 y + \tfrac{2}{3}x^2 y + \tfrac{1}{4}x^3 y + \tfrac{1}{3}x^2 y + y^2 = x^3 y + x^2 y + y^2,$$

in agreement with (2–31).

Throughout we have assumed that P and Q have continuous first partial derivatives for all (x, y). If there is a discontinuity at a single point (a, b), we cannot, in general, find one function $u(x, y)$ whose differential is $P\,dx +$

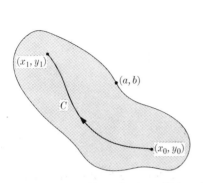

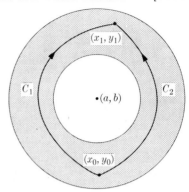

Fig. 2–5.   Domain in which solution of exact equation is given by line
integral.

Fig. 2–6.   Domain in which line integral fails to give solution of exact
equation.

$Q \, dy$ in the domain (Section 2–1) consisting of the xy-plane without the
point (a, b). We can, however, find such a function in each domain which
does not "surround" (a, b); such a domain is shown in Fig. 2–5. For
each point (x_1, y_1) of the domain we can evaluate u by (2–43), where C
is an arbitrary path from a fixed point (x_0, y_0) [replacing $(0, 0)$] to (x_1, y_1),
and not leaving the domain. The same procedure for the domain of Fig.
2–6 can be attempted, but in general two paths such as C_1, C_2 will yield
different values for u, so that no definite function u is obtained.

For a complete discussion of the above case, see pp. 225–258 of Reference 4 at the end of this chapter.

Solution of initial value problem. If the general solution of an exact
equation is obtained in the form $u(x, y) = c$, then the solution through
(x_0, y_0) is obtained by choosing c so that $u(x_0, y_0) = c$. Hence the
equation of the particular solution is

$$u(x, y) - u(x_0, y_0) = 0. \tag{2–45}$$

(The locus (2–45) may consist of several curves, one of which will define
the particular solution sought.) If u is obtained as a line integral on a path
C, as in Fig. 2–5, then (2–45) becomes

$$\int_{(x_0, y_0)}^{(x, y)} P \, dx + Q \, dy = 0. \tag{2–46}$$

When the variables are separable, (2–46) reduces to

$$\int_{x_0}^{x} P(x) \, dx + \int_{y_0}^{y} Q(y) \, dy = 0. \tag{2–47}$$

PROBLEMS

1. Show that the following equations are exact, and find all solutions:

(a) $2xy\, dx + (x^2 + 1)\, dy = 0$,

(b) $(2x + y)\, dx + (x - 2y)\, dy = 0$,

(c) $(15x^2y^2 - y^4)\, dx + (10x^3y - 4xy^3 + 5y^4)\, dy = 0$,

(d) $e^{x^2 y}(1 + 2x^2y)\, dx + x^3 e^{x^2 y}\, dy = 0$,

(e) $[x \cos (x + y) + \sin (x + y)]\, dx + x \cos (x + y)\, dy = 0$,

(f) $3xy(1 + x^2)^{1/2}\, dx + [(1 + x^2)^{3/2} + \sin y]\, dy = 0$,

(g) $(x^2 + y^2)^2 (x\, dx + y\, dy) + 2\, dx + 3\, dy = 0$,

(h) $[(y\, dx - x\, dy)/y^2] + x\, dx = 0$,

(i) $(2xy^2 + 2xye^{2x} + e^{2x}y)\, dx + (2x^2y + xe^{2x})\, dy = 0$,

(j) $[(y - x)\, dx - 2x\, dy]/(x + y)^3 = 0$.

2. Find the particular solution specified:

(a) $(2x + y + 1)\, dx + (x + 3y + 2)\, dy = 0$; $y = 0$ when $x = 0$;

(b) $(3x + y + 5)^3(3\, dx + dy) + 2\, dx - dy = 0$; $y = 1$ when $x = -2$;

(c) $x^2\, dx + ye^y\, dy = 0$; $y = 1$ when $x = 0$;

(d) $e^{x^2}\, dx + \sin (1 + y^2)\, dy = 0$; $y = 0$ when $x = 0$;

(e) $ye^{-x^2}\, dx + [\int_0^x e^{-t^2}\, dt + y]\, dy = 0$; $y = 1$ when $x = 1$.

3. Find the general solution of the equation

$$(x + y)\, dx + x\, dy = 0,$$

(a) by means of (2–38);   (b) by means of (2–39);   (c) by means of (2–44).

4. Let $P(x, y)$ and $Q(x, y)$ be homogeneous of degree n: $P(tx, ty) = t^n P(x, y)$, $Q(tx, ty) = t^n Q(x, y)$. Show with the aid of Eq. (2–44) that if the equation $P\, dx + Q\, dy = 0$ is exact, then its general solution is $xP(x, y) + yQ(x, y) = c$.

5. Obtain the general solutions of the following parts of Prob. 1, with the aid of the theorem of Prob. 4: (a) part (b), (b) part (c).

6. Obtain the general solution of the exact equation

$$\frac{y\, dx - x\, dy}{x^2 + y^2} = 0,$$

with the aid of a line integral (2–42) from $(1, 0)$ to a general point (x_1, y_1). Use as path C a circular arc $x = \cos t$, $y = \sin t$, $0 \le t \le \theta_1$, plus the line segment from $(\cos \theta_1, \sin \theta_1)$ to (x_1, y_1), where θ_1 is the polar angle of (x_1, y_1). Show that the solution obtained is not single-valued if (x_1, y_1) is not restricted to lie in a domain which does not surround the origin.

ANSWERS

1. (a) $x^2y + y = c$, (b) $x^2 + xy - y^2 = c$, (c) $5x^3y^2 - xy^4 + y^5 = c$, (d) $xe^{x^2 y} = c$, (e) $x \sin (x + y) = c$, (f) $y(1 + x^2)^{3/2} - \cos y = c$, (g)

$(x^2 + y^2)^3 + 12x + 18y = c,$ (h) $2x + x^2y = cy,$ (i) $x^2y^2 + xe^{2x}y = c,$
(j) $x = c(x + y)^2.$

2. (a) $2x^2 + 2xy + 3y^2 + 2x + 4y = 0,$ (b) $(3x + y + 5)^4 + 4(2x - y + 5) = 0,$ (c) $x^3 + 3e^y(y - 1) = 0,$ (d) $\int_0^x e^{t^2} dt + \int_0^y \sin(1 + t^2)\, dt = 0,$
(e) $2y \int_0^x e^{-t^2} dt + y^2 = 2 \int_0^1 e^{-x^2} dx + 1.$

3. $x^2 + 2xy = c.$

6. Solutions are $\theta = \tan^{-1}(y/x) = c.$

2–7 Integrating factors.

If an equation

$$P(x, y)\, dx + Q(x, y)\, dy = 0 \qquad (2\text{–}48)$$

is not exact, we can attempt to make it exact by multiplying by a suitable function of x and y. For example, the equation

$$y\, dx - x\, dy = 0 \qquad (2\text{–}49)$$

is not exact, but it becomes exact when multiplied by y^{-2}, for

$$\frac{y\, dx - x\, dy}{y^2} \equiv d\left(\frac{x}{y}\right). \qquad (2\text{–}50)$$

We say that y^{-2} is an *integrating factor* for Eq. (2–49). Similarly,

$$2y\, dx + x\, dy = 0 \qquad (2\text{–}51)$$

becomes exact on multiplication by x:

$$2xy\, dx + x^2\, dy \equiv d(x^2y) \equiv 0, \qquad (2\text{–}51\text{a})$$

so that x is an integrating factor.

It should be noted that multiplication of the differential equation by the integrating factor may introduce new discontinuities in the coefficients and may also introduce extraneous solutions (curves along which the integrating factor is zero). In the case of (2–49), the factor y^{-2} is discontinuous when $y = 0$. In the case of (2–51), the factor x is zero when $x = 0$, so that $x = 0$ is a solution of (2–51a); it happens to be a solution of (2–51) also. In general, the differential equation is changed when multiplied by the factor and a careful analysis is needed to ensure that no solutions are lost or gained in the process.

If an integrating factor has been found, then the given equation is replaced by an exact equation which can be solved by the methods of the preceding sections. Hence, except for the difficulties mentioned in the preceding paragraph, the general solution of the given equation is known.

EXAMPLE 1. $(xy + y - 1) \, dx + x \, dy = 0.$

Solution. Here e^x is an integrating factor, for

$$e^x(xy + y - 1) \, dx + xe^x \, dy = 0 \qquad\qquad (2\text{--}52)$$

is exact:

$$\frac{\partial}{\partial y} \, e^x(xy + y - 1) \equiv e^x(x + 1) \equiv \frac{\partial}{\partial x} \, (xe^x).$$

We readily find the solutions of the new equation

$$xye^x - e^x = c.$$

Since e^x is never 0, the integrating factor introduces no complications and (2–52) gives the general solution of the given equation; [there is one singular point: (0, 1), through which two solutions pass: $x = 0$ and $xy = 1 - e^{-x}$].

EXAMPLE 2. $(3xy + y^2) \, dx + (3xy + x^2) \, dy = 0.$ Here the equation is not exact, but becomes exact when multiplied by $(x + y)$. For

$$(x + y)(3xy + y^2) \, dx + (x + y)(3xy + x^2) \, dy = 0$$

is exact, as is readily verified. In fact, the equation can be written

$$d\,[(x + y)^2 xy] = 0,$$

so that the solutions of the new equation are

$$(x + y)^2 xy = c.$$

Now the line $x + y = 0$ appears as a solution; however, it is not extraneous, since it is a solution of the original differential equation.

These two examples suggest the very great variety of integrating factors that may occur. It is natural to ask whether every equation $P \, dx + Q \, dy = 0$ has an integrating factor. It can be proved that an integrating factor does always exist, provided attention is restricted to a suitably small domain (Section 12–11); however, this fact does not help to find integrating factors for particular problems (any more than the existence theorem of Section 1–8 helps us to find solutions).

The difficult problem of finding integrating factors for particular equations challenges the ingenuity. Here are some suggestions which are often helpful.

First note four commonly occurring exact differentials:

$$d(xy) = y\,dx + x\,dy, \tag{2-53}$$

$$d\left(\frac{y}{x}\right) = \frac{x\,dy - y\,dx}{x^2}, \tag{2-54}$$

$$d \arctan \frac{y}{x} = \frac{x\,dy - y\,dx}{x^2 + y^2}, \tag{2-55}$$

$$d\tfrac{1}{2} \log (x^2 + y^2) = \frac{x\,dx + y\,dy}{x^2 + y^2}. \tag{2-56}$$

Next observe that *if $du = P\,dx + Q\,dy$, then $P\,dx + Q\,dy$ remains an exact differential when multiplied by any continuous function of the function u.* For example,

$$2x\,dx + 2y\,dy = d(x^2 + y^2) = du,$$

$$(x^2 + y^2)(2x\,dx + 2y\,dy) = u\,du = d\left(\frac{u^2}{2}\right) \qquad (u = x^2 + y^2),$$

$$\frac{2x\,dx + 2y\,dy}{x^2 + y^2} = \frac{du}{u} = d \log u \qquad (u = x^2 + y^2).$$

The above explains (2–56), and we see that (2–55) is obtained from (2–54) by the reasoning:

$$\frac{x\,dy - y\,dx}{x^2 + y^2} = \frac{1}{1 + (y^2/x^2)} \frac{x\,dy - y\,dx}{x^2} = \frac{1}{1 + u^2}\,du = d \arctan u,$$

where $u = y/x$.

Given a particular equation, we can try to recognize an *exact portion* of the expression $P\,dx + Q\,dy$. We then try to multiply this *part* by an appropriate function $f(u)$ (where u is the function whose differential appears) in order to make the remaining portion exact.

EXAMPLE 3. $(xy^2 + y)\,dx + x\,dy = 0$. This can be written

$$xy^2\,dx + d(xy) = 0.$$

We can multiply by a function of xy without spoiling the exactness of the second group. If we multiply by $(xy)^{-2}$ we remove y^2 from the first term:

$$\frac{dx}{x} + \frac{1}{(xy)^2}\,d(xy) = 0,$$

$$\log|x| - \frac{1}{xy} = c.$$

EXAMPLE 4. $(x + 3y^2)\,dx + 2xy\,dy = 0$. Here $x\,dx$ is exact. We try to multiply by some function of x which will make $3y^2\,dx + 2xy\,dy$ exact. If we call this function $v(x)$, then we want

$$3y^2 v\,dx + 2xvy\,dy = 0$$

to be an exact differential equation. Hence, by Eq. (2–33),

$$6yv \equiv 2y(xv' + v)$$

is the condition imposed. Accordingly, for $y \neq 0$,

$$6v = 2xv' + 2v, \qquad xv' - 2v = 0, \qquad x\frac{dv}{dx} - 2v = 0.$$

By separation of variables, we find that $v = x^2$ is a solution. Hence x^2 is the integrating factor sought. The new equation and general solution are found to be

$$(x^3 + 3x^2y^2)\,dx + 2x^3y\,dy = 0,$$

$$\frac{x^4}{4} + x^3y^2 = c.$$

EXAMPLE 5. $(3y + 8xy^2)\,dx + (2x + 6x^2y)\,dy = 0$. Here we consider the group $3y\,dx + 2x\,dy$, which is not exact, but becomes so after multiplication by $(xy)^{-1}$, which has the effect of separating the variables:

$$(xy)^{-1}(3y\,dx + 2x\,dy) = \frac{3\,dx}{x} + \frac{2\,dy}{y}$$

$$= d(3\log x + 2\log y) = d(\log x^3y^2).$$

If the whole equation is multiplied by $(xy)^{-1}$, it becomes

$$3\frac{dx}{x} + 2\frac{dy}{y} + 8y\,dx + 6x\,dy = 0.$$

The first two terms together are exact and will remain exact when multiplied by any function of $\log x^3y^2$ or by any function of x^3y^2. We let $z = x^3y^2$ and try to choose $v(z)$ so that $v(z)(8y\,dx + 6x\,dy)$ is exact. Hence, by Eq. (2–33),

$$\frac{\partial}{\partial y}(8yv) = \frac{\partial}{\partial x}(6xv),$$

$$8v + 8y\frac{\partial v}{\partial y} = 6v + 6x\frac{\partial v}{\partial x},$$

$$8v + 8yv'(z)2x^3y = 6v + 6xv'(z)3x^2y^2,$$

$$2zv'(z) - 2v = 0.$$

Therefore we can choose $v = z$. The equation now becomes

$$zd\log z + 8x^3y^3\,dx + 6x^4y^2\,dy = 0,$$

$$z\frac{dz}{z} + d(2x^4y^3) = 0,$$

$$x^3y^2 + 2x^4y^3 = c.$$

The methods illustrated may not always be successful, but they are worth trying.

2–8 Linear equation of first order. A differential equation of first order is said to be linear if it can be written in the form

$$y' + p(x)y = q(x). \tag{2–57}$$

In differential form this equation is

$$[p(x)y - q(x)]\,dx + dy = 0.$$

The term $q(x)\,dx$ by itself is exact and will remain so after multiplication by a function of x, $v(x)$. We try to make the group $p(x)y\,dx + dy$ exact by multiplying by $v(x)$; that is,

$$v(x)p(x)y\,dx + v(x)\,dy = 0$$

is to be an exact equation. Accordingly, by Eq. (2–33),

$$v(x)p(x) = v'(x), \qquad p(x) \, dx = \frac{dv}{v},$$

and we can choose $v = e^{\int p \, dx}$. The original equation becomes

$$v'(x)y \, dx + v(x) \, dy - v(x)q(x) \, dx = 0;$$

therefore

$$yv(x) - \int v(x)q(x) \, dx = c,$$

$$ye^{\int p \, dx} - \int e^{\int p \, dx}q(x) \, dx = c,$$

$$y = e^{-\int p \, dx}\int e^{\int p \, dx}q \, dx + ce^{-\int p \, dx}. \tag{2–58}$$

Equation (2–58) is a general formula for the solutions of the linear equation (2–57). Because of its importance, we state the result as a theorem:

THEOREM. *Let $p(x)$ and $q(x)$ be continuous for $a < x < b$. Let particular choices of the indefinite integrals $\int p \, dx$ and $\int e^{\int p \, dx}q \, dx$ be made. Then (2–58) gives all solutions of (2–57) over the interval $a < x < b$. In particular, for each point (x_0, y_0), $a < x_0 < b$, c can be chosen uniquely so that (2–58) passes through (x_0, y_0).*

Proof. We let $v = e^{\int p \, dx}$, so that v is continuous and

$$v' = e^{\int p \, dx}p(x) = pv.$$

If $y(x)$ is a solution of (2–57) for $a < x < b$, then successively

$$y' + py = q,$$

$$vy' + pvy = qv,$$

$$vy' + v'y = qv,$$

$$\frac{d}{dx}(vy) = qv,$$

$$vy = \int qv \, dx + c,$$

where the indefinite integral can be chosen as the given indefinite integral of qv. Now $v(x) \neq 0$. Hence we can divide by v:

$$y = v^{-1}\int qv \, dx + cv^{-1}; \tag{2–59}$$

therefore y has form (2–58). Next let y have form (2–58), so that (2–59) holds. We now reverse the steps, and arrive finally at the equation $y' + py = q$. Since v is continuous and $v \neq 0$, all the steps are reversible, and we conclude that every function (2–58) defines a solution of (2–57).

We can write (2–58) as

$$y = r(x) + \frac{c}{v(x)}.$$

The condition that when $x = x_0$, $y = y_0$, leads to the equation

$$y_0 = r(x_0) + \frac{c}{v(x_0)} \qquad [v(x_0) \neq 0].$$

Hence $c = v(x_0)[y_0 - r(x_0)]$. With this value of c (and only with this value) the initial condition is satisfied.

EXAMPLE. $y' + xy = x$. Here $p = x$ and $v = e^{x^2/2}$. Multiplying by v, we find

$$e^{x^2/2}y' + xe^{x^2/2} y = xe^{x^2/2},$$

$$e^{x^2/2}y = \int xe^{x^2/2}\, dx = e^{x^2/2} + c,$$

$$y = 1 + ce^{-x^2/2}.$$

The linear equation is of fundamental importance for applications. Accordingly, the whole of the following chapter is devoted to linear equations, and examples are given of typical applications.

PROBLEMS

1. Find an integrating factor for each of the following differential equations and obtain the general solution:

(a) $(x + 2y)\, dx + x\, dy = 0$,
(b) $(x + 3y)\, dx + x\, dy = 0$,
(c) $y\, dx + (y - x)\, dy = 0$,
(d) $2y^2\, dx + (2x + 3xy)\, dy = 0$,
(e) $(x^2 + y^2 + x)\, dx + y\, dy = 0$,
(f) $y\, dx + (x + x^2y^4)\, dy = 0$,
(g) $(2 + 2y^3)\, dx + 3xy^2\, dy = 0$,
(h) $y\, dx + (y^3 - 2x)\, dy = 0$,
(i) $(3y + 3e^x y^{2/3})\, dx + x\, dy = 0$,
(j) $(x + x^2y + y^3)\, dx + (y - x^3 - xy^2)\, dy = 0$,
(k) $(xy + y^2)\, dx + (xy - x^2)\, dy = 0$,
(l) $(y^3 - 2x^2y)\, dx + (2xy^2 - x^3)\, dy = 0$,
(m) $(5y - 6x)\, dx + x\, dy = 0$,
(n) $(\sin y + x^2 + 2x)\, dx + \cos y\, dy = 0$,
(o) $(3x - y^2)\, dx - 4xy\, dy = 0$.

2. Find the general solution of each of the following linear differential equations:

(a) $(dy/dx) + 3y = x$,

(b) $(dy/dx) + [y/(x + 1)] = \sin x$,

(c) $(\sin^2 x - y) \, dx - \tan x \, dy = 0$,

(d) $(y^2 - 1) \, dx + (y^3 - y + 2x) \, dy = 0$,

(e) $(dx/dt) + x = e^{2t}$.

3. Find the particular solution indicated:

(a) $(3xy + 2) \, dx + x^2 \, dy = 0$; $y = 1$ when $x = 1$;

(b) $xy' + 2y = 2x \cos 2x + 2 \sin 2x$; $y = 1$ when $x = \pi$.

4. (a) An equation of form

$$y' + p(x)y = q(x)y^n \qquad (n \neq 1)$$

is called a *Bernoulli equation*. Show that the substitution $u = y^{1-n}$ reduces the equation to a linear equation for u as function of x. Obtain a formula analogous to (2–58) for the general solution.

(b) Show that the equation

$$[p(x)y^{k+1} + q(x)y^m] \, dx + r(x)y^k \, dy = 0 \qquad (m \neq k + 1)$$

can be written as a Bernoulli equation.

(c) Determine which of the equations of Prob. 1 are Bernoulli equations for y as a function of x or for x as a function of y.

5. Show that $v(x, y)$ is an integrating factor of the differential equation $P \, dx + Q \, dy = 0$ if and only if

$$Q \frac{\partial v}{\partial x} - P \frac{\partial v}{\partial y} = v \left(\frac{\partial P}{\partial y} - \frac{\partial Q}{\partial x} \right).$$

(*Additional problems* are given at the end of the chapter.)

ANSWERS

1. (a) $x^3 + 3x^2y = c$;   (b) $x^4 + 4x^3y = c$;   (c) $x + y \log |y| = cy$ and $y = 0$; (d) $y \log |x^2y^3| - 2 = cy$, $x = 0$ and $y = 0$; (e) $2x + \log (x^2 + y^2) = c$; (f) $xy^4 - 3 = 3cxy$, $y = 0$, $x = 0$; (g) $x^2(y^3 + 1) = c$; (h) $x + y^3 = cy^2$, $y = 0$;   (i) $xy^{1/3} + e^x = c$;   (j) $y = x \tan \{ - [2(x^2 + y^2)]^{-1} + c\}$;   (k) $x + y \log |xy| = cy$, $y = 0$, $x = 0$;   (l) $x^2y^4 - x^4y^2 = c$;   (m) $x^5y = x^6 + c$; (n) $e^x(\sin y + x^2) = c$;   (o) $x(x - y^2)^2 = c$.

2. (a) $9y = 3x - 1 + ce^{-3x}$;   (b) $(x + 1)y = \sin x - (x + 1) \cos x + c$; (c) $3y \sin x = \sin^3 x + c$; (d) $2x(y - 1) = (y + 1)[4y - y^2 - \log (y + 1)^4 + c]$, $y = -1$; (e) $3x = e^{2t} + ce^{-t}$.

3. (a) $x^3y + x^2 = 2$, (b) $x^2y = x^2 \sin 2x + \pi^2$.

4. (a) $y = v^{-1}[(1 - n)\int v^{1-n}q \, dx + c]^{1/(1-n)}$, $v = e^{\int p \, dx}$;   (c) parts (a), (b), (c), (e), (f), (g), (h), (i), (m), (o).

2–9 Method of substitution. One very general way to attack the first order equation $P\,dx + Q\,dy = 0$ is to introduce new variables u, v by the equations

$$x = \phi(u, v), \qquad y = \psi(u, v). \tag{2–60}$$

We shall consider the process formally, give examples, and then point out assumptions which should be satisfied to ensure that the steps to be taken have meaning.

From (2–60) we obtain

$$dx = \frac{\partial \phi}{\partial u}\,du + \frac{\partial \phi}{\partial v}\,dv, \qquad dy = \frac{\partial \psi}{\partial u}\,du + \frac{\partial \psi}{\partial v}\,dv. \tag{2–61}$$

Hence x, y, dx, dy are expressed in terms of u, v, du, dv, and the differential equation $P\,dx + Q\,dy = 0$ becomes a differential equation in u, v:

$$P[\phi(u, v), \psi(u, v)]\left(\frac{\partial \phi}{\partial u}\,du + \frac{\partial \phi}{\partial v}\,dv\right)$$

$$+ Q[\dots, \dots]\left(\frac{\partial \psi}{\partial u}\,du + \cdots\right) = 0. \tag{2–62}$$

When we collect terms, (2–62) becomes

$$P_1(u, v)\,du + Q_1(u, v)\,dv = 0. \tag{2–63}$$

Let us suppose that the general solution of (2–63) has been found in form

$$U(u, v) = c. \tag{2–64}$$

From (2–60) we obtain u, v as functions of x, y,

$$u = F(x, y), \qquad v = G(x, y). \tag{2–65}$$

Substitution in (2–64) yields

$$U[F(x, y), \quad G(x, y)] = c, \tag{2–66}$$

which represents the general solution of $P\,dx + Q\,dy = 0$.

EXAMPLE 1. $3x^5\,dx - y(y^2 - x^3)\,dy = 0$.

Solution. We set

$$u = x^3, \qquad v = y^2, \qquad du = 3x^2\,dx, \qquad dv = 2y\,dy.$$

The equation becomes homogeneous:

$$u \, du - \tfrac{1}{2}(v - u) \, dv = 0.$$

Solution by the method of Section 2–4 yields

$$(v - 2u)(v + u)^2 = c;$$

hence

$$(y^2 - 2x^3)(y^2 + x^3)^2 = c$$

is the general solution of the given equation.

EXAMPLE 2.

$$(x\sqrt{x^2 + y^2} + x^2y + y^2 + y^3) \, dx$$
$$+ (y\sqrt{x^2 + y^2} - x^3 - xy - xy^2) \, dy = 0.$$

Solution. We introduce polar coordinates:

$$x = r \cos \theta, \qquad y = r \sin \theta, \qquad dx = \cos \theta \, dr - r \sin \theta \, d\theta,$$
$$dy = \sin \theta \, dr + \cos \theta \, d\theta.$$

The equation becomes

$$(r^2 \cos \theta + r^3 \sin \theta \cos^2 \theta + r^2 \sin^2 \theta + r^3 \sin^3 \theta)(\cos \theta \, dr - r \sin \theta \, d\theta)$$
$$+ (r^2 \sin \theta - r^3 \cos^3 \theta - r^2 \sin \theta \cos \theta - r^3 \sin^2 \theta \cos \theta)$$
$$\times (\sin \theta \, dr + r \cos \theta) \, d\theta = 0,$$
$$r^2 \, dr - (r^3 \sin \theta + r^4) \, d\theta = 0,$$

which is a Bernoulli equation (Prob. 4 in Section 2–8) that becomes linear after the substitution $u = r^{-1}$. The solutions are found to be

$$re^{\cos \theta} \left(c - \int e^{-\cos \theta} \, d\theta \right) = 1.$$

The solution can be rewritten in rectangular coordinates if desired. If this is to be done, it is simpler to replace the indefinite integral of $e^{-\cos \theta}$ by

$$\int_0^\theta e^{-\cos t} \, dt$$

and then to set $\theta = \tan^{-1}(y/x)$.

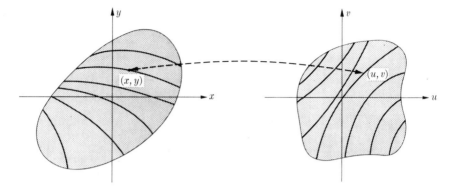

FIG. 2–7.   Change of variables in first order equation.

Discussion of the method.   The equations (2–60) can be regarded as a transformation, or mapping, of a region in the uv-plane into the xy-plane. We want to be able to solve for u, v in terms of x, y as in (2–65), which means that *we must assume a one-to-one correspondence between the points of a domain in the uv-plane and those of a domain in the xy-plane*, as suggested in Fig. 2–7. Under such a correspondence the family of solution curves of the given differential equation is replaced by a curve family in the uv-plane. If the functions (2–60) and (2–65) have continuous partial derivatives, then it follows from calculus that each curve in the uv-domain satisfies (2–63); conversely, each solution of (2–63) corresponds to a solution of $P\,dx + Q\,dy = 0$.

The requirement that the correspondence be one-to-one is ordinarily satisfied only if attention is restricted to sufficiently small domains in the xy-plane and the uv-plane. In Example 1, the transformation is given by $u = x^3$, $v = y^2$; this correspondence is one-to-one if, for example, $y > 0$ and $v > 0$. The inverse functions are $x = u^{1/3}$, $y = v^{1/2}$. To ensure that these have continuous partial derivatives, we must make a further restriction; for example, $u > 0$, $x > 0$. Thus we apparently obtain only the solutions in the first quadrant: $x > 0$, $y > 0$. However, a similar analysis applies to each of the other quadrants. The lines $x = 0$, $y = 0$ are solution curves; the origin is a singular point.

In Example 2, the transformation is given by $x = r \cos \theta$, $y = r \sin \theta$ and is one-to-one if, for example, $r > 0$ and $-\pi < \theta < \pi$. Under these restrictions the solutions are meaningful and provide the solutions in the xy-plane without the negative x-axis. To study the solutions near the omitted line, we can require $r > 0$, $0 < \theta < 2\pi$, and the same formula is obtained.

In certain cases we do not need to assume the transformation to be one-to-one. A much simpler reasoning is sufficient. Given the differential equation $P\,dx + Q\,dy = 0$, we suppose the functions $u(x, y)$, $v(x, y)$ to be such that (in a certain domain)

$$P_1(u, v)\,du + Q_1(u, v)\,dv \equiv P\,dx + Q\,dy$$

when u, v, $du = (\partial u/\partial x)\,dx + (\partial u/\partial y)\,dy$, $dv = (\partial v/\partial x)\,dx + (\partial v/\partial y)\,dy$ are expressed in terms of x and y. Suppose further that we have found an integrating factor V for the equation $P_1\,du + Q_1\,dv = 0$, that is, that

$$dU(u, v) = VP_1\,du + VQ_1\,dv,$$

so that the solutions are given by $U(u, v) = c$. Then we can also write

$$dU = V(P_1\,du + Q_1\,dv) = V(P\,dx + Q\,dy);$$

that is, when u and v are expressed in terms of x and y, $V(u, v) = V[u(x, y), v(x, y)]$ becomes an integrating factor for $P\,dx + Q\,dy = 0$ and the solutions are given by $U[u(x, y), v(x, y)] = c$. The crucial fact is that *it may not be necessary to solve for x and y in terms of u and v.*

Example 1 fits the conditions described. For, with $u = x^3$, $v = y^2$, we have

$$u\,du - \tfrac{1}{2}(v - u)\,dv \equiv 3x^5\,dx - y(y^2 - x^3)\,dy.$$

PROBLEMS

1. Apply the given transformation to obtain a differential equation in the new variables; then obtain the general solution of the given differential equation:

(a) $(2x + y)\,dx + (x + 5y)\,dy = 0$; $u = x - y, v = x + 2y$;

(b) $3x^2 ye^y\,dx + x^3 e^y(y + 1)\,dy = 0$; $u = x^3$, $v = ye^y$;

(c) $(2x - 3y)\,dx - 3x\,dy = 0$; $u = x^2 - 3xy, v = x + y$;

(d) $(x + 2y)\,dx + (y - 2x)\,dy = 0$; $x = r\cos\theta, y = r\sin\theta$;

(e) $(3x^3 + xy^2 - x^2y^2 - y^4)\,dx + (3x^2y + y^3 + x^3y + xy^3)\,dy = 0$; $x = r\cos\theta, y = r\sin\theta$.

2. Find an appropriate transformation of variables and obtain the general solution:

(a) $e^x \sin y\,dx + e^x \cos y\,dy = 0$;

(b) $(x^2 + y^2)^3(y\,dx + x\,dy) + 6xy(x^2 + y^2)^2(x\,dx + y\,dy) = 0$;

(c) $2(x^3 - xy^3)\,dx + 3(x^2y^2 + y^5)\,dy = 0$.

3. Show that an appropriate translation of axes: $x = u + h$, $y = v + k$, converts

$$(a_1x + b_1y + c_1)\,dx + (a_2x + b_2y + c_2)\,dy = 0$$

(a_1, b_1, c_1, a_2, b_2, c_2 constants) into a homogeneous equation in u, v, provided

$a_1b_2 - a_2b_1 \neq 0$. Show that, when $a_1b_2 - a_2b_1 = 0$, the equation can be written in the form

$$[k_1(\alpha x + \beta y) + c_1]\, dx + [k_2(\alpha x + \beta y) + c_2]\, dy = 0.$$

Show that, if $\alpha^2 + \beta^2 \neq 0$, the substitution $x = \alpha u + \beta v$, $y = \beta u - \alpha v$ now leads to a separation of variables.

4. Find all solutions of the following equations (see Prob. 3):

(a) $(x + 2y - 1)\, dx + (2x - y - 7)\, dy = 0$;
(b) $(x + y + 1)\, dx + (2x + 2y + 1)\, dy = 0$;
(c) $(3x - 3y + 2)\, dx + (2x - 2y + 1)\, dy = 0$;
(d) $(2x + 3y)\, dx + (3x + 2y + 1)\, dy = 0$.

5. Show that the introduction of polar coordinates: $x = r \cos\theta$, $y = r \sin\theta$ leads to a separation of variables in a homogeneous equation $y' = F(x, y)$.

(*Additional problems* are given at the end of the chapter.)

ANSWERS

1. (a) $u\, du + v\, dv = 0$, $2x^2 + 2xy + 5y^2 = c$;   (b) $u\, dv + v\, du = 0$, $x^3 y e^y = c$;
(c) $du = 0$, $x^2 - 3xy = c$;   (d) $r\, dr - 2r^2\, d\theta = 0$, $r = ce^{2\theta}$;   (e) $r^3(1 + 2\cos^2\theta)\, dr + r^5 \sin\theta\, d\theta = 0$, $r = c \exp [(\sqrt{2}/2) \tan (\sqrt{2}\cos\theta)]$.

2. (a) $e^x \sin y = c$,   (b) $xy(x^2 + y^2)^3 = c$,   (c) $\log (x^4 + y^6) + 2 \operatorname{arc\,tan} (x^2/y^3) = c$.

4. (a) $x^2 + 4xy - y^2 - 2x - 14y = c$;   (b) $x + 2y + \log |x + y| = c$ and $x + y = 0$;   (c) $\log |5x - 5y + 3| + 15x + 10y = c$, $5x - 5y + 3 = 0$;   (d) $x^2 + 3xy + y^2 + y = c$.

2–10 Orthogonal trajectories. Let there be given a family of curves in a domain D in the xy-plane. We assume that each curve has a well-defined tangent at each point and that exactly one curve passes through each point of D, as suggested in Fig. 2–8.

Now let us suppose that a second family of curves is given in D, with similar properties, and that each curve of the second family meets each curve of the first family at *right angles*. Then we say that the second family forms a set of *orthogonal trajectories* of the first family; similarly, the first family is a set of orthogonal trajectories of the second. (The word *trajectory* is synonymous with *solution curve;* the term arose in mechanics.)

If the first family consists of the solutions of a differential equation

$$P(x, y)\, dx + Q(x, y)\, dy = 0, \tag{2-67}$$

then, under reasonable assumptions, the second family is *uniquely* defined

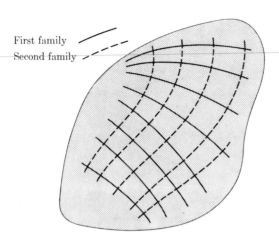

First family

Second family

FIG. 2–8.   Orthogonal trajectories.

as the family of solutions of the differential equation

$$Q(x, y) \, dx - P(x, y) \, dy = 0 \qquad (2\text{–}68)$$

in D.   For example, if $P^2 + Q^2 \neq 0$ in D and $P(x, y), Q(x, y)$ have continuous first partial derivatives in D, then the existence theorem (Section 1–8) guarantees a unique solution of (2–67) or of (2–68) through each point of D.   The corresponding slopes are

$$\frac{dy}{dx} = -\frac{P}{Q}, \qquad \frac{dy}{dx} = \frac{Q}{P}. \qquad (2\text{–}69)$$

Since these are negative reciprocals of each other (one of the two may be ∞), the corresponding curves are orthogonal.

If the first family is given as the primitive of (2–67) in the form $F(x, y, c) = 0$, then Eq. (2–67) can be obtained as in Section 1–10 by differentiating and eliminating c.   From (2–67) we obtain (2–68) by interchanging coefficients and changing sign.   The orthogonal trajectories are then the solutions of (2–68).

EXAMPLE 1.   Find the orthogonal trajectories of the circles $x^2 + y^2 = c$.

Solution.
$$2x \, dx + 2y \, dy = 0,$$

$$y \, dx - x \, dy = 0,$$

$$\frac{dx}{x} - \frac{dy}{y} = 0,$$

$$y = cx, \qquad x = 0.$$

The orthogonal trajectories are the rays through the origin. The origin itself is a singular point for both families.

EXAMPLE 2. Find the orthogonal trajectories of the family $x^2 - cx + 4y = 0$.

Solution.
$$2x - c + 4y' = 0, \tag{2-70}$$

$$x^2 - x(2x + 4y') + 4y = 0, \tag{2-71}$$

$$4xy' - 4y + x^2 = 0. \tag{2-72}$$

We have used derivatives instead of differentials; hence we replace y' by $-1/y'$ to obtain the orthogonal trajectories:

$$\frac{-4x}{y'} - 4y + x^2 = 0,$$

$$-4x\,dx + (x^2 - 4y)\,dy = 0.$$

The substitution $u = x^2$ leads to a linear equation for u as a function of y. The solutions are found to be

$$(x^2 - 4y - 8)e^{-y/2} = c.$$

There are two common errors in finding orthogonal trajectories. The first is to fail to eliminate the constant from the given primitive; in Example 2 this would lead to the use of (2-70) instead of (2-72) as the differential equation of the given family. The second error is to forget to replace y' by $-1/y'$, so that the given family is obtained instead of the orthogonal trajectories.

* 2–11 Other applications of the first order equation. Applications of linear equations are given in the next chapter; here we confine ourselves to applications of nonlinear equations.

A. *Particle moving on a line.* Let a particle of mass m move on the x-axis subject to a force F which depends only on the position of the particle. Newton's second law gives the differential equation

$$m\frac{d^2x}{dt^2} = F(x). \tag{2-73}$$

Now if $v = dx/dt$, then

$$\frac{d^2x}{dt^2} = \frac{dv}{dt} = \frac{dv}{dx}\frac{dx}{dt} = v\frac{dv}{dx}. \tag{2-74}$$

Hence Eq. (2–73) becomes a first
order equation:

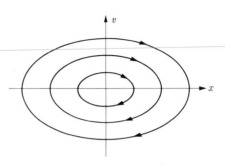

$$mv \frac{dv}{dx} = F(x). \qquad (2\text{--}75)$$

The solutions are obtained by separation of variables:

$$mv \, dv = F(x) \, dx,$$

$$m \frac{v^2}{2} = \int F(x) \, dx + c_1. \qquad (2\text{--}76)$$

Let

$$U(x) = -\int F(x) \, dx, \qquad (2\text{--}77)$$

FIG. 2–9. Velocity versus displacement for motion subject to Hooke's
law.

for some particular choice of the indefinite integral. Then (2–76) becomes

$$m \frac{v^2}{2} + U(x) = c_1. \qquad (2\text{--}78)$$

The term $\frac{1}{2}mv^2$ represents the *kinetic energy* of the particle; the function
$U(x)$ (determined up to an additive constant) is called the *potential
energy*. Hence (2–78) states:

$$\text{kinetic energy} + \text{potential energy} = \text{const} \qquad (2\text{--}79)$$

for each motion of the system. Equation (2–79) is the *law of conservation
of energy* for the case considered. The constant on the right is called the
total energy.

If, for example, $F(x) = -kx$ (Hooke's law), where $k > 0$, then the
curves (2–78) in the xv-plane are ellipses (Fig. 2–9):

$$\tfrac{1}{2}mv^2 + \tfrac{1}{2}kx^2 = c_1. \qquad (2\text{--}80)$$

When $v > 0$, x increases as t increases, so that the motion is to the right;
when $v < 0$, x decreases as t increases. The motion of the particle on the
x-axis is a projection of the motion on the ellipse; hence the particle oscil-
lates back and forth, with maximum displacement $\sqrt{2c_1/k}$.

The time t can be found in terms of x from (2–78), for (2–78) gives

$$\frac{dx}{dt} = v = \pm \sqrt{\frac{2}{m}[c_1 - U(x)]}, \quad \pm \sqrt{\frac{2}{m}} \, dt = \frac{dx}{\sqrt{c_1 - U(x)}},$$

$$\pm \sqrt{\frac{2}{m}} \, t = \int \frac{dx}{\sqrt{c_1 - U(x)}} + c_2. \qquad (2\text{--}81)$$

The plus sign applies to the upper half-plane ($v > 0$), the minus sign to the lower half.

If the particle is subject to a friction which depends only on velocity, in addition to the force F, then Eq. (2–75) is replaced by

$$mv \frac{dv}{dx} = F(x) + G(v). \tag{2-82}$$

This is a first order equation, to which some of the methods of this chapter may be applicable. When the force depends in any way on position and velocity we are led to a first order equation:

$$mv \frac{dv}{dx} = F(x, v). \tag{2-83}$$

When the force depends only on velocity v and time t, the differential equation can be written

$$m \frac{dv}{dt} = F(v, t), \tag{2-84}$$

which is again of first order. If the general solution is obtained in the form

$$\phi(v, t) = c_1, \tag{2-85}$$

then replacement of v by dx/dt leads to a first order equation for x in terms of t.

B. *Motion of the planets.* It can be shown that, to a first approximation, each planet moves in a plane as though it were attracted by a fixed mass at the center of the sun. (See Chapter 5 of Reference 6 at the end of this chapter.) We choose coordinate axes with origin at the sun and introduce polar coordinates relative to these axes. We denote by **F** the force acting on the planet and by F_r, F_θ its components along the r- and θ-directions, as indicated in Fig. 2–10. Then

$$F_r = -\frac{k}{r^2}, \qquad F_\theta = 0, \tag{2-86}$$

since all the force is toward the sun and since the force is proportional to the inverse square of the distance.

It is shown in the calculus that the components a_r, a_θ of the acceleration vector are

$$a_r = \frac{d^2r}{dt^2} - r\omega^2, \qquad a_\theta = 2\omega \frac{dr}{dt} + r \frac{d\omega}{dt}, \tag{2-87}$$

where $\omega = d\theta/dt$ is the angular velocity. (See p. 360 of Reference 7 at the end of this chapter.) Since $ma_r = F_r$, $ma_\theta = F_\theta$, where m is the mass of the planet, we conclude that

$$m\left(\frac{d^2r}{dt^2} - r\omega^2\right) = -\frac{k}{r^2}, \qquad m\left(2\omega\frac{dr}{dt} + r\frac{d\omega}{dt}\right) = 0. \qquad (2\text{–}88)$$

These differential equations govern the motion.

From the second part of (2–88) we obtain the first order equation

$$2\omega\,dr + r\,d\omega = 0, \qquad\qquad (2\text{–}89)$$

for which r is an integrating factor, and we find

$$r^2\omega = h, \qquad\qquad (2\text{–}90)$$

where h is a constant. The quantity $mr^2\omega$ is termed the angular momentum (more precisely, the xy-component of angular momentum) and Eq. (2–90) expresses the *conservation of angular momentum*. If $h = 0$, then either $r \equiv 0$ or

$$\omega = \frac{d\theta}{dt} \equiv 0.$$

This implies that θ is constant and that the planet moves on a straight line through the sun. Such a path can be followed by a meteor which falls into the sun.

For fixed nonzero h, the first part of (2–88) becomes

$$m\left(\frac{d^2r}{dt^2} - \frac{h^2}{r^3}\right) = -\frac{k}{r^2}. \qquad\qquad (2\text{–}91)$$

This second order equation governs the variation of r with time t. To study this, we introduce the new variable

$$u = \frac{1}{r} \qquad\qquad (2\text{–}92)$$

and seek a differential equation for u in terms of θ, rather than t. Now, by Eq. (2–90),

$$\frac{dr}{dt} = -\frac{1}{u^2}\frac{du}{dt} = -\frac{1}{u^2}\frac{du}{d\theta}\frac{d\theta}{dt} = -\frac{\omega}{u^2}\frac{du}{d\theta} = -h\frac{du}{d\theta}.$$

Similarly,

$$\frac{d^2r}{dt^2} = \frac{d}{dt}\left(-h\frac{du}{d\theta}\right) = \frac{d}{d\theta}\left(-h\frac{du}{d\theta}\right)\frac{d\theta}{dt} = -h\frac{d^2u}{d\theta^2}\,hu^2.$$

Accordingly, (2–91) becomes

$$m\left(-h^2u^2\frac{d^2u}{d\theta^2} - h^2u^3\right) = -ku^2$$

or, with the abbreviation $K = k/mh^2$, $(h \neq 0)$,

$$\frac{d^2u}{d\theta^2} + u = K. \tag{2–93}$$

This equation has the same form as (2–73) and we analyze it in the same way:

$$v = \frac{du}{d\theta}, \qquad v\frac{dv}{du} + u = K, \qquad \int v\,dv + \int (u - K)\,du = \frac{c_1^2}{2},$$

$$v^2 + (u - K)^2 = c_1^2, \qquad v = \pm\sqrt{c_1^2 - (u - K)^2},$$

$$\frac{du}{\sqrt{c_1^2 - (u - K)^2}} = \pm d\theta, \qquad \cos^{-1}\frac{u - K}{c_1} = \pm\theta - c_2,$$

$$u - K = c_1\cos(\pm\theta - c_2). \tag{2–94}$$

The $\pm$ sign can be absorbed in the constants c_1, c_2. With $r = 1/u$, we finally find

$$r = \frac{1}{K + c_1\cos(\theta - c_2)}. \tag{2–95}$$

It is shown in analytic geometry that the equation of a conic section with a focus at the origin is

$$r = \frac{l}{1 + e\cos(\theta - \beta)}, \tag{2–96}$$

where e is the eccentricity, $\theta = \beta$ is the axis of symmetry through the focus, and l is one-half the latus rectum (see Fig. 2–10). From (2–95) we conclude that each planet moves in an orbit which is a conic section, clearly an ellipse ($e < 1$). Comets follow approximately parabolic orbits ($e = 1$); bodies passing the solar system at great speed follow hyperbolic orbits ($e > 1$).

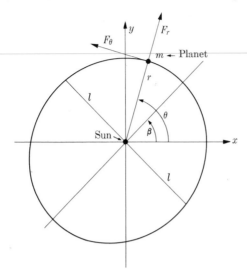

FIG. 2–10.   Motion of a planet.

C. *Chemical processes.* We visualize first a process in which compound C is formed from compounds A and B, and we assume that the rate of formation of C is proportional to the amounts of A and B present. If x denotes the amount of C (e.g., in grams), and a, b denote the amounts of A and B, then we have

$$\frac{dx}{dt} = kab. \qquad (2\text{--}97)$$

Now the amount of A present at time t equals an initial amount a_0 minus the amount used up to form C; the latter amount will be αx, where α is a constant. Similar reasoning applies to compound B. Hence

$$a = a_0 - \alpha x, \qquad b = b_0 - \beta x. \qquad (2\text{--}98)$$

The constants α and β are such that x grams of C are formed of αx grams of A and βx grams of B. From (2–97) and (2–98) we obtain the differential equation

$$\frac{dx}{dt} = k(a_0 - \alpha x)(b_0 - \beta x), \qquad (2\text{--}99)$$

which can be solved by separation of variables (Prob. 10 below).

Now we discuss the qualitative properties of the solutions of (2–99). We note, first of all, that particular solutions are given by

$$x = \frac{a_0}{\alpha}, \qquad x = \frac{b_0}{\beta}. \qquad (2\text{--}100)$$

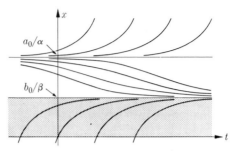

FIG. 2–11. Approach to equilibrium in a chemical process.

If $a_0/\alpha > b_0/\beta$, then these are distinct lines parallel to the t-axis (Fig. 2–11). Furthermore, since dx/dt is positive below the line $x = b_0/\beta$, negative between the two lines, and positive above the line a_0/α, we deduce that the solutions have the appearance of Fig. 2–11. We note that the isoclines are lines $x = $ const, from which it follows that a solution curve remains such after translation to the left or right. (In the diagram of Fig. 2–11 only the shaded portion has physical meaning, for x must be positive, and so also must $a = a_0 - \alpha x$, $b = b_0 - \beta x$.)

In more complicated processes, where compound C may be formed from three or more substances, Eq. (2–99) is replaced by the general equation

$$\frac{dx}{dt} = k(a_1 - \alpha_1 x)(a_2 - \alpha_2 x) \cdots (a_n - \alpha_n x). \qquad (2\text{–}101)$$

The qualitative analysis is similar to that just given (Prob. 11 below).

D. *Growth of population.* The rate of growth of any population (human, animal, bacterial) is determined by the number of individuals present at a given time and by the environment. For example, the maximum food supply will limit the number of individuals who can live in a given region. When the population is very small, the environment has little effect (e.g., the food supply appears unlimited). If x is a measure of the population at time t, the simplest law postulated is

$$\frac{dx}{dt} = kx \ (k = \text{const}). \qquad (2\text{–}102)$$

It would be meaningless to choose x, which is an integer, as the actual population; for then dx/dt would be either zero or infinite (Fig. 2–12); but we can choose a smooth function which approximates the true population and for which the differential equation (2–102) has meaning.

The effect of the environment can be taken into account in many ways. We can assume that the maximum allowable value of x (e.g., because of exhaustion of the food supply) is x_0, and postulate the differential equation

$$\frac{dx}{dt} = kx(x_0 - x), \qquad x \leqq x_0.$$
$$(2\text{--}103)$$

FIG. 2–12. Population growth approximated by a smooth curve.

An interesting variant of (2–103) is obtained by assuming that because of technological improvements, x_0 is a slowly increasing function of t. For example, we might assume

$$\frac{dx}{dt} = kx(at + b - x),$$

$$x < at + b, \qquad (2\text{--}104)$$

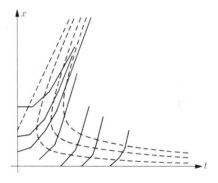

FIG. 2–13. Population growth with improving environment.

where a, b are constants. Analysis of the solutions by isoclines leads to the diagram of Fig. 2–13. Equation (2–104) is reduced to a linear equation by the substitution $y = 1/x$ (Prob. 12 below), and hence the solutions can be obtained explicitly.

E. *A nonlinear electric circuit.* To a first approximation, electric circuits are described by linear differential equations (see Chapters 3–6). Here, however, we consider a circuit containing a nonlinear element, for which the analysis is considerably more difficult.

The circuit chosen (Fig. 2–14) contains an inductance L, resistance R, and applied emf $v(t)$. In linear theory the current I obeys the differential equation

$$L\frac{dI}{dt} + RI = v(t), \qquad (2\text{--}105)$$

but we must now take into account nonlinearities in the inductance term, which are very pronounced if the coil contains an iron bar. In this case, (2–105) is replaced by the equation

$$\frac{d\Lambda(I)}{dt} + RI = v(t), \qquad (2\text{--}106)$$

where $\Lambda(I)$ is the magnetic flux caused by current I. The function $\Lambda(I)$

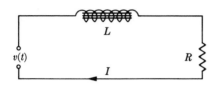

FIG. 2–14. Electric circuit with non-linear inductance.

is given empirically by experiments with different materials. It is, in fact, a two-valued function (Fig. 2–15), with one value for increasing I and another for decreasing I; its graph is the "hysteresis loop."

We shall approximate $\Lambda(I)$ by a single-valued function, which does fit the average values fairly well:

$$\Lambda(I) = b \sinh^{-1}(aI), \tag{2–107}$$

where a and b are constants. Since

$$\frac{d\Lambda(I)}{dt} = \frac{d\Lambda}{dI}\frac{dI}{dt}, \tag{2–108}$$

Eq. (2–106) becomes

$$\frac{ab}{\sqrt{1 + a^2I^2}}\frac{dI}{dt} + RI = v(t), \tag{2–109}$$

which is a nonlinear differential equation for I in terms of t.

If the applied voltage v is constant, the variables in (2–109) can be separated (Prob. 13 below). The solutions have the qualitative appearance of Fig. 2–16. If $v(t)$ is sinusoidal,

$$v = v_0 \sin \omega t, \tag{2–110}$$

the equation is difficult to solve. Graphical solution leads to a diagram such as that of Fig. 2–17 (Prob. 13 below).

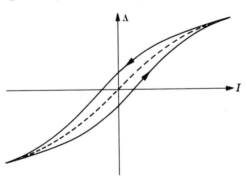

FIG. 2–15. Magnetic flux versus current.

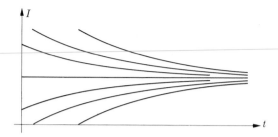

FIG. 2-16. Current in nonlinear circuit with constant applied emf.

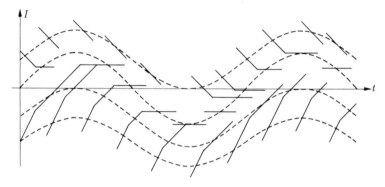

FIG. 2-17. Current in nonlinear circuit with sinusoidal applied emf.

Remarks. The examples chosen are only a few of the many types which occur in all the sciences. They illustrate the type of reasoning by which one may gain information on the physical system from the differential equation itself.

PROBLEMS

1. Find the family of orthogonal trajectories for each of the following families of curves, and graph both families in each case:

(a) $y^2 = 4cx$,   (b) $x^2 + y^2 + cx = 0$,

(c) $x^2 + y^2 + 2cy - 1 = 0$,   (d) $y = ce^{-2x}$,

(e) the family of all circles through $(1, 1)$ and $(-1, -1)$,

(f) the family of similar ellipses: $(x^2/a^2) + (y^2/4a^2) = 1$.

2. (a) Let a family of curves be the family of solutions of a differential equation $y' = f(x, y)$. Let a second similar family have the property that at each point (x, y) the angle from the tangent to the curve of the first family through (x, y) to the tangent of the second family through (x, y) is α (rad). Show that the second family satisfies the differential equation

$$y' = \frac{f(x, y) + \tan \alpha}{1 - f(x, y) \tan \alpha}.$$

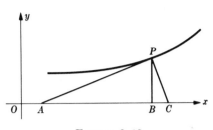

FIGURE 2-18

(b) Find the family of curves whose tangents form the angle α with the circles $x^2 + y^2 = c^2$.

(c) Find the family of curves whose tangents form the angle $\pi/4$ with the hyperbolas $xy = c$.

3. In Fig. 2–18 a general curve $y = f(x)$ is shown; $P(x, y)$ is an arbitrary point on the curve, AP is tangent to the curve, and PC is normal to the curve. Find $f(x)$ in each of the following cases:

(a) $OA = BP$ for all P; when $x = 2, y = 1$;
(b) $OA = AB$ for all P; when $x = 3, y = 1$;
(c) $(BP)^2 = BC$ for all P; when $x = 1, y = 1$.

4. Let a particle of mass m move on the x-axis subject to a force $F(x, v)$, so that $m(d^2x/dt^2) = F$. Let $x = 0$ and $v = v_0$ when $t = 0$. Find v in terms of x and t in terms of x for each of the following cases:

(a) $F = -mg$ (falling body),
(b) $F = -kx - mg$ (mass suspended by a spring),
(c) $F = -kx - ax^3$ (nonlinear spring),
(d) $F = -kmb/(x - b)^2$ (attraction by inverse square law),
(e) $F = -bv - hv^3$ (nonlinear friction).

Throughout, m, g, b, k, a, b, h denote positive constants.

5. Let a particle of mass m move on the x-axis subject to a force $F = -kx + g(v)$, where k is a positive constant and $g(v) = -bv^2$ when $v \geqq 0$, $g(v) = bv^2$ when $v \leqq 0$, $b = \text{const} > 0$. Find v in terms of x and plot the corresponding curves in the xv-plane. Give a physical interpretation.

6. Let a particle of mass m move on the x-axis subject to a force F; let $x = x_0$, $v = v_0$ when $t = 0$. Find x and v in terms of t for each of the following cases (m, g, b, k are constants):

(a) $F = -mg - bv$,
(b) $F = -bv + k \sin t$,
(c) $F = bv^3 e^{-kt}$.

7. The motion of a *simple pendulum* (Fig. 2–19) can be described by equating the tangential force $-mg \sin \theta$ to m times the acceleration component a_θ. Hence, by (2–87),

$$mL \frac{d\omega}{dt} = -mg \sin \theta.$$

Show that $d\omega/dt = \omega(d\omega/d\theta)$, find ω in terms of θ, and analyze the motion graphically in the $\omega\theta$-plane.

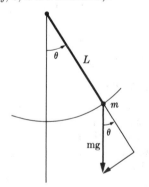

FIG. 2–19. Simple pendulum.

8. Show that, for a planet moving about the sun, conservation of angular momentum (Eq. 2–90) implies that the radius vector of the planet sweeps out area at a constant rate.

9. Find the path of a planet if the gravitational force is replaced by an attractive force proportional to distance.

10. Find the general solution of Eq. (2–99).

11. Discuss with the aid of isoclines the qualitative nature of the solutions of Eq. (2–101).

12. Solve each of the following equations and interpret the results in terms of population growth: (a) Eq. (2–102), (b) Eq. (2–103), (c) Eq (2–104). [*Hint* for (c): Set $y = 1/x$.]

13. (a) Find the solutions of Eq. (2–109) when v is a constant. (b) Analyze the solutions of Eq. (2–109) graphically with the aid of isoclines for the case $v = v_0 \sin \omega t$ (see Fig. 2–17).

ANSWERS

1. (a) $2x^2 + y^2 = c^2$, (b) $x^2 + y^2 + cy = 0$, (c) $c(x^2 + y^2) - x + c = 0$, (d) $y^2 = x + c$, (e) $c(x^2 + y^2) - x - y + 2c = 0$, (f) $y^4 = cx$, $x = 0$.

2. (b) $r = ce^{-\theta \tan \alpha}$ (in polar coordinates), (c) $x^2 - 2xy + y^2 = c$.

3. (a) $2y - y \log |y| = x$, (b) $9y = x^2$, (c) $y = e^{x-1}$.

4. (a) $v^2 - v_0^2 = -2gx$, $x = v_0 t - \frac{1}{2}gt^2$; (b) $\frac{1}{2}m(v^2 - v_0^2) = -mgx - \frac{1}{2}kx^2$, $x = (mg/k) \cos \beta t + (v_0/\beta) \sin \beta t - (mg/k), \beta = (k/m)^{1/2}$; (c) $\frac{1}{2}m(v^2 - v_0^2) = -\frac{1}{2}kx^2 - \frac{1}{4}ax^4$, $t = \pm \int_0^x (v_0^2 - km^{-1}u^2 - \frac{1}{2}am^{-1}u^4)^{-1/2} \, du$; (d) $\frac{1}{2}(v^2 - v_0^2) = kx(x - b)^{-1}$, $t = \pm \int_0^x [v_0^2 + 2ku(u - b)^{-1}]^{-1/2} \, du$; (e) $v = [cv_0 - \tan (bcx/m)]/[c + c^2 v_0 \tan (bcx/m)]$, $t = \int_0^x [c + c^2 v_0 \tan (bcu/m)]/[cv_0 - \tan (bcu/m)] \, du$.

5. $(v^2 - \frac{1}{2}mkb^{-2} \pm kb^{-1}x)e^{\pm 2bx/m} = $ const, plus for $v \geqq 0$, minus for $v \leqq 0$.

6. (a) $v = e^{-bt/m}(v_0 + gmb^{-1}) - gmb^{-1}$, $x = x_0 - mgtb^{-1} + (1 - e^{-bt/m})$ $b^{-2}(m^2g + mbv_0)$; (b) $v = (b^2 + m^2)^{-1}[\{(b^2 + m^2)v_0 + km\}e^{-bt/m} + k(b \sin t - m \cos t)]$, $x = (b^3 + m^2b)^{-1}[\{-mv_0(b^2 + m^2) - km^2\}e^{-bt/m} - kb(b \cos t + m \sin t)] + x_0 + (mv_0 + k)/b$; (c) $v^2 = kmv_0^2/[2bv_0^2(e^{-kt} - 1) + km]$, $x = x_0 + v_0\int_0^t [km/\{2bv_0^2(e^{-ku} - 1) + km\}]^{1/2} \, du$.

7. $L\omega^2 = 2g \cos \theta + c$.

9. $r^2[c_1 + (c_1^2 - K)^{1/2} \sin (2\theta + c_2)] = 1$, $K = km^{-1}h^{-2}$. By converting to rectangular coordinates, we can verify that the path is an ellipse with *center* at the origin.

10. $x = (a_0 - b_0 ce^{qt})(\alpha - \beta ce^{qt})^{-1}$, $q = k(a_0\beta - b_0\alpha)$; $x = b_0/\beta$, $x = a_0/\alpha$.

12. (a) $x = ce^{kt}$;   (b) $x = cx_0 e^{kx_0 t}(1 + ce^{kx_0 t})^{-1}$; (c) $x = -e^{pt^2 + qt}[k\int e^{pt^2 + qt} \, dt + c]^{-1}$, $p = \frac{1}{2}ka$, $q = kb$.

13. (a) $I = [ke^{2\beta t} - 2ce^{\beta t} - kc^2][a(e^{2\beta t} + 2kce^{\beta t} - c^2)]^{-1}$, $\beta t > \log |c|$, $k = av_0/R$, $\beta = Rq/(ab)$, $q = (1 + k^2)^{1/2}$.

MISCELLANEOUS PROBLEMS

For each of the following first order differential equations determine whether the equation has separable variables, is homogeneous, is linear, or is exact, and find the general solution.

1. $y' = (x + 1)/y$.
2. $y' + y = 2x + 1$.
3. $(2xy - y + 2x) \, dx + (x^2 - x) \, dy = 0$.
4. $y' = (x^2 - 1)/(y^2 + 1)$.
5. $[\{y/(xy + 1)\} + x^2] \, dx + [x \, dy/(xy + 1)] = 0$.
6. $y \sin \log x \, dx - \tan y \, dy = 0$.
7. $y' = [(x + \sqrt{x^2 - y^2})/y]$.
8. $(2x \sin xy + x^2 y \cos xy) \, dx + x^3 \cos xy \, dy = 0$.
9. $y' = y + e^y$.
10. $(2x - y) \, dx + (x + 2y) \, dy = 0$.
11. $(2x + y + 1) \, dx + (x + 3y + 2) \, dy = 0$.
12. $y' = xy^2 + 2xy$.
13. $(y - x^2 - 2xy) \, dx + (x^2 - x) \, dy = 0$.
14. $y' = (2xe^{-2x} - 2y^3)/3y^2$.
15. $(y + y^2) \, dx + (x - 2y - 4y^2 - 2y^3) \, dy = 0$.
16. $(dy/dx) + [(4x^3y + y^4)/(x^4 + 4xy^3)] = 0$.
17. $dy/dx = [2x(4x^2 - y^3)]/[3y^2(2x^2 + y^3)]$.
18. $(2y^2 + xy) \, dx + (y^2 - x^2 - xy) \, dy = 0$.
19. $(2x^3y^2 + 3x^2y^3 - 1) \, dx + (2x^4y + x^3y^2 - 1) \, dy = 0$.
20. $(x^3 + xy^2 - x^2y - y^3 + y^2) \, dx + (x^2y + y^3 + x^3 + xy^2 - xy) \, dy = 0$.
21. $t(dx/dt) + x = t^3$.
22. $(1 + 3x + 3y - x^2 + 2xy - y^2) \, dx$
     $+ (1 - 3x - 3y + x^2 - 2xy + y^2) \, dy = 0$.
23. $(3xe^{3x} \sin 2y + e^{3x} \sin 2y) \, dx + (2xe^{3x} \cos 2y + 2y) \, dy = 0$.
24. $y' = x^2y^2 + xy^2 - x^2y - y^2 - xy + y$.
25. $(5x^2y^3 + 4xy^2 + 3y) \, dx + (4x^3y^2 + 3x^2y + 2x) \, dy = 0$.

ANSWERS

1. $y^2 = (x + 1)^2 + c$.   2. $y = 2x - 1 + ce^{-x}$.   3. $x^2y - xy + x^2 = c$.
4. $y^3 + 3y = x^3 - 3x + c$.   5. $(xy + 1)^3 = ce^{-x^3}$.   6. $\int \sin \log x \, dx -$
$\int (\tan y/y) \, dy = c$. 7. $x + \sqrt{x^2 - y^2} = c$. 8. $x^2 \sin xy = c$. 9. $\int [dy/(y + e^y)] =$
$x + c$.   10. $r = ce^{-\theta/2}$ (polar coords.).   11. $2x^2 + 2xy + 3y^2 + 2x + 4y = c$.
12. $\log |y/(y + 2)| = x^2 + c$, $y = 0$, $y = -2$.   13. $x + y = c(x^2 + y)$.
14. $y^3e^{2x} = x^2 + c$. 15. $y^2 - [xy/(1 + y)] = c$. 16. $x^4y + xy^4 = c$. 17. $(y^3 +$
$4x^2)^2(y^3 - x^2)^3 = c$. 18. $(x + y)e^{x/y} = c$. 19. $x^3y^2 + 1 = c(x + y)$. 20. $r =$

$ce^{-\theta} + \frac{1}{2}(\sin\theta - \cos\theta)$(polar coords.).   21. $4xt - t^4 = c.$ 22. $9(x - y)^2 +$
$21y - 33x + 2 = ce^{3(y-x)}.$ 23. $y^2 + xe^{3x}\sin 2y = c.$ 24. $y(1 -$
$ce^{(2x^3+3x^2-6x)/6}) = 1.$ 25. $x^5y^4 + x^4y^3 + x^3y^2 = c.$

SUGGESTED REFERENCES

1. AGNEW, RALPH, *Differential Equations*. New York: McGraw-Hill, 1942.
2. FORSYTH, A. R., *Theory of Differential Equations*, Vols. 1–6. Cambridge, Eng.: Cambridge University Press, 1890–1906.
3. KAMKE, E., *Differentialgleichungen, Lösungsmethoden und Lösungen*, Vol. 1, 2nd ed. Leipzig: Akademische Verlagsgesellschaft, 1943.
4. KAPLAN, WILFRED, *Advanced Calculus*. Reading, Mass.: Addison-Wesley, 1952.
5. MCLACHLAN, N. W., *Ordinary Non-linear Differential Equations in Engineering and Physical Sciences*. London: Oxford University Press, 1950.
6. MOULTON, F. R., *Celestial Mechanics*, 2nd ed. New York: Macmillan, 1914.
7. THOMAS, GEORGE B., *Calculus*. Reading, Mass.: Addison-Wesley, 1953.

CHAPTER 3

PROPERTIES OF SOLUTIONS OF THE LINEAR
EQUATION OF FIRST ORDER

3-1 Input and output. We shall now consider the first order linear equation in the form

$$a(t) \, D_t x + x = F(t), \tag{3-1}$$

where $D_t x$ denotes the derivative dx/dt. The independent variable is the *time*, t. The dependent variable x suggests a *displacement;* in particular applications, x may represent a measure of temperature, current, charge, velocity, mass, or some other physical quantity.

Equation (3-1) can be reduced to standard form by division by $a(t)$:

$$D_t x + p(t)x = q(t), \qquad p(t) = \frac{1}{a(t)}, \qquad q(t) = \frac{F(t)}{a(t)}. \tag{3-2}$$

We shall generally assume $a(t) \neq 0$, so that Eqs. (3-1) and (3-2) are equivalent. Conversely, if $p(t) \neq 0$, Eq. (3-2) can be reduced to form (3-1) by division by $p(t)$.

In Eq. (3-1) we term $F(t)$ the *input* and think of it as describing a law of variation which we are trying to force x to follow. If $a(t)$ were 0, then x would be exactly $F(t)$; therefore the term $a(t) \, D_t x$ describes an *obstacle* which prevents x from equaling $F(t)$. A particular solution $x(t)$ of Eq. (3-1) will be called an *output*. It will be seen that in many cases, after a certain *transient* period has elapsed, there is effectively only one output function $x(t)$; we can then speak of *the* output.

A physical model for the relation between input and output is provided by the simple experiment of measurement of the temperature of the atmosphere. The input is the actual temperature; the output is the instantaneous reading of a thermometer. If the thermometer is brought outdoors from a heated house, the output will not agree with the input until some time (transient period) has elapsed.

Another physical example is that of the acceleration of an automobile, where x is the speed at time t. For a given setting of the throttle there is a corresponding maximum speed on a level road. If we press the accelerator pedal down suddenly, the car does not instantly attain the corresponding maximum speed; the actual speed (output) does not immediately agree with the desired speed (input) because of the obstacle represented by the

78

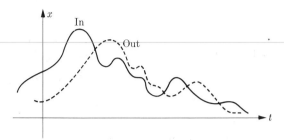

FIG. 3-1.   Input and output.

inertia of the car. A typical variation of input and output is shown in Fig. 3-1.

A great variety of other examples can be given. Indeed, our daily life is filled with examples of our wishing something to happen in a prescribed manner (input) and then finding that events (output) are not exactly as we wished them to be, because certain obstacles prevented realization of our plans. In some situations the fact that the output is a modification of the input may be desirable; in fact, considerable effort may be expended to design a mechanism which modifies the input in a prescribed manner. For example, the springs and shock absorber of an automobile are designed to transform the input (shaking of the car by a bumpy road) into a quite different output (the smooth ride enjoyed by the passengers).

The relation between input and output cannot always be described by a linear equation (3-1). When the obstacles are complicated, an equation of higher order, a nonlinear equation, or some other mathematical formalism is needed. We emphasize the linear case because of its simplicity and its wide applicability, and because of the insight it gives into more complicated problems.

In describing physical systems, engineers often make use of a "block diagram," a simple version of which is shown in Fig. 3-2. The physical system is considered as a "black box," about whose interior we know little. We know only that for each input $F(t)$ there is a corresponding output $x(t)$ (for given initial conditions) and we wish to study the relation between F and x. More complicated systems can be described by several black boxes; the input can be transmitted to Box 1, whose output may be transmitted to Box 2, and so on.

An example of a black box, with a diagram of what is inside, is given in Fig. 3-3, which illustrates a simple control system. The input $f(t)$ leads to an output $x(t)$; we are trying to

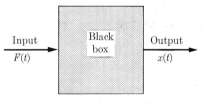

FIG. 3-2.   Simple block diagram.

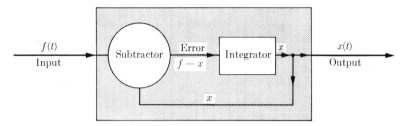

FIG. 3–3. Simple control system.

make output equal input. At each instant, output and input are compared; both are fed into the subtractor, which computes the error $f(t) - x(t)$. The error is then fed to an amplifying integrator which produces an output

$$x = \frac{1}{a} \int [f(t) - x(t)]\, dt;$$

that is,

$$a\frac{dx}{dt} = f(t) - x(t),$$

so that Eq. (3–1) holds. Since this system continually feeds the output back to influence the input, we say that the system has "feedback." A simple illustration is provided by a spectator trying to keep up with a procession; if he is lagging, he observes the error and accelerates accordingly.

3–2 Exponential decay. We now consider Eq. (3–1) and assume that $a(t)$ is a positive constant, denoted simply by a. We seek the output which corresponds to *zero input:* $F(t) \equiv 0$. The equation reads:

$$a\,D_t x + x = 0. \tag{3–3}$$

The general solution is found (e.g., by separation of variables) to be

$$x = ce^{-t/a}. \tag{3–4}$$

When $t = 0$, $x = c$, so that c can be interpreted as x_0, the initial value of the output, and Eq. (3–4) becomes

$$x = x_0 e^{-t/a}. \tag{3–5}$$

If $x_0 = 0$, the output x is identically 0, so that output equals input. If $x_0 \neq 0$, each solution x approaches 0 as t increases, as shown in Fig. 3–4

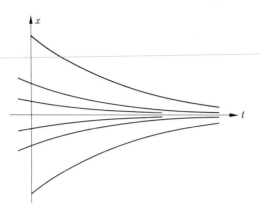

FIG. 3–4. Exponential decay.

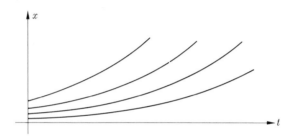

FIG. 3–5. Exponential growth.

The phenomenon described by Eq. (3–5) is known as *exponential decay*. The physical quantity measured by x has one preferred value, namely $x = 0$. If we attempt to depart from this value by choosing an initial x_0 different from 0, the value of x will gradually come back to 0 and, after a transient period, will effectively coincide with 0. We call the solutions (3–5) themselves *transients*, since they all approach 0 as $t \to +\infty$.

When one possible output is $x \equiv$ const, as here, we term that output an *equilibrium solution*. The equilibrium solution is called *stable* if all outputs (or at least those whose initial values are close enough to the equilibrium value) approach the equilibrium solution as $t \to +\infty$. Thus $x \equiv 0$ is a stable equilibrium solution in the case at hand.

The rapidity of approach to the equilibrium value 0 is determined by the size of the constant a; the larger a is, the more slowly x approaches 0. Physically a has the dimension of *time* and is called the *time constant* (other terms used are *solution time, relaxation time*). We note that when $t = a$, $x = x_0 e^{-1}$; when $t = 2a$, $x = x_0 e^{-2} = x_0(e^{-1})^2$; when $t = 3a$, $x = x_0 e^{-3} = x_0(e^{-1})^3$. Thus the values of x at the equally spaced times $t = 0,\ a,\ 2a,\ 3a,\ \dots$ form a *decreasing geometric progression* with ratio

$e^{-1} = 0.367879$. The time constant can be interpreted as the time necessary to reduce the deviation from equilibrium to 37% (approximately)
of its initial value. In general, for a sequence of t values which form an
arithmetic progression with difference d, the corresponding x values form
a geometric progression with ratio $e^{-d/a}$.

If the constant a is negative rather than positive, then the solutions (3–5)
describe increasing exponential functions. The direction of the time axis
of Fig. 3–4 must therefore be reversed, and we obtain the diagram of
Fig. 3–5. The phenomenon is now called *exponential growth*. We still have
the particular output $x = 0$; however, a slight departure from 0 in the
initial value x_0 leads to a solution which departs further and further from
0 (approaching $+\infty$ or $-\infty$) as t increases. We say that the equilibrium
$x = 0$ is *unstable*.

3–3 Constant input. If we now assume that the input $F(t)$ is constant,
equal to a fixed number F_0, and that a is a positive constant, the differential
equation becomes

$$a\,D_t x + x = F_0. \tag{3–6}$$

The solutions are found to be

$$x = F_0 + ce^{-t/a}. \tag{3–7}$$

They are plotted in Fig. 3–6.

The effect of the constant is thus simply to replace the equilibrium
solution $x = 0$ by $x = F_0$. For each solution, the deviation from the
desired value (the "error") is $x - F_0$; this quantity decays exponentially
to 0 as $t \to +\infty$ and is hence called a *transient*. Again, then, the equilibrium is stable. When a is negative, the direction of time is reversed and
the equilibrium becomes unstable.

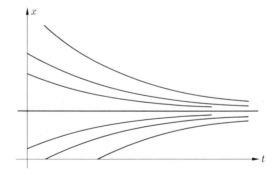

Fig. 3–6. Response to constant input.

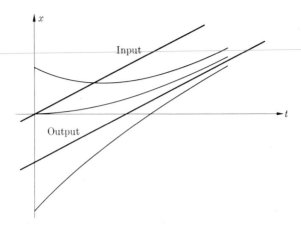

FIG. 3–7. Response to ramp input.

3–4 Ramp input. We again assume a to be a positive constant, but that

$$F(t) = kt, \qquad (3\text{--}8)$$

where k is constant. We term $F(t)$ a *ramp input*. The differential equation becomes

$$a\,D_t x + x = kt. \qquad (3\text{--}9)$$

By the general method for linear equations, the solutions are found to be

$$x = k(t - a) + ce^{-t/a}. \qquad (3\text{--}10)$$

Input and output are shown in Fig. 3–7. We recognize the transient term $ce^{-t/a}$, which decays exponentially as $t \to +\infty$. The first term $k(t - a)$ is no longer constant, but can be considered as a "steady state." It is a particular output, and every output approaches this steady state as $t \to +\infty$; that is, the deviation of any output from $k(t - a)$ is a transient. We accordingly say that the steady state is *stable*.

We note that the steady-state output $k(t - a)$ *lags the input* by time a, the time constant. (This can be used as a definition of the time constant or as a way of measuring it.) Thus the output imitates the input but is never able to reproduce it instantaneously.

It is somewhat arbitrary to call $x = k(t - a)$ the steady state here, for any one output could be called the steady state (indeed, the difference between any two outputs must be a transient); however, $x = k(t - a)$ is certainly the simplest output.

3–5 General input. Superposition principle. For constant a (>0) and general $F(t)$, our equation and solutions are

$$a\, D_t x + x = F(t), \tag{3-11}$$

$$x = \frac{1}{a} e^{-t/a} \int F(t) e^{t/a}\, dt + c e^{-t/a}. \tag{3-12}$$

If we assume $F(t)$ to be continuous for all t, then Eq. (3–12) defines a family of solutions $x(t)$ for all t (Section 2–8). In Section 3–6 it will be seen that Eq. (3–12) can be applied much more generally.

In Eq. (3–12) we denote the first term by $G(t)$:

$$G(t) = \frac{1}{a} e^{-t/a} \int F(t) e^{t/a}\, dt, \tag{3-13}$$

where some one choice of the indefinite integral is made. Then the outputs are

$$x = G(t) + c e^{-t/a}. \tag{3-14}$$

It is tempting to call $G(t)$ the steady state, for then the general output differs from the steady state by a transient (the difference decays exponentially as $t \to +\infty$), and the steady state is stable. It must, however, be emphasized that the choice of $G(t)$ depends on which indefinite integral is chosen in Eq. (3–13); changing the choice of indefinite integral adds the term $\text{const} \times e^{-t/a}$ to $G(t)$. Thus we see that the steady state is not precisely defined. However, we can agree to consider two solutions as effectively the same if they differ by a transient; then there is effectively only one steady state and, in fact, only one output.

Now let $F_1(t)$, $F_2(t)$ be inputs, continuous for all t. Let $G_1(t)$, $G_2(t)$ be corresponding outputs, defined as in (3–13). Then an *output corresponding to a linear combination* $c_1 F_1(t) + c_2 F_2(t)$ *is the linear combination* $c_1 G_1(t) + c_2 G_2(t)$. This is the *superposition principle*, which we justify by remarking that

$$\frac{1}{a} e^{-t/a} \int [c_1 F_1(t) + c_2 F_2(t)] e^{t/a}\, dt = c_1 \frac{1}{a} e^{-t/a} \int F_1(t) e^{t/a}\, dt$$

$$+ c_2 \frac{1}{a} e^{-t/a} \int F_2(t) e^{t/a}\, dt,$$

for appropriate choices of the indefinite integrals.

EXAMPLE 1. $3\, D_t x + x = 5 + 2t$. By Section 3–3, an output corresponding to the constant input 5 is $x = 5$; by Section 3–4, an output corresponding to the ramp input $2t$ is $2(t - 3)$; hence an output corresponding to the input $5 + 2t$ is $5 + 2(t - 3)$. The general output is

$$x = 5 + 2(t - 3) + c e^{-t/3}.$$

EXAMPLE 2. $3\,D_t x + x = 4t + 3e^{2t}$.   An output corresponding to e^{2t} is

$$\frac{1}{3}\,e^{-t/3}\int e^{2t}e^{t/3}\,dt = \frac{e^{2t}}{7}.$$

For the ramp input $4t$ plus the input $3e^{2t}$, we thus have an output

$$4(t - 3) + \tfrac{3}{7}e^{2t}.$$

Ignoring transients, we can say simply: multiplication of input by a constant multiplies the output by a constant; addition of two inputs produces the sum of the corresponding outputs.

PROBLEMS

1. Determine the input and the time constant for each of the following equations:

(a) $D_t x + 3x = e^t$,   (b) $5\,D_t x + 2x = \sin 2t$,
(c) $2\,D_t x + x = te^{-t}$,   (d) $D_t x + x = t^2$.

2. Find the output which is 0 for $t = 0$ and compare with the input graphically:

(a) $D_t x + 2x = 0$,   (b) $D_t x + 2x = 1$,
(c) $2\,D_t x + x = 3$,   (d) $2\,D_t x + x = t$,
(e) $2\,D_t x + x = \sin t$,   (f) $2\,D_t x + x = \sin 5t$,
(g) $10\,D_t x + x = \sin t$,   (h) $2\,D_t x + 5x = \sin t$,
(i) $2\,D_t x + x = 1 + t$,   (j) $2\,D_t x - x = t$.

3. (a) Verify that for the equation $5\,D_t x + x = F(t)$, the following list of inputs and outputs is correct:

Input	Output	Input	Output
1	1	e^t	$e^t/6$
t	$t - 5$	$\sin t$	$(\sin t - 5\cos t)/26$
t^2	$t^2 - 10t + 50$	$\cos t$	$(\cos t + 5\sin t)/26$

(b) For the equation of part (a) and with the aid of the list given, find outputs for the following inputs:

(i) $3t + 2$,   (ii) $2t^2 - t + 3$,
(iii) $3e^t - 5$,   (iv) $2t - 13\sin t$,
(v) $7\sin t + 9\cos t$,   (vi) $4\sin[t + (\pi/3)]$.

4. Prove that for the general equation $a(t)\,D_t x + x = F(t)$, with $a(t) \neq 0$, an output can be identical with an input only when the input is a constant.

5. Figure 3–8 shows part of a transient curve $ce^{-t/a}$. Show how the rest

FIGURE 3–8

of the curve can be found by graphical means alone. Also, show how the other curves $c'e^{-t/a}$ can be found for $c' \neq c$.

6. Give an example which illustrates the concepts of input and output in each of the following contexts:

(a) turning on a lamp,   (b) using a thermostat,

(c) steering a car,   (d) carrying the ball in a football game,

(e) conducting an orchestra,   (f) tracing a curve on paper,

(g) roasting a chicken,   (h) using a telephone,

(i) listening to a radio,   (j) conducting military maneuvers.

ANSWERS

1. Inputs: (a) $\frac{1}{3}e^t$, (b) $\frac{1}{2}\sin 2t$, (c) te^{-t}, (d) t^2. Time constants: (a) $\frac{1}{3}$, (b) $\frac{5}{2}$, (c) 2, (d) 1.

2. (a) 0, (b) $\frac{1}{2}(1 - e^{-2t})$, (c) $3(1 - e^{-t/2})$, (d) $t - 2 + 2e^{-t/2}$, (e) $\frac{1}{5}(-2\cos t + \sin t + 2e^{-t/2})$, (f) $\frac{1}{101}(-10\cos 5t + \sin 5t + 10e^{-t/2})$, (g) $\frac{1}{101}(-10\cos t + \sin t + 10e^{-t/10})$, (h) $\frac{1}{29}(-2\cos t + 5\sin t + 2e^{-5t/2})$, (i) $t - 1 + e^{-t/2}$, (j) $2e^{t/2} - t - 2$.

3. (b) (i) $3t - 13$, (ii) $2t^2 - 21t + 108$, (iii) $\frac{1}{2}e^t - 5$, (iv) $2t - 10 - \frac{1}{2}(\sin t - 5\cos t)$, (v) $2\sin t - \cos t$, (vi) $\frac{1}{13}[(1 + 5\sqrt{3})\sin t + (\sqrt{3} - 5)\cos t]$.

3–6 Discontinuous inputs. It is convenient to allow $F(t)$ to have jump discontinuities, as shown in Fig. 3–9. At each discontinuity, $F(t)$ has limits from the left and from the right; in each interval at most a finite number of such discontinuities occur. We call such a function "piecewise continuous." It can be integrated without difficulty.

In particular,

$$G(t) = \int_0^t F(u)\, du$$

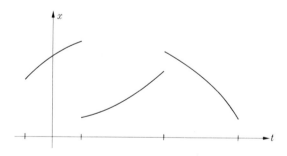

FIG. 3–9. Piecewise-continuous function.

is well defined for all t, and furthermore, $G(t)$ is continuous. For by the law of the mean, for h sufficiently small in absolute value,

$$G(t + h) - G(t) = \int_t^{t+h} F(u)\, du = hF(t_1), \qquad (3\text{–}15)$$

where t_1 is between t and $t + h$; this holds even if F is discontinuous at t, for F will then be continuous between t and $t + h$, when h is sufficiently small. If we let h approach 0 through positive values in Eq. (3–15), then $F(t_1)$ approaches its right-hand limit at t, and hence $G(t + h) - G(t) \to 0$; similarly, as h approaches 0 through negative values, $G(t + h) - G(t) \to 0$. Accordingly, $G(t)$ is continuous.

From Eq. (3–15) we reason also that

$$\frac{G(t + h) - G(t)}{h} = F(t_1). \qquad (3\text{–}16)$$

If F is continuous at t, we obtain a limit as $h \to 0$:

$$G'(t) = \lim_{h \to 0} \frac{G(t + h) - G(t)}{h} = F(t). \qquad (3\text{–}17)$$

If F is discontinuous, we can let h approach 0 through positive values or through negative values:

$$G'_+(t) = \lim_{h \to 0+} \frac{G(t + h) - G(t)}{h} = \lim_{t_1 \to t+} F(t_1),$$

$$\qquad (3\text{–}18)$$

$$G'_-(t) = \lim_{h \to 0-} \frac{G(t + h) - G(t)}{h} = \lim_{t_1 \to t-} F(t_1).$$

Thus G has left- and right-hand derivatives, equal to the left- and right-hand limits of F at t.

EXAMPLE 1. Let $F(t) = 0$ for $t < 0$, $F(t) = 1$ for $0 < t < 2$, and $F(t) = 0$ for $t \geqq 2$. Then $G(t)$ represents the area under the graph of $F(t)$ between 0 and t. Therefore $G(t) = 0$ for $t \leqq 0$, $G(t) = t$ for $0 \leqq t \leqq 2$, and $G(t) = 2$ for $t \geqq 2$. Thus $G(t)$ is continuous; $G'(t) = F(t)$ except at $t = 0$ and at $t = 2$, where $G(t)$ has "corners"; the two slopes at each corner are the two limits of $F(t)$, from the left and from the right. (See Fig. 3–10.)

If we use such a piecewise-continuous function as input in the equation

$$a\frac{dx}{dt} + x = F(t), \qquad (3\text{–}19)$$

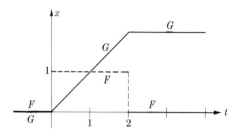

FIG. 3–10. Integral of a piecewise-continuous function.

then we can again obtain the solutions by the formula

$$x = \frac{1}{a} e^{-t/a} \int F(t)e^{t/a}\, dt + ce^{-t/a}. \tag{3-20}$$

The product $F(t)e^{t/a}$ is also piecewise continuous, and its indefinite integrals are well defined by the reasoning above:

$$\int F(t)e^{t/a}\, dt = \int_0^t F(u)e^{u/a}\, du + C. \tag{3-21}$$

If we substitute (3–20) in the differential equation (3–19), we verify (as in Section 2–8) that the equation is satisfied *except* at the jumps of $F(t)$. At each such point, x will remain continuous and have a corner; the right-(left-) hand derivative will satisfy Eq. (3–19) with F interpreted as its right- (left-) hand limit. This follows from our discussion of $G(t)$ above.

We now generalize the concept of solution of a differential equation by calling the function $x(t)$ a solution of (3–19) if $x(t)$ is continuous and satisfies the differential equation wherever $F(t)$ is continuous. The reasoning of Section 2–8 shows that all such solutions are given by Eq. (3–20) and that there will be precisely one solution which satisfies an initial condition: $x = x_0$ when $t = t_0$.

EXAMPLE 2. $3 D_t x + x = F(t)$, where $F(t)$ is the function of Example 1 (Fig. 3–10) above. We apply (3–20). To find an indefinite integral, we use (3–21):

$$\int F(t)e^{t/3}\, dt = \int_0^t F(u)e^{u/3}\, du = \begin{cases} 0, & t \leq 0, \\ 3(e^{t/3} - 1), & 0 \leq t \leq 2, \\ 3(e^{2/3} - 1), & t \geq 2. \end{cases}$$

Accordingly,

$$x = ce^{-t/a} + \phi(t),$$

$$\phi(t) = \begin{cases} 0, & t \leqq 0, \\ 1 - e^{-t/3}, & 0 \leqq t \leqq 2, \\ (e^{2/3} - 1)e^{-t/3}, & t \geqq 2. \end{cases}$$

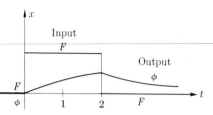

The input $F(t)$ and output $\phi(t)$ are shown in Fig. 3–11. The general output differs from $\phi(t)$ by a transient.

Fig. 3–11. Response to a step-function input.

We can obtain the output of Fig. 3–11 by the following reasoning. We start at large negative t with $x = 0$. Since the input is 0, the output remains at the equilibrium value 0 until $t = 0$. At this point the input jumps to 1 (new instructions are given); the output now behaves as though $F(t)$ were going to remain equal to 1 for all t; the output cannot guess that new orders will be given at $t = 2$! With a constant input of 1 and initial value 0 at $t = 0$, the output follows an exponential decay curve (time constant 3) relative to the equilibrium value 1. At $t = 2$ new orders instruct the output to behave as though $F(t)$ would remain 0 forever; the initial value at $t = 2$ is $1 - e^{-2/3}$. Accordingly, the output follows an exponential decay curve relative to the equilibrium value 0.

3–7 Step-function inputs. A piecewise-continuous function $F(t)$ which is constant between jumps is called a *step function* (Fig. 3–12). We can obtain the output for an equation $a\, D_t x + x = F(t)$ by the reasoning above (Section 3–6). For example, if $x = 0$ for $t = 0$ and we are interested only in the future ($t > 0$), we can obtain the output by following a succession of exponential decay curves relative to the momentary equilibrium values. Between $t = 0$ and the first jump at t_1 the error will be reduced

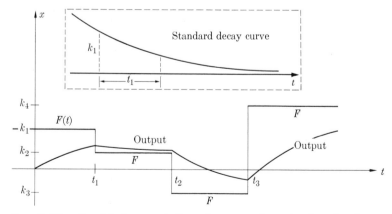

Fig. 3–12. Graphical computation of response to a step-function input.

from its initial value of k_1 by a factor of $e^{-t_1/a}$. At $t = t_1$ the new equilibrium value k_2 is introduced; the output differs from k_2 by an error E_2 [equal to $k_1(1 - e^{-t_1/a}) - k_2$]. Between $t = t_1$ and $t = t_2$ this error is reduced by a factor $e^{-(t_2-t_1)/a}$. This leads to an error E_3 at $t = t_2$, relative to the new equilibrium value k_3, and so on. The output continually pursues the input; when it gets close to its goal, the input jumps to a new value.

It should be remarked that the separate parts of the output in Fig. 3–12 are all portions of the single curve $x = e^{-t/a}$, also shown in the figure. Thus the curve between $t = 0$ and $t = t_1$ is the same as the portion of the standard curve which starts at the point where the value is k_1 and continues for t_1 units. *When this standard curve is available, the whole output can be found graphically.* The graphical process is made easier if the input is traced on transparent paper and the tracing is then placed in position over the proper portion of the standard graph.

The time constant a plays a crucial role in determining the nature of the output. If a is large, the mechanism is sluggish and the output will be unable to follow rapid changes in the input; if a is small, the output will very closely resemble the input.

It does not follow that a mechanism with the smallest possible value of a gives the optimum performance. Very often the orders given to a mechanism contain unintentional irregularities. A "wise" mechanism will ignore these, and a large time constant helps in this regard. For example, if an automobile responded instantly to every slight variation in pressure on the accelerator, it would make travel a very jerky process.

An equivalent description of the dependence on choice of time constant is obtained by keeping a fixed and varying the frequency of the jumps in the input. If $F(t)$ jumps back and forth between two values, the output will swing back and forth between intermediate values (Fig. 3–13). As the rapidity of oscillation of $F(t)$ is increased, the output will oscillate over

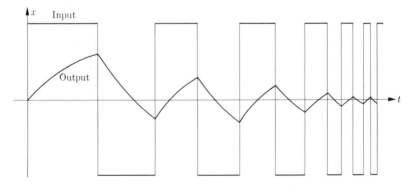

Fig. 3–13.   Response to oscillatory input.

a narrower and narrower range. For extremely rapid fluctuation of F, the output is approximately constant. (This phenomenon is the basis of motion pictures; the input fluctuates violently 32 times per second, but the human eye notices only a slight fluctuation and sees a picture which changes only gradually.)

Throughout this section we have assumed a to be positive (stable case) and have singled out one output function. Other output functions differ from the one chosen by transients and hence, after an initial period has elapsed, effectively agree with the output chosen.

3–8 Step-function approximation to arbitrary input. If $F(t)$ is a general continuous or piecewise-continuous input, we can approximate $F(t)$ arbitrarily closely by a step function $F_1(t)$, as suggested in Fig. 3–14. We can write

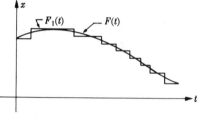

$$F(t) = F_1(t) + E(t), \quad (3\text{--}22)$$

where $E(t)$ is the error in the approximation. By the superposition principle, the actual output consists of the output from input $F_1(t)$ plus that from $E(t)$. It can be shown (Prob. 5 below) that if $|E(t)| < \epsilon$ (except perhaps at discontinuity points), then the output x_E from $E(t)$ satisfies the same inequality:

Fig. 3–14.   Approximation by step-function.

$|x_E| < \epsilon$ for all t, after an initial transient period. Hence by making the error $E(t)$ small, we obtain a good approximation to the true output by replacing $F(t)$ by the step function $F_1(t)$.

EXAMPLE 1. $2\,D_t x + x = F(t)$, where $F(t) = 0$ for $t \leq 0$, $F(t) = 2t - t^2$ for $0 \leq t \leq 2$, and $F(t) = 0$ for $t \geq 2$. We choose $F_1(t)$ to be 0 for $t < \frac{1}{2}$ and $t > \frac{3}{2}$ and to be 1 for $\frac{1}{2} \leq t \leq \frac{3}{2}$. The two inputs and corresponding outputs (for 0 initial value) are shown in Fig. 3–15. The

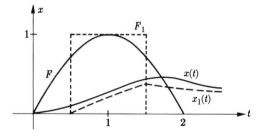

Fig. 3–15.   Response to parabolic pulse versus response to approximating square pulse.

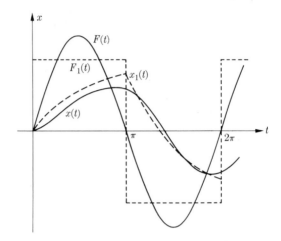

FIG. 3–16.   Response to sinusoidal wave versus response to approximating square wave.

error $E(t)$ has a maximum of $\frac{3}{4}$, but the outputs never differ by more than 0.2.

EXAMPLE 2. $2 D_t x + x = \sin t$.   We approximate the input by a *square wave:* $F_1(t) = 0.75$ for $0 < t < \pi$, $F_1(t) = -0.75$ for $\pi < t < 2\pi$, ... The two outputs for zero initial value are shown in Fig. 3–16. Again the approximation is strikingly good.

PROBLEMS

1. For each of the following functions $F(t)$, find $\int_0^t F(u)\, du = G(t)$; graph both $F(t)$ and $G(t)$.

(a) $F(t) = 0$ for $t < 0$; $F(t) = 1$ for $t \geqq 0$.

(b) $F(t) = 0$ for $t < 0$; $F(t) = \sin t$ for $0 \leqq t \leqq \pi$; $F(t) = 0$ for $t \geqq \pi$.

(c) $F(t) = [t]$, where $[t]$ is the integer n such that $n \leqq t < n + 1$; thus $[3.562] = 3$, $[-2.567] = -3$, $[-5.00] = -5$.

(d) $F(t) = (-1)^{[t]}$ [see part (c)].

(e) $F(t) = t - [t]$ [see part (c)].

(f) $F(t) = e^{t-[t]}$ [see part (c)].

2. For each of the following differential equations, obtain the general output, and graph input and output for several choices of initial value:

(a) $2 D_t x + x = F(t)$, $F(t) = 0$ for $t < 0$, $F(t) = 1$ for $t \geqq 0$;

(b) $2 D_t x + x = F(t)$, $F(t)$ as in Prob. 1(b);

(c) $2 D_t x + x = F(t)$, $F(t)$ as in Prob. 1(c);

(d) $2 D_t x + x = F(t)$, $F(t)$ as in Prob. 1(d);

(e) $5 D_t x + x = F(t)$, $F(t)$ as in Prob. 1(d).

3. (a) Graph the function $x = e^{-t/4}$ carefully for $-4 \leqq t \leqq 8$.

(b) With the aid of the graph of part (a), find graphically the output, with initial value 0, for the differential equation $4\,D_t x + x = F(t)$, if $F(t) = 0$ for $t < 0$, $F(t) = 1$ for $0 \leq t < 2$, $F(t) = 0$ for $2 \leq t < 3$, $F(t) = 2$ for $3 \leq t < 5$, $F(t) = 0$ for $t \geq 5$.

(c) Repeat part (b), with initial value -1.

4. For the differential equation $D_t x + x = F(t)$, obtain the output with initial value 0 for the following choices of $F(t)$ and compare with the output for initial value 0 when $F(t)$ is replaced by the step function $F_1(t)$ given. The output corresponding to $F_1(t)$ is to be obtained graphically, as in Prob. 3.

(a) $F(t) = 0, t < 0; F(t) = e^t - 1, 0 \leq t \leq 1; F(t) = (e^2 - e)e^{-t}, t \geq 1;$ $F_1(t) = 0, t < \frac{1}{2}; F_1(t) = 1, \frac{1}{2} \leq t < \frac{3}{2}; F_1(t) = 0, t \geq \frac{3}{2}.$

(b) $F(t) = 0, t < 0; F(t) = \cos t, 0 \leq t < \pi; F(t) = 0, t \geq \pi; F_1(t) = 0,$ $t < 0; F_1(t) = 1, 0 \leq t < \pi/6; F_1(t) = 0.7, \pi/6 \leq t < \pi/3; F_1(t) = 0,$ $\pi/3 \leq t \leq 2\pi/3; F_1(t) = -0.7, 2\pi/3 \leq t < 5\pi/6; F_1(t) = -1, 5\pi/6 \leq t$ $< \pi, F_1(t) = 0, t \geq \pi.$

5. (a) Prove that if $E(t)$ is piecewise continuous and $-\epsilon < E(t) < \epsilon$ for $t > 0$, and if a is a positive constant, then

$$-a\epsilon\, e^{t/a} < \int_0^t E(u)e^{u/a}\, du < a\epsilon\, e^{t/a}.$$

(b) With the aid of the result of part (a), show that if $|E(t)| < \epsilon$ for $t > 0$, then the output $x(t)$ with initial value 0 for the equation $a\,D_t x + x = E(t)$ satisfies the inequality $|x(t)| < \epsilon$ for $t > 0$.

ANSWERS

1. (a) $G(t) = 0, t < 0; G(t) = t, t \geq 0;$ (b) $G(t) = 0, t < 0; G(t) = 1 - \cos t, 0 \leq t \leq \pi; G(t) = 2, t \geq \pi;$ (c) $G(t) = nt - \frac{1}{2}n(n+1)$ for $n \leq t \leq n+1;$ (d) $G(t) = t - 2n,$ for $2n \leq t \leq 2n+1; G(t) = 2n + 2 - t$ for $2n+1 \leq t \leq 2n+2;$ (e) $G(t) = \frac{1}{2}[n + (t-n)^2]$ for $n \leq t \leq n+1;$ (f) $e^{t-[t]} + (e-1)[t] - 1.$

2. (a) $x = ce^{-t/2} + G(t),$ $G = 0$ for $t \leq 0,$ $G = 1 - e^{-t/2}, t \geq 0;$ (b) $x = ce^{-t/2} + G(t),$ $G = 0,$ $t \leq 0,$ $G = \frac{1}{5}(\sin t - 2\cos t + 2e^{-t/2}),$ $0 \leq t \leq \pi,$ $G = \frac{2}{5}(e^{\pi/2} + 1)e^{-t/2},$ $t \geq \pi;$ (c) $x = ce^{-t/2} + G(t),$ $G = n - e^{-t/2}(e^{1/2} - e^{(n+1)/2})/(1 - e^{1/2}),$ $n \leq t \leq n+1;$ (d) $x = ce^{-t/2} + G(t),$ $G = (-1)^n - e^{(1-t)/2}[2(-1)^n e^{n/2} + e^{-1/2} - 1]/(1 + e^{1/2}),$ $n \leq t \leq n+1;$ (e) $x = ce^{-t/5} + G(t),$ $G = (-1)^n - e^{(1-t)/5}[2(-1)^n e^{n/5} + e^{-1/5} - 1]/(1 + e^{1/5}).$

4. Actual outputs: (a) $x = 0, t < 0;$ $x = \frac{1}{2}(e^t + e^{-t}) - 1, 0 \leq t \leq 1,$ $x = \frac{1}{2}(e - 1)e^{-t}(2te - 1 - e), t > 1;$ (b) $x = 0, t < 0;$ $x = \frac{1}{2}(\sin t + \cos t - e^{-t}), 0 \leq t \leq \pi, x = -\frac{1}{2}(e^{\pi} + 1)e^{-t}, t \geq \pi.$

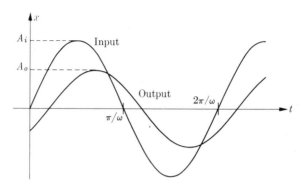

FIG. 3–17.   Response to sinusoidal input.

3–9 Sinusoidal inputs. The step-function analysis of Section 3–7 has given some insight into the type of output to be expected from an *oscillatory* input. As will be seen, a much more complete description is obtained by considering the case of a *sinusoidal* input

$$F(t) = A_i \sin \omega t \qquad (\omega > 0). \tag{3-23}$$

Here A_i is the *input amplitude* and ω is the *input frequency* (radians per unit time); these quantities are indicated in Fig. 3–17.

The output for $a\, D_t x + x = F(t)$, $(a > 0)$, is given by

$$x = \frac{1}{a} e^{-t/a} \int A_i \sin \omega t \; e^{t/a} \, dt + ce^{-t/a}. \tag{3-24}$$

On carrying out the integration, we find

$$x = \frac{A_i}{1 + a^2\omega^2} (\sin \omega t - a\omega \cos \omega t) + ce^{-t/a}, \tag{3-25}$$

which can be written

$$x = A_o \sin (\omega t - \alpha) + ce^{-t/a}, \tag{3-26}$$

where

$$A_o = \frac{A_i}{\sqrt{1 + a^2\omega^2}}, \qquad \tan \alpha = a\omega, \quad 0 < \alpha < \tfrac{1}{2}\pi. \tag{3-27}$$

The output can be considered as a *steady-state* term

$$x_1 = A_o \sin (\omega t - \alpha) \tag{3-28}$$

plus a transient. We term A_o the *output amplitude*. The output frequency is again ω. The subtraction of α represents a *lag in phase;* the output

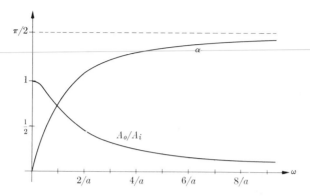

FIG. 3-18. Amplification and phase lag for sinusoidal input.

oscillation lags the input oscillation by α radians. Thus $\alpha = \frac{1}{2}\pi$ signifies a lag of one-fourth of an oscillation, $\alpha = \pi/3$ signifies a lag of one-sixth of an oscillation. This is shown in Fig. 3-17, in which α is approximately $\frac{1}{4}\pi$.

From (3-27) it follows that the output amplitude is less than the input amplitude; the *amplification factor* A_o/A_i is less than 1. The value of the amplification factor depends on the input frequency ω (a is fixed); the more rapid the input oscillations, the more the amplitude is decreased. Furthermore, as $\omega \to +\infty$, $A_o/A_i \to 0$, as shown graphically in Fig. 3-18. In the same figure the variation of the phase lag α with ω is also shown; α goes from 0 to $\frac{1}{2}\pi$ as ω goes from 0 to $+\infty$. From (3-27) it follows that fixing ω and varying a is equivalent to fixing a and varying ω; multiplying a by a constant k has the same effect on A_o as multiplying ω by k.

If the input $F(t)$ is taken as $A_i \cos \omega t$, the output is found to be

$$x = A_o \cos (\omega t - \alpha) + ce^{-t/a}, \tag{3-29}$$

where A_o and α are again given by (3-27) (Prob. 2 below). By the superposition principle it follows that the output corresponding to

$$F(t) = c_1 \cos \omega t + c_2 \sin \omega t \tag{3-30}$$

is

$$x = \frac{1}{\sqrt{1 + a^2\omega^2}} [c_1 \cos (\omega t - \alpha) + c_2 \sin (\omega t - \alpha)] + ce^{-t/a}. \tag{3-31}$$

We can write (3-30) and (3-31) as follows:

$$F(t) = A_i \sin (\omega t + \beta), \qquad A_i = \sqrt{c_1^2 + c_2^2}, \qquad \tan \beta = \frac{c_1}{c_2}, \tag{3-32}$$

$$x = A_o \sin (\omega t + \beta - \alpha) + ce^{-t/a}, \tag{3-33}$$

(Prob. 3 below); A_o and α are still given by (3–27). Thus the input has a phase shift of β and amplitude A_i; the output has a phase shift of $\beta - \alpha$ and amplitude A_o. Input and output have the *same frequency.*

3–10 Fourier series. If n is a positive integer and a_n, b_n are constants, the function

$$a_n \cos n\omega t + b_n \sin n\omega t \qquad (\omega = \text{const} > 0)$$

represents an oscillation of frequency $n\omega$. The *period* of each oscillation is $2\pi/n\omega$; hence in a time interval of length $2\pi/\omega = \tau$, the function completes n oscillations. If we form a linear combination of such oscillations for $n = 1, 2, \ldots, N$:

$$\sum_{n=1}^{N} (a_n \cos n\omega t + b_n \sin n\omega t),$$

then we obtain a function $F(t)$ which repeats itself after each interval of length $\tau = 2\pi/\omega$; that is,

$$F(t + \tau) = F(t) \qquad \text{for all } t, \tag{3–34}$$

for this relation holds for each term of the sum. We say that $F(t)$ *has period* τ. If we let N become infinite, we obtain an infinite series

$$\sum_{n=1}^{\infty} (a_n \cos n\omega t + b_n \sin n\omega t). \tag{3–35}$$

If the series converges for all t, then its sum $F(t)$ again satisfies (3–34) and has period τ. Now it can be shown that *every* function $F(t)$ which has period $\tau = 2\pi/\omega$ and satisfies certain continuity conditions can be represented as such a series plus a constant:

$$F(t) = \frac{a_0}{2} + \sum_{n=1}^{\infty} (a_n \cos n\omega t + b_n \sin n\omega t). \tag{3–36}$$

(The notation $a_0/2$ for the constant term is chosen to simplify the general law of coefficients given below.) The representation (3–36) holds, in particular, if $F(t)$ has a continuous derivative $F'(t)$ for all t. (See Reference 1 and Reference 3, Chapter 7 at the end of this chapter.)

Let us assume that $F(t)$ can be represented by (3–36). We then multiply both sides of the equation by $\cos m\omega t$ and integrate from 0 to τ:

$$\int_0^\tau F(t) \cos m\omega t \, dt = \int_0^\tau \frac{a_0}{2} \cos m\omega t \, dt$$

$$+ \sum_{n=1}^\infty \left[a_n \int_0^\tau \cos n\omega t \cos m\omega t \, dt + b_n \int_0^\tau \sin n\omega t \cos m\omega t \, dt \right]. \quad (3\text{--}37)$$

The term-by-term integration of the infinite series can be justified if, for example, $F(t)$ is continuous. Now for $m = 1, 2, \ldots$,

$$\int_0^\tau \cos n\omega t \cos m\omega t \, dt = \begin{cases} 0, & n \neq m, \\ \frac{1}{2}\tau, & n = m, \end{cases} \quad (3\text{--}38)$$

$$\int_0^\tau \sin n\omega t \cos m\omega t \, dt = 0, \quad (3\text{--}39)$$

(Prob. 4 below). Hence (3–37) becomes

$$\int_0^\tau F(t) \cos m\omega t \, dt = \frac{\tau}{2} a_m, \qquad m = 1, \ 2, \ \ldots. \quad (3\text{--}40)$$

Similarly, we find that (3–40) is correct for $m = 0$ and that

$$\int_0^\tau F(t) \sin m\omega t \, dt = \frac{\tau}{2} b_m, \qquad m = 1, \ 2, \ \ldots. \quad (3\text{--}41)$$

We obtain, accordingly, the *rule of coefficients:*

$$a_n = \frac{2}{\tau} \int_0^\tau F(t) \cos n\omega t \, dt, \qquad b_n = \frac{2}{\tau} \int_0^\tau F(t) \sin n\omega t \, dt, \quad (3\text{--}42)$$

where $\tau = 2\pi/\omega$. The constants a_n, b_n defined by (3–42) are called the *Fourier coefficients* of $F(t)$ for the interval 0 to τ, and the corresponding series (3–36) is called the *Fourier series* of $F(t)$ over the same interval. If $F(t)$ is given only between 0 and τ, and $F(t)$ is piecewise continuous, then the Fourier series of $F(t)$ is well defined. If $F(t)$ is given as a periodic function of period τ, and $F(t)$ is piecewise continuous, then the coefficients can also be computed by the formulas

$$a_n = \frac{2}{\tau} \int_{-\tau/2}^{\tau/2} F(t) \cos n\omega t \, dt, \qquad b_n = \frac{2}{\tau} \int_{-\tau/2}^{\tau/2} F(t) \sin n\omega t \, dt. \quad (3\text{--}43)$$

Indeed, the integral of $F(t) \cos n\omega t$ has the same value over any interval of length τ. The same statement applies to $F(t) \sin n\omega t$ (Prob. 5 below).

EXAMPLE. Let $F(t)$ have period 2π and be equal to 1 for $0 < t < \pi$, and equal to -1 for $-\pi < t < 0$. At the jump points we assign the value 0, which is the average of limits from the left and from the right. (See Fig. 3–19.) The function $F(t)$ is a *square wave*.

For this function, $\tau = 2\pi$ and $\omega = 2\pi/\tau = 1$. Formulas (3–43) give

$$a_n = -\frac{1}{\pi} \int_{-\pi}^{0} \cos nt \, dt + \frac{1}{\pi} \int_{0}^{\pi} \cos nt \, dt = 0,$$

$$b_n = -\frac{1}{\pi} \int_{-\pi}^{0} \sin nt \, dt + \frac{1}{\pi} \int_{0}^{\pi} \sin nt \, dt = \begin{cases} 0, & n = 2, 4, \ldots, \\ \dfrac{4}{n\pi}, & n = 1, 3, 5, \ldots \end{cases}$$

Accordingly, the Fourier series of $F(t)$ is

$$\frac{4}{\pi} \sin t + \frac{4}{3\pi} \sin 3t + \cdots = \frac{4}{\pi} \sum_{n=1}^{\infty} \frac{\sin (2n - 1)t}{2n - 1}.$$

It can be proved that this series converges to $F(t)$ for all t. At the jump points each term of the series is 0, so that the sum is 0, in agreement with the assignment of values of $F(t)$. Such a condition is typical; at a jump discontinuity the series gives the average value:

$$\frac{1}{2} \left[\lim_{h \to 0+} F(t + h) + \lim_{h \to 0-} F(t + h) \right].$$

At a general t, a partial sum of the series gives an approximation to $F(t)$.

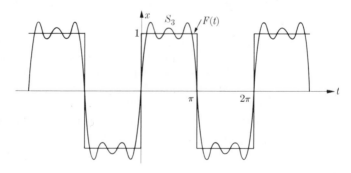

FIG. 3–19. Approximation of square wave by partial sum of Fourier series.

In Fig. 3–19 such an approximation is shown, namely that of the partial sum

$$S_3 = \frac{4}{\pi}\left(\sin t + \frac{\sin 3t}{3} + \frac{\sin 5t}{5}\right).$$

3–11 General oscillatory input. To determine the output of the differential equation

$$a\,D_t x + x = F(t) \qquad (a > 0),\tag{3–44}$$

when the input $F(t)$ is a periodic function of period τ, we need only represent $F(t)$ by its Fourier series:

$$F(t) = \frac{a_0}{2} + \sum_{n=1}^{\infty}(a_n \cos n\omega t + b_n \sin n\omega t).$$

Then

$$
\begin{aligned}
x &= \frac{1}{a}e^{-t/a}\int e^{t/a}\left[\frac{a_0}{2} + \sum_{n=1}^{\infty}(a_n \cos n\omega t + b_n \sin n\omega t)\right]dt + ce^{-t/a}\\
&= \frac{a_0}{2} + \sum_{n=1}^{\infty}\frac{(a_n - an\omega b_n)\cos n\omega t + (b_n + an\omega a_n)\sin n\omega t}{1 + a^2 n^2 \omega^2} + ce^{-t/a}
\end{aligned}
$$

$$\tag{3–45}$$

(the term-by-term integration can be justified if, for example, $F(t)$ is continuous). We thus obtain a Fourier series which represents a periodic output; the general output is equal to this periodic function or "steady state" plus a transient.

The relation between input and output is more easily seen in terms of amplitude and phase. We write

$$F(t) = \frac{a_0}{2} + \sum_{n=1}^{\infty}A_n^i \sin(n\omega t + \beta_n).\tag{3–46}$$

Then, as in Section 3–9, the steady-state output is

$$x(t) = \frac{a_0}{2} + \sum_{n=1}^{\infty}A_n^o \sin(n\omega t + \beta_n - \alpha_n),$$

$$\tag{3–47}$$

$$A_n^o = \frac{A_n^i}{\sqrt{1 + n^2\omega^2 a^2}}, \qquad \tan\alpha_n = an\omega, \quad 0 < \alpha_n < \tfrac{1}{2}\pi.$$

From (3–47) we can draw important conclusions. As n increases, the radical in the denominator increases, so that the successive terms of the

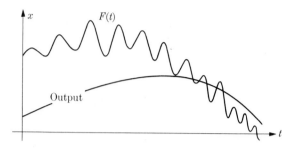

FIG. 3–20. Smoothing effect for first order linear equation.

input become progressively less important in the output. The output depends principally on the constant term, the fundamental oscillation ($n = 1$), and the first few "higher harmonics" ($n = 2, 3, \ldots$). The constant term is simply the average value of $F(t)$:

$$\frac{a_0}{2} = \frac{1}{\tau} \int_0^\tau F(t)\, dt.$$

Hence output and input have the same average value over a period. When large in amplitude, the higher harmonics give $F(t)$ a ragged appearance, as in Fig. 3–20; however, the output tends to ignore the higher harmonics and is therefore much smoother. The system represented by the linear differential equation $a\, D_t x + x = F(t)$ can thus be regarded as a *filter* which allows constants or low-frequency oscillations to pass through with only slight change, but which cuts out high-frequency oscillations.

EXAMPLE. Let $F(t)$ be the square wave of Fig. 3–19 above, and let $a = 1$. The differential equation is

$$D_t x + x = \frac{4}{\pi} \sum_{n=1}^{\infty} \frac{\sin (2n - 1)t}{2n - 1}.$$

Here $\omega = 1$, and (3–45), (3–47) give

$$x = \frac{4}{\pi} \sum_{n=1}^{\infty} \frac{-(2n - 1) \cos (2n - 1)t + \sin (2n - 1)t}{[1 + (2n - 1)^2](2n - 1)}$$

$$= \frac{4}{\pi} \sum_{n=1}^{\infty} \frac{\sin [(2n - 1)t - \alpha_n]}{(2n - 1)\sqrt{1 + (2n - 1)^2}}, \qquad \tan \alpha_n = 2n - 1,$$

for the steady-state output.

Explicit formula for the periodic output. If we write the general output of (3–44) as

$$x = G(t) + ce^{-t/a},$$

$$G(t) = \frac{1}{a} e^{-t/a} \int_0^t e^{u/a} F(u) \, du,$$

then $G(t)$ will not generally represent a periodic function. However, for appropriate choice of c, the output must agree with the steady-state output found above. To find c, we impose the condition

$$x(\tau) - x(0) = 0.$$

Accordingly,

$$G(\tau) + ce^{-\tau/a} - G(0) - c = 0,$$

$$c = \frac{G(\tau)}{1 - e^{-\tau/a}},$$

since $G(0) = 0$. Therefore, the periodic solution sought is

$$x = G(t) + \frac{G(\tau)}{1 - e^{-\tau/a}} e^{-t/a}, \tag{3–48}$$

$$G(t) = \frac{1}{a} e^{-t/a} \int_0^t e^{u/a} F(u) \, du.$$

We have required only that $x(0) = x(\tau)$, but it can be verified (Prob. 8 below) that (3–48) defines a function which is periodic for all t.

PROBLEMS

1. Find the periodic output and compare with input for each of the following equations:
   (a) $D_t x + x = 3 \sin t + \sin 5t$,
   (b) $10 \, D_t x + x = 3 \sin t + \sin 5t$,
   (c) $D_t x + x = 3 \sin t + \sin 5t + 5 \sin 100t$,
   (d) $D_t x + x = \cos t + 3 \sin t$,
   (e) $D_t x + x = \sin^3 t$.

2. Prove that (3–29) is the general output of the equation $a \, D_t x + x = A_i \cos \omega t$.

3. (a) Prove that (3–30) can be written in form (3–32). Explain how the quadrant of β is determined.
   (b) Prove that the general output of $a \, D_t x + x = A_i \sin (\omega t + \beta)$ is given by (3–33).

4. Prove (3–38) and (3–39).

5. Prove that if $f(t)$ has period τ and $f(t)$ is piecewise continuous, then

$$\int_c^{c+\tau} f(t)\, dt = \int_0^\tau f(t)\, dt$$

for every value of c. [*Hint:* Interpret the integrals as areas.]

6. Expand each of the following functions in a Fourier series. Graph the first few partial sums and compare with the function.

(a) $F(t) = t, 0 \leq t \leq \pi, F(t) = 2\pi - t$ for $\pi \leq t \leq 2\pi$, period 2π.
(b) $F(t) = t, -\pi \leq t \leq \pi$, period 2π.
(c) $F(t) = t^2, -1 \leq t \leq 1$, period 2.
(d) $F(t) = e^t, 0 \leq t \leq 1$, period 1.

7. By using the functions (a), (b), (c), (d) of Prob. 6 as inputs, find the corresponding periodic outputs of $D_t x + x = F(t)$ as Fourier series.

8. Prove that if $F(t)$ is piecewise continuous and has period τ, then (3–48) defines x as a periodic function of t. [*Hint:* Show that $x(t + \tau) - x(t) = \frac{1}{a} e^{-(t+\tau)/a} \int_\tau^{t+\tau} e^{u/a} F(u)\, du - \frac{1}{a} e^{-t/a} \int_0^t e^{u/a} F(u)\, du$. Set $v = u - \tau$ in the first integral.]

9. Obtain the periodic solutions of Prob. 7(a), (d) by means of (3–48).

ANSWERS

1. (a) $(3/\sqrt{2}) \sin (t - \arctan 1) + (1/\sqrt{26}) \sin (5t - \arctan 5)$,
(b) $(3/\sqrt{101}) \sin (t - \arctan 10) + (1/\sqrt{2501}) \sin (5t - \arctan 50)$,
(c) $(3/\sqrt{2}) \sin (t - \arctan 1) + (1/\sqrt{26}) \sin (5t - \arctan 5)$
$$+ (5/\sqrt{10,001}) \sin (100t - \arctan 100),$$
(d) $(1/\sqrt{2}) \cos (t - \arctan 1) + (3/\sqrt{2}) \sin (t - \arctan 1)$,
(e) $(3\sqrt{2}/8) \sin (t - \arctan 1) - (\sqrt{10}/40) \sin (3t - \arctan 3)$.

6. (a) $\dfrac{\pi}{2} + \dfrac{2}{\pi} \displaystyle\sum_{n=1}^{\infty} \dfrac{(-1)^n - 1}{n^2} \cos nt,$ (b) $-2 \displaystyle\sum_{n=1}^{\infty} (-1)^n \dfrac{\sin nt}{n},$

(c) $\dfrac{1}{3} + \dfrac{4}{\pi^2} \displaystyle\sum_{n=1}^{\infty} \dfrac{(-1)^n \cos n\pi t}{n^2},$

(d) $e - 1 + 2 \displaystyle\sum_{n=1}^{\infty} \dfrac{(e - 1) \cos 2\pi nt - 2\pi(e - 1)n \sin 2\pi nt}{1 + 4\pi^2 n^2}.$

7. (a) $\dfrac{\pi}{2} + \dfrac{2}{\pi} \displaystyle\sum_{n=1}^{\infty} \dfrac{(-1)^n - 1}{n^2(1 + n^2)} (\cos nt + n \sin nt),$

(b) $-2 \displaystyle\sum_{n=1}^{\infty} \dfrac{(-1)^n \sin (nt - \arctan n)}{n\sqrt{1 + n^2}},$

(c) $\dfrac{1}{3} + \dfrac{4}{\pi^2} \displaystyle\sum_{n=1}^{\infty} \dfrac{(-1)^n}{n^2\sqrt{1 + \pi^2 n^2}} \cos{(n\pi t - \arctan{n\pi})},$

(d) $e - 1 + 2 \displaystyle\sum_{n=1}^{\infty} \dfrac{[4\pi^2 e^2 n^2 + (e - 1)^2]^{1/2}}{(1 + 4\pi^2 n^2)^{3/2}} \sin{(2\pi nt + \beta_n - \alpha_n)},$

$\tan{\beta_n} = -2(e - 1)/(4\pi en), \tfrac{1}{2}\pi < \beta_n < \pi, \tan{\alpha_n} = 2\pi n, 0 < \alpha_n < \tfrac{1}{2}\pi.$

9. (a) $x = t - 1 + 2e^{-t}/(1 + e^{-\pi}),$ $0 \leqq t \leqq \pi,$ $x = 2\pi - t + 1$
$- 2e^{\pi-t}/(1 + e^{-\pi}), \pi \leqq t \leqq 2\pi;$ (d) $x = \tfrac{1}{2}[e^t + e^{1-t}], 0 \leqq t \leqq 1.$

3-12 Memory interpretation of output.

The general output of

$$a\,D_t x + x = F(t) \qquad (a = \text{const} > 0) \tag{3-49}$$

can be written

$$x = \frac{1}{a}\,e^{-t/a} \int_{t_0}^{t} F(u)e^{u/a}\,du + ce^{-t/a}. \tag{3-50}$$

This expression differs from the form (3–12) only in that the first term gives the particular solution such that $x = 0$ for $t = t_0$. We can even allow t_0 to be $-\infty$:

$$x = \frac{1}{a}\,e^{-t/a} \int_{-\infty}^{t} F(u)e^{u/a}\,du + ce^{-t/a}. \tag{3-51}$$

The integral from $-\infty$ to t can be written as the integral from $-\infty$ to 0 plus the integral from 0 to t. If the improper integral

$$\int_{-\infty}^{0} F(u)e^{u/a}\,du \tag{3-52}$$

exists, it is simply a constant. Thus Eq. (3–51) reads

$$x = \frac{1}{a}\,e^{-t/a} \left(\text{const} + \int_{0}^{t} F(u)e^{u/a}\,du \right) + ce^{-t/a},$$

which is clearly equivalent to (3–50) with $t_0 = 0$. To ensure existence of the improper integral (3–52), $F(t)$ must be reasonably small for large negative t. For most applications, $F(t)$ can be considered to be 0 for large negative t, and often for all negative t, so that (3–52) does exist.

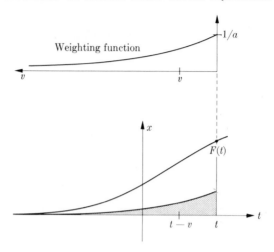

FIG. 3–21. Output as weighted average of past input.

We can write the particular output represented by the first term of (3–51) in other ways:

$$x = \frac{1}{a} e^{-t/a} \int_{-\infty}^{t} F(u) e^{u/a} \, du, \qquad x = \frac{1}{a} \int_{-\infty}^{t} F(u) e^{(u-t)/a} \, du,$$

$$x = \int_{0}^{\infty} F(t - v) \left(\frac{1}{a} e^{-v/a} \right) dv. \tag{3–53}$$

The last equation is obtained from the previous one by the substitution $v = t - u$.

Equation (3–53) can be interpreted as a weighted average over the past of $F(t)$. The value of F a time v ago [the value $F(t - v)$] receives weight $e^{-v/a}/a$. The total weight is

$$\int_{0}^{\infty} \frac{1}{a} e^{-v/a} \, dv = 1. \tag{3–54}$$

Thus the value of the output at time t can be computed graphically as shown in Fig. 3–21, where the graph of the weighting function $e^{-v/a}/a$ has been reversed and placed above the graph of $F(t)$, with the origin above the value of t here considered. The value of F at time $t - v$ multiplied by the weight $e^{-v/a}/a$ yields a new curve. The area under this curve from $-\infty$ to t is the value of x sought.

It is clear that the recent values of $F(t)$ receive most weight. Hence (except for a transient), the output is mainly affected by the instantaneous value of the input F. However, since the values in the past do receive some weight, we can say that the output does "remember" the recent past and reproduces a certain average between past and present. Thus the output has a "memory" of the past values of the input and uses all of these; the memory becomes weaker and weaker, approaching zero, as values further in the past are recalled. Values of F in the very distant past are, in effect, "forgotten."

The shape of the weighting curve depends on the time constant a; the larger a is, the flatter the curve and the better the memory. For very small a, the curve drops to zero very quickly as v increases, and the memory is very poor.

3–13 Effect of negative a and variable a. Stationarity. Throughout most of the preceding discussion, the coefficient a has been assumed to be a positive constant. If a is a negative constant, all the conclusions with respect to *increasing* time t must be replaced by conclusions with respect to *decreasing* time t. The transient term $ce^{-t/a}$ can be considered negligible only as t decreases. If we follow increasing time, this term grows indefinitely and approaches $\pm\infty$ as $t \to +\infty$. The particular outputs discussed above have limited practical value, for they are *unstable:* the slightest change in the initial conditions will introduce a term $ce^{-t/a}$ which will grow and completely change the appearance of the output as t increases.

The laws of growth of populations (human, animal, bacterial) are expressed by equations which, to first approximation, are of form $a\,D_t x + x = F(t)$, with a negative (Section 2–11). Accordingly, such growth has the unstable character described above. (Perhaps that is the reason why life on this planet is so difficult to manage!) Other similar processes, such as the growth of money at compound interest, some chemical reactions (e.g., explosions), storms in the ocean or atmosphere, spread of epidemics, and the spreading of rumors are also cases in which the output is relatively unaffected by the input. The phenomenon is exceedingly difficult to control.

If the coefficient a varies with time t, we can still obtain the solution by the general formula for linear equations (Section 2–8):

$$x(t) = e^{-\int dt/a(t)} \int \frac{F(t)}{a(t)} e^{\int dt/a(t)}\, dt + ce^{-\int dt/a(t)}. \tag{3–55}$$

Without looking at this formula in detail, we can describe each solution qualitatively. If $a(t)$ remains positive, then the output $x(t)$ still attempts to follow the input; so long as a is large and positive, the follow-up will be sluggish; if a becomes small and positive, the follow-up will become very rapid. If a crosses continuously from positive values to negative values,

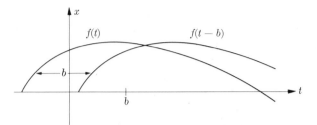

FIG. 3–22.   Effect of replacement of t by $t - b$.

then trouble arises, for the integrals with $a(t)$ in the denominator become improper and, in fact, the existence theorem of Chapter 1 is not applicable. Each such case requires special analysis. If a remains negative, the solutions are unstable. (See also Probs. 8–10 below.)

There is one fundamental difference between the case of variable a and that of constant a: *if a is constant and $x = f(t)$ is a solution of the differential equation $a\, D_t x + x = F(t)$, then for every choice of the constant b, $x = f(t - b)$ is a solution of $a\, D_t x + x = F(t - b)$.* For $a\, D_t f(t) + f(t) = F(t)$ implies $a\, D_t f(t - b) + f(t - b) = F(t - b)$. Replacement of t by $t - b$ has the effect of translating the graph of $f(t)$ by b units, as in Fig. 3–22. The assertion can thus be interpreted as follows: delay of the input by time b results in delay of the output by time b. We can also say that our mechanism is *stationary*: it responds to a given type of input in the same way at all times. When a is variable, the above reasoning fails, and the mechanism is changing its character as t varies.

3–14 Step-by-step integration. Several graphical procedures which have been suggested for the case of constant a (Fig. 3–12 and Fig. 3–21) can be adapted to the case of variable a (see Prob. 6 below). For example, $F(t)$ and $a(t)$ can be approximated by step functions as in Section 3–8. If a is variable (and positive), the procedure of Fig. 3–12 requires different exponential decay curves for different intervals of t. This procedure closely resembles the method of step-by-step integration (Section 1–5), which is itself very easy to apply. The crucial formula is

$$\Delta x = \frac{x - F(t)}{a(t)}\,\Delta t.$$

If $F(t)$ is itself a very complicated function, or is given only in tabular or graphical form, some such numerical or graphical procedure is advisable.

An alternative numerical procedure is to employ (3–55). The indefinite integrals can be replaced by definite integrals from t_0 to t and then evaluated for each t by the trapezoidal rule or Simpson's rule.

PROBLEMS

1. Find the solution $x(t)$ such that $x = 0$ for $t = 0$ for each of the following equations. Compare output with input, and discuss the extent to which the output remembers the input:

(a) $D_t x + x = F(t)$, $F(t) = 0$ for $t < 0$, $F(t) = t - t^2$ for $0 \leq t \leq 1$, $F(t) = 0$ for $t > 1$;

(b) $5 D_t x + x = F(t)$, $F(t)$ as in (a);

(c) $D_t x + x = F(t)$, $F(t) = 0$ for $t < 0$, $F(t) = 10 \sin t$ for $0 \leq t \leq \pi$, $F(t) = \sin t$ for $t > \pi$.

2. Graph the weighting function $e^{-v/a}/a$, $v \geq 0$, for $a = 0.1$, $a = 1$, $a = 10$.

3. For the following equations find the output graphically, as in Fig. 3–21:

(a) $D_t x + x = F(t)$, $F(t) = 0$ for $t < 0$, $F(t) = 1$ for $0 \leq t < 2$, $F(t) = 0$ for $t \geq 2$;

(b) $D_t x + x = F(t)$, $F(t) = 0$ for $t < 0$, $F(t) = t$ for $t \geq 0$.

4. Compare output with input for the following equations with given initial conditions:

(a) $-2 D_t x + x = 1 - t$, $x = 0$ for $t = 0$;

(b) $-2 D_t x + x = \sin t$, $x = 1$ for $t = 0$;

(c) $(1 + t) D_t x + x = 1 - t$, $x = 0$ for $t = 0$;

(d) $(1 + t) D_t x + x = \sin t$, $x = 0$ for $t = 0$;

(e) $(1 - t^2) D_t x + x = 1 - t^2$, $x = 1$ for $t = 0$.

5. Solve by step-by-step integration with $\Delta t = 0.1$ from $t = 0$ to $t = 0.5$. Compare with the exact solution.

(a) $5 D_t x + x = 1 - t$, $x = 2$ for $t = 0$;

(b) $2 D_t x + x = \sin t$, $x = 1$ for $t = 0$;

(c) $(1 + t) D_t x + x = t^2$, $x = 0$ for $t = 0$;

(d) $D_t x - x = 1$, $x = 0$ for $t = 0$.

6. Show that Eq. (3–53) can be generalized for the equation with variable coefficient: $a(t) D_t x + x = F(t)$ to provide an output of form

$$x = \int_0^\infty F(t - v) H_t(v)\, dv,$$

where the weighting function $H_t(v)$ depends on t.

7. Show that if $a(t)$ is continuous and never 0, the substitution

$$v = \int_0^t \frac{du}{a(u)}$$

converts $a(t) D_t x + x = F(t)$ into an equation with constant coefficients. The variable v can be interpreted as a special scale on the t-axis.

8. Show that the equilibrium solution $x \equiv 0$ of each of the following equations is unstable, even though $a(t) > 0$:

(a) $(1 + t^2) D_t x + x = 0,$ (b) $e^t D_t x + x = 0.$

9. Show that if $a(t)$ is continuous and greater than 0 for all t, then the equilibrium solution $x \equiv 0$ of the equation $a(t) D_t x + x = 0$ is stable if and only if

$$\int_0^\infty \frac{dt}{a(t)} = \infty.$$

Relate this result to Prob. 7.

10. Show that even though $a(t)$ is negative for all t, each solution of the following equations is bounded for all t, so that the equilibrium solution has a kind of stability:

(a) $-(1 + t^2) D_t x + x = 0,$ (b) $-\cosh t\, D_t x + x = 0.$

ANSWERS

1. (a) $x = 0, t < 0$; $x = 3e^{-t} - t^2 + 3t - 3, 0 \le t \le 1$; $x = 0.282e^{-t}$, $t \ge 1$;   (b) $x = 0,\ t < 0$; $x = 55e^{-t/5} - t^2 + 11t - 55,\ 0 \le t \le 1$; $x = 0.037e^{-t/5},\ t \ge 1$;   (c) $x = 0,\ t < 0$; $x = 5e^{-t} + 5\,(\sin t - \cos t),$ $0 \le t \le \pi$; $x = \frac{1}{2}(\sin t - \cos t) + 109e^{-t}, t \ge \pi.$

4. Outputs: (a) $e^{t/2} - t - 1$; (b) $0.2\,(\sin t + 2 \cos t + 3e^{t/2})$; (c) $(2t - t^2)/(2 + 2t), t > -1$; (d) $(1 - \cos t)/(1 + t), t > -1$; (e) $x = t - 1 + \sqrt{(1 - t)/(1 + t)}\,(\arcsin t + 2),\ -1 < t \le 1$; $x = t - 1 + \sqrt{(t - 1)/(t + 1)}\,[\log(t + \sqrt{t^2 - 1}) + c], t \ge 1,\ c$ undetermined.

5. Computed values at $t = 0.5$: (a) 1.8843, (b) 0.8206, (c) 0.0272, (d) 0.6105. Exact values: (a) 1.8807, (b) 0.8352, (c) 0.0278, (d) 0.6487.

6. $H(t, v) = b(t - v) \exp \int_t^{t-v} b(u)du,$ where $b(t) = 1/a(t).$

3–15 Physical applications. We give here several specific examples of phenomena whose laws are expressed by linear equations of first order. In each case the equation must be considered as an idealization of what is observed. For restricted ranges of the variables, the accuracy is in general very good.

A. *Heat conduction.* If a body at temperature x is immersed in a medium at temperature F, then the temperature of the body will approach temperature F at a rate which is proportional to the difference between x and F:

$$D_t x = k(F - x) \qquad (k > 0), \tag{3–56}$$

or

$$\frac{1}{k} D_t x + x = F. \tag{3–56a}$$

If F also varies with time, then $F(t)$ serves as input, the actual temperature as output. (See Reference 5, p. 221.)

B. *Motion against friction.* If a particle of mass m moves with velocity $v = D_t x$ along a line (the x-axis), and is subject to no force other than frictional resistance proportional to velocity, then the variation of velocity is governed by the linear equation

$$m\, D_t v + kv = 0. \qquad (3\text{–}57)$$

(An example is that of a boy sliding on ice.) If, in addition, a force $f(t)$ is applied, the equation becomes

$$m\, D_t v + kv = f(t). \qquad (3\text{–}58)$$

In particular, $f(t)$ may be constant; this case is illustrated by a body, (e.g., a parachute) falling against air resistance.

C. *Electric circuits.* (See Section 2–11.) If a circuit contains only a resistance R, a capacitance C, and an emf ε, as in Fig. 3–23, then the charge on the capacitor varies with time t according to the linear equation:

$$R\, D_t q + \frac{1}{C}\, q = \varepsilon. \qquad (3\text{–}59)$$

Here the emf ε may depend on time t and, in particular, be sinusoidal:

$$\varepsilon = \varepsilon_0 \sin \omega t. \qquad (3\text{–}60)$$

The current I in the circuit is simply the rate of change of the charge q:

$$D_t q = I. \qquad (3\text{–}61)$$

Hence the current I obeys the equation

$$R\, D_t I + \frac{1}{C}\, I = D_t \varepsilon. \qquad (3\text{–}62)$$

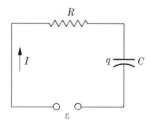

FIG. 3–23. εRC-circuit.

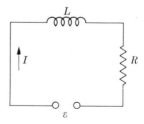

FIG. 3–24. εRL-circuit.

If the circuit contains only an inductance L, a resistance R, and an emf $\mathcal{E}$, as in Fig. 3–24, then the current obeys the linear equation:

$$L\, D_t I + RI = \mathcal{E} \tag{3–63}$$

(see Reference 6).

D. *Radioactive decay.* A substance such as radium disintegrates at a rate proportional to the mass remaining. If x is the mass at time t, we have

$$D_t x = -kx \qquad (k > 0). \tag{3–64}$$

If more of the substance is being produced (e.g., in a nuclear reactor) at rate $f(t)$, then the total mass x obeys the equation

$$D_t x + kx = f(t). \tag{3–65}$$

E. *Growth of population.* (See Section 2–11.) Consider, for example, a population in a friendly environment that permits unlimited expansion. The number of offspring and the number of deaths in a short time interval are then approximately proportional to the size of the population at the beginning of the interval. Idealizing the process as a continuous one, we obtain the differential equation

$$D_t x = kx \qquad (k > 0), \tag{3–66}$$

where x is a number which measures the size of the population. If, in addition to the natural growth, there are gains or losses in population due to immigration or emigration, then (3–66) becomes

$$D_t x = kx + f(t). \tag{3–67}$$

In all five examples, the principles discussed in this chapter find immediate application. All but the last example represent stable systems, and the output can be considered as some sort of imitation of the input, but lagging input and ignoring some of its irregularities.

PROBLEMS

1. A thermometer which reads 75°F indoors is taken outdoors. After five minutes it reads 65°; after another five minutes it reads 60°. What is the outdoor temperature?

2. If the temperature of a body is changing rapidly, the instantaneous reading of a thermometer in the body will not agree with that of the body. Let us assume the reading r to be related to the actual temperature x by a linear equation

$a\,D_t r + r = x$, $(a > 0)$. Discuss the manner in which r will vary when the temperature of the medium surrounding the body varies.

3. Mr. Smith and Mr. Brown both order coffee at the lunch counter and receive their cups at the same time. Mr. Smith adds cream at once but does not drink his coffee until five minutes later; Mr. Brown waits five minutes, then adds cream and begins to drink. Who drinks the hotter coffee?

4. (a) A man weighing (with his load) 200 lb makes a parachute jump and reaches a steady velocity of 10 mi/hr. Find the force due to air resistance when his speed is 5 mi/hr.

(b) If the same man jumps again from a height of h feet and takes 10 sec to open his parachute, how low can h be if he is to strike the ground with a speed less than 20 mi/hr.

5. In an RC-circuit, as in Fig. 3–23, if the applied emf $\mathcal{E}$ is irregular, what can be said about the relative smoothness of $\mathcal{E}$, I, and q?

6. The half-life of a radioactive substance is the time needed for one-half of a given amount to disintegrate. Express this in terms of the time constant.

7. The population of the United States from 1790 on, at ten-year intervals, was as follows, in millions: 3.9, 5.3, 7.2, 9.6, 12.9, 17, 23, 31, 39, 50, 63, 76, 92, 108, 122, 135, 150. If the growth obeys the law $D_t x = k(t)x$, discuss the behavior of $k(t)$ and its deviation from constancy. Choose some reasonable average value for k and predict the population in the years 2000, 2100, 2500, and 3000.

ANSWERS

1. 55°. 3. Mr. Smith. 4. (a) 100 lb, (b) 1754 ft. 6. $a \log 2$.

SUGGESTED REFERENCES

1. CHURCHILL, RUEL V., *Fourier Series and Boundary Value Problems.* New York: McGraw-Hill, 1941.

2. DRAPER, CHARLES S., McKAY, WALTER, and LEES, SIDNEY, *Instrument Engineering*, Vols. 1 and 2 (esp. Vol. 2, Chap. 18). New York: McGraw-Hill, 1952–1953.

3. KAPLAN, W., *Advanced Calculus.* Reading, Mass.: Addison-Wesley, 1952.

4. LAWDEN, DEREK F., *Mathematics of Engineering Systems.* New York: John Wiley and Sons, Inc., 1954.

5. LEMON, HARVEY B., and FERENCE, MICHAEL, JR., *Analytical Experimental Physics.* Chicago: University of Chicago Press, 1943.

6. SKILLING, HUGH H., *Transient Electric Currents.* New York: McGraw-Hill, 1937.

7. TRIMMER, JOHN D., *Response of Physical Systems.* New York: John Wiley and Sons, Inc., 1950.

CHAPTER 4

LINEAR EQUATIONS OF ARBITRARY ORDER

4–1 Linear differential equations. An ordinary linear differential equation of order n is a differential equation of form

$$a_0(x)D_x^n y + a_1(x)D_x^{n-1}y + \cdots + a_{n-1}(x)D_x y + a_n(x)y = Q(x), \quad (4\text{--}1)$$

where $D_x y = y'$, $D_x^2 y = y''$, ... It will be assumed here that the coefficients $a_0(x)$, $a_1(x)$, ..., $a_n(x)$, and the right-hand member $Q(x)$ are defined and continuous in some interval (perhaps infinite) of the x-axis and that $a_0(x) \neq 0$ in this interval. The fundamental theorem of Section 1–8 then implies that there is one and only one solution $y = y(x)$ of (4–1) that satisfies given initial conditions

$$y(x_0) = y_0, \qquad y'(x_0) = y_0', \qquad \ldots, \qquad y^{(n-1)}(x_0) = y_0^{(n-1)} \quad (4\text{--}2)$$

at a point x_0 of the interval.

The following are examples of linear equations:

$$y'' + y = \sin 2x, \qquad (4\text{--}3)$$

$$x^2 y'' - xy' + e^x y = \log x \qquad (x > 0), \qquad (4\text{--}4)$$

$$D_x^5 y - x D_x^3 y + x^3 D_x y = 0, \qquad (4\text{--}5)$$

$$y'' + y = 0, \qquad (4\text{--}6)$$

$$y' + x^2 y = e^x. \qquad (4\text{--}7)$$

If $Q(x) \equiv 0$, the equation (4–1) is called a *homogeneous linear equation* (not to be confused with the homogeneous first order equation of Section 2–4). Thus Eqs. (4–5), (4–6) are homogeneous; the other examples are *nonhomogeneous*. If we replace $Q(x)$ by 0 in a general equation (4–1), we obtain a new homogeneous equation, which we call the *related homogeneous equation* of Eq. (4–1). Thus Eq. (4–6) is the related homogeneous equation of Eq. (4–3).

If the coefficients $a_0(x)$, $a_1(x)$, ..., $a_n(x)$ are all constants, hence independent of x, the equation (4–1) is said to have *constant coefficients*, even though $Q(x)$ may depend on x. Thus (4–3) and (4–6) have constant coefficients; the other examples do not.

112

4–2 Linear independence. We call n functions $y_1(x), \ldots, y_n(x)$ *linearly independent* over a given interval if no one function is expressible as a linear combination of the others, with constant coefficients, over the interval. If the functions are not linearly independent over the interval, they are termed *linearly dependent.* For example, the functions

$$y_1(x) = \cos 2x, \qquad y_2(x) = \cos^2 x, \qquad y_3(x) = \sin^2 x$$

are linearly dependent for all x, since $y_1(x) \equiv y_2(x) - y_3(x)$ for all x. The functions $y_1 = 1$, $y_2 = x$, $y_3 = x^2$ are linearly independent for all x. For if, for example,

$$x^2 \equiv A \cdot 1 + B \cdot x$$

with constant A, B, then the quadratic function $x^2 - Bx - A$ would be identically 0; but the quadratic equation $x^2 - Bx - A = 0$ has at most two roots, so that there is a contradiction. Similarly, relations of form $x \equiv A \cdot 1 + B \cdot x^2$ and $1 \equiv A \cdot x + B \cdot x^2$ are found to be impossible.

The definition of linear independence can be rephrased as follows: *the functions $y_1(x), \ldots, y_n(x)$ are linearly independent over a given interval, if an identity*

$$k_1 y_1(x) + \cdots + k_n y_n(x) \equiv 0 \tag{4–8}$$

with constant k_1, k_2, ... can hold over the given interval only if $k_1 = 0$, $k_2 = 0, \ldots, k_n = 0$. If the new condition holds, then no one function is expressible as a linear combination of the others; if, for example,

$$y_n(x) \equiv A_1 y_1(x) + \cdots + A_{n-1} y_{n-1}(x), \tag{4–9}$$

then Eq. (4–8) holds with $A_1 = k_1$, $A_2 = k_2, \ldots, A_{n-1} = k_{n-1}$, $-1 = k_n$, so that not all k's are zero. If the new condition fails, then one function is expressible as a linear combination of the others; for if (4–8) holds and, for example, $k_n \neq 0$, then (4–8) can be solved for $y_n(x)$ to give a relation (4–9) with $A_1 = -(k_1/k_n), \ldots, A_{n-1} = -(k_{n-1}/k_n)$. Thus the two definitions of linear independence are equivalent.

The importance of the concept of linear independence for linear differential equations can be illustrated by considering the example

$$y''' - y'' - y' + y = 0. \tag{4–10}$$

We verify that $y = e^x$ is a solution of this equation; from the form of the equation, $y = c_1 e^x$ will also be a solution for each choice of the constant c_1. Similarly, $y = c_2 x e^x$ and $y = c_3 e^{-x}$ are found to be solutions, with c_2, c_3

constant. Finally, we observe that the sum of such solutions is a solution; that is,

$$y = c_1 e^x + c_2 x e^x + c_3 e^{-x} \tag{4–11}$$

is a solution. Since (4–11) contains three arbitrary constants, it appears to be the general solution. By the existence theorem of Section 1–8 we can prove that (4–11) is the general solution of Eq. (4–10) by showing that the c's can be chosen so that (4–11) satisfies *prescribed initial conditions:*

$$y(x_0) = y_0, \qquad y'(x_0) = y'_0, \qquad y''(x_0) = y''_0. \tag{4–12}$$

From (4–11) we find that (4–12) will be satisfied if

$$y_0 = c_1 e^{x_0} + c_2 x_0 e^{x_0} + c_3 e^{-x_0},$$

$$y'_0 = c_1 e^{x_0} + c_2 e^{x_0}(x_0 + 1) - c_3 e^{-x_0}, \tag{4–13}$$

$$y''_0 = c_1 e^{x_0} + c_2 e^{x_0}(x_0 + 2) + c_3 e^{-x_0}.$$

These are simultaneous linear equations for c_1, c_2, c_3. The determinant of coefficients is

$$\begin{vmatrix} e^{x_0} & x_0 e^{x_0} & e^{-x_0} \\ e^{x_0} & e^{x_0}(x_0 + 1) & -e^{-x_0} \\ e^{x_0} & e^{x_0}(x_0 + 2) & e^{-x_0} \end{vmatrix} = e^{4x_0} \neq 0.$$

Hence Eqs. (4–13) can be solved uniquely for c_1, c_2, c_3 and (4–11) is the general solution of (4–10).

Now the functions e^x, e^{-x}, $\cosh x$ are also solutions of Eq. (4–10), so that the same reasoning would lead us to the expression

$$y = c_1 e^x + c_2 e^{-x} + c_3 \cosh x \tag{4–14}$$

for the general solution. If we attempt to prove that (4–14) is the general solution by showing that arbitrary initial conditions can be satisfied, we are led to contradictory equations. For example, the initial conditions

$$\text{for } x = 0: \quad y = 1, \qquad y' = 0, \qquad y'' = 0$$

cannot be satisfied. They lead to the contradictory equations

$$1 = c_1 + c_2 + c_3, \qquad 0 = c_1 - c_2, \qquad 0 = c_1 + c_2 + c_3.$$

What is the reason for the difficulty? A close inspection of Eq. (4–14) reveals that it contains effectively only *two* arbitrary constants. For $\cosh x = \frac{1}{2}(e^x + e^{-x})$, so that (4–14) can be written

$$y = c_1 e^x + c_2 e^{-x} + \tfrac{1}{2}c_3(e^x + e^{-x}) = C_1 e^x + C_2 e^{-x},$$

where $C_1 = c_1 + \frac{1}{2}c_3$, $C_2 = c_2 + \frac{1}{2}c_3$. Since there are only *two* arbitrary constants at our disposal, it is clear why the *three* initial conditions cannot be satisfied.

The reduction of the number of arbitrary constants in (4–14) was possible because the three functions e^x, e^{-x}, $\cosh x$ are *linearly dependent*, for

$$\cosh x \equiv \tfrac{1}{2}e^x + \tfrac{1}{2}e^{-x}.$$

A similar reduction cannot be made in (4–11), since the functions e^x, xe^x, e^{-x} are *linearly independent*. It will be seen that the general solution of a linear differential equation has the form

$$y = y^*(x) + c_1 y_1(x) + \cdots + c_n y_n(x), \tag{4–15}$$

where $y_1(x)$, $\ldots$, $y_n(x)$ are *linearly independent* solutions of the related homogeneous equation and $y^*(x)$ is one particular solution of the given equation. However, if $y_1(x)$, $\ldots$, $y_n(x)$ are linearly dependent, the number of arbitrary constants in (4–15) can be reduced and (4–15) fails to provide the general solution.

4–3 Fundamental theorem. Test for linear independence. We can now state a general existence theorem for linear differential equations.

THEOREM 1. *Let the linear differential equation*

$$a_0(x)D_x^n y + a_1(x)D_x^{n-1} y + \cdots + a_n(x)y = Q(x) \tag{4–16}$$

be given, with the related homogeneous equation

$$a_0(x)D_x^n y + a_1(x)D_x^{n-1} y + \cdots + a_n(x)y = 0. \tag{4–17}$$

Let $a_0(x)$, $\ldots$, $a_n(x)$, $Q(x)$ be defined and continuous over an interval of the x-axis and let $a_0(x) \neq 0$ on this interval.

There exists a set of n functions which are solutions of Eq. (4–17) and are linearly independent over the given interval. If $y_1(x)$, $\ldots$, $y_n(x)$ is such a set of linearly independent solutions of Eq. (4–17) over the given *interval, then*

$$y = c_1 y_1(x) + \cdots + c_n y_n(x) \tag{4-18}$$

is the general solution of Eq. (4–17).

There exist solutions of the nonhomogeneous equation Eq. (4–16) over the given interval. If $y = y^*(x)$ is one such solution and $y_1(x), \ldots, y_n(x)$ are chosen as above, then

$$y = y^*(x) + c_1 y_1(x) + \cdots + c_n y_n(x) \tag{4-19}$$

is the general solution of Eq. (4–16).

The proof of this theorem is given in Chapter 12. One striking difference from the general existence theorem of Section 1–8 is that *all solutions of the linear equation* (4–16) *are valid over the whole interval under consideration.* For a nonlinear equation this need not be so (Prob. 6 below).

Remark. Theorem 1 implies that we cannot find $n + 1$ linearly independent solutions of the homogeneous equation (4–17). For let $y_1(x), \ldots,$ $y_{n+1}(x)$ be such linearly independent solutions. Then $y_1(x), \ldots, y_n(x)$ are also linearly independent; indeed, a relation

$$k_1 y_1(x) + \cdots + k_n y_n(x) \equiv 0,$$

with not all k's equal to zero, is a special case of a relation of form

$$k_1 y_1(x) + \cdots + k_n y_n(x) + k_{n+1} y_{n+1}(x) \equiv 0,$$

with not all k's equal to zero. Hence the general solution of (4–17) is given by

$$y = c_1 y_1(x) + \cdots + c_n y_n(x).$$

Now $y_{n+1}(x)$ is also a solution of (4–17). Therefore by Theorem 1, for some choice of the c's,

$$y_{n+1}(x) \equiv c_1 y_1(x) + \cdots + c_n y_n(x).$$

This makes $y_1(x), \ldots, y_{n+1}(x)$ linearly dependent, contrary to assumption.
We proceed to illustrate the theorem by examples.

EXAMPLE 1. $y'' + y = 0$. The functions $\sin x$ and $\cos x$ are solutions for all x, since $(\sin x)'' = -\sin x$, $(\cos x)'' = -\cos x$. The two functions are linearly independent, since neither one is a constant times the other. Hence by Theorem 1 the general solution is

$$y = c_1 \sin x + c_2 \cos x. \tag{4-20}$$

To check the correctness of this result, we replace y by $c_1 \sin x + c_2 \cos x$ in the differential equation. The left-hand side becomes

$$-c_1 \sin x - c_2 \cos x + (c_1 \sin x + c_2 \cos x)$$

or

$$c_1 (-\sin x + \sin x) + c_2 (-\cos x + \cos x). \qquad (4\text{-}21)$$

Since this is identically 0, the solutions do check. The terms were grouped in a special way in (4–21) to illustrate the fact that replacing y by $c_1 \sin x + c_2 \cos x$ is the same as

$$(c_1 \text{ times the result of replacing } y \text{ by } \sin x)$$

$$\text{plus}$$

$$(c_2 \text{ times the result of replacing } y \text{ by } \cos x).$$

In other words, the left-hand side of the differential equation "operates" on each term of (4–20) separately and constants factor out. Since $\sin x$ and $\cos x$ are solutions, each leads to a zero term, and the final result is zero, as desired. This reasoning can be extended to the general case of (4–17). Since $y_1(x), \ldots, y_n(x)$ are solutions, each gives rise to a zero term when (4–18) is substituted in (4–17). The constants factor out and the result is zero.

It remains to verify that the constants c_1 and c_2 can be chosen to match the given initial conditions

$$\text{for } x = x_0: \quad y = y_0, \quad y' = y_0'. \qquad (4\text{-}22)$$

We must solve the equations

$$y_0 = c_1 \sin x_0 + c_2 \cos x_0,$$
$$y_0' = c_1 \cos x_0 - c_2 \sin x_0. \qquad (4\text{-}23)$$

The unique solution is

$$c_1 = y_0 \sin x_0 + y_0' \cos x_0, \qquad c_2 = y_0 \cos x_0 - y_0' \sin x_0.$$

Thus, for this example, the theorem above (for the homogeneous case) is completely verified.

Let us re-examine the last steps. Why was it possible to solve (4–23) for c_1, c_2? Equations (4–23) are linear equations in the unknowns c_1, c_2, and the determinant of the coefficients is

$$\begin{vmatrix} \sin x_0 & \cos x_0 \\ \cos x_0 & -\sin x_0 \end{vmatrix} = -\sin^2 x_0 - \cos^2 x_0 = -1.$$

This determinant is never 0, hence solution is always possible. If we write the general solution as $c_1 y_1(x) + c_2 y_2(x)$, the equations (4–23) become

$$y_0 = c_1 y_1(x_0) + c_2 y_2(x_0),$$

$$y_0' = c_1 y_1'(x_0) + c_2 y_2'(x_0),$$

and the determinant of the coefficients is

$$W = \begin{vmatrix} y_1(x_0) & y_2(x_0) \\ y_1'(x_0) & y_2'(x_0) \end{vmatrix}$$

If the determinant is not equal to zero for any value of x_0 in the interval considered, then initial conditions can be satisfied. The determinant W is called the *Wronskian determinant* of the functions $y_1(x)$, $y_2(x)$ and the fact that it is not zero is *equivalent to linear independence of $y_1(x)$, $y_2(x)$*:

THEOREM 2. *Let $y_1(x)$, . . . , $y_n(x)$ be solutions of the homogeneous equation (4–17) over an interval. Then $y_1(x)$, . . . , $y_n(x)$ are linearly independent over the interval if and only if $W(x) \neq 0$ on the interval, where $W(x)$ is the Wronskian determinant:*

$$W(x) = \begin{vmatrix} y_1(x) & y_2(x) & \cdots & y_n(x) \\ y_1'(x) & y_2'(x) & & y_n'(x) \\ \vdots & & & \vdots \\ y_1^{(n-1)}(x) & y_2^{(n-1)}(x) & \cdots & y_n^{(n-1)}(x) \end{vmatrix} \qquad (4\text{–}24)$$

Theorem 2 is a consequence of Theorem 1; for if $y_1(x)$, . . . , $y_n(x)$ are linearly independent, then $y = c_1 y_1(x) + \cdots + c_n y_n(x)$ is the general solution of (4–17). Hence it must be possible to satisfy arbitrary initial conditions. This leads as above to simultaneous linear equations for c_1, . . . , c_n, whose determinant is $W(x_0)$. Since these can be solved for c_1, . . . , c_n for arbitrary choice of y_0, y_0', . . . , $y_0^{(n-1)}$, the determinant $W(x_0)$ must be different from 0. Conversely, if $W(x) \neq 0$, the functions must be linearly independent; otherwise, for appropriate constants $k_1, \ldots, k_n$ not all 0,

$$k_1 y_1(x) + \cdots + k_n y_n(x) \equiv 0.$$

We differentiate this relation $n - 1$ times:

$$k_1 y_1'(x) + \cdots + k_n y_n'(x) \equiv 0,$$
$$\vdots$$
$$k_1 y_1^{(n-1)}(x) + \cdots + k_n y_n^{(n-1)}(x) \equiv 0.$$

Thus the k's satisfy n homogeneous equations. Since the k's are not all 0, the determinant of the coefficients must be zero for every x. But the determinant is $W(x)$, which we have assumed to be different from 0. Since this is a contradiction, the functions must be linearly independent.

EXAMPLE 2. $y'' + y = 2e^x$. Here $y^*(x) = e^x$ is a particular solution, for $(e^x)'' + e^x = 2e^x$. Hence, by the result of Example 1,

$$y = e^x + c_1 \cos x + c_2 \sin x \tag{4-25}$$

is the general solution. We can verify without difficulty that (4–25) always satisfies the differential equation and that the initial conditions can always be met. The only difference between this and the preceding example is that while the terms $c_1 \cos x$ and $c_2 \sin x$ give rise to zero terms, the term e^x gives rise to $e^x + e^x$, which just matches the right-hand side.

The "term-by-term" analyses here are a reflection of the basic fact that the differential equation is *linear*. The left-hand side acts on a given $y(x)$ as a *linear operator*. This point of view will be further discussed below.

EXAMPLE 3. $y''' - 6y'' + 11y' - 6y = 4 - 12x$. Here we verify that $y_1 = e^x$, $y_2 = e^{2x}$, $y_3 = e^{3x}$ are solutions of the related homogeneous equation. They are linearly independent, since

$$W(x) = \begin{vmatrix} e^x & e^{2x} & e^{3x} \\ e^x & 2e^{2x} & 3e^{3x} \\ e^x & 4e^{2x} & 9e^{3x} \end{vmatrix} = 2e^{6x} \neq 0.$$

We also verify that $y^* = 2x + 3$ is a solution of the nonhomogeneous equation. Hence

$$y = 2x + 3 + c_1 e^x + c_2 e^{2x} + c_3 e^{3x}$$

is the general solution of the given equation.

In all the above examples, we have had the particular solutions given. Most of the rest of this chapter is devoted to methods for finding particular solutions.

PROBLEMS

1. Given the differential equation $x^2 y'' + 4xy' + 2y = 0$,

(a) verify that $y = c_1 x^{-2} + c_2 x^{-1}$, $x > 0$, satisfies the equation for every choice of c_1 and c_2;

(b) verify that c_1 and c_2 can be chosen uniquely to satisfy the initial conditions $y = y_0$, $y' = y_0'$ for $x = x_0$, $(x_0 > 0)$.

Hence $y = c_1 x^{-2} + c_2 x^{-1}$ is the general solution for $x > 0$.

2. Follow the procedure of Prob. 1 to show that $y = x^2 + e^{-x}(c_1 \cos 2x + c_2 \sin 2x)$ is the general solution of the differential equation

$$y'' + 2y' + 5y = 5x^2 + 4x + 2.$$

3. Show that the following sets of functions are linearly independent for all x:
(a) e^{ax}, e^{bx}, e^{cx} (a, b, c distinct);   (b) x, x^2, x^3;
(c) $\sin x, \cos x, \sin 2x$;   (d) $e^x, xe^x, \sinh x$.

4. Determine which of the following sets of functions are linearly independent for all x:
(a) $\sinh x, e^x, e^{-x}$;   (b) $\sin 3x, \sin x, \sin^3 x$;
(c) $1 + x, 1 + 2x, x^2$;   (d) $x^2 - x + 1, x^2 - 1, 3x^2 - x - 1$.

5. Let $y_1(x)$ and $y_2(x)$ both be solutions of the differential equation

$$y'' + p(x)y' + q(x)y = 0, \qquad a < x < b.$$

Let $W(x) = y_1(x)y_2'(x) - y_1'(x)y_2(x)$ be the Wronskian determinant of the two solutions.
(a) Prove that $W'(x) = -p(x)W(x)$.
(b) Prove that $W(x) = ce^{-\int p(x)\,dx}$ for some constant c.
(c) Prove that if $W(x_0) = 0$ for some one $x_0, a < x_0 < b$, then $W(x) \equiv 0$, for $a < x < b$.
(d) Show that the result in (c) follows from Theorem 2.

6. Show that every solution of the *nonlinear* differential equation

$$y' = 1 + y^2$$

is defined over an interval of length at most π.

7. Let $y_1(x), y_2(x), \ldots, y_n(x)$ be linearly independent over a certain interval. Let $c_1, \ldots, c_n, C_1, \ldots, C_n$ be constants such that

$$c_1 y_1(x) + \cdots + c_n y_n(x) \equiv C_1 y_1(x) + \cdots + C_n y_n(x)$$

over the interval. Show that $c_1 = C_1, \ldots, c_n = C_n$.

ANSWERS

4. (a), (b), (d) are linearly dependent. 6. Solutions are $y = \tan(x + c)$.

4–4 Differential operators. Superposition principle. It is convenient to write Dy for $D_x y$, $D^2 y$ for $D_x^2 y$, and in general

$$D^n y \equiv D_x^n y \equiv \frac{d^n y}{dx^n} \qquad (n = 1, 2, \ldots). \tag{4–26}$$

Furthermore we can write $y'' + y \equiv D^2 y + y \equiv (D^2 + 1)y$, $x^2 y'' - x^3 y' + 3y \equiv x^2 D^2 y - x^3 D y + 3y \equiv (x^2 D^2 - x^3 D + 3)y$, and in general

TABLE 4–1

y	$u = Ly, L = D^2 + 2D - 1$
e^{5x}	$34e^{5x}$
x^2	$2 + 4x - x^2$
$\sin x$	$-2 \sin x + 2 \cos x$
$2e^{5x} + x^2$	$68e^{5x} + 2 + 4x - x^2$

$$a_0(x) D_x^n y + a_1(x) D_x^{n-1} y + \cdots$$

$$+ a_n(x) y \equiv [a_0(x) D^n + a_1(x) D^{n-1} + \cdots + a_n(x)] y. \quad (4\text{--}27)$$

An expression

$$a_0(x) D^n + a_1(x) D^{n-1} + \cdots + a_n(x) \quad (4\text{--}28)$$

by itself will now be termed a (linear) *differential operator*. It describes an operation to be carried out on some function, but does not tell which function is "operated on." For example, $D^2 + 2D - 1$ is a differential operator; it stands for "formation of the second derivative of plus twice the first derivative of minus one times . . . " When the blank is filled in, we can compute the result of the operation. For example,

$$(D^2 + 2D - 1)e^{5x} = D^2 e^{5x} + 2De^{5x} - e^{5x}$$

$$= 25e^{5x} + 10e^{5x} - e^{5x} = 34e^{5x}.$$

We can abbreviate a given differential operator (4–28) by a single letter, e.g., L. Then we write Ly for the result of applying L to a particular function y of x. For example, if $L = D^2 + 2D - 1$, then $L(e^{5x}) = 34e^{5x}$ as above. Accordingly, the operator L resembles a function, except that the independent and dependent variables are functions; if y is a function of x, then $Ly = u$ is another function. We can emphasize the similarity of operator to function by forming a table of values, such as Table 4–1.

Each operator L of form (4–28) is *linear*. By this we mean that if $y_1(x)$ and $y_2(x)$ are functions defined for $a < x < b$ and are such that Ly_1, Ly_2 have meaning, then for each pair of constants c_1, c_2,

$$L[c_1 y_1(x) + c_2 y_2(x)] = c_1 L[y_1(x)] + c_2 L[y_2(x)]. \quad (4\text{--}29)$$

For example, for the operator of Table 4–1,

$$L(2e^{5x} + x^2) = 2L(e^{5x}) + L(x^2);$$

this is in agreement with the table.

To prove (4–29) in general, we proceed as follows:

$$\{a_0(x)D^n + \cdots + a_n(x)\}[c_1y_1(x) + c_2y_2(x)]$$

$$= a_0(x)D^n[c_1y_1(x) + c_2y_2(x)] + \cdots + a_n(x)[c_1y_1(x) + c_2y_2(x)]$$

$$= a_0(x)[c_1D^ny_1(x) + c_2D^ny_2(x)] + \cdots + [c_1a_n(x)y_1(x) + c_2a_n(x)y_2(x)]$$

$$= c_1[a_0(x)D^ny_1 + \cdots + a_n(x)y_1] + c_2[a_0(x)D^ny_2 + \cdots + a_n(x)y_2]$$

$$= c_1\{a_0(x)D^n + \cdots + a_n(x)\}y_1 + c_2\{a_0(x)D^n + \cdots + a_n(x)\}y_2$$

$$= c_1Ly_1 + c_2Ly_2.$$

By repeated application of (4–29) we obtain the rule

$$L[c_1y_1(x) + \cdots + c_ny_n(x)] = c_1L[y_1(x)] + \cdots + c_nL[y_n(x)]. \qquad (4\text{–}30)$$

A linear differential equation

$$a_0(x)D^ny + \cdots + a_n(x)y = Q \qquad (4\text{–}31)$$

can be written in the abbreviated form

$$Ly = Q, \qquad (4\text{–}32)$$

where $L = a_0(x)D^n + \cdots + a_n(x)$. If $y_1(x), \ldots, y_n(x)$ are solutions of the related homogeneous equation, then $Ly_1 = 0, \ldots, Ly_n = 0$. Hence, by (4–30),

$$L(c_1y_1 + \cdots + c_ny_n) = 0 \qquad (4\text{–}33)$$

for every choice of the constants $c_1, \ldots, c_n$. If $y^*(x)$ is a solution of the nonhomogeneous equation (4–31), then $Ly^* = Q$. Hence by (4–30) and (4–33)

$$L(y^* + c_1y_1 + \cdots + c_ny_n) = Q. \qquad (4\text{–}34)$$

Thus $y = y^* + c_1y_1 + \cdots + c_ny_n$ is a solution of (4–32) for every choice of $c_1, \ldots, c_n$. Furthermore, if $y(x)$ is any solution of the nonhomogeneous equation, then $L(y - y^*) = Ly - Ly^* = Q(x) - Q(x) = 0$. Hence $y - y^*$ is a solution of the homogeneous equation; that is, the general solution of the nonhomogeneous equation equals a particular solution $y^*(x)$ plus the general solution of the homogeneous equation. These remarks explain the *form* of the general solution of (4–31). (To complete the proof of Theorem 1, we would have to establish the *existence* of the solutions $y^*, y_1, \ldots, y_n$; this is done in Chapter 12.)

Now let $y_1^*(x)$ satisfy (4–31) when $Q = Q_1(x)$, and let $y_2^*(x)$ satisfy (4–31) when $Q = Q_2(x)$; then $Ly_1^* = Q_1$, $Ly_2^* = Q_2$ and by (4–29),

$$L(c_1 y_1^* + c_2 y_2^*) = c_1 Q_1 + c_2 Q_2.$$

Hence $c_1 y_1^* + c_2 y_2^*$ is a solution of (4–31) when Q is replaced by $c_1 Q_1 + c_2 Q_2$. This result, which is known as the *superposition principle* (Section 3–5), can be stated as follows: *if the right-hand member of a nonhomogeneous linear differential equation is a linear combination of two functions Q_1, Q_2, then a particular solution is obtainable as the same linear combination of two solutions of the equation with Q_1, Q_2, respectively as right-hand members.* The principle can be extended to linear combinations of n functions $Q_1(x), \ldots, Q_n(x)$.

4–5 Linear differential equations with constant coefficients. Homogeneous case. Theorem 1 of Section 4–3 shows us how to form the general solution of a linear differential equation when we know a set of linearly independent solutions $y_1(x), \ldots, y_n(x)$ of the related homogeneous equation and one solution $y^*(x)$ of the nonhomogeneous equation. However, we as yet have no information on how to find $y_1(x), \ldots, y_n(x)$ and $y^*(x)$. For the general linear equation, obtaining this information presents difficulties, although infinite power series prove to be very helpful (Chapter 9).

For the special case of *constant coefficients* the problem is reducible to a standard problem of algebra and can be considered to be completely solved. Here we give the solution for the homogeneous case.

Let the given differential equation be

$$a_0 D^n y + \cdots + a_{n-1} Dy + a_n y = 0 \tag{4–35}$$

or, in operator notation,

$$(a_0 D^n + \cdots + a_n)y = 0. \tag{4–36}$$

The coefficients $a_0, \ldots, a_n$ are assumed to be constants and $a_0 \neq 0$.

We now seek a solution of the form $y = e^{rx}$, where r is constant. Substitution of this expression for y in the left-hand side of (4–35) yields

$$a_0 r^n e^{rx} + \cdots + a_{n-1} r e^{rx} + a_n e^{rx} \equiv f(r) e^{rx}, \tag{4–37}$$

where

$$f(r) = a_0 r^n + a_1 r^{n-1} + \cdots + a_{n-1} r + a_n. \tag{4–38}$$

We call $f(r)$ the *characteristic polynomial* associated with the given differential equation. We see that e^{rx} *will be a solution of the differential equation if r is chosen as a root of*

$$a_0 r^n + a_1 r^{n-1} + \cdots + a_{n-1}r + a_n = 0, \qquad (4\text{--}39)$$

which we call the *characteristic equation* associated with the differential equation; it is simply the equation $f(r) = 0$. We note that if we write

$$L = a_0 D^n + \cdots + a_n, \qquad (4\text{--}40)$$

then the differential equation is simply the equation

$$Ly = 0. \qquad (4\text{--}41)$$

Equation (4–37) states that

$$L(e^{rx}) = f(r)e^{rx}, \qquad (4\text{--}42)$$

and $f(r)$ is obtained from L by replacing D by r.

If now equation (4–39) has n distinct real roots $r_1, \ldots, r_n$, then the functions

$$y_1 = e^{r_1 x}, \qquad y_2 = e^{r_2 x}, \qquad \ldots, \qquad y_n = e^{r_n x} \qquad (4\text{--}43)$$

are solutions of the differential equation. These functions are linearly independent for all x (Prob. 4 below). Hence the general solution sought is

$$y = c_1 e^{r_1 x} + \cdots + c_n e^{r_n x}. \qquad (4\text{--}44)$$

It is shown in algebra that an equation of degree n has n roots. However, some of the n roots may be complex and some may be equal. It will be seen that Eq. (4–44) can be modified appropriately to cover these cases.

For example, the equation

$$y'' + y = 0 \qquad \text{or} \qquad (D^2 + 1)y = 0 \qquad (4\text{--}45)$$

has the characteristic equation

$$r^2 + 1 = 0 \qquad (4\text{--}46)$$

with roots $\pm i$, $(i = \sqrt{-1})$. Proceeding formally, we obtain the "solutions" e^{ix} and e^{-ix}. From these imaginary solutions we can form linear combinations which are real:

$$\cos x = \frac{e^{ix} + e^{-ix}}{2}, \qquad \sin x = \frac{e^{ix} - e^{-ix}}{2i}, \qquad (4\text{--}47)$$

as follows from the Euler identity

$$e^{ix} = \cos x + i \sin x. \tag{4-48}$$

(See p. 534 of Reference 10 at the end of the chapter.) The functions

$$y_1(x) = \cos x, \qquad y_2(x) = \sin x,$$

are indeed linearly independent solutions of Eq. (4–45), and hence

$$y = c_1 \cos x + c_2 \sin x$$

is the general solution.

A similar analysis applies to complex roots in general. As is shown in algebra, for equations with real coefficients such roots come in pairs $a \pm bi$ (complex conjugates). From the two functions

$$e^{(a \pm bi)x} = e^{ax} (\cos bx \pm i \sin bx)$$

we obtain the two real functions

$$e^{ax} \cos bx, \quad e^{ax} \sin bx \tag{4-49}$$

as solutions of the corresponding differential equation. A proof is given in Section 4–8.

If the characteristic equation has a multiple real root r, of multiplicity k, then it can be shown (Prob. 5 below) that the functions

$$e^{rx}, \quad xe^{rx}, \quad \ldots, \quad x^{k-1}e^{rx} \tag{4-50}$$

are solutions. If $a \pm bi$ is a pair of multiple complex roots, then we obtain as solutions the functions (4–49) and the functions obtained by multiplying these by $x, x^2, \ldots, x^{k-1}$.

We can summarize the rules as follows.

THEOREM 3. *Let there be given a homogeneous linear differential equation with constant coefficients:*

$$(a_0 D^n + \cdots + a_n)y = 0 \qquad (a_0 \neq 0) \tag{4-51}$$

with the characteristic equation

$$a_0 r^n + \cdots + a_n = 0. \tag{4-52}$$

After finding the n roots of (4–52), we assign

(i) *to each simple real root r the function e^{rx};*

(ii) *to each pair $a \pm bi$ of simple complex roots the functions $e^{ax} \cos bx$, $e^{ax} \sin bx$;*

(iii) *to each real root r of multiplicity k the functions e^{rx}, xe^{rx}, ..., $x^{k-1}e^{rx}$;*

(iv) *to each pair $a \pm bi$ of complex roots of multiplicity k the functions*

$$e^{ax} \cos bx, \qquad xe^{ax} \cos bx, \qquad ..., \qquad x^{k-1}e^{ax} \cos bx,$$

$$e^{ax} \sin bx, \qquad xe^{ax} \sin bx, \qquad ..., \qquad x^{k-1}e^{ax} \sin bx.$$

The n functions $y_1(x)$, ..., $y_n(x)$ thus obtained are linearly independent solutions of (4–51) for all x, and

$$y = c_1 y_1(x) + \cdots + c_n y_n(x) \tag{4–53}$$

is the general solution of (4–51).

A proof of this theorem is given in Section 4–8.

EXAMPLE 1. $y'' - y = 0$. The characteristic equation $r^2 - 1 = 0$ has the distinct roots 1, -1; hence the general solution is

$$y = c_1 e^x + c_2 e^{-x}.$$

EXAMPLE 2. $(D^3 - D^2 + D - 1)y = 0$. The characteristic equation $r^3 - r^2 + r - 1 = 0$ has the distinct roots 1, i, $-i$; hence the general solution is

$$y = c_1 e^x + c_2 \cos x + c_3 \sin x.$$

EXAMPLE 3. $(D^2 + D + 1)y = 0$. The characteristic equation is $r^2 + r + 1 = 0$, with roots $\frac{1}{2}(-1 \pm \sqrt{3}i)$. Hence by part (ii) of the above theorem the general solution is

$$y = e^{-x/2}\left(c_1 \cos \frac{\sqrt{3}}{2}x + c_2 \sin \frac{\sqrt{3}}{2}x\right).$$

EXAMPLE 4. $(D^6 + 8D^4 + 16D^2)y = 0$. The characteristic equation is $r^6 + 8r^4 + 16r^2 = 0$ or

$$r^2(r^2 + 4)^2 = 0.$$

The roots are 0, 0, $\pm 2i$, $\pm 2i$. By parts (iii) and (iv) of Theorem 3 the general solution is

$$y = c_1 + c_2 x + c_3 \cos 2x + c_4 \sin 2x + c_5 x \cos 2x + c_6 x \sin 2x.$$

It thus appears that the homogeneous linear equation with constant coefficients is completely solved; once we have solved a certain algebraic equation, we can write down all solutions of the differential equation. However, in a practical sense this method is not always satisfactory, for the solution of an algebraic equation is not always simple or brief. In Section 4–6 we shall discuss further the solution of algebraic equations.

PROBLEMS

1. Find the general solution of each of the following equations:

(a) $(D^2 - 4)y = 0$,
(b) $(D^2 + 4D)y = 0$,
(c) $(D^3 - 3D^2 + 3D - 1)y = 0$,
(d) $(D^4 + 1)y = 0$,
(e) $(D^5 + 2D^3 + D)y = 0$,
(f) $D_t x + 3x = 0$,
(g) $D_t^2 x + D_t x + 7x = 0$,
(h) $D^5 y = 0$,
(i) $(D^5 + 1)y = 0$,
(j) $(D^6 + 64)y = 0$,
(k) $(D^8 - 256)y = 0$,
(l) $(36D^3 - 144D^2 + 191D - 84)y = 0$,
(m) $(D^3 + 5D^2 - 1)y = 0$,
(n) $(D^4 - 5D^2 + D + 4)y = 0$,
(o) $(D^4 + D - 1)y = 0$,
(p) $(D^4 + D^3 + D^2 + D + 1)y = 0$.

2. Find the solution which satisfies the given initial conditions:

(a) $(D^2 + 1)y = 0$; for $x = 0$: $y = 1$ and $y' = 0$;
(b) $D_t^2 x + D_t x - 3x = 0$; $x = 0$ and $D_t x = 1$ for $t = 0$.

3. Find a solution satisfying the given boundary conditions:

(a) $y'' - y' - 6y = 0$; $y = 1$ for $x = 0$, $y = 0$ for $x = 1$;
(b) $y'' + y = 0$; $y = 1$ for $x = 0$, $y = 2$ for $x = \pi/2$;
(c) $y'' + y = 0$; $y = 0$ for $x = 0$, $y = 0$ for $x = \pi$.

4. Let $r_1, \ldots, r_n$ be distinct real numbers. Prove that the functions $e^{r_1 x}, \ldots, e^{r_n x}$ are linearly independent for all x. Two possible methods are the following.

(a) Assume a relation $k_1 e^{r_1 x} + \cdots + k_n e^{r_n x} = 0$ to hold for all x, with not all $k_i = 0$. Assume the r's to be numbered so that $r_1 < r_2 \cdots < r_n$. Multiply by $e^{-r_n x}$ and let $x \to +\infty$ to conclude that $k_n = 0$. Then multiply by $e^{-r_{n-1} x}$ and let $x \to +\infty$ to conclude that $k_{n-1} = 0$; thus finally show that all k's must be zero.

(b) The Wronskian determinant is $W_n e^{(r_1 + \cdots + r_n)x}$, where

$$W_n = \begin{vmatrix} 1 & 1 & \ldots & 1 \\ r_1 & r_2 & & r_n \\ \vdots & & & \vdots \\ r_1^{n-1} & r_2^{n-1} & \ldots & r_n^{n-1} \end{vmatrix}$$

We can show by induction that so long as the r's are distinct, W_n cannot be 0.

This is clearly true for $n = 1$. If it is true for $n = k$, then form W_n for $n = k + 1$:

$$W_{k+1} = \begin{vmatrix} 1 & 1 & \cdots & 1 \\ r_1 & r_2 & & r_{k+1} \\ \vdots & & & \vdots \\ r_1^k & r_2^k & \cdots & r_{k+1}^k \end{vmatrix}$$

If we replace r_{k+1} by a variable r, then W_{k+1} becomes a polynomial, $P(r)$; the coefficient of r^k in $P(r)$ is a determinant W_k and, by induction assumption, is not 0. Hence $P(r)$ has degree k. Show that $P(r_1) = 0$, $P(r_2) = 0, \ldots,$ $P(r_k) = 0$ and that $P(r) \neq 0$ for r distinct from $r_1, \ldots, r_k$. Therefore $W_{k+1} = P(r_{k+1}) \neq 0$, and the induction is complete.

Remark. The determinant W_n is known as a *Vandermonde determinant.* An extension of the argument of part (b) shows that

$$W_n = (r_2 - r_1)(r_3 - r_1)(r_3 - r_2) \cdots (r_n - r_{n-1}).$$

5. Let $L = a_0 D^n + \cdots + a_n$, where $a_0, \ldots, a_n$ are constants. Let $f(r) = a_0 r^n + \cdots + a_n$.

(a) Prove: $L(xe^{rx}) = f(r)xe^{rx} + f'(r)e^{rx}$. [*Hint:* It is sufficient to prove the rule when L consists of one term $a_k D^k$; the general case then follows by addition. For this case, $D^k(xe^{rx})$ can be found by the Leibnitz rule for differentiation of products:

$$D^k(uv) = uD^k v + kDuD^{k-1}v + \frac{k(k-1)}{2!} D^2 u D^{k-2}v + \cdots + (D^k u)v. \Big]$$

(b) Prove that for $m = 1, 2, 3, \ldots,$

$$L(x^m e^{rx}) = e^{rx}\left[f(r)x^m + mf'(r)x^{m-1} + \frac{m(m-1)}{2!} f''(r)x^{m-2} + \cdots \right].$$

[*Hint:* Proceed as in part (a).]

(c) If r_1 is a double root of $f(r)$, then $f(r_1) = 0$, $f'(r_1) = 0$. Set $r = r_1$ in the conclusion of part (a) to conclude that $xe^{r_1 x}$ must then be a solution of the differential equation $Ly = 0$.

(d) Reason as in part (c) to show that if r_1 is a root of multiplicity k of $f(r)$, then $e^{r_1 x}, xe^{r_1 x}, \ldots, x^{k-1}e^{r_1 x}$ are solutions of the differential equation $Ly = 0$. Another proof of this result is given in Prob. 7 following Section 4–13.

ANSWERS

1. (a) $c_1 e^{2x} + c_2 e^{-2x}$;   (b) $c_1 + c_2 e^{-4x}$;   (c) $c_1 e^x + c_2 x e^x + c_3 x^2 e^x$; (d) $e^{ax}(c_1 \cos ax + c_2 \sin ax) + e^{-ax}(c_3 \cos ax + c_4 \sin ax)$, $a = \sqrt{2}/2$; (e) $c_1 + c_2 \cos x + c_3 \sin x + c_4 x \cos x + c_5 x \sin x$;   (f) $c_1 e^{-3t}$;

(g) $e^{-t/2}(c_1 \cos bt + c_2 \sin bt)$, $b = 3\sqrt{3}/2$;   (h) $c_1 + c_2 x + c_3 x^2 + c_4 x^3 + c_5 x^4$;   (i) $c_1 e^{-x} + e^{0.809x}(c_2 \cos 0.588x + c_3 \sin 0.588x) + e^{-0.309x}(c_4 \cos 0.951x + c_5 \sin 0.951x)$;   (j) $c_1 \cos 2x + c_2 \sin 2x + e^{1.732x}(c_3 \cos x + c_4 \sin x) + e^{-1.732x}(c_5 \cos x + c_6 \sin x)$;   (k) $c_1 e^{2x} + c_2 e^{-2x} + c_3 \cos 2x + c_4 \sin 2x + e^{ax}(c_5 \cos ax + c_6 \sin ax) + e^{-ax}(c_7 \cos ax + c_8 \sin ax)$, $a = \sqrt{2}$;   (l) $c_1 e^{(3/2)x} + c_2 e^{(4/3)x} + c_3 e^{(7/6)x}$;   (m) $c_1 e^{0.429x} + c_2 e^{-4.96x} + c_3 e^{-0.470x}$;   (n) $c_1 e^{1.22x} + c_2 e^{1.78x} + c_3 e^{-0.86x} + c_4 e^{-2.15x}$;   (o) $c_1 e^{-1.22x} + c_2 e^{0.73x} + e^{0.25x}(c_3 \cos 1.03x + c_4 \sin 1.03x)$;   (p) $e^{0.309x}(c_1 \cos 0.951x + c_2 \sin 0.951x) + e^{-0.809x}(c_1 \cos 0.588x + c_2 \sin 0.588x)$.

2. (a) $\cos x$,   (b) $(2/\sqrt{13})e^{-t/2} \sinh(\sqrt{13}t/2)$.

3. (a) $(e^{3x} - e^{5-2x})/(1 - e^5)$,   (b) $\cos x + 2 \sin x$,   (c) $c \sin x$.

*4–6 Solution of algebraic equations. In this section we shall review briefly some of the methods available for finding roots of algebraic equations. References which supply a more complete treatment of the subject are given at the end of the section.

We write

$$a_0 z^n + \cdots + a_{n-1} z + a_n = 0 \qquad (a_0 \neq 0) \qquad (4\text{--}54)$$

and seek the real or complex numbers z which satisfy the equation. We write $z = x + iy$ so that when z is real, z reduces to the real number x, and x satisfies the equation

$$a_0 x^n + \cdots + a_{n-1} x + a_n = 0. \qquad (4\text{--}55)$$

We shall assume throughout that the coefficients $a_0, \ldots, a_n$ are real; much of what is described will be applicable to the complex case also.

We know from algebra that the equation (4–54) has n roots, some of which may coincide. Since the coefficients are real, the complex roots come in conjugate pairs $a \pm bi$. If z_1 is a root of (4–54), then $z - z_1$ is a factor of the polynomial

$$f(z) = a_0 z^n + \cdots + a_{n-1} z + a_n. \qquad (4\text{--}56)$$

Division of $f(z)$ by $z - z_1$ yields a polynomial of degree $n - 1$. Hence *for each root found we can lower the degree by 1.*

The division by $z - z_1$ can be carried out by synthetic division (see Reference 3, p. 13). If $a \pm bi$ is a pair of complex roots, then we can remove two roots at once by dividing by

$$(z - a - bi)(z - a + bi) \equiv z^2 - 2az + a^2 + b^2.$$

If roots are known only approximately (as will almost always be the case), the division process will have a small remainder. In replacing this by zero, we introduce an error in the roots to be found; at times, the error can be significant, even though the root being factored out is known very accurately. Accordingly, a certain amount of rechecking is necessary.

Finding real roots. Consider Eq. (4–55) and the related graph of

$$f(x) = a_0 x^n + \cdots + a_{n-1} x + a_n. \tag{4-57}$$

For large positive or negative x the term of highest degree dominates; hence if n is odd there must be at least one real root; if n is even there may be none.

The Descartes rule of signs gives us more information. We write out the sequence of signs of the coefficients (omitting zero terms); for example, for $x^3 - x + 1 = 0$ we write $+ - +$. We then count the number of *alternations of sign;* $x^3 - x + 1 = 0$ has two alternations. According to the rule, the number of positive roots is at most equal to the number of alternations and differs from that number by an even number; in the example considered there are either two positive roots or no positive roots. The number of negative roots can be estimated similarly by replacing x by $-x$ and then counting alternations: $x^3 - x + 1 = 0$ becomes $-x^3 + x + 1 = 0$, with one alternation and hence one negative root.

It can be shown that each root x of (4–55) satisfies the inequality

$$|x| \leqq \frac{M}{|a_0|} + 1; \qquad M = \max\left(|a_1|, \ldots, |a_n|\right). \tag{4-58}$$

(See Reference 3, pp. 21–23.) Thus the search for roots should be confined to the interval $-b \leqq x \leqq b$, where $b = 1 + M/|a_0|$. It is usually a simple matter to study the graph of $f(x)$ in this interval, with the aid of the derivative $f'(x)$, and hence to obtain approximate location of roots. If x_0 is an approximate root, then a better approximation x_1 can be obtained by *Newton's method:*

$$x_1 = x_0 - \frac{f(x_0)}{f'(x_0)}, \tag{4-59}$$

provided $f(x_0)f''(x_0) > 0$ and we know that x_0 lies within an interval containing a root of $f(x)$ but none of $f'(x)$ or $f''(x)$. Under the conditions mentioned, repeated application of (4–59) yields a sequence of numbers which converges to the root sought.

EXAMPLE 1. $x^3 + x + 1 = 0$. The rule of signs shows that there are no positive roots and one negative root. The inequality (4–58) becomes

$|x| \leqq 2$. We need graph $f(x)$ only between -2 and 2. Since $f'(x) = 3x^2 + 1$, there are no critical points. The graph thus has the appearance of Fig. 4–1; there is only one real root, close to -1. Application of Newton's method yields the successive approximations -0.75, -0.686, -0.68234. Division by $x + 0.68234$ yields the new equation

$$x^2 - 0.68234x + 1.4655878 = 0.$$

Its roots are complex: $0.34117 \pm 1.16155i$.

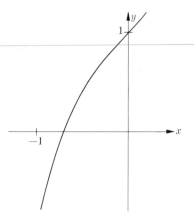

FIG. 4–1. $y = x^3 + x + 1$.

Finding complex roots. For a cubic equation we can always proceed as in the preceding example; there is always at least one real root which can be factored out to leave a quadratic equation. A fourth degree equation which has two real roots can be analyzed similarly, as can every equation having only one pair of complex roots. When there are two or more pairs of complex roots, a direct attack on the complex roots is needed. One method, based on *moments*, is described here.

We consider the equation (4–54) and assume that $a_0 = 1$. Let the roots be $z_1, z_2, \ldots, z_n$ and let

$$s_k = z_1^k + z_2^k + \cdots + z_n^k \qquad (k = 0, 1, 2, \ldots). \qquad (4\text{–}60)$$

We call s_k the kth moment of the roots. It can be proved (Reference 3, pp. 134–136) that the s_k are obtainable from the coefficients $a_1, a_2, \ldots, a_n$ by the following formulas:

$$s_0 = n, \qquad s_1 + a_1 = 0, \qquad s_2 + a_1 s_1 + 2a_2 = 0, \qquad \ldots,$$

$$s_k + a_1 s_{k-1} + \cdots + a_{k-1} s_1 + ka_k = 0$$

$$(k = 1, \ldots, n - 1), \qquad (4\text{–}61)$$

$$s_k + a_1 s_{k-1} + \cdots + a_{n-1} s_{k-n+1} + a_n s_{k-n} = 0$$

$$(k \geqq n).$$

For example, for the equation

$$z^5 - 4z^3 + 2z^2 - z + 1 = 0,$$

we have

$$a_1 = 0, \qquad a_2 = -4, \qquad a_3 = 2, \qquad a_4 = -1, \qquad a_5 = 1,$$

$$s_0 = 5, \qquad s_1 + 0 = 0, \qquad s_2 + 0s_1 - 8 = 0,$$

$$s_3 + 0s_2 - 4s_1 + 6 = 0,$$

$$s_4 + 0s_3 - 4s_2 + 2s_1 - 4 = 0,$$

$$s_k + 0s_{k-1} - 4s_{k-2} + 2s_{k-3} - s_{k-4} + s_{k-5} = 0,$$

for $k = 5, 6, \ldots$ Accordingly,

$$s_0 = 5, \qquad s_1 = 0, \qquad s_2 = 8, \qquad s_3 = -6, \qquad s_4 = 36,$$

$$s_5 = 4s_3 - 2s_2 + s_1 - s_0 = -45, \qquad s_6 = 164, \qquad \ldots$$

We can easily compute s_5, s_6, . . . in tabular form as suggested in Fig. 4–2. The coefficients -1, 1, -2, 4, 0 are written vertically on a slip of paper. The slip is placed next to the corresponding values of the s_k's. The two columns are multiplied, the results are added, and the next s_k is obtained.

Now let us suppose that the roots of the equation are distributed as in Fig. 4–3, so that the roots farthest from the origin are a pair of conjugate complex roots z_1, $\bar{z}_1$. *Then for large k, s_k can be approximated by $z_1^k + \bar{z}_1^k$.* For by De Moivre's formula,

$$z_1^k = r_1^k(\cos k\theta_1 + i \sin k\theta_1);$$

in general the kth powers of the roots have distances from the origin that are the kth powers of the original distances. Hence for large k the roots z_1, $\bar{z}_1$ with largest distance will contribute proportionately more and more to s_k. If we assume that $s_k = z_1^k + \bar{z}_1^k$, then s_k would coincide with the kth moment for a certain quadratic equation $az^2 + bz + c = 0$ whose roots are z_1, $\bar{z}_1$. Hence (for $k \geqq 2$) by the formulas (4–61),

$$s_k + \frac{b}{a} s_{k-1} + \frac{c}{a} s_{k-2} = 0,$$

or

$$as_k + bs_{k-1} + cs_{k-2} = 0. \tag{4–62}$$

This should hold approximately for large k. If we compute s_k for $k = 1$, $2, \ldots$, we can then determine a, b, c by forming Eq. (4–62) for two successive large values of k. This leads us to a system of three equations:

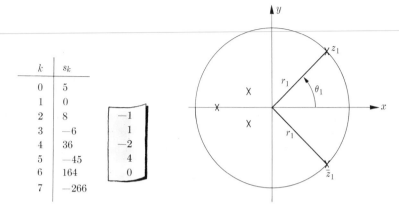

k	s_k
0	5
1	0
2	8
3	−6
4	36
5	−45
6	164
7	−266

−1
1
−2
4
0

FIG. 4–2.   Computation of moments.

FIG. 4–3   Finding complex roots farthest from origin.

$$az^2 + bz + c = 0,$$

$$as_k + bs_{k-1} + cs_{k-2} = 0, \qquad (4\text{–}63)$$

$$as_{k+1} + bs_k + cs_{k-1} = 0.$$

Elimination of a, b, c yields the desired quadratic equation in determinant form:

$$\begin{vmatrix} z^2 & z & 1 \\ s_k & s_{k-1} & s_{k-2} \\ s_{k+1} & s_k & s_{k-1} \end{vmatrix} = 0. \qquad (4\text{–}64)$$

The roots of this quadratic, for large k, approximate the roots z_1, $\bar{z}_1$.

If the root farthest from the origin is real, then the same reasoning as above leads us to the linear equation $az + b = 0$ and to the system

$$az + b = 0, \qquad as_k + bs_{k-1} = 0. \qquad (4\text{–}65)$$

Hence for large k

$$z = \frac{s_k}{s_{k-1}} \qquad (4\text{–}66)$$

is (approximately) the real root sought.

It is important to know in advance whether there is a pair of complex roots, or a single real root, or some other combination of roots at maximum distance from the origin. We can obtain a clue from the moments themselves. When there is a single real root at maximum distance, the s_k will eventually either have constant sign (for a positive root) or will alternate in sign (for a negative root); this cannot happen in the complex case. However when there are three or more roots at equal maximum distance from the origin, the method will fail. The difficulty can almost always be avoided by a simple substitution such as $z' = z + 1$; the new equation in z' will generally not have the complication described. Similar remarks apply when the roots at maximum distance are $\pm a$, where a is real.

If the roots at maximum distance are only $x_1 \pm iy_1$, but these roots are double, triple, or of higher multiplicity, then it can be verified that these roots satisfy the quadratic equation (4–64), approximately, for large k. The fact that the roots are multiple is revealed in a root of the derivative $f'(z)$ at each point:

$$f'(z) = na_0 z^{n-1} + (n-1)a_1 z^{n-2} + \cdots + a_{n-1}.$$

A similar remark applies when there is a multiple real root at maximum distance.

EXAMPLE 2. $z^3 + z + 1 = 0$. This is, in effect, the same as Example 1. The formulas (4–61) give the following sequence of values for s_k for $k = 0$, $1, 2, \ldots, 21$:

$$3, 0, -2, -3, 2, 5, 1, -7, -6, 6, 13, 0, -19,$$
$$-13, 19, 32, -6, -51, -26, 57, 77, -31.$$

The signs do not suggest real roots; hence we assume a pair of complex roots and form the equation (4–64). For $k = 10$ this becomes

$$\begin{vmatrix} z^2 & z & 1 \\ 13 & 6 & -6 \\ 0 & 13 & 6 \end{vmatrix} = 0,$$

$$114z^2 - 78z + 169 = 0,$$

$$z = 0.342 \pm 1.17i.$$

For $k = 20$ we should obtain a better approximation:

$$\begin{vmatrix} z^2 & z & 1 \\ 77 & 57 & -26 \\ -31 & 77 & 57 \end{vmatrix} = 0,$$

$$5251z^2 - 3583z + 7696 = 0,$$

$$z = 0.34117 \pm 1.16156i.$$

The method can also be applied to finding the root or roots nearest to the origin. We set $w = 1/z$ and obtain an equation for w: if

$$z = r\,(\cos\theta + i\sin\theta),$$

then

$$w = (\cos\theta - i\sin\theta)/r = [\cos(-\theta) + i\sin(-\theta)]/r.$$

Hence the roots w farthest from the origin correspond to the roots z nearest to the origin. For the equation of Example 2 we find $w^3 + w^2 + 1 = 0$, and s_k has the following values for $k = 0, 1, 2, \ldots, 21$:

$$3,\ -1,\ 1,\ -4,\ 5,\ -6,\ 10,\ -15,\ 21,\ -31,\ 46,\ -67,\ 98,$$

$$-144,\ 211,\ -309,\ 453,\ -664,\ 973,\ -1426,\ 2090,\ -3063.$$

The alternating signs indicate a negative root. By (4–66), with $k = 21$, $w = -3063/2090$ and $z = -2090/3063 = -0.68234$.

The roots found agree with those found previously (to five significant figures).

For further discussion of algebraic equations the reader is referred to the books by Dickson, Scarborough, and Willers listed at the end of the chapter (References 3, 9, and 11).

PROBLEMS

1. Find all roots of the following equations:

(a) $z^3 - z + 1 = 0$, (b) $z^3 + z^2 - z + 1 = 0$,
(c) $z^4 + 4z^3 + z - 1 = 0$, (d) $z^4 - z^3 + 1 = 0$,
(e) $3z^5 + z^4 - 5z^3 - 17z^2 - 7z - 5 = 0$,
(f) $24z^6 + 16z^5 + 58z^4 + 12z^3 - 18z - 45 = 0$.

2. (a) Prove: if $f(z) = a_0 z^n + \cdots + a_n$, $a_0 \neq 0$, and s_k is defined as in the text, then for sufficiently large $|z|$

$$\frac{zf'(z)}{f(z)} = s_0 + \frac{s_1}{z} + \frac{s_2}{z^2} + \cdots + \frac{s_n}{z^n} + \cdots$$

[*Hint:* Let $f(z) = a_0(z - z_1)(z - z_2) \cdots (z - z_n)$. Show that

$$\frac{zf'(z)}{f(z)} = \frac{z}{z - z_1} + \frac{z}{z - z_2} + \cdots + \frac{z}{z - z_n}$$

$$= \frac{1}{1 - (z_1/z)} + \frac{1}{1 - (z_2/z)} + \cdots$$

Replace each term by the corresponding geometric series

$$1/(1 - r) = 1 + r + \cdots + r^n + \cdots, \qquad |r| < 1.]$$

(b) Obtain the moments s_k for $z^3 + z + 1 = 0$ by applying the result of (a). The series can be obtained by dividing $zf'(z)$ by $f(z)$ by long division.

ANSWERS

1. (a) -1.3247, $0.6624 \pm 0.5623i$;
   (b) -1.8393, $0.41964 \pm 0.60625i$;
   (c) -4.0750, 0.48578, $-0.20539 \pm 0.68044i$;
   (d) $1.0189 \pm 0.60260i$, $-0.51892 \pm 0.66631i$;
   (e) 2.09454, $-1.04727 \pm 1.13594i$, $-0.16667 \pm 0.05528i$;
   (f) ± 0.86603, $\pm 1.2247i$, $-0.33333 \pm 1.2472i$.

2. (b) $zf'(z)/f(z) = 3 - 2z^{-2} - 3z^{-3} + 2z^{-4} + \cdots$; $s_0 = 3$, $s_1 = 0$, $s_2 = -2$, $s_3 = -3$, $s_4 = 2, \ldots$

*4–7 Complex form of solution of homogeneous linear differential equations. We return to the discussion of Section 4–5. When the characteristic equation has complex roots $a \pm bi$, we can proceed formally and introduce the terms

$$c_1 e^{(a+bi)x} + c_2 e^{(a-bi)x} \tag{4–67}$$

in the solution. From the identity

$$e^{(a\pm bi)x} = e^{ax}(\cos bx \pm i \sin bx) \tag{4–68}$$

the real solution can then be recovered, for (4–67) becomes

$$e^{ax}[(c_1 + c_2) \cos bx + i(c_1 - c_2) \sin bx]. \tag{4–69}$$

If we now introduce new arbitrary constants:

$$C_1 = c_1 + c_2, \qquad C_2 = i(c_1 - c_2), \tag{4–70}$$

then (4–69) becomes

$$e^{ax}[C_1 \cos bx + C_2 \sin bx], \tag{4–71}$$

as in Section 4–5.

From (4–70) we find

$$c_1 = \tfrac{1}{2}(C_1 - iC_2), \qquad c_2 = \tfrac{1}{2}(C_1 + iC_2). \tag{4–72}$$

Now the constants C_1, C_2 are to be real, so that (4–71) gives the usual real solution. Hence (4–72) shows that c_1, c_2 are *conjugate complex numbers:* $c_2 = \bar{c}_1$. If c_1, c_2 are any pair of conjugate complex constants, then $C_1 = c_1 + c_2$ and $C_2 = i(c_1 - c_2)$ will be real, and (4–67) is equivalent to the real solution (4–71). Thus we can write our real solutions as

$$c_1 e^{(a+bi)x} + \bar{c}_1 e^{(a-bi)x} \qquad \text{or} \qquad c_1 e^{rx} + \bar{c}_1 e^{\bar{r}x},$$

where $r = a + bi$, $\bar{r} = a - bi$.

EXAMPLE 1. $y'' + y = 0$. The characteristic equation has roots $\pm i$. The general solution can be written in the two forms

$$y = C_1 \cos x + C_2 \sin x, \tag{4–73}$$

$$y = c e^{ix} + \bar{c} e^{-ix}. \tag{4–74}$$

Here C_1, C_2 are arbitrary *real* constants, while c is an arbitrary *complex* constant. If the characteristic equation has several complex roots, possibly multiple, as in Example 2, a similar procedure can be followed.

EXAMPLE 2. $(D^6 + 2D^5 + 4D^4 + 4D^3 + 5D^2 + 2D + 2)y = 0$. The characteristic equation can be written as $(r^2 + 1)^2(r^2 + 2r + 2) = 0$, so that the roots are $\pm i$, $\pm i$, $-1 \pm i$. The solutions can be written in the form

$$y = C_1 \cos x + C_2 \sin x + C_3\, x \cos x + C_4\, x \sin x$$
$$+ \, e^{-x}(C_5 \cos x + C_6 \sin x)$$

or in the form

$$y = c_1 e^{ix} + \bar{c}_1 e^{-ix} + c_2 x e^{ix} + \bar{c}_2 x e^{-ix}$$
$$+ \, e^{-x}(c_3 e^{ix} + \bar{c}_3 e^{-ix})$$
$$= e^{ix}(c_1 + c_2 x + c_3 e^{-x}) + e^{-ix}(\bar{c}_1 + \bar{c}_2 x + \bar{c}_3 e^{-x}),$$

where c_1, c_2, c_3 are arbitrary complex constants.

EXAMPLE 3. $(D^3 - 3D^2 + 7D - 5)y = 0$. The characteristic roots are $1, 1 \pm 2i$. The solutions can be written

$$y = c_1 e^x + e^x(c_2 e^{2ix} + \bar{c}_2 e^{-2ix}),$$

where c_1 is real, but c_2 is complex.

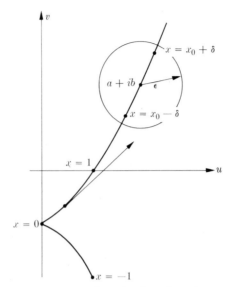

FIG. 4–4. Complex function of x.

*4–8 General theory of complex solutions. The preceding discussion has been purely formal, and a complex solution such as (4–74) appears simply as another way of writing the real solution (4–73). However, we can take a more general point of view which enables us to calculate freely with complex solutions.

By a *complex function* of a *real variable* is meant a function $F(x)$ which assigns a complex number to each x (of a certain interval). Thus

$$F(x) = x^2 + i(x^3 - 1)$$

is such a function; typical values are as follows:

$$F(0) = -i, \qquad F(1) = 1, \qquad F(2) = 4 + 7i.$$

Other examples of complex functions are the following:

$$F(x) = \cos x + i \sin x = e^{ix},$$

$$F(x) = e^{2x} \cos x + ie^{2x} \sin x = e^{(2+i)x}.$$

The value of a complex function $F(x)$ for each x is a complex number $w = u + iv$, where u and v are real. We call u the *real part* of w, and v

the *imaginary part* of w; in symbols,

$$w = u + iv, \quad u = \text{Re } (w), \quad v = \text{Im } (w). \quad (4\text{--}75)$$

The complex number w can be represented graphically as a point in the uv-plane, as in Fig. 4–4.

If a complex function $F(x)$ is given, then as x varies, $w = F(x)$ in general traces a *curve* in the uv-plane, as in Fig. 4–4, for which

$$w = x^2 + i(x^3 - 1); \quad (4\text{--}76)$$

indeed, Eq. (4–76) is the same as the two equations

$$u = x^2, \quad v = x^3 - 1,$$

which can be interpreted as parametric equations of a curve in the uv-plane. In general, a complex function $F(x)$ can be written as

$$F(x) = f(x) + ig(x),$$

where f and g are real, and the equation $w = F(x)$ is equivalent to the two real equations

$$u = f(x), \quad v = g(x).$$

Operations on complex functions are similar to those for real functions; we can add, subtract, multiply, and divide complex functions defined over the same interval, except for division by $0 = 0 + i0$. We can also take limits; the equation

$$\lim_{x \to x_0} F(x) = a + ib \quad (4\text{--}77)$$

means that $F(x)$ is to be arbitrarily close to $a + ib$ for x sufficiently close to x_0. More precisely, for each $\epsilon > 0$, we must be able to choose $\delta > 0$ so that when $0 < |x - x_0| < \delta$, $F(x)$ is within distance ϵ of $a + ib$ (Fig. 4–4). It is easily seen that (4–77) is equivalent to the two equations

$$\lim_{x \to x_0} f(x) = a, \quad \lim_{x \to x_0} g(x) = b, \quad (4\text{--}78)$$

if $F(x) = f(x) + ig(x)$. From the definition or from (4–78) we can verify that the theorem on limit of sum, product, and quotient remains valid.

The *derivative* of a complex function is defined exactly as for real functions:

$$F'(x) = \lim_{\Delta x \to 0} \frac{F(x + \Delta x) - F(x)}{\Delta x}. \quad (4\text{--}79)$$

Since we can take limits in real and imaginary parts separately, (4–79) is the same as

$$F'(x) = f'(x) + ig'(x), \qquad (4\text{–}80)$$

wherever both $f'(x)$ and $g'(x)$ exist.

EXAMPLE 1. If $w = F(x) = x^2 + i(x^3 - 1)$, then $F'(x) = 2x + 3x^2 i$.

We also use other standard notations for the derivative: w', dw/dx, $D_x w$. Higher derivatives are defined as usual, and we find

$$F''(x) = f''(x) + ig''(x), \qquad \ldots \qquad (4\text{–}81)$$

The first and second derivatives can be interpreted geometrically. For the complex number $f'(x) + ig'(x)$ can be interpreted as the *velocity vector* of the moving point whose position at *time* x is $u + iv$ (see Fig. 4–4). As shown in calculus, this vector is tangent to the path. Similarly, the second derivative $f''(x) + ig''(x)$ can be interpreted as the *acceleration vector*.

The basic rules of calculus continue to hold:

$$[F(x) + G(x)]' = F'(x) + G'(x), \qquad (F \cdot G)' = FG' + F'G,$$

$$\left(\frac{F}{G}\right)' = \frac{GF' - FG'}{G^2}, \qquad (4\text{–}82)$$

provided F and G have derivatives over the interval considered and there is no division by zero. The proofs are left as exercises (Prob. 4 below).

EXAMPLE 2. If $w = e^{ix} = \cos x + i \sin x$, then $D_x w = -\sin x + i \cos x = i(\cos x + i \sin x) = ie^{ix}$, $D_x^2 w = iD_x e^{ix} = i^2 e^{ix} = -e^{ix}$.

EXAMPLE 3. If $w = e^{(a+bi)x} = e^{ax}(\cos bx + i \sin bx)$, where a and b are real constants, then

$$D_x w = e^{ax}[(a \cos bx - b \sin bx) + i(a \sin bx + b \cos bx)]$$

$$= (a + bi)e^{ax}(\cos bx + i \sin bx)$$

$$= (a + bi)e^{(a+bi)x},$$

$$D_x^2 w = (a + bi) D_x e^{(a+bi)x} = (a + bi)^2 e^{(a+bi)x},$$

and in general

$$D_x^n w = (a + bi)^n e^{(a+bi)x} \qquad (n = 1, 2, \ldots). \qquad (4\text{–}83)$$

A *definite integral* can also be defined as a limit or, more simply, by the equation

$$\int_a^b F(x)\, dx = \int_a^b [f(x) + ig(x)]\, dx = \int_a^b f(x)\, dx + i \int_a^b g(x)\, dx. \quad (4\text{–}84)$$

The indefinite integral is defined as usual:

$$\int F(x)\, dx = G(x) + C \quad \text{if} \quad G'(x) = F(x), \quad\quad (4\text{–}85)$$

where C is an arbitrary *complex* constant. The integrals have the familiar properties. In particular,

$$\int_a^b F'(x)\, dx = F(b) - F(a) = F(x)\Big|_a^b. \quad\quad (4\text{–}86)$$

Let a linear differential equation be given:

$$(a_0 D_x^n + \cdots + a_n)y = Q(x). \quad\quad (4\text{–}87)$$

A complex function $F(x) = f(x) + ig(x)$ is said to be a *solution* of the equation over some interval if

$$a_0 D_x^n F(x) + \cdots + a_n F(x) \equiv Q(x) \quad\quad (4\text{–}88)$$

over the interval. The coefficients $a_0, a_1, \ldots$ may be functions of x; in fact, the coefficients and $Q(x)$ may be complex functions.

EXAMPLE 4. $y = e^{ix}$ is a solution of $y'' + y = 0$ (see Example 2 above).

EXAMPLE 5. $y = e^{(1+2i)x}$ is a solution of the equation $y'' - 2y' + 5y = 0$. For

$$(D_x^2 - 2D_x + 5)e^{(1+2i)x} \equiv [(1 + 2i)^2 - 2(1 + 2i) + 5]e^{(1+2i)x} \equiv 0.$$

EXAMPLE 6. $y = e^{ix}$ is a solution of $(ix - 1)y'' + xy' - y = 0$.

EXAMPLE 7. $y = 1 - ix + ce^{-ix}$ is a solution of $y' + iy = x$, for every choice of the complex constant c. This solution can be obtained by the general method for first order linear equations (Section 2–8). For e^{ix} is the integrating factor, and

$$e^{ix}y = \int xe^{ix}\, dx = \frac{xe^{ix}}{i} - \int \frac{e^{ix}}{i}\, dx = \frac{xe^{ix}}{i} + e^{ix} + c$$

by integration by parts. Hence $y = 1 - ix + ce^{-ix}$, and this is the general complex solution.

If the coefficients $a_0, \ldots$ and the function $Q(x)$ are real, there are, in general, complex solutions of Eq. (4–87); see Examples 4 and 5 above. The real part of every such complex solution is a real solution. For if

$$a_0(D_x^n f + iD_x^n g) + \cdots + a_n(f + ig) = Q,$$

then we can compare real parts on both sides:

$$a_0 D_x^n f + \cdots + a_n f = Q;$$

that is, $f = \mathrm{Re}\,(f + ig)$ is a solution. Every real solution f can be considered as the real part of a complex solution $f + i0$. Hence *if all complex solutions are known, all real solutions are obtained as all real parts of the complex solutions.*

A set of n complex functions is said to be *linearly independent* (with respect to complex coefficients) over a given interval if no one function can be expressed as a linear combination, with complex coefficients, of the others; this is equivalent to the statement that an identity

$$k_1 F_1(x) + \cdots + k_n F_n(x) \equiv 0,$$

where $k_1, \ldots, k_n$ are complex constants, can hold only if $k_1 = 0, k_2 = 0,$ $\ldots, k_n = 0$. A theorem analogous to Theorem 1 (Section 4–2) holds for complex functions: the general solution of (4–84) has the form

$$y = y*(x) + c_1 F_1(x) + \cdots + c_n F_n(x),$$

where $F_1(x), \ldots, F_n(x)$ are linearly independent complex solutions of the homogeneous equation. A proof is given in Chapter 12. When the coefficients are constant, the following basic theorem applies.

THEOREM 4. *Let there be given a homogeneous linear differential equation with constant coefficients (possibly complex),*

$$(a_0 D_x^n + \cdots + a_n)y = 0 \qquad (a_0 \neq 0). \qquad (4\text{–}89)$$

Let $r_1, \ldots, r_n$ be the roots of the characteristic equation

$$a_0 r^n + \cdots + a_n = 0. \qquad (4\text{–}90)$$

To each root r of multiplicity k assign the functions $e^{rx}, xe^{rx}, \ldots, x^{k-1}e^{rx}$.

Then the n functions so obtained are linearly independent complex solutions of (4–89), so that the general solution is

$$y = e^{r_1 x}(c_1 + c_2 x + \cdots + c_{k_1} x^{k_1 - 1}) + \cdots \qquad (4\text{-}91)$$

Proof: The fact that the functions $e^{r_1 x}$, $x e^{r_1 x}$, ... are solutions is demonstrated as in the real case (Prob. 5 following Section 4–5; see also Prob. 7 following Section 4–13). To show that the functions are linearly independent, we assume that some linear combination of them is identically 0; that is, that (4–91) is identically 0 for some choice of c_1, c_2 ... (not all 0). This is a relation of form

$$P_1(x)e^{r_1 x} + P_2(x)e^{r_2 x} + \cdots + P_m(x)e^{r_m x} \equiv 0, \qquad (4\text{-}92)$$

where r_1, r_2, ... are distinct and $P_1(x)$, ..., $P_m(x)$ are polynomials of degrees k_1, ..., k_m.

We now prove by induction on m that such a relation (4–92) is impossible. For $m = 1$ we have the relation

$$P_1(x)e^{r_1 x} \equiv 0;$$

hence $P_1(x) \equiv 0$. This is impossible, since a polynomial of degree k_1 has at most k_1 roots and $e^{(a+bi)x} \not\equiv 0$ (Prob. 5 below).

Now we suppose the relation proved to be impossible when there are $m - 1$ terms and demonstrate its impossibility when there are m terms. Let us suppose that (4–92) holds. Then we multiply by $e^{-r_m x}$ to obtain the relation

$$P_1(x)e^{(r_1 - r_m)x} + P_2(x)e^{(r_2 - r_m)x} + \cdots + P_{m-1}(x)e^{(r_{m-1} - r_m)x} + P_m(x) \equiv 0.$$

If we differentiate this relation, we obtain a similar one in which P_m is replaced by a polynomial of degree $k_m - 1$ but each of the polynomials $P_1(x)$, ..., $P_{m-1}(x)$ is replaced by a polynomial of the *same* degree:

$$D[(ax^k + \cdots)e^{bx}] = e^{bx}[abx^k + x^{k-1}(ka + \cdots) + \cdots].$$

By repeated differentiation we eventually replace $P_m(x)$ by 0 and have a relation

$$Q_1(x)e^{(r_1 - r_m)x} + \cdots + Q_{m-1}(x)e^{(r_{m-1} - r_m)x} \equiv 0,$$

where $Q_1(x)$, ..., $Q_{m-1}(x)$ have degrees k_1, ..., k_{m-1}. By the induction hypothesis, such a relation is impossible. Hence the assertion is established for all m, and the functions $e^{r_1 x}$, $xe^{r_1 x}$, $\ldots$, $e^{r_2 x}$, $\ldots$ are linearly independent.

We can now prove Theorem 3 of Section 4–5. When the coefficients are real, the complex roots are in conjugate pairs: $r_2 = \bar{r}_1$, $r_4 = \bar{r}_3$, $\ldots$ The real solutions are obtained as the real parts of the complex solutions (4–91). But if r is complex, $r = a + bi$, then

$$\mathrm{Re}\,[c_1 e^{rx} + c_2 e^{\bar{r}x}] = \mathrm{Re}\,[c_1 e^{ax}\,(\cos bx + i\sin bx) + c_2 e^{ax}\,(\cos bx - i\sin bx)]$$

$$= e^{ax}[C_1 \cos bx + C_2 \sin bx],$$

where $C_1 = \mathrm{Re}\,(c_1 + c_2)$, $C_2 = \mathrm{Im}\,(c_2 - c_1)$. Similarly,

$$\mathrm{Re}\,[c_1 x^h e^{rx} + c_2 x^h e^{\bar{r}x}] = x^h e^{ax}[C_1 \cos bx + C_2 \sin bx]$$

and, if r is real,

$$\mathrm{Re}\,[c_1 x^h e^{rx}] = C_1 x^h e^{rx},$$

where $C_1 = \mathrm{Re}\,(c_1)$. Hence the general real solution has the form described in Theorem 3. Since this is the general solution, the n functions concerned must be linearly independent real functions.

Remark. For the real case we now have three ways of writing the solutions. We illustrate these for the equation $y'' + y = 0$:

$$y = C_1 \cos x + C_2 \sin x,$$

$$y = ce^{ix} + \bar{c}e^{-ix}, \tag{4–93}$$

$$y = \mathrm{Re}\,(c_1 e^{ix} + c_2 e^{-ix}).$$

Here C_1, C_2 are arbitrary real constants; c, c_1, and c_2 are arbitrary complex constants.

PROBLEMS

1. Write the general real solution of each of the following equations in terms of complex constants and their conjugates:

(a) $y'' + 4y = 0$, (b) $y'' + 4y' + 5y = 0$,
(c) $(D^5 + 6D^3 + 9D)y = 0$, (d) $(D^5 + D^4 + D^3 + D^2 + D + 1)y = 0$.

2. Differentiate each of the following functions:

(a) $xe^{(1+i)x}$, (b) $(e^{ix} - e^{-ix})/(e^{ix} + e^{-ix})$,
(c) $(e^{ix} - e^{-ix})^3$, (d) $(1 + ix)^4$.

3. Graph as curves in the uv-plane ($w = u + iv$):

(a) $w = xe^{ix}$, (b) $w = x + e^{ix}$.

4. Prove the validity of the following rules for complex functions $F(x)$, $G(x)$, assuming existence of the derivatives concerned over some interval:

(a) $[F(x) + G(x)]' = F'(x) + G'(x)$ [Hint: Use (4–80).],

(b) $[F(x)G(x)]' = F(x)G'(x) + F'(x)G(x)$,

(c) $\left[\dfrac{1}{G(x)}\right]' = -\dfrac{G'(x)}{[G(x)]^2}$ $[G(x) \neq 0]$,

(d) $\left[\dfrac{F(x)}{G(x)}\right]' = \dfrac{G(x)F'(x) - F(x)G'(x)}{[G(x)]^2}$ $[G(x) \neq 0]$.

[Hint: For (d) apply the results of parts (b) and (c).]

5. Prove that $e^{a+bi} \neq 0$ for every complex number $a + bi$.

6. Find the general complex solution of each of the following:

(a) $(D^2 + 4)y = 0$,   (b) $(D^2 - 3iD - 2)y = 0$.

7. Obtain the general real solution as the real part of the general complex solution:

(a) $(D^2 + 9)y = 0$,   (b) $(D^2 + 4D + 5)y = 0$,

(c) $(D^4 + 18D^2 + 81)y = 0$,   (d) $(D^3 - 6D^2 + 10D)y = 0$.

ANSWERS

1. (a) $ce^{2ix} + \bar{c}e^{-2ix}$;   (b) $ce^{(-2+i)x} + \bar{c}e^{(-2-i)x}$;   (c) $C_1 + c_2e^{rx} + \bar{c}_2e^{\bar{r}x} + c_3xe^{rx} + \bar{c}_3xe^{\bar{r}x}$, $r = \sqrt{3}i$, and C_1 is real; (d) $C_1e^{-x} + c_2e^{rx} + \bar{c}_2e^{\bar{r}x} + c_3e^{sx} + \bar{c}_3e^{\bar{s}x}$, where $r = e^{\pi i/3}$, $s = e^{2\pi i/3}$, and C_1 is real.

2. (a) $(1 + x + ix)e^{(1+i)x}$, (b) $4i/(e^{ix} + e^{-ix})^2$, (c) $3i(e^{ix} - e^{-ix})^2(e^{ix} + e^{-ix})$, (d) $4i(1 + ix)^3$.

6. (a) $c_1e^{2ix} + c_2e^{-2ix}$, (b) $c_1e^{ix} + c_2e^{2ix}$.

7. (a) Re $(c_1e^{3ix} + c_2e^{-3ix})$, (b) Re $(c_1e^{(-2+i)x} + c_2e^{(-2-i)x})$, (c) Re $(c_1e^{3ix} + c_2e^{-3ix} + c_3xe^{3ix} + c_4xe^{-3ix})$, (d) Re $(c_1 + c_2e^{(3+i)x} + c_3e^{(3-i)x})$.

4–9 The nonhomogeneous linear equation.   Method of variation of parameters. By Theorem 1 of Section 4–3 the general solution of a linear equation such as

$$(a_0D^n + \cdots + a_n)y = Q(x) \tag{4–94}$$

has the form

$$y = y^*(x) + c_1y_1(x) + \cdots + c_ny_n(x), \tag{4–95}$$

where $y^*(x)$ is a particular solution of the nonhomogeneous equation (4–94)

and $c_1y_1(x) + \cdots + c_ny_n(x)$ is the general solution of the related homogeneous equation

$$(a_0D^n + \cdots + a_n)y = 0;$$

the expression $c_1y_1(x) + \cdots + c_ny_n(x)$ is called the *complementary function*. At least when the coefficients are constant, we know how to find the
complementary function, but we have as yet no inkling of how to find the
particular solution $y^*(x)$ of the nonhomogeneous equation. Now, however, we discuss one method for finding $y^*(x)$. This method, called *variation of parameters*, is applicable to equations with variable or constant
coefficients, but the complementary function must be known before the
method can be applied. Other methods for finding $y^*(x)$ are described in
succeeding sections.

We illustrate our procedure for an equation of second order:

$$a_0y'' + a_1y' + a_2y = Q(x). \tag{4-96}$$

The complementary function $c_1y_1(x) + c_2y_2(x)$ is assumed known. We
consider the two equations

$$y = c_1y_1(x) + c_2y_2(x),$$
$$y' = c_1y_1'(x) + c_2y_2'(x),$$

which are valid for solutions of the homogeneous equation. We now replace the constants c_1, c_2 by variables $v_1(x)$, $v_2(x)$ and consider the equations

$$y = v_1(x)y_1(x) + v_2(x)y_2(x),$$
$$y' = v_1(x)y_1'(x) + v_2(x)y_2'(x) \tag{4-97}$$

as defining a *change* of variables. Instead of a second order equation for y,
we obtain a system of two first order equations for v_1 and v_2. To find the
system, we differentiate the first equation of (4–97):

$$y' = v_1y_1' + v_2y_2' + v_1'y_1 + v_2'y_2.$$

Subtraction of the second equation of (4–97) yields

$$v_1'y_1 + v_2'y_2 = 0, \tag{4-98}$$

which is the *first of the two equations sought*. Differentiation of the second
of (4–97) yields

$$y'' = v_1y_1'' + v_2y_2'' + v_1'y_1' + v_2'y_2'. \tag{4-99}$$

We replace y, y', y'' by the expressions (4–97), (4–99) in the given differential equation (4–96):

$$a_0(v_1y_1'' + v_2y_2'' + v_1'y_1' + v_2'y_2') + a_1(v_1y_1' + v_2y_2')$$
$$+ a_2(v_1y_1 + v_2y_2) = Q(x).$$

This can be written as

$$v_1(a_0y_1'' + a_1y_1' + a_2y_1) + v_2(a_0y_2'' + a_1y_2' + a_2y_2)$$
$$+ a_0(v_1'y_1' + v_2'y_2') = Q(x).$$

But $y_1(x)$ is a solution of the homogeneous equation; hence the coefficient of v_1 is 0. For the same reason, the coefficient of v_2 is 0. Accordingly, we obtain the *second desired equation:*

$$a_0(v_1'y_1' + v_2'y_2') = Q(x). \tag{4–100}$$

Equations (4–98) and (4–100) are simultaneous equations for v_1', v_2'. They can be solved for v_1', v_2', and v_1, v_2 can then be obtained by integration.

EXAMPLE 1. $y'' - y = e^{3x}$. The complementary function is $c_1e^x + c_2e^{-x}$. Equations (4–98), (4–100) become

$$v_1'e^x + v_2'e^{-x} = 0, \qquad v_1'e^x - v_2'e^{-x} = e^{3x}.$$

Hence $2v_1' = e^{2x}$, $2v_2' = -e^{4x}$, and we can choose v_1 as $e^{2x}/4$, v_2 as $-e^{4x}/8$. Only one particular solution is needed in each case, hence

$$y^* = v_1e^x + v_2e^{-x} = \frac{1}{4}e^{3x} - \frac{1}{8}e^{3x} = \frac{e^{3x}}{8}$$

is the particular solution sought, and the general solution is

$$y = \tfrac{1}{8}e^{3x} + c_1e^x + c_2e^{-x}.$$

The method carries over to higher order equations.

THEOREM 5. *Let the function $a_0(x)$, $\ldots$, $a_n(x)$, $Q(x)$ be continuous in some interval of x, within which $a_0(x) \neq 0$. Let $c_1y_1(x) + \cdots + c_ny_n(x)$ be the complementary function for the differential equation*

$$[a_0(x)D^n + \cdots + a_n(x)]y = Q(x).$$

Then a particular solution $y^*(x)$ is given by

$$y^*(x) = v_1(x)y_1(x) + \cdots + v_n(x)y_n(x),$$

where $v_1(x), \ldots, v_n(x)$ are chosen to satisfy the equations

$$v_1'(x)y_1(x) + \cdots + v_n'(x)y_n(x) = 0,$$
$$v_1'(x)y_1'(x) + \cdots + v_n'(x)y_n'(x) = 0,$$
$$\vdots \qquad\qquad\qquad\qquad (4\text{-}101$$
$$v_1'(x)y_1^{(n-2)}(x) + \cdots + v_n'(x)y_n^{(n-2)}(x) = 0,$$
$$a_0[v_1'(x)y_1^{(n-1)}(x) + \cdots + v_n'(x)y^{(n-1)}(x)] = Q(x).$$

Proof. The justification follows the reasoning for $n = 2$. The equations

$$y = v_1y_1 + \cdots + v_ny_n,$$
$$y' = v_1y_1' + \cdots + v_ny_n',$$
$$\vdots \qquad\qquad\qquad\qquad (4\text{-}102)$$
$$y^{(n-1)} = v_1y_1^{(n-1)} + \cdots + v_ny_n^{(n-1)}$$

are considered as defining the change of variables. Differentiation of the first and subtraction of the second yields the first of (4-101); differentiation of the second of (4-102) and subtraction of the third leads to the second of (4-101), and so on. Differentiation of the last of (4-102) gives

$$y^{(n)} = v_1y_1^{(n)} + \cdots + v_ny_n^{(n)} + v_1'y_1^{(n-1)} + \cdots + v_n'y_n^{(n-1)}. \qquad (4\text{-}103)$$

Multiplication of the equations (4-102) by $a_n(x), a_{n-1}(x), \ldots, a_1(x)$ and of (4-103) by $a_0(x)$ and addition of all n equations yields the last of (4-101) by virtue of the fact that $y_1(x), \ldots, y_n(x)$ are solutions of the homogeneous equation.

The equations (4-101) are n simultaneous linear equations for the unknowns $v_1'(x), \ldots, v_n'(x)$. The determinant of the coefficients is

$$a_0(x)\begin{vmatrix} y_1(x) & \cdots & y_n(x) \\ \vdots & & \vdots \\ y_1^{(n-1)}(x) & \cdots & y_n^{(n-1)}(x) \end{vmatrix} = a_0(x)W(x),$$

where $W(x)$ is the Wronskian determinant of the functions $y_1(x), \ldots, y_n(x)$. Since these functions must be linearly independent, Theorem 2 shows that $W(x) \neq 0$. We have assumed $a_0(x) \neq 0$; therefore the determinant is not 0 and equations (4–101) have a unique solution for $v_1'(x), \ldots, v_n'(x)$. The solution can be expressed in terms of determinants:

$$v_1'(x) = \begin{vmatrix} 0 & y_2(x) & \cdots & y_n(x) \\ \vdots & & & \vdots \\ Q(x) & y_2^{(n-1)}(x) & \cdots & y_n^{(n-1)}(x) \end{vmatrix} \div [a_0(x)W(x)], \cdots$$

From this we see that $v_1'(x)$ equals a continuous function of x; hence integration is possible and we can obtain $v_1(x)$ (up to an arbitrary constant). Similarly, $v_2(x), \ldots, v_n(x)$ can be found. Since $v_1(x), \ldots, v_n(x)$ satisfy (4–101), it follows at once that $y^*(x) = v_1 y_1(x) + \cdots + v_n y_n(x)$ satisfies (4–102) and the given differential equation.

EXAMPLE 2. $y'' + 2y' + 2y = e^{5x}$. The characteristic equation has the complex roots $-1 \pm i$. We apply the method of variation of parameters in terms of *complex* functions. If we obtain a complex particular solution, its real part will be a real particular solution (Section 4–8). We thus choose $y_1(x)$ to be $e^{(-1+i)x}$ and $y_2(x)$ to be $e^{(-1-i)x}$. Equations (4–101) become

$$v_1' e^{(-1+i)x} + v_2' e^{(-1-i)x} = 0,$$

$$v_1'(-1+i)e^{(-1+i)x} + v_2'(-1-i)e^{(-1-i)x} = e^{5x}.$$

Elimination gives the equations

$$v_1' = -\tfrac{1}{2}ie^{(6-i)x}, \qquad v_2' = \tfrac{1}{2}ie^{(6+i)x}.$$

Hence we can choose

$$v_1 = -\frac{1}{2}\frac{i}{6-i}e^{(6-i)x}, \qquad v_2 = \frac{1}{2}\frac{i}{6+i}e^{(6+i)x},$$

$$y^* = v_1 y_1 + v_2 y_2 = e^{5x}\left(-\frac{1}{2}\frac{i}{6-i} + \frac{1}{2}\frac{i}{6+i}\right) = \frac{e^{5x}}{37}.$$

EXAMPLE 3. $(D^3 - 7D^2 + 14D - 8)y = \log x$, $x > 0$. The complementary function is $c_1 e^x + c_2 e^{2x} + c_3 e^{4x}$. The equations (4–101) become

$$v_1' e^x + v_2' e^{2x} + v_3' e^{4x} = 0,$$

$$v_1' e^x + 2v_2' e^{2x} + 4v_3' e^{4x} = 0,$$

$$v_1' e^x + 4v_2' e^{2x} + 16v_3' e^{4x} = \log x.$$

We can consider $v_1'e^x$, $v_2'e^{2x}$, $v_3'e^{4x}$ as the unknowns. We solve by determinants:

$$v_1'e^x = \begin{vmatrix} 0 & 1 & 1 \\ 0 & 2 & 4 \\ \log x & 4 & 16 \end{vmatrix} \div \begin{vmatrix} 1 & 1 & 1 \\ 1 & 2 & 4 \\ 1 & 4 & 16 \end{vmatrix} = \frac{\log x}{3}.$$

Hence $v_1' = (e^{-x}\log x)/3$. Similarly, we find $v_2' = -(e^{-2x}\log x)/2$, $v_3' = (e^{-4x}\log x)/6$; accordingly,

$$y^* = v_1 y_1 + v_2 y_2 + v_3 y_3$$

$$= \tfrac{1}{3}e^x \int e^{-x}\log x\, dx - \tfrac{1}{2}e^{2x}\int e^{-2x}\log x\, dx + \tfrac{1}{6}e^{4x}\int e^{-4x}\log x\, dx.$$

PROBLEMS

1. Find the general solution of each of the following by the method of variation of parameters:

(a) $y'' - y = e^x$;   (b) $y''' - 6y'' + 11y' - 6y = e^{4x}$;
(c) $y'' + y = \cot x$;   (d) $y'' + 4y = \sec 2x$;
(e) $y'' - y = 1/x$, $x > 0$;   (f) $y'' + 4y' + 5y = xe^x$.

2. Solve the first order linear equation $y' + p(x)y = Q(x)$ by first solving the related homogeneous equation and then obtaining a particular solution by variation of parameters.

3. Verify that $y = c_1 x + c_2 x^2$ is the general solution of the equation $x^2 y'' - 2xy' + 2y = 0$ and find the general solution of the equation $x^2 y'' - 2xy' + 2y = x^3$.

4. Verify that $y = c_1 e^x + c_2 x^{-1}$ is the complementary function for

$$x(x+1)y'' + (2 - x^2)y' - (2+x)y = (x+1)^2$$

and find the general solution.

5. Let an equation of second order

$$y'' + a_1 y' + a_2 y = Q(x)$$

have constant coefficients a_1, a_2 and distinct characteristic roots r_1, r_2. Apply the method of variation of parameters to obtain the particular solution

$$y = \frac{e^{r_1 x}}{r_1 - r_2} \int e^{-r_1 x} Q(x)\, dx + \frac{e^{r_2 x}}{r_2 - r_1} \int e^{-r_2 x} Q(x)\, dx.$$

ANSWERS

1. (a) $c_1 e^x + c_2 e^{-x} + \frac{1}{2} x e^x$,   (b) $c_1 e^x + c_2 e^{2x} + c_3 e^{3x} + \frac{1}{6} e^{4x}$,
(c) $c_1 \cos x + c_2 \sin x - \sin x \log |\csc x + \cot x|$,
(d) $c_1 \cos 2x + c_2 \sin 2x + \frac{1}{4} \cos 2x \log |\cos 2x| + \frac{1}{2} x \sin 2x$,
(e) $c_1 e^x + c_2 e^{-x} + \frac{1}{2} e^x \int e^{-x} x^{-1}\, dx - \frac{1}{2} e^{-x} \int e^x x^{-1}\, dx$,
(f) $(5x - 3)e^x/50 + c e^{(-2+i)x} + \bar{c} e^{(-2-i)x}$.

3. $c_1 x + c_2 x^2 + \frac{1}{2} x^3$.

4. $c_1 e^x + c_2 x^{-1} - \frac{1}{2}(x + 2)$.

4–10 Method of undetermined coefficients. Consider a linear equation with constant coefficients:

$$(a_0 D^n + \cdots + a_n)y = Q(x). \tag{4–104}$$

We seek a particular solution and do not assume the complementary function to be known. The following examples illustrate the method of undetermined coefficients.

EXAMPLE 1. $(D^2 - 3D + 7)y = 10e^{2x}$. We try to choose k so that $k e^{2x}$ is a solution. Substitution yields the equations

$$4k e^{2x} - 6k e^{2x} + 7k e^{2x} = 10e^{2x}, \qquad 5k = 10, \qquad k = 2.$$

Hence $2e^{2x}$ is a particular solution. The complementary function is found to be

$$e^{3x/2}\left(c_1 \cos \frac{\sqrt{19}}{2} x + c_2 \sin \frac{\sqrt{19}}{2} x\right).$$

The general solution is the sum of the complementary function and $2e^{2x}$.

EXAMPLE 2. $(D^2 - 3D + 7)y = 10xe^{2x}$. If we set $y = kxe^{2x}$, we find it impossible to choose k so that the equation is satisfied, since $(D^2 - 3D + 7)xe^{2x}$ is of form $e^{2x}(ax + b)$. This suggests trying $y = e^{2x}(k_1 x + k_2)$. The substitution in the equation can be arranged as follows:

$$
\begin{array}{r|l}
7 & y \;\; = e^{2x}(k_1 x + k_2) \\
-3 & y' \;\, = e^{2x}(2k_1 x + 2k_2 + k_1) \\
1 & y'' = e^{2x}(4k_1 x + 4k_2 + 4k_1) \\
\hline
\end{array}
$$

$$y'' - 3y' + 7y = e^{2x}(5k_1 x + 5k_2 + k_1) \equiv 10xe^{2x}.$$

Hence $5k_1x + 5k_2 + k_1 \equiv 10x$, $5k_1 = 10$, $5k_2 + k_1 = 0$, and

$$k_1 = 2, \qquad k_2 = -\tfrac{2}{5}, \qquad y = e^{2x}(2x - \tfrac{2}{5}).$$

(See Prob. 8 below.)

The above examples suggest that, in general, if $Q = e^{ax}(b_0x^m + b_1x^{m-1} + \cdots)$, our trial function should have the similar form

$$y = e^{ax}(k_1x^m + k_2x^{m-1} + \cdots).$$

The following example shows that this form will not always succeed.

EXAMPLE 3. $y'' - y = xe^x$. The substitution of $y = e^x(k_1x + k_2)$ leads to an impossible equation:

$$y'' - y = 2e^xk_1 \equiv xe^x.$$

The trouble can be traced to the fact that e^x is a solution of the homogeneous equation. We discover by experimenting that multiplication by x eliminates the difficulty; that is, we set $y = xe^x(k_1x + k_2) = e^x(k_1x^2 + k_2x)$ and can then determine the coefficients:

$$
\begin{array}{r|l}
-1 & y = e^x(k_1x^2 + k_2x) \\
0 & y' = e^x[k_1x^2 + (2k_1 + k_2)x + k_2] \\
1 & y'' = e^x[k_1x^2 + (4k_1 + k_2)x + 2k_1 + 2k_2] \\
\hline
\end{array}
$$

$$y'' - y = e^x(4k_1x + 2k_1 + 2k_2) \equiv xe^x.$$

Hence

$$4k_1x + 2k_1 + 2k_2 \equiv x, \qquad k_1 = \tfrac{1}{4}, \qquad k_2 = -\tfrac{1}{4},$$

$$y^* = e^x[(x^2/4) - (x/4)].$$

EXAMPLE 4. $y'' + 2y' + 2y = \sin 3x$. The substitution $y = k \sin 3x$ is ineffective, but $y = k_1 \cos 3x + k_2 \sin 3x$ leads to a solution:

$$
\begin{array}{r|l}
2 & y = k_1 \cos 3x + k_2 \sin 3x \\
2 & y' = 3k_2 \cos 3x - 3k_1 \sin 3x \\
1 & y'' = -9k_1 \cos 3x - 9k_2 \sin 2x \\
\hline
\end{array}
$$

$$y'' + 2y' + 2y = \cos 3x\,(6k_2 - 7k_1) + \sin 3x\,(-7k_2 - 6k_1) \equiv \sin 3x,$$

$$6k_2 - 7k_1 = 0, \qquad -7k_2 - 6k_1 = 1,$$

$$k_1 = -\frac{6}{85}, \qquad k_2 = -\frac{7}{85},$$

$$y^* = \frac{1}{85}(-6\cos 3x - 7\sin 3x),$$

$$y = e^{-x}(c_1\cos x + c_2\sin x) + \frac{1}{85}(-6\cos 3x - 7\sin 3x).$$

We now describe the method precisely; the rules given are suggested by the preceding examples. A proof is given in Section 4–12.

Let $Q(x)$ have the form

$$e^{ax}\cos bx\,(p_0 x^m + p_1 x^{m-1} + \cdots + p_m) + e^{ax}\sin bx\,(q_0 x^m + \cdots + q_m),$$

where some of the constants a, b, p_0, p_1, ..., q_0, q_1, ... may be 0. If $a \pm bi$ is not a root of the characteristic equation

$$a_0 r^n + \cdots + a_n = 0,$$

then there is a solution y^* of the differential equation (4–104) of form similar to that of Q:

$$y^* = e^{ax}\cos bx\,(k_1 x^m + k_2 x^{m-1} + \cdots)$$
$$+ e^{ax}\sin bx\,(l_1 x^m + l_2 x^{m-1} + \cdots).$$

The coefficients k_1, k_2, ..., l_1, l_2, ... are determined by substitution of the expression for y^* in the differential equation and choosing the coefficients so that the equation becomes an identity. If $a \pm bi$ is a root of multiplicity h of the characteristic equation, then the trial function must be multiplied by x^h.

When the right-hand member is formed of two or more groups of terms, each associated with a different choice of $a \pm bi$, the particular solution is a sum of those for each group. This follows from the superposition principle (Section 4–4).

EXAMPLE 5. $y'' + y = x + 2\cos x + \sin x$. Here the term x corresponds to $a \pm bi$ equal to 0:

$$x = xe^{0x}\cos 0x,$$

while the terms $2\cos x + \sin x$ correspond to $a \pm bi$ equal to $\pm i$. Now 0 is not a root of the characteristic equation, but $\pm i$ are roots. Hence our trial function is

$$y^* = k_1 x + k_2 + x(k_3\cos x + k_4\sin x).$$

Upon substituting, we find that $k_1 = 1$, $k_2 = 0$, $k_3 = -\frac{1}{2}$, $k_4 = 1$, and

$$y = c_1 \cos x + c_2 \sin x + x + x \sin x - \tfrac{1}{2}x \cos x$$

is the general solution.

PROBLEMS

1. Find the general solution of each of the following equations:

(a) $(D^2 - 9)y = e^x$,   (b) $(D^2 + 4)y = e^{2x}$,

(c) $(D^2 + 2D + 1)y = 5$,   (d) $(D^2 + 2D + 1)y = x + e^{4x}$,

(e) $(D^2 + 2D + 1)y = e^{-x} + e^x$,   (f) $(D^2 + 2D + 1)y = x^2 e^{-x}$,

(g) $(D^2 + 4)y = 5 \sin 3x + \cos 3x + \sin 2x$,

(h) $(D^2 + 4)y = x \cos 2x + 3e^{2x} \sin 2x$,

(i) $(D^3 + 3D^2 + 3D + 1)y = e^{-x} \sin x$,

(j) $(D^2 + 2D + 5)y = e^{-x} \cos 2x$.

2. Find a particular solution of each of the following:

(a) $(D^3 + D + 1)y = e^{3x}$,

(b) $(D^4 - D^3 + D - 1)y = e^x + 3x - 2$,

(c) $(D^5 + D^3 + D + 1)y = x^2 - x - 2$,

(d) $(D^4 + D^3 + D^2)y = 3x + 2$,

(e) $(D^4 + D + 1)y = \cos 2x + 3$.

3. Obtain the solution satisfying the given initial conditions:

(a) $y'' + y = \sin x$; $y = 1$ and $y' = 0$ for $x = 0$;

(b) $y'' + 2y' + y = \cos x$; $y = 0$ and $y' = 0$ for $x = 0$.

4. Prove that a particular solution of the equation $(a_0 D^n + \cdots + a_n)y = e^{bx}$ is given by $e^{bx}/f(b)$, where $f(r) = a_0 r^n + \cdots + a_n$, provided $f(b) \neq 0$.

5. Prove that if the coefficients $a_0, \ldots, a_n$ are real and $y(x)$ is a complex solution of

$$(a_0 D^n + \cdots + a_n)y = F(x) = f(x) + ig(x),$$

then Re $[y(x)]$ is a solution of

$$(a_0 D^n + \cdots + a_n)y = f(x) = \text{Re } [F],$$

while Im $[y(x)]$ is a solution of

$$(a_0 D^n + \cdots + a_n)y = g(x) = \text{Im } [F].$$

6. Use the rule of Prob. 5 to obtain particular solutions of the following:

(a) $(D^2 + D + 2)y = \cos x = \text{Re } [e^{ix}]$,

(b) $(D^2 + D + 2)y = \sin x = \text{Im } [e^{ix}]$,

[*Hint:* Obtain a particular solution of $(D^2 + D + 2)y = e^{ix}$, then take real and imaginary parts.]

(c) $(D^2 + D + 2)y = e^{2x}\cos 3x = \operatorname{Re}[e^{(2+3i)x}]$,

(d) $(D^2 + D + 2)y = xe^{2x}\cos 3x = \operatorname{Re}[xe^{(2+3i)x}]$.

7. Show that if $h > 0$, $\lambda > 0$, $\omega > 0$, then a particular solution of

$$(D^2 + 2hD + \lambda^2)y = A\sin\omega x$$

is given by

$$y = \frac{A\sin\omega(x-\alpha)}{[(\lambda^2-\omega^2)^2 + 4h^2\omega^2]^{1/2}}, \qquad \alpha = \tan^{-1}\frac{2h\omega}{\omega^2-\lambda^2},$$

where $0 < \alpha < \pi$.

8. Example 2 in the text above leads to the identity

$$5k_1xe^{2x} + (5k_2 + k_1)e^{2x} \equiv 10xe^{2x} + 0e^{2x}.$$

Show that the relations $5k_1 = 10$, $5k_2 + k_1 = 0$ follow from the *linear independence* of the functions xe^{2x}, e^{2x}. (See Prob. 7 following Section 4–3.)

ANSWERS

1. (a) $c_1e^{3x} + c_2e^{-3x} - (e^x/8)$, (b) $c_1\cos 2x + c_2\sin 2x + e^{2x}/8$,
(c) $c_1e^{-x} + c_2xe^{-x} + 5$, (d) $c_1e^{-x} + c_2xe^{-x} + x - 2 + e^{4x}/25$, (e) $c_1e^{-x} +$
$c_2xe^{-x} + x^2e^{-x}/2 + e^x/4$,   (f) $c_1e^{-x} + c_2xe^{-x} + x^4e^{-x}/12$,   (g) $c_1\cos 2x +$
$c_2\sin 2x - \sin 3x - (\cos 3x)/5 - (x\cos 2x)/4$,   (h) $c_1\cos 2x + c_2\sin 2x -$
$(2x^2\sin 2x + x\cos 2x)/16 + (3/20)e^{2x}(\sin 2x - 2\cos 2x)$,   (i) $c_1e^{-x} +$
$c_2xe^{-x} + c_3x^2e^{-x} + e^{-x}\cos x$, (j) $e^{-x}(c_1\cos 2x + c_2\sin 2x + \frac{1}{4}x\sin 2x)$.

2. (a) $e^{3x}/31$, (b) $\frac{1}{2}xe^x - 3x - 1$, (c) $x^2 - 3x + 1$,
(d) $\frac{1}{2}(x^3 - x^2)$, (e) $3 + (17\cos 2x + 2\sin 2x)/293$.

3. (a) $\cos x + \frac{1}{2}(\sin x - x\cos x)$, (b) $\frac{1}{2}(\sin x - xe^{-x})$.

6. (a) $\frac{1}{2}(\cos x + \sin x)$, (b) $\frac{1}{2}(\sin x - \cos x)$,
(c) $e^{2x}(15\sin 3x - \cos 3x)/226$,
(d) $e^{2x}[x(-226\cos 3x + 3390\sin 3x) - 1194\sin 3x + 1300\cos 3x]/226^2$.

4–11 Algebra of differential operators. Let $L_1 = a_0D^n + \cdots + a_n$, $L_2 = b_0D^m + \cdots + b_m$ be two differential operators; the coefficients may depend on x. The sum $L_1 + L_2$ is defined as the operator which, when applied to y, yields $L_1y + L_2y$. Thus

$$(L_1 + L_2)y = L_1y + L_2y$$

$$= (a_0D^n + \cdots + a_n)y + (b_0D^m + \cdots + b_m)y.$$

The last two expressions can be combined into one by collecting terms of the same degree. Hence we conclude that differential operators are added as if they were polynomials in D; for example,

$$(xD^2 - D + 2) + (xD + x) = xD^2 + (x - 1)D + x + 2.$$

Multiplication of L_1 *on the left* by a constant k or by a function of x is defined similarly by the equations

$$(kL_1)y = k(L_1y), \qquad [g(x)L_1]y = g(x)(L_1y),$$

and we again conclude that the operation is the same as for polynomials. From the definitions given, these algebraic rules follow:

$$L_1 + L_2 = L_2 + L_1, \qquad L_1 + (L_2 + L_3) = (L_1 + L_2) + L_3,$$

$$g(x)(L_1 + L_2) = g(x)L_1 + g(x)L_2, \qquad k(L_1 + L_2) = kL_1 + kL_2.$$

We now define multiplication of operators. We define L_1L_2 as the operator such that

$$(L_1L_2)y = L_1(L_2y) \qquad\qquad (4\text{--}105)$$

wherever the right-hand side has meaning. For example, $(D + 1)(D - 1)$ is evaluated as follows:

$$[(D + 1)(D - 1)]y = (D + 1)[(D - 1)y] = (D + 1)(Dy - y)$$
$$= D(Dy - y) + Dy - y$$
$$= D^2y - Dy + Dy - y = D^2y - y$$
$$= (D^2 - 1)y.$$

Accordingly, $(D + 1)(D - 1) = D^2 - 1$, just as for polynomials in D. However, the results are different *when the coefficients are variable.* Thus

$$[(xD + 1)(xD - 1)]y = (xD + 1)[(xD - 1)y]$$
$$= (xD + 1)(xDy - y)$$
$$= xD(xDy - y) + xDy - y$$
$$= x(xD^2y + Dy - Dy) + xDy - y$$
$$= x^2D^2y + xDy - y.$$

Accordingly,

$$(xD + 1)(xD - 1) = x^2D^2 + xD - 1,$$

in disagreement with formal algebraic procedure. If we analyze the above example, we see that the difficulty arises from the application of D to xDy; the product rule is used, and a new term Dy is introduced because of the variable coefficient x. This complication cannot arise when the coefficients are constant, and we have the general rule:

Operators with constant coefficients can be multiplied as if they were polynomials in the variable D.

To prove this rule, we remark that each operator $a_0 D^n + \cdots + a_n = L$ with constant coefficients can be completely specified by giving its characteristic polynomial $f(r) = a_0 r^n + \cdots + a_n$; furthermore, $f(r)$ can be found from the equation

$$L(e^{rx}) = f(r)e^{rx}. \tag{4–106}$$

Now if $L_1 = a_0 D^n + \cdots + a_n$, $L_2 = b_0 D^m + \cdots + b_m$, then let $f(r) = a_0 r^n + \cdots + a_n$, $g(r) = b_0 r^m + \cdots + b_m$ be the associated characteristic polynomials. From the above definition of multiplication of operators, it follows that $L_1 L_2$ is an operator with constant coefficients. To find the associated characteristic polynomial, we evaluate

$$(L_1 L_2)e^{rx} = L_1(L_2 e^{rx}) = L_1[g(r)e^{rx}] = g(r)L_1(e^{rx}) = g(r)f(r)e^{rx}.$$

Hence $L_1 L_2$ has the characteristic polynomial $f(r)g(r)$, which shows that $L_1 L_2$ is obtained from L_1, L_2 by multiplying them as if they were polynomials in D.

Operators with *constant coefficients* can now be added, multiplied, and multiplied by constants, just as polynomials in D. In particular, multiplication is *commutative, associative,* and *distributive:*

$$L_1 L_2 = L_2 L_1, \qquad L_1(L_2 L_3) = (L_1 L_2)L_3,$$
$$L_1(L_2 + L_3) = L_1 L_2 + L_1 L_3. \tag{4–107}$$

Associativity and distributivity also hold for operators with variable coefficients, but commutativity fails in general (Probs. 2, 3 following Section 4–12).

4–12 Factorization of operators. Associated with each operator $a_0 D^n + \cdots + a_n$ with constant coefficients is the characteristic polynomial $f(r) = a_0 r^n + \cdots + a_n$. It is convenient to denote the operator itself by $f(D)$:

$$f(D) = a_0 D^n + \cdots + a_n. \tag{4–108}$$

Of course, $f(D)$ is not a function of D in the ordinary sense.

If the roots r_1, r_2, ... of the characteristic equation are known, then there is a corresponding factorization of $f(r)$:

$$f(r) = a_0(r - r_1)(r - r_2) \cdots (r - r_n). \tag{4-109}$$

Accordingly, we obtain a *factorization of $f(D)$*:

$$f(D) = a_0 D^n + \cdots + a_n = a_0(D - r_1)(D - r_2) \cdots (D - r_n). \tag{4-110}$$

For we are allowed to multiply the expressions on the right of (4-110) as though D were a variable; (4-109) shows that the result is $f(D)$.

Factorization provides us with a new method for finding a particular solution of a nonhomogeneous equation.

EXAMPLE 1. $(D^2 - 1)y = e^{-x}$. We write the equation in factored form:

$$[(D + 1)(D - 1)]y = e^{-x}$$

or

$$(D + 1)[(D - 1)y] = e^{-x}.$$

This suggests introduction of an auxiliary variable:

$$(D - 1)y = u, \qquad (D + 1)u = e^{-x}.$$

The second equation is a first order linear equation for u; when u has been found, we treat the first equation as a first order equation for y:

$$u = e^{-x} \int e^x e^{-x} \, dx = xe^{-x},$$

$$y = e^x \int e^{-x} u \, dx = e^x \int xe^{-2x} \, dx$$

$$= e^x \left(\frac{xe^{-2x}}{-2} + \frac{e^{-2x}}{-4} \right) = -\frac{1}{2} xe^{-x} - \frac{1}{4} e^{-x}.$$

We have carried no arbitrary constants, since we seek only a particular solution. Furthermore, $-\frac{1}{4}e^{-x}$ is part of the complementary function, so that $-\frac{1}{2}xe^{-x}$ is itself a particular solution.

EXAMPLE 2. $(D^2 - 1)y = 2 (\sec^3 x - \sec x)$. The procedure used in Example 1 leads to the equations

$$(D - 1)y = u, \qquad (D + 1)u = 2 (\sec^3 x - \sec x),$$

$$u = 2e^{-x}\int e^x (\sec^3 x - \sec x)\, dx, \qquad y = e^x \int e^{-x} u\, dx.$$

It happens that the integrals can be expressed in terms of elementary functions:

$$\int e^x (\sec^3 x - \sec x)\, dx = \tfrac{1}{2}e^x (\sec x \tan x - \sec x),$$

$$u = \sec x \tan x - \sec x,$$

$$y = e^x \int e^{-x} (\sec x \tan x - \sec x)\, dx$$

$$= e^x (e^{-x} \sec x) = \sec x.$$

EXAMPLE 3. $(D^2 + 1)^3 y = \sin 2x$.   Here the equation is given in factored form. We now introduce two auxiliary variables:

$$(D^2 + 1)y = u, \qquad (D^2 + 1)u = v, \qquad (D^2 + 1)v = \sin 2x.$$

We could factor the quadratic expressions into linear factors:

$$D^2 + 1 = (D + i)(D - i)$$

and split the equations further; however, this procedure is not needed in this case. By undetermined coefficients we find, successively,

$$v = \frac{\sin 2x}{-3}, \qquad u = \frac{\sin 2x}{9}, \qquad y = \frac{\sin 2x}{-27}.$$

Justification of the method of undetermined coefficients. We now give a proof that the rule of Section 4–10 does provide a particular solution of a nonhomogeneous equation

$$Ly \equiv a_0 y^{(n)} + \cdots + a_n = Q(x) \equiv p(x)e^{ax} \cos bx + q(x)e^{ax} \sin bx,$$

where $p(x)$ and $q(x)$ are polynomials of degree at most m. Let the characteristic roots be $r_1, \ldots, r_n$, so that

$$L = a_0(D - r_1) \cdots (D - r_n).$$

From the form given, $Q(x)$ is a particular solution of an equation $L_1 y = 0$, where

$$L_1 = [(D - a)^2 + b^2]^{m+1}.$$

This we see by writing out the general solution of the equation $L_1 y = 0$. Now let y be a solution of the equation $Ly = Q$. We multiply by L_1 on both sides:

$$L_1 Ly = L_1 Q = 0.$$

Hence y is also a solution of the equation $L_1 Ly = 0$; that is,

$$a_0[(D - a)^2 + b^2]^{m+1}(D - r_1) \cdots (D - r_n)y = 0.$$

The general solution of this equation consists of the general solution of the equation $Ly = 0$, that is, the complementary function $y_c(x)$, plus additional terms arising from the new roots $a \pm bi$. The additional terms have the form described for the trial function in the rule of Section 4–10; we denote the sum of these terms by $y_k(x)$. Thus $y = y_c(x) + y_k(x)$. Substitution in the equation $Ly = Q$ gives

$$L[y_c(x) + y_k(x)] = Q.$$

But $L[y_c(x)] = 0$, and hence

$$L[y_k(x)] = Q.$$

Therefore, for appropriate choice of the k's, $y_k(x)$ is a particular solution.

PROBLEMS

1. Combine and simplify the following:
   (a) $x^2(D^2 - D + 1) + x[xD^2 + (x + 1)D - 2]$,
   (b) $(D - 1)^2 + (D + 1)^2$,
   (c) $(D - 1)(D^2 + D + 1)$,
   (d) $(xD - 1)(D + 2)$,
   (e) $(x^2 D^2 - e^x)(e^x D + x)$.

2. Verify the following inequalities:
   (a) $(xD + x)(xD - 1) \neq (xD - 1)(xD + x)$,
   (b) $(D + e^x)(e^x D + 1) \neq (e^x D + 1)(D + e^x)$.

3. Let L_1, L_2, L_3 be arbitrary differential operators with coefficients which have derivatives of all orders over an interval.
   (a) Prove the associative law: $L_1(L_2 L_3) = (L_1 L_2)L_3$.
   (b) Prove the distributive law: $L_1(L_2 + L_3) = L_1 L_2 + L_1 L_3$.

4. Find a particular solution by factoring the operator in each of the following equations:

(a) $(D - 1)^2 y = e^x$, (b) $(D^2 - 2D - 3)y = e^{5x}$,
(c) $(D^4 + 5D^2 + 4)y = \sin 3x$, (d) $(D + 1)^5 y = xe^{-x}$,
(e) $(D - 1)^2 y = e^{e^x}(e^{2x} - e^x + 1)$,
(f) $(D^2 + 3D + 2)y = (2/x^3) - (3/x^2) + (2/x)$.

ANSWERS

1. (a) $2x^2 D^2 + xD + x^2 - 2x$, (b) $2D^2 + 2$, (c) $D^3 - 1$, (d) $xD^2 + (2x - 1)D - 2$, (e) $x^2 e^x D^3 + (2x^2 e^x + x^3)D^2 + (x^2 e^x + 2x^2 - e^{2x})D - xe^x$.

4. (a) $x^2 e^x/2$, (b) $e^{5x}/12$, (c) $(\sin 3)x/40$, (d) $x^6 e^{-x}/720$, (e) e^{e^x}, (f) $1/x$.

4–13 Inverse operators. Since the reciprocal of a polynomial is not a polynomial, we cannot interpret $1/(a_0 D^n + \cdots)$ as a differential operator. Instead, we interpret it as the inverse of the operator; that is,

$$h(x) = \frac{1}{a_0 D^n + \cdots + a_n} Q(x)$$

if

$$(a_0 D^n + \cdots a_n)h(x) = Q(x).$$

In other words, $1/(a_0 D^n + \cdots + a_n)$ applied to $Q(x)$ yields a solution $y(x)$ of the differential equation

$$(a_0 D^n + \cdots a_n)y = Q(x).$$

Since there are many such solutions and we have not specified any initial conditions, the inverse operator is ambiguously defined. For example,

$$\frac{1}{D + 2} e^x$$

stands for a solution of $(D + 2)y = e^x$. Possible choices are $e^x/3$, $(e^x/3) + e^{-2x}$, $(e^x/3) + 2e^{-2x}$, etc. We shall write, for example,

$$\frac{1}{D + 2} e^x = \frac{e^x}{3},$$

with the understanding that we have selected only one of many possible choices. We are interested mainly in obtaining particular solutions; other solutions are obtained by adding the complementary function.

Throughout we restrict attention to operators with *constant coefficients*. The following rules will simplify the evaluations:

$$\frac{1}{f(D)} [c_1 Q_1(x) + c_2 Q_2(x)] = c_1 \frac{1}{f(D)} Q_1(x) + c_2 \frac{1}{f(D)} Q_2(x), \quad (4\text{-}111)$$

$$\frac{1}{f(D)g(D)} Q(x) = \frac{1}{f(D)} \left[\frac{1}{g(D)} Q(x) \right] = \frac{1}{g(D)} \left[\frac{1}{f(D)} Q(x) \right]. \quad (4\text{-}112)$$

Equation (4-111) is simply the superposition principle (Section 4-4). Equation (4-112) follows from the definition of multiplication and from the commutative law (4-107); it is equivalent to the factorization method of Section 4-12.

EXAMPLE 1. Evaluate $[1/(D - a)]e^{ax}$. We must find a solution of the differential equation

$$(D - a)y = e^{ax}.$$

The method of undetermined coefficients leads to the trial function kxe^{ax}; k is found to be 1. Hence

$$\frac{1}{D - a} e^{ax} = xe^{ax}. \quad (4\text{-}113)$$

EXAMPLE 2. Evaluate $[1/f(D)]e^{ax}$, where $f(D) = a_0 D^n + \cdots + a_n$, $f(a) \neq 0$. The differential equation is

$$f(D)y = e^{ax}.$$

Since $f(a) \neq 0$, we seek a solution of form ke^{ax}:

$$f(D)ke^{ax} = kf(a)e^{ax} = e^{ax}.$$

Hence $k = 1/f(a)$ and

$$\frac{1}{f(D)} e^{ax} = \frac{e^{ax}}{f(a)} \quad [f(a) \neq 0]. \quad (4\text{-}114)$$

EXAMPLE 3. Evaluate $[1/(D - a)^m]e^{ax}$, where $m = 1, 2, 3, \ldots$ We are led to the equation

$$(D - a)^m y = e^{ax}$$

and the trial function $kx^m e^{ax}$. Now

$$(D - a)x^m e^{ax} = mx^{m-1}e^{ax},$$
$$(D - a)^2 x^m e^{ax} = m(m - 1)x^{m-2}e^{ax},$$
$$\vdots$$
$$(D - a)^m x^m e^{ax} = m!e^{ax}.$$

Hence $k = 1/m!$ and

$$\frac{1}{(D-a)^m} e^{ax} = \frac{x^m e^{ax}}{m!}. \tag{4–115}$$

EXAMPLE 4. Show that

$$\frac{1}{(D-a)^m f(D)} e^{ax} = \frac{x^m e^{ax}}{m! f(a)}, \qquad [f(a) \neq 0], \tag{4–116}$$

where $m = 1, 2, 3, \ldots$ We can reason:

$$\frac{1}{(D-a)^m f(D)} e^{ax} = \frac{1}{(D-a)^m} \left[\frac{1}{f(D)} e^{ax} \right]$$

$$= \frac{1}{(D-a)^m} \frac{e^{ax}}{f(a)} = \frac{x^m e^{ax}}{m! f(a)},$$

by (4–113) and (4–115) above. The first step is based on (4–112).

EXAMPLE 5. Evaluate $[1/(D^2 + a^2)] \sin bx$, $a \neq b$. The differential equation $(D^2 + a^2)y = \sin bx$ can be solved by undetermined coefficients. An alternative procedure is to find a solution y of the equation $(D^2 + a^2)y = e^{bix}$; the *imaginary part* of y will then satisfy the equation with $\sin bx = \text{Im}\,(e^{bix})$ on the right. (See Probs. 5, 6 following Section 4–10.) Hence, by Example 2,

$$\frac{1}{D^2 + a^2} \sin bx = \text{Im}\left(\frac{1}{D^2 + a^2} e^{ibx} \right) = \text{Im}\left(\frac{e^{ibx}}{a^2 - b^2} \right),$$

$$\frac{1}{D^2 + a^2} \sin bx = \frac{\sin bx}{a^2 - b^2} \qquad (a \neq b). \tag{4–117}$$

In the same way we show that

$$\frac{1}{D^2 + a^2} \cos bx = \frac{\cos bx}{a^2 - b^2} \qquad (a \neq b). \tag{4–118}$$

The rules thus far established permit us to obtain solutions of particular equations with great ease.

EXAMPLE 6. $(3D^2 - 2D + 5)y = e^{2x}$. By (4–114) a particular solution is

$$y = \frac{1}{3D^2 - 2D + 5} e^{2x} = \frac{e^{2x}}{13}.$$

EXAMPLE 7. $(D^3 - D)y = e^x$. By (4–116) a particular solution is

$$y = \frac{1}{(D-1)(D^2 + D)} e^x = \frac{x e^x}{2}.$$

4–14 A table of inverse operators. Table 4–2 systematically lists rules for inverse operators. Rules 1 through 4 and 10 through 11 are established in the preceding section, and Rules 5 through 9 are obtained by similar methods. The remaining rules require discussion.

Rule 12 merely states that integration is the inverse of differentiation. Rule 13 is the expression for a solution of the first order linear equation

$$(D - a)y = Q(x)$$

(see Section 2–8).

Rule 14 is a rearrangement of the result of repeated application of Rule 13. First, we make a special choice of the indefinite integral in Rule 13:

$$\frac{1}{D - a} Q(x) = e^{ax} \int_c^x e^{-au} Q(u) \, du;$$

here c is any convenient value within the interval in which $Q(x)$ is given (and continuous). Next, by Rule 11

$$\frac{1}{(D - a)^2} Q(x) = \frac{1}{D - a} P(x); \qquad P(x) = e^{ax} \int_c^x e^{-au} Q(u) \, du,$$

$$\frac{1}{(D - a)^2} Q(x) = e^{ax} \int_c^x e^{-av} P(v) \, dv = e^{ax} \int_c^x \int_c^v e^{-au} Q(u) \, du \, dv$$

$$= e^{ax} \int_c^x \int_u^x e^{-au} Q(u) \, dv \, du$$

$$= e^{ax} \int_c^x e^{-au} Q(u)(x - u) \, du.$$

A similar procedure justifies the rule for general m.

The proofs of Rules 15 and 16 are left as exercises (Probs. 5, 6 below). Rule 16 is known as the *exponential shift*, and we can illustrate it by an example:

$$\frac{1}{D^2 + 2D + 5} e^{-x} \sin 2x = e^{-x} \frac{1}{(D - 1)^2 + 2(D - 1) + 5} \sin 2x$$

$$= e^{-x} \frac{1}{D^2 + 4} \sin 2x = \frac{-xe^{-x} \cos 2x}{4}.$$

For the last step, Rule 5 is applied.

TABLE 4–2

RULES FOR INVERSE OPERATORS

No.	Expression	Value (one choice)
1	$\dfrac{1}{f(D)} e^{ax}$	$\dfrac{e^{ax}}{f(a)}$ $[f(a) \neq 0]$
2	$\dfrac{1}{(D-a)^m} e^{ax}$	$\dfrac{x^m e^{ax}}{m!}$ $(m = 1, 2, \ldots)$
3	$\dfrac{1}{(D-a)^m f(D)} e^{ax}$	$\dfrac{x^m e^{ax}}{m! f(a)}$ $[m = 1, 2, \ldots, f(a) \neq 0]$
4	$\dfrac{1}{D^2 + a^2} \sin bx$	$\dfrac{\sin bx}{a^2 - b^2}$ $(a \neq b)$
5	$\dfrac{1}{D^2 + a^2} \sin ax$	$\dfrac{-x \cos ax}{2a}$
6	$\dfrac{1}{D^2 + a^2} \cos bx$	$\dfrac{\cos bx}{a^2 - b^2}$ $(a \neq b)$
7	$\dfrac{1}{D^2 + a^2} \cos ax$	$\dfrac{x \sin ax}{2a}$
8	$\dfrac{1}{aD^2 + bD + c} \sin \omega x$	$\dfrac{(c - a\omega^2) \sin \omega x - b\omega \cos \omega x}{(c - a\omega^2)^2 + b^2 \omega^2}$ (denom $\neq 0$)
9	$\dfrac{1}{aD^2 + bD + c} \cos \omega x$	$\dfrac{(c - a\omega^2) \cos \omega x + b\omega \sin \omega x}{(c - a\omega^2)^2 + b^2 \omega^2}$ (denom $\neq 0$)
10	$\dfrac{1}{f(D)} [c_1 Q_1(x) + c_2 Q_2(x)]$	$c_1 \dfrac{1}{f(D)} Q_1(x) + c_2 \dfrac{1}{f(D)} Q_2(x)$
11	$\dfrac{1}{f(D) g(D)} Q(x)$	$\dfrac{1}{f(D)} \left[\dfrac{1}{g(D)} Q(x) \right]$ or $\dfrac{1}{g(D)} \left[\dfrac{1}{f(D)} Q(x) \right]$
12	$\dfrac{1}{D} Q(x)$	$\displaystyle\int Q(x)\, dx$

(continued)

No.	Expression	Value (one choice)
13	$\dfrac{1}{D-a}Q(x)$	$e^{ax}\displaystyle\int e^{-ax}Q(x)\,dx$
14	$\dfrac{1}{(D-a)^m}Q(x)$ $(m = 1, 2, \ldots)$	$\dfrac{e^{ax}}{(m-1)!}\displaystyle\int_c^x e^{-au}(x-u)^{m-1}Q(u)\,du$ (c is arbit.)
15	$\dfrac{1}{(D-a)^2+b^2}Q(x)$ $(b \neq 0)$	$\dfrac{e^{ax}}{b}\displaystyle\int_c^x e^{-au}\sin b(x-u)\,Q(u)\,du$ (c is arbit.)
16	$\dfrac{1}{f(D)}e^{ax}Q(x)$	$e^{ax}\dfrac{1}{f(D+a)}Q(x)$
17	$\dfrac{1}{f(D)}e^{ax}P(x)$ $[f(a) \neq 0,$ $P(x)$ a polyn. of deg. $N]$	$e^{ax}\left[g(a)P(x)+\dfrac{g'(a)}{1!}P'(x)+\cdots+\dfrac{g^{(N)}(a)}{N!}P^{(N)}(x)\right]$ $[g(r) = 1/f(r)]$
18	$\dfrac{1}{(D-a)^m f(D)}e^{ax}P(x)$ $[f(a) \neq 0, m = 1, 2, \ldots,$ P a polyn. of deg. $N]$	$e^{ax}\left[g(a)Q(x)+\dfrac{g'(a)}{1!}Q'(x)+\cdots+\dfrac{g^{(N)}(a)}{N!}Q^{(N)}(x)\right]$ $[g(r) = 1/f(r), Q(x) = (1/D^m)P(x)]$
19	$\dfrac{1}{f(D)}Q(x)$ $[f(r) = a_0(r-r_1)\ldots$ $(r-r_n),$ $r_1, \ldots, r_n$ distinct]	$\displaystyle\sum_{k=1}^n \dfrac{1}{f'(r_k)}\dfrac{1}{D-r_k}Q(x)$
20	$\dfrac{1}{f(D)}Q(x)$	$\displaystyle\int_c^x Q(u)W(x-u)\,du$, where $y = W(x)$ is solution of $f(D)y \equiv (a_0 D^n + \cdots)y = 0$ such that $W(0) = 0$, $W'(0) = 0, \ldots, W^{(n-2)}(0) = 0, W^{(n-1)}(0) = 1/a_0$
21	$\dfrac{1}{f(D)}Q(x-x_0)$	$\phi(x-x_0)$, where $\phi(x) = \dfrac{1}{f(D)}Q(x)$

To obtain Rule 17, we can reason as follows. We first assert that if

$$\frac{1}{f(D)} Q(x, a) = y(x, a),$$

then

$$\frac{1}{f(D)} \frac{\partial Q}{\partial a} = \frac{\partial y}{\partial a} \qquad (4\text{–}119)$$

(all derivatives concerned are assumed continuous). For if

$$f(D)y(x, a) = Q(x, a),$$

then

$$f(D) \frac{\partial y}{\partial a} = (a_0 D^n + \cdots + a_n) \frac{\partial y}{\partial a} = a_0 \frac{\partial^{n+1} y}{\partial x^n \partial a} + \cdots$$

$$= \frac{\partial}{\partial a} (a_0 D^n y + \cdots) = \frac{\partial}{\partial a} f(D)y = \frac{\partial Q}{\partial a},$$

by interchange of order of differentiation. We now apply (4–119) with $Q = e^{ax}$:

$$\frac{1}{f(D)} e^{ax} = \frac{e^{ax}}{f(a)} = g(a)e^{ax},$$

$$\frac{1}{f(D)} xe^{ax} = \frac{\partial}{\partial a} [g(a)e^{ax}] = e^{ax}[xg(a) + g'(a)],$$

and by repeated differentiation with respect to a (with the aid of Leibnitz' rule—see Prob. 5(a) following Section 4–5), we obtain

$$\frac{1}{f(D)} x^m e^{ax} = e^{ax} \left[x^m g(a) + mx^{m-1} \frac{g'(a)}{1!} + \cdots + m! \frac{g^{(m)}(a)}{m!} \right].$$

This shows the correctness of Rule 17 when $P(x) = x^m$. By the superposition principle (Rule 10) we obtain the rule for a general polynomial.

Rule 18 can be deduced from Rule 17 and Rules 11 and 16. For

$$\frac{1}{(D - a)^m f(D)} e^{ax} P(x) = \frac{1}{(D - a)^m} \left[\frac{1}{f(D)} e^{ax} P(x) \right]$$

$$= \frac{1}{(D - a)^m} e^{ax} \left[g(a)P(x) + g'(a)P'(x) + \cdots + \frac{g^{(N)}(a)}{N!} P^{(N)}(x) \right]$$

(continued)

$$= e^{ax} \frac{1}{D^m} [g(a)P(x) + g'(a)P'(x) + \cdots]$$

$$= e^{ax} \left[g(a)Q(x) + g'(a)Q'(x) + \cdots + \frac{g^{(N)}(a)}{N!} Q^{(N)}(x) \right],$$

where $Q(x) = (1/D^m)P(x)$ is an m-fold integral of $P(x)$.

EXAMPLE A.

$$\frac{1}{D^2 + 5} [(2x + 1)e^{3x}] = e^{3x} \left[\frac{2x + 1}{14} - \frac{3}{49} \right].$$

Here $a = 3, g(r) = (r^2 + 5)^{-1}, g'(r) = -2r(r^2 + 5)^{-2}, P(x) = 2x + 1$.

EXAMPLE B.

$$\frac{1}{D^2 - 9} [(2x + 1)e^{3x}] = e^{3x} \left[\frac{x^2 + x}{6} - \frac{2x + 1}{36} \right].$$

Here Rule 18 applies: $f(D) = D + 3, g(r) = (r + 3)^{-1}, a = 3, P(x) = 2x + 1, Q(x) = x^2 + x$.

By allowing a to be complex (that is, $a = \alpha + \beta i$) Rules 17 and 18 can be extended to functions $e^{\alpha x} \cos \beta x \, P(x)$, $e^{\alpha x} \sin \beta x \, P(x)$. *Hence Rules 17 and 18 completely replace the method of undetermined coefficients.*

Rule 19 is known as the *Heaviside expansion.* It is equivalent to the statement that (when the characteristic roots are distinct) $1/f(D)$ can be replaced by its *partial-fraction* expansion; for example,

$$\frac{1}{(D - 2)(D - 3)} = \frac{1}{D - 3} - \frac{1}{D - 2}.$$

In the general case, $1/f(r)$ has the partial-fraction expansion

$$\frac{1}{f(r)} = \sum_{k=1}^{n} \frac{A_k}{r - r_k}.$$

Each A_k is equal to $1/f'(r_k)$. For if we multiply both sides by $r - r_k$ and then let $r \to r_k$, the right-hand side has the limit A_k and the left-hand side has the limit

$$\lim_{r \to r_k} \frac{r - r_k}{f(r)} = \lim_{r \to r_k} \frac{r - r_k}{f(r) - f(r_k)} = \frac{1}{f'(r_k)},$$

since $f(r_k) = 0$. Hence Rule 19 is simply a partial-fraction expansion of the inverse operator.

To prove the validity of Rule 19, we apply induction. For $n = 1$ the rule is an identity. We can also verify it for $n = 2$ as follows. Since

$$\frac{1}{a_0(r - r_1)(r - r_2)} = \frac{1}{a_0}\left[\frac{1}{r_1 - r_2}\frac{1}{r - r_1} + \frac{1}{r_2 - r_1}\frac{1}{r - r_2}\right],$$

the rule states that

$$\frac{1}{a_0(D - r_1)(D - r_2)}Q$$

$$= \frac{1}{a_0}\left[\frac{1}{r_1 - r_2}\frac{1}{D - r_1}Q(x) + \frac{1}{r_2 - r_1}\frac{1}{D - r_2}Q(x)\right];$$

it can be proved to be correct by the method of variation of parameters (see Prob. 5 following Section 4–9). Now we suppose the rule to have been proved correct when $f(r)$ has degree n. We then proceed to prove it true when $f(r)$ has degree $n + 1$. We thus assume

$$\frac{1}{a_0(D - r_1) \cdots (D - r_n)}Q = \sum_{k=1}^{n}\frac{A_k}{D - r_k}Q(x),$$

where the A_k are the coefficients of the partial-fraction expansion. We apply the operator $1/(D - r_{n+1})$ to both sides:

$$\frac{1}{a_0(D - r_1) \cdots (D - r_n)(D - r_{n+1})}Q$$

$$= \sum_{k=1}^{n}\frac{A_k}{(D - r_{n+1})(D - r_k)}Q(x).$$

By the case $n = 2$ verified above, each term on the right can be replaced by its partial-fraction expansion. If this is done and terms are collected, the (unique) partial-fraction expansion of the left-hand side is obtained. Hence the rule is true for degree $n + 1$ and, by induction, for all n.

Remark. We can prove in the same way that $1/f(D)$ can be replaced by its partial-fraction expansion even when the characteristic roots are not distinct.

Rule 20 gives the solution of the nonhomogeneous equation with $y = 0$, $y' = 0, \ldots, y^{(n-1)} = 0$ for $x = c$. (For a full discussion see Reference 8.) The rule can be proved by means of the partial-fraction expansion. For example, if $f(D) = a_0 D^2 + a_1 D + a_2$ and the roots r_1, r_2 are distinct, then, as above,

$$
y = \frac{1}{a_0} \left[\frac{1}{r_1 - r_2} \frac{1}{D - r_1} Q + \frac{1}{r_2 - r_1} \frac{1}{D - r_2} Q \right]
$$

$$
= \frac{1}{a_0} \left[\frac{e^{r_1 x}}{r_1 - r_2} \int_c^x e^{-r_1 u} Q(u)\, du + \frac{e^{r_2 x}}{r_2 - r_1} \int_c^x e^{-r_2 u} Q(u)\, du \right]
$$

$$
= \int_c^x Q(u) \frac{e^{r_1(x-u)} - e^{r_2(x-u)}}{a_0(r_1 - r_2)}\, du.
$$

Since

$$
W(x) = \frac{e^{r_1 x} - e^{r_2 x}}{a_0(r_1 - r_2)}
$$

is a solution of the homogeneous equation and $W(0) = 0$, $W'(0) = 1/a_0$, the rule is verified. A general proof is outlined in Prob. 9 below.

Rule 21 is a *principle of stationarity*, discussed for the first order equation in Section 3–13. If $Q(x)$ is regarded as an input, then $[1/f(D)]\, Q(x)$ is an output. The principle states that delay of the input by "time" x_0 has the effect of delaying the output by the same time. For the validity of the principle it is crucial that the coefficients be *constant*. To justify the rule, we remark that

$$
D\phi(x - x_0) = \phi'(x - x_0), \qquad D^2\phi(x - x_0) = \phi''(x - x_0), \qquad \ldots,
$$

and hence that if $f(D)\phi(x) = Q(x)$,

$$
f(D)\phi(x - x_0) = Q(x - x_0).
$$

The operational rules listed and their elaborations are known as the *Heaviside calculus* (after Oliver Heaviside (1850–1925); see Reference 6 at the end of this chapter). The most striking characteristic is the correctness of the results obtained by manipulating operational expressions as though D were a numerical variable. The procedure is not infallible and every new rule must be tested to establish its validity. An alternative method, the results of which parallel those of Heaviside, is based on the *Laplace transform*, which is described briefly in the next section.

PROBLEMS

1. Find a particular solution of each of the following equations, with the aid of the operational rules (numbers of rules suggested are given in parentheses):

(a) $(D^2 - 7D + 2)y = e^{5x}$ (1),

(b) $(D^2 - 3D + 2)y = e^{2x}$ (3),

(c) $(D^2 - 3D + 2)y = \cos 4x$ (1 or 9),

(d) $(D^3 + 4D)y = \sin 3x$ (11, 12, 6),

(e) $(D - 2)(D^2 + 3D + 2)y = e^{2x} \cos 3x$ (1 alone or 11, 16, 12, 8),

(f) $(D^3 + D + 1)y = e^{4x}(2x + 3)$ (17),

(g) $(D^3 + D^2 + 1)y = (x^2 - 1) \cos x$ (17),

(h) $(D - 1)^2(D^3 + D^2 + 1)y = xe^x$ (18),

(i) $(D - 2)^4 y = e^{2x}x^3$ (14 or 18),

(j) $(D^2 - 3D + 2)y = \log x$ (19, 13),

(k) $(D^2 + 2D + 2)y = 1/(1 + x^2)$ (15).

2. (a) ... (j). With the aid of operational rules, find a particular solution of each of the equations of Prob. 1 following Section 4–10.

3. (a) ... (e). With the aid of operational rules, find a particular solution of each of the equations of Prob. 2 following Section 4–10.

4. Extend the proof of Rule 14 to the operator $1/(D - a)^3$.

5. Prove Rule 15. [*Hint:* Write the result as $y(x) = (e^{ax}/b)[p(x) \sin bx - q(x) \cos bx]$, where

$$p(x) = \int_c^x \cos bu\, e^{-au} Q(u)\, du, \qquad q(x) = \int_c^x \sin bu\, e^{-au} Q(u)\, du,$$

so that $p'(x) = \cos bx\, e^{-ax}Q(x)$, $q'(x) = \sin bx\, e^{-ax}Q(x)$. Show that $y(x)$ satisfies the differential equation $[(D - a)^2 + b^2]y = Q(x)$.]

6. (a) Prove: $[1/(D - b)]e^{ax}Q(x) = e^{ax}[1/(D + a - b)]Q(x)$. [*Hint:* Apply Rule 13 to both sides.]

(b) Prove Rule 16. [*Hint:* Write $f(D)$ in factored form and then apply the result of part (a) to shift e^{ax} past each factor in turn:

$$\frac{1}{a_0(D - r_1) \cdots (D - r_n)} e^{ax}Q(x)$$

$$= \frac{1}{a_0(D - r_1) \cdots (D - r_{n-1})} e^{ax}\left[\frac{1}{(D + a - r_n)} Q(x)\right]$$

$$= \frac{1}{a_0(D - r_1) \cdots (D - r_{n-2})} e^{ax}\left[\frac{1}{(D + a - r_{n-1})(D + a - r_{n-2})} Q(x)\right]$$

$$= \cdots]$$

7. (a) Prove: $(D - b)[e^{ax}\phi(x)] = e^{ax}[(D + a - b)\phi(x)]$.

(b) Prove: $f(D)[e^{ax}\phi(x)] = e^{ax}[f(D + a)\phi(x)]$. [*Hint:* Proceed as in the proof of Rule 16 in Prob. 6.]

(c) Prove that Rule 16 follows from the result of part (b).

(d) Prove: $(D - a)^m[e^{ax}\phi(x)] = e^{ax}D^m\phi(x)$.

(e) Apply the rule of part (d) to show that if r is a root of multiplicity k of the equation $f(r) = 0$, then $e^{rx}, xe^{rx}, \ldots, x^{k-1}e^{rx}$ are solutions of the equation $f(D)y = 0$. [*Hint:* $f(D) = (D - r)^k g(D)$.]

8. Let $f(r) = a_0 r^n + \cdots$, $g(r) = 1/f(r)$. If $f(a) \neq 0$, then $g(r)$ can be expanded in a Taylor series about $r = a$:

$$\frac{1}{f(r)} = g(r) = g(a) + g'(a)(r - a) + \cdots$$

$$+ \frac{g^{(m)}}{m!} (r - a)^m + \cdots$$

This suggests the operator identity

$$\frac{1}{f(D)} = g(a) + g'(a)(D - a) + \cdots$$

$$+ \frac{g^{(m)}(a)}{m!} (D - a)^m + \cdots \qquad (*)$$

Use this identity to evaluate $[1/f(D)]e^{ax}P(x)$, where $P(x)$ is a polynomial, and show, with the aid of the result of Prob. 7(d), that Rule 17 is obtained. [The identity $(*)$ is valid only for operation on a restricted class of functions, which includes those of form $e^{ax}P(x)$.]

9. *Proof of Rule* 20. It is shown in advanced calculus (see, for example, Reference 7, pp. 218–222) that

$$\frac{\partial}{\partial x} \int_c^x F(x, u) \, du = F(x, x) + \int_c^x \frac{\partial F}{\partial x} (x, u) \, du,$$

provided F and $\partial F/\partial x$ are continuous. Apply this rule to

$$y = \int_c^x Q(u) W(x - u) \, du,$$

where $Q(x)$ is continuous for all x and $W(x)$ satisfies the conditions described in Rule 20. Conclude successively that

$$y' = \int_c^x Q(u) W'(x - u) \, du,$$

$$y'' = \int_c^x Q(u) W''(x - u) \, du, \qquad \dots,$$

$$y^{(n-1)} = \int_c^x Q(u) W^{(n-1)}(x - u) \, du,$$

$$y^{(n)} = \frac{Q(x)}{a_0} + \int_c^x Q(u) W^{(n)}(x - u) \, du,$$

and hence that

$$f(D)y = Q(x) + \int_c^x Q(u)[a_0 W^{(n)}(x - u) + \cdots + a_n W(x - u)] \, du$$

$$= Q(x).$$

Show also that $y(c) = 0$, $y'(c) = 0, \ldots, y^{(n-1)}(c) = 0$.

10. (a) Apply Rule 20 to evaluate $[1/(D - a)^m]Q(x)$ and show that Rule 14 is obtained.

(b) Apply Rule 20 to evaluate $[1/(D^2 - 2aD + a^2 + b^2)]Q(x)$, $b \neq 0$, and show that Rule 15 is obtained.

ANSWERS

1. (a) $-e^{5x}/8$, (b) xe^{2x}, (c) $(-7 \cos 4x - 6 \sin 4x)/170$, (d) $(\cos 3x)/15$,
(e) $e^{2x}(\sin 3x - 7 \cos 3x)/450$, (f) $e^{4x}(138x + 109)/4761$,
(g) $(-6x - 22) \cos x - (x^2 + 4x - 5) \sin x$, (h) $e^x(x^3 - 5x^2)/18$,
(i) $x^7 e^{2x}/840$, (j) $e^{2x}\int e^{-2x} \log x \, dx - e^x \int e^{-x} \log x \, dx$,
(k) $e^{-x}\int_0^x e^u \sin (x - u)(1 + u^2)^{-1} \, du$.

4–15 Laplace transforms. Let $f(x)$ be defined for $0 \leq x < \infty$. The Laplace transform of $f(x)$ is defined as the function $F(s)$, where

$$F(s) = \int_0^\infty f(x)e^{-sx} \, dx. \qquad (4\text{--}120)$$

We write $F(s) = \mathcal{L}[f]$ and consider $F(s)$ as the result of applying an operator $\mathcal{L}$ to f. Various assumptions on $f(x)$ will ensure that the integral has meaning for some values of s; in general, we allow s to be complex, so that $F(s)$ is a complex-valued function of a complex variable. Here we proceed formally in order to indicate briefly the properties of the Laplace transform and their utilization in solving differential equations. (For a detailed treatment the reader is referred to References 2 and 8 at the end of this chapter.)

If $f(x) \equiv 1$, then

$$\mathcal{L}[1] = F(s) = \int_0^\infty e^{-sx} \, dx = \frac{e^{-sx}}{-s}\Big|_0^\infty = \frac{1}{s}. \qquad (4\text{--}121)$$

The evaluation of the improper integral is correct if $s > 0$ or, if s is complex, when Re $(s) > 0$. If $f(x) \equiv x^n$, where n is a positive integer, then

$$\mathcal{L}[f] = F(s) = \int_0^\infty x^n e^{-sx} \, dx = \frac{x^n e^{-sx}}{-s}\Big|_0^\infty + \frac{n}{s} \int_0^\infty x^{n-1} e^{-sx} \, dx,$$

and hence, for Re $(s) > 0$

$$\mathcal{L}[x^n] = \frac{n}{s} \mathcal{L}[x^{n-1}]. \tag{4-122}$$

Repeated application of this rule, together with (4–121), gives the conclusion

$$\mathcal{L}[x^n] = \frac{n!}{s^{n+1}} \qquad (n = 0, 1, 2, \ldots). \tag{4-123}$$

A similar analysis gives the rule

$$\mathcal{L}[x^n e^{ax}] = \frac{n!}{(s - a)^{n+1}} \qquad (n = 0, 1, 2, \ldots). \tag{4-124}$$

Further transform pairs $f(x)$, $F(s)$ can be obtained from some general rules. The Laplace transform is *linear:* if c_1, c_2 are constants, then

$$\mathcal{L}[c_1 f_1 + c_2 f_2] = c_1 \mathcal{L}[f_1] + c_2 \mathcal{L}[f_2]; \tag{4-125}$$

this follows directly from the definition. Under appropriate assumptions,

$$\mathcal{L}[f'(x)] = \int_0^\infty f'(x)e^{-sx}\,dx = f(x)e^{-sx}\Big|_0^\infty + s\int_0^\infty f(x)e^{-sx}\,dx$$
$$= -f(0) + s\mathcal{L}[f];$$

$f(x)$ must, in particular, satisfy the condition $f(x)e^{-sx} \to 0$ as $x \to \infty$, for some values of s. By repeated application of the conclusion we obtain the general rules for *transformation of derivatives:*

$$\mathcal{L}[f'] = s\mathcal{L}[f] - f(0),$$
$$\mathcal{L}[f''] = s^2\mathcal{L}[f] - [f'(0) + sf(0)],$$
$$\mathcal{L}[f'''] = s^3\mathcal{L}[f] - [f''(0) + sf'(0) + s^2 f(0)], \tag{4-126}$$
$$\vdots$$
$$\mathcal{L}[f^{(n)}] = s^n\mathcal{L}[f] - [f^{(n-1)}(0) + sf^{(n-2)}(0) + \cdots + s^{n-1}f(0)].$$

Finally, a function f (under certain restrictions) is uniquely determined by its Laplace transform:

$$\mathcal{L}[f] = \mathcal{L}[g] \qquad \text{implies} \qquad f = g. \tag{4-127}$$

EXAMPLES.

$$\mathcal{L}\left[\sin ax\right] = \mathcal{L}\left[\frac{e^{aix} - e^{-aix}}{2i}\right] = \frac{1}{2i}\,\mathcal{L}[e^{aix}] - \frac{1}{2i}\,\mathcal{L}[e^{-aix}]$$

$$= \frac{1}{2i}\,\frac{1}{s - ai} - \frac{1}{2i}\,\frac{1}{s + ai} = \frac{a}{s^2 + a^2}\,. \tag{4-128}$$

Since $(\sin ax)' = a\cos ax$, we conclude from (4–126):

$$\mathcal{L}\left[\cos ax\right] = \frac{s}{s^2 + a^2}\,. \tag{4-129}$$

The Laplace transform can be used to obtain the solution for $x \geqq 0$ of a linear differential equation with constant coefficients, satisfying prescribed initial conditions at $x = 0$. We illustrate the procedure by an example. Given the equation with initial conditions

$$\frac{d^2y}{dx^2} + 4y = e^{3x}, \qquad y = a, \quad y' = b \quad \text{when} \quad x = 0,$$

we seek $\mathcal{L}[y]$. From the given equation,

$$\mathcal{L}[y'' + 4y] = \mathcal{L}[e^{3x}] = \frac{1}{s - 3}\,.$$

Hence by (4–126)

$$s^2\mathcal{L}[y] - (b + as) + 4\mathcal{L}[y] = \frac{1}{s - 3}\,,$$

$$\mathcal{L}[y] = \frac{1}{(s^2 + 4)(s - 3)} + \frac{as + b}{s^2 + 4}$$

$$= \frac{1}{13}\,\frac{1}{s - 3} - \frac{1}{13}\,\frac{s + 3}{s^2 + 4} + \frac{as + b}{s^2 + 4}\,.$$

By Eqs. (4–124), (4–128), (4–129), and linearity we can write

$$\mathcal{L}[y] = \frac{1}{13}\,\mathcal{L}[e^{3x}] - \frac{1}{13}\,\mathcal{L}\left[\cos 2x\right] - \frac{3}{26}\,\mathcal{L}\left[\sin 2x\right]$$

$$+ a\mathcal{L}\left[\cos 2x\right] + \frac{b}{2}\,\mathcal{L}\left[\sin 2x\right]$$

$$= \mathcal{L}\left[\frac{e^{3x}}{13} - \frac{\cos 2x}{13} - \frac{3\sin 2x}{26} + a\cos 2x + \frac{b}{2}\sin 2x\right],$$

$$y = \frac{e^{3x}}{13} - \frac{\cos 2x}{13} - \frac{3\sin 2x}{26} + a\cos 2x + \frac{b}{2}\sin 2x.$$

The effectiveness of the method depends upon availability of a good table of transform pairs. Such tables closely resemble the table of inverse operators in the preceding section. For lists of Laplace transforms and inverse transforms see References 1, 4, 5, and 8.

PROBLEMS

1. Find the Laplace transforms of the following functions:

(a) $\cosh x$,   (b) $\sinh x$,

(c) $x \sin x$,   (d) $x^2 \sin x$,

(e) $f(x) = 1, 0 \leq x < 1, f(x) = 0, x \geq 1$,

(f) $f(x) = x^2 - x, 0 \leq x \leq 1, f(x) = 0, x > 1$,

(g) $e^{ax} \sin bx$,   (h) $e^{ax} \cos bx$.

2. Find the Laplace transforms of the following functions, with the aid of results given above and linearity:

(a) $x^4 + 2$,   (b) $xe^x - 1$,

(c) $\sin 3x - \cos 5x$,   (d) $(1 - x^2)e^{2x}$.

3. With the aid of the Laplace transform, find the solutions that satisfy the given initial conditions:

(a) $y' + y = e^{-x}; y = 1$ for $x = 0$;

(b) $(D^2 + 3D + 2)y = x$; for $x = 0: y = 1$ and $y' = -1$;

(c) $(D + 1)^3 y = e^{-x}, y = 0$; for $x = 0: y' = 0, y'' = 0$;

(d) $(D^2 + 2D + 2)y = \sin x$; for $x = 0: y = 0, y' = 1$.

[*Hint:* For (d) use the results of Probs. 1(g) and 1(h).]

ANSWERS

1. (a) $\dfrac{s}{s^2 - 1}$,   (b) $\dfrac{1}{s^2 - 1}$,   (c) $\dfrac{2s}{(s^2 + 1)^2}$,   (d) $\dfrac{2 - 6s^2}{(s^2 + 1)^3}$,

(e) $\dfrac{1 - e^{-s}}{s}$,   (f) $\dfrac{1 + e^{-s}}{-s^2} + \dfrac{2}{s^2}(1 - e^{-s})$,   (g) $\dfrac{b}{(s - a)^2 + b^2}$,

(h) $\dfrac{s - a}{(s - a)^2 + b^2}$.

2. (a) $\dfrac{24}{s^5} + \dfrac{2}{s}$,   (b) $\dfrac{1}{(s - 1)^2} - \dfrac{1}{s}$,   (c) $\dfrac{3}{s^2 + 9} - \dfrac{s}{s^2 + 25}$,

(d) $\dfrac{1}{s - 2} - \dfrac{2}{(s - 2)^3}$.

3. (a) $e^{-x}(x + 1)$,   (b) $\frac{1}{4}(8e^{-x} - e^{-2x} - 3 + 2x)$,   (c) $x^3 e^{-x}/6$,

(d) $\frac{1}{5}(-2 \cos x + \sin x + 2e^{-x} \cos x + 6e^{-x} \sin x)$.

SUGGESTED REFERENCES

1. *C. R. C. Standard Mathematical Tables*, 11th ed. Cleveland: Chemical Rubber Pub. Co., 1957.

2. CHURCHILL, RUEL V., *Modern Operational Mathematics in Engineering*. New York: McGraw-Hill, 1944.

3. DICKSON, LEONARD E., *First Course in the Theory of Equations*. New York: John Wiley and Sons, Inc., 1922.

4. ERDÉLYI, A., editor, *Tables of Integral Transforms*, Vol. I, compiled by the staff of the Bateman Manuscript Project. New York: McGraw-Hill, 1954.

5. GARDNER, M. F., and BARNES, J. L., *Transients in Linear Systems*, Vol. I. London: Chapman and Hall, 1942.

6. HEAVISIDE, OLIVER, *Electromagnetic Theory*. New York: Dover, 1950.

7. KAPLAN, WILFRED, *Advanced Calculus*. Reading, Mass.: Addison-Wesley, 1952.

8. KAPLAN, WILFRED, *Operational Methods for Linear Systems*. Reading, Mass.: Addison-Wesley (to appear).

9. SCARBOROUGH, JAMES B., *Numerical Mathematical Analysis*, 2nd ed. Baltimore: Johns Hopkins Press, 1950.

10. THOMAS, GEORGE B., *Calculus*. Reading, Mass.: Addison-Wesley, 1953.

11. WILLERS, F. A., *Practical Analysis*, transl. by R. T. Beyer. New York: Dover, 1948.

CHAPTER 5

PROPERTIES OF SOLUTIONS OF LINEAR
DIFFERENTIAL EQUATIONS

5–1 Input-output analysis. The purpose of this chapter is to extend the analysis of Chapter 3 to linear equations of arbitrary order; the conclusions reached are similar to those of Chapter 3.

We again let time t be the independent variable and let x be the dependent variable. The equation to be discussed is the general linear equation

$$a_0(t) \frac{d^n x}{dt^n} + \cdots + a_{n-1}(t) \frac{dx}{dt} + a_n(t)x = F(t), \tag{5-1}$$

with $a_0(t) \neq 0$. We shall use the operator notation of Chapter 4, with $D = d/dt$, $D^2 = d^2/dt^2$, ... Accordingly, Eq. (5–1) can be written

$$[a_0(t)D^n + \cdots + a_{n-1}(t)D + a_n(t)]x = F(t). \tag{5-1a}$$

In view of the physical applications, we shall also assume $a_n(t) \neq 0$. By division by $a_n(t)$, the equation can then be written in the form (5–1a), with the coefficient of x equal to 1; thus we can consider the equation in the form

$$[a_0(t)D^n + \cdots + a_{n-1}(t)D + 1]x = F(t). \tag{5-2}$$

Throughout most of the chapter the coefficients $a_0(t), \ldots, a_{n-1}(t)$ will be assumed to be constants. In any case, $a_0(t), \ldots, a_{n-1}(t), F(t)$ will be assumed to be continuous for all t, unless otherwise noted.

The motivation for writing the equation in form (5–2) lies in the approximation which it suggests. *If $a_0(t), \ldots, a_{n-1}(t)$ are all small, then, approximately, $x = F(t)$.* We think of nonzero coefficients $a_0(t), \ldots, a_{n-1}(t)$ as describing the presence of some mechanism which prevents x from coinciding with $F(t)$. We term $F(t)$ the *input*, and we call an actual solution $x(t)$ of (5–2) an *output*. Thus, to a first approximation,

$$\text{output} \sim \text{input.} \tag{5-3}$$

Our problem is to discuss the *nature* of the deviation of output from input.

We could, of course, leave the equation in form (5–1), call $F(t)$ the input and a solution $x(t)$ an output. If $a_n(t) \neq 0$, the approximation (5–3) is replaced by

$$\text{output} \sim \text{input} \div a_n(t). \tag{5-3a}$$

We prefer to work with (5–2) because of the greater simplicity of (5–3) as compared with (5–3a). If $a_n(t) \equiv 0$, the division by $a_n(t)$ is impossible. In this case, if $a_{n-1}(t) \neq 0$, we introduce a new unknown $y = dx/dt$ and consider y as output. For example, the equation $(3D^2 + D)x = F(t)$ is replaced by $(3D + 1)y = F(t)$; if the term in Dx is also missing, we set y equal to $D^k x$, where $D^k x$ is the derivative of lowest order appearing. The final output x is obtained from y by repeated integration.

5–2 Stability. Transients. By the theory of Chapter 4 the general solution of Eq. (5–2) has the form

$$x = x^*(t) + c_1 x_1(t) + \cdots + c_n x_n(t), \qquad (5\text{–}4)$$

where $c_1, \ldots, c_n$ are arbitrary constants. The equation (5–2) is said to be *stable* if for every choice of $c_1, \ldots, c_n$, the complementary function $c_1 x_1(t) + \cdots + c_n x_n(t)$ approaches 0 as $t \to +\infty$. In other words, Eq. (5–2) is stable if every solution of the related homogeneous equation

$$[a_0(t)D^n + \cdots + a_{n-1}(t)D + 1]x = 0 \qquad (5\text{–}5)$$

approaches 0 as $t \to \infty$. In particular then, $x_1(t), \ldots, x_n(t)$ must approach 0 as $t \to +\infty$; conversely, if $x_1(t) \to 0, \ldots, x_n(t) \to 0$ as $t \to +\infty$, then every linear combination $c_1 x_1(t) + \cdots + c_n x_n(t)$ must approach 0 as $t \to +\infty$, and the equation is stable. If at least one solution of the homogeneous equation fails to approach 0 as $t \to +\infty$, the equation (5–2) is said to be *unstable*.

The significance of stability is that there is effectively a unique output for a given input. For if the equation is stable, then any output $x(t)$ differs from any other [for example, $x^*(t)$] by a *transient*, that is, by a term which approaches 0 as $t \to \infty$. Indeed, from Eq. (5–4),

$$x(t) - x^*(t) = c_1 x_1(t) + \cdots + c_n x_n(t) \to 0 \qquad \text{as} \qquad t \to \infty.$$

The choice of the particular solution $x^*(t)$ is irrelevant here. If $x^{**}(t)$ is any other particular solution, then it is also a possible output, so that $x^{**}(t) - x^*(t) \to 0$ as $t \to +\infty$; hence also

$$x(t) - x^{**}(t) = [x(t) - x^*(t)] + [x^*(t) - x^{**}(t)] \to 0.$$

The appearance of the outputs for a stable equation with given input is shown in Fig. 5–1. All solutions come together as t increases. If two solutions are far apart for $t = 0$, it takes a relatively long time for them to come together. More generally, if at $t = 0$ the initial values of $x, D_t x, \ldots,$ $D_t^{n-1} x$ are not close together for the two solutions, then it will take rela-

tively longer for the two to come together. The differences in the initial
values give rise to a "transient error," which tends to disappear as t
increases. This is illustrated by a
teacher summoning her young pupils
to enter the school building after a
recess. The children may be widely
scattered when she calls (the initial
values vary widely), but after some
time all are passing through the
doorway and are then in the school-

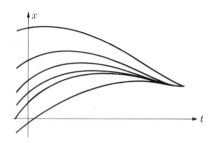

FIG. 5-1. Output of stable equation.

room. If the school discipline is very poor, some do not heed the call
at all (the initial errors do not all disappear) and the system is unstable.

Now let the coefficients $a_0(t), \ldots, a_{n-1}(t)$ be *constants*, $(a_0 \neq 0)$.
Then the nature of the solutions of Eq. (5–5) is determined by the characteristic equation

$$a_0 r^n + \cdots + a_{n-1} r + 1 = 0. \tag{5–6}$$

The differential equation (5–2) *is stable if all roots of the characteristic equation have negative real parts; otherwise, the equation is unstable.* For if r_1 is
a real root of multiplicity k of the characteristic equation, then $e^{r_1 t}$,
$t e^{r_1 t}, \ldots, t^{k-1} e^{r_1 t}$ are solutions of the homogeneous differential equation (5–5); if $r_1 < 0$, then each of these solutions approaches 0 as $t \to +\infty$;
if $r_1 \geqq 0$, then none of them approaches 0 as $t \to +\infty$. If $a \pm bi$ is a
pair of complex roots of multiplicity k, then we have solutions of form
$t^h e^{at} \cos bt$, $t^h e^{at} \sin bt$, $(h = 0, 1, \ldots, k - 1)$; if $a < 0$, all these solutions are transients; if $a \geqq 0$, none is a transient (Prob. 3 below). Hence
when all roots have negative real parts, all solutions of the homogeneous
equation are transients; when at least one root has positive or zero real
part, a nontransient solution of the homogeneous equation exists.

Remarks. If $\pm bi$ is a pair of simple roots, the corresponding functions
$\cos bt$, $\sin bt$ are not transients; however, they do not become arbitrarily
large as $t \to +\infty$. We sometimes call an equation (5–2) stable if all
solutions of the homogeneous equation are either transients or, at worst,
remain *bounded* as $t \to +\infty$. With this definition, the output of a stable
equation is not uniquely determined as above; however, if the initial
conditions are confined to a narrow range, the possible outputs never
differ greatly.

In mathematical literature one usually refers to a "stable solution" of a
differential equation, rather than to a "stable differential equation." A
solution $x = x^*(t)$, $(t_0 \leqq t < \infty)$, of a differential equation is termed
asymptotically stable if for every solution $x(t)$ whose initial conditions are

close enough to those of $x^*(t)$ when $t = t_0$, $\lim [x(t) - x^*(t)] = 0$ as $t \to +\infty$. A stable linear equation (5–1), as defined above, is simply an equation for which every solution is asymptotically stable for every choice of $F(t)$. The preceding remarks are somewhat oversimplified. For more details the reader is referred to References 2, 3, and 7, and to Chapter 11.

5–3 Testing for stability. We shall consider only equations with constant coefficients. By the criterion of the preceding section, testing for stability requires examination of the roots of an algebraic equation and determination of whether there are roots with positive or zero real parts.

EXAMPLE 1. $\frac{2}{5}D^2x + \frac{3}{5}Dx + x = F(t)$. The characteristic equation is

$$2r^2 + 3r + 5 = 0.$$

The roots are

$$\frac{-3 \pm \sqrt{-31}}{4} = -\frac{3}{4} \pm \frac{\sqrt{31}}{4} i.$$

The real parts are equal to $-\frac{3}{4}$; the equation is stable.

The example suggests a general rule for the equation of second order,

$$(aD^2 + bD + c)x = F(t). \tag{5-7}$$

The equation is stable if a, b, and c all have the same sign; it is unstable otherwise. This is equivalent to the statement that the roots of

$$ar^2 + br + c = 0$$

have negative real parts when $a > 0$, $b > 0$, $c > 0$ or $a < 0$, $b < 0$, $c < 0$, and not otherwise. The proof is left as an exercise (Prob. 4 below).

EXAMPLE 2. $(D^3 + \frac{1}{2}D^2 + \frac{1}{2}D + 1)x = F(t)$.

The characteristic equation is

$$2r^3 + r^2 + r + 2 = 0.$$

The roots are -1, $\frac{1}{4}(1 \pm \sqrt{15}i)$. The equation is unstable.

The second example shows that positiveness of the coefficients does not ensure stability for a third order equation. However, it can be shown that if the coefficients are not all of the same sign, then the equation is unstable.

Criteria have been developed for testing the stability of equations of arbitrary order with constant coefficients; see References 4, 5, and 7 at the end of this chapter.

PROBLEMS

1. Test the following equations for stability:

(a) $(3D^2 + 2D + 1)x = F(t)$, (b) $(D^3 + D^2 + D + 1)x = F(t)$,

(c) $(D^3 + 2D^2 + 2D + 1)x = F(t)$,

(d) $(D^3 + 5D^2 + D + 1)x = F(t)$.

2. For each of the following equations obtain the general solution and plot input and output for several choices of initial conditions:

(a) $(D^2 + 3D + 2)x = t^2 - t$, (b) $(D^2 + 2D + 2)x = t^2 - t$,

(c) $(D^2 - 3D + 2)x = t^2 - t$.

3. (a) Prove: if $a < 0$, $h = 1, 2, \ldots$, then $t^h e^{at} \to 0$ as $t \to +\infty$, but if $a \geqq 0$ then $t^h e^{at}$ does not approach 0 as $t \to +\infty$.

(b) Prove: if $a < 0$, then $f(t) = t^h e^{at} \cos bt \to 0$ and $g(t) = t^h e^{at} \sin bt \to 0$ as $t \to +\infty$. [$Hint:$ $|f(t)| \leqq t^h e^{at}$, $|g(t)| \leqq t^h e^{at}$.]

(c) Prove: if $a \geqq 0$, $b > 0$, then the functions $f(t)$, $g(t)$ of part (b) do not approach 0 as $t \to +\infty$. [$Hint:$ Consider the values of $f(t)$ for $bt = 2n\pi$ and of $g(t)$ for $bt = \frac{1}{2}\pi + 2n\pi$, $(n = 1, 2, \ldots)$.]

4. Prove: if $a > 0$, then the roots of the quadratic equation $ar^2 + br + c = 0$ have negative real parts if and only if $b > 0$ and $c > 0$.

ANSWERS

1. (a) Stable, (b) unstable, (c) stable, (d) stable.

2. General solutions:

(a) $\frac{1}{2}(t^2 - 4t + 5) + c_1 e^{-t} + c_2 e^{-2t}$,

(b) $\frac{1}{2}(t^2 - 3t + 2) + e^{-t}(c_1 \cos t + c_2 \sin t)$,

(c) $\frac{1}{2}(t^2 + 2t + 2) + c_1 e^t + c_2 e^{2t}$.

5–4 Study of transients for second order equations. We consider a second order equation with constant coefficients:

$$a_0 D^2 x + a_1 Dx + x = F(t). \tag{5–8}$$

We assume that the equation is stable, so that

$$a_0 > 0, \quad a_1 > 0, \tag{5–9}$$

and proceed to consider the output corresponding to *zero input*. The output is therefore a solution of the homogeneous equation

$$a_0 D^2 x + a_1 Dx + x = 0 \tag{5–10}$$

and is itself a *transient*. In effect, we are thus simply studying the transients of Eq. (5–8).

We introduce the new constants

$$h = \frac{1}{2}\frac{a_1}{a_0} > 0, \qquad \lambda = \sqrt{\frac{1}{a_0}} > 0. \tag{5-11}$$

Then Eq. (5–10) can be written

$$D^2x + 2hDx + \lambda^2 x = 0, \tag{5-12}$$

and the characteristic equation is

$$r^2 + 2hr + \lambda^2 = 0. \tag{5-13}$$

We now consider three cases:

> Case 1. $0 < h < \lambda$.   *Underdamped case.*
>
> Case 2. $h = \lambda$.   *Critically damped case.*
>
> Case 3. $h > \lambda$.   *Overdamped case.*

In Case 1 the characteristic roots are complex: $r = -h \pm i\sqrt{\lambda^2 - h^2}$. The solutions are

$$x = e^{-ht}(c_1 \cos \beta t + c_2 \sin \beta t), \qquad \beta = \sqrt{\lambda^2 - h^2}. \tag{5-14}$$

By trigonometry we can write

$$c_1 \cos \beta t + c_2 \sin \beta t = A \sin (\beta t + \alpha),$$
$$A \cos \alpha = c_2, \qquad A \sin \alpha = c_1. \tag{5-15}$$

Thus A and α are polar coordinates of the point whose rectangular coordinates are c_2, c_1 (Fig. 5–2). The solution (5–14) becomes

$$x = Ae^{-ht} \sin (\beta t + \alpha). \tag{5-16}$$

Since $\sin (\beta t + \alpha)$ varies between -1 and $+1$, the solution varies between the two curves $x = Ae^{-ht}$, $x = -Ae^{-ht}$, touching these curves when $\beta t + \alpha$ is an odd multiple of $\pi/2$. The solution is called a *damped oscillation.* (See Fig. 5–3.)

In Case 2 the characteristic roots are real and equal: $r = -h, -h$. The solutions are

$$x = c_1 e^{-ht} + c_2 t e^{-ht}. \tag{5-17}$$

If $c_2 = 0$, this represents exponential decay (Section 3–2). If $c_2 \neq 0$, the solution approaches 0 as $t \to +\infty$, but is permitted one change of sign (*overshooting*), as in Fig. 5–4.

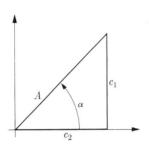

FIGURE 5–2

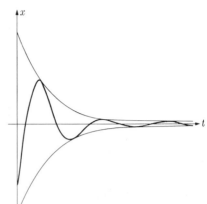

FIG. 5–3. Damped oscillation.

FIG. 5–4. Critical damping.

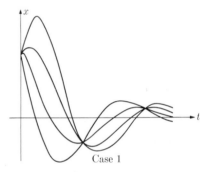

Case 1

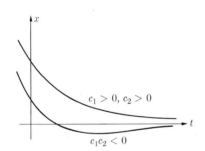

$c_1 > 0, c_2 > 0$

$c_1 c_2 < 0$

FIG. 5–5. Overcritical damping.

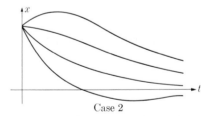

Case 2

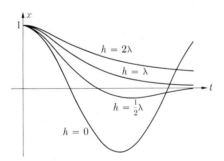

$h = 2\lambda$

$h = \lambda$

$h = \frac{1}{2}\lambda$

$h = 0$

FIG. 5–7. Solutions with fixed λ.

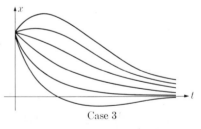

Case 3

FIG. 5–6. Solutions with fixed initial value of x.

In Case 3 the characteristic roots are distinct, real, and negative: $r = -h \pm \sqrt{h^2 - \lambda^2}$. The solutions are

$$x = c_1 e^{-at} + c_2 e^{-bt},$$

$$-a = -h - \mu, \quad -b = -h + \mu, \tag{5-18}$$

$$\mu = \sqrt{h^2 - \lambda^2} = \frac{a - b}{2}.$$

Thus $-a$ and $-b$ are negative. If c_1 and c_2 have the same sign, x can be considered as an average of two exponential decay functions; when c_1 and c_2 have opposite signs, there is conflict between the two and a change of sign occurs. The two cases are illustrated in Fig. 5-5.

We can study the three cases in greater detail by considering the solutions with the initial values $x = x_0$, $D_t x = v_0$ when $t = 0$. The solutions are then found to be (Prob. 3 below) as follows:

Case 1: $x = e^{-ht}\{x_0 \cos \beta t + [(v_0 + hx_0)/\beta] \sin \beta t\}$,

Case 2: $x = e^{-ht}[x_0 + (v_0 + hx_0)t]$,   (5-19)

Case 3: $x = [1/(a - b)][(v_0 + ax_0)e^{-bt} - (v_0 + bx_0)e^{-at}]$.

If we now fix x_0 (e.g., at 1) and vary v_0, we obtain in each case a family of solutions, as shown in Fig. 5-6. These are drawn for a particular choice of x_0, h, β, and μ; however the appearance is similar for other values. Changing x_0, $(x_0 \neq 0)$, while keeping h, β, μ fixed has the effect of changing the vertical scale. When $x_0 = 0$, the solutions are as follows:

Case 1: $x = (v_0/\beta)e^{-ht} \sin \beta t$,

Case 2: $x = v_0 t e^{-ht}$,   (5-20)

Case 3: $x = [v_0/(a - b)](e^{-bt} - e^{-at})$.

These are easily pictured (Prob. 4 below); changing v_0 in (5-20) merely varies the vertical scale.

We can also compare the three cases by keeping λ fixed and letting h vary, $(0 < h < \infty)$. Typical solutions for $x_0 = 1$, $v_0 = 0$ are shown in Fig. 5-7 for $t \geq 0$. The figure also includes the limiting case $h = 0$, in which the solutions are

$$x = c_1 \cos \lambda t + c_2 \sin \lambda t,$$

i.e., pure sinusoidal oscillations. They are not transients but, as remarked at the end of Section 5-2, are sometimes included with transients in the definition of stability. Figure 5-7 clearly shows the effect of increasing h; the oscillations are slowed down and decreased in amplitude until they disappear and are replaced by exponential decay. In physical applications, h is a measure of friction or some similar loss of energy. When h is 0, the system is "conservative" and does not lose energy; when h is positive, a loss of energy occurs and slows down the motion.

Time constant. The definition of a time constant is more complicated for a second order equation because of the three types of solutions. However, in all three cases an exponential factor governs the rate at which the transient approaches 0. When the transient is underdamped, the factor is e^{-ht} and we call $1/h$ the time constant τ. When the solution is overdamped, there are two exponentials, but the larger of the two is e^{-bt}, where $b = h - \sqrt{h^2 - \lambda^2}$; we call $1/b$ the time constant τ; the same definition applies in Case 2. In general, the roots of the characteristic equation are $-h \pm \sqrt{h^2 - \lambda^2}$ and we choose the time constant τ as minus the reciprocal of the real part of the root having the larger real part:

$$\tau = \frac{1}{\mathrm{Re}\,(h - \sqrt{h^2 - \lambda^2})} = \begin{cases} \dfrac{1}{h} & (0 < h \leq \lambda), \\[3mm] \dfrac{1}{h - \sqrt{h^2 - \lambda^2}} & (h \geq \lambda). \end{cases} \tag{5-21}$$

Large values of τ indicate a very slow approach of the transient to 0; small values signify a rapid decay of the transient.

We can rephrase our conclusions in terms of the original constants a_0, a_1:

Case 1. Damped oscillations. $a_1^2 < 4a_0$, $\tau = 2a_0/a_1$.

Case 2. Critical damping. $a_1^2 = 4a_0$, $\tau = 2a_0/a_1$.

Case 3. Overcritical damping $a_1^2 > 4a_0$, $\tau = \dfrac{2a_0}{a_1 - \sqrt{a_1^2 - 4a_0}}$.
(no oscillations).

PROBLEMS

1. Let $x = e^{-t} \sin 2t$. Make a careful graph, showing maximum and minimum points and points of contact with the curves $x = \pm e^{-t}$.

2. Graph the solutions of $(D^2 + 3D + 2)x = 0$ for which $x = 1$ when $t = 0$. For what initial values of $v = dx/dt$ does the solution never cross the t-axis for $-\infty < t < \infty$?

3. Verify that (5–19) provides the solution of (5–12) with prescribed initial values in (a) Case 1, (b) Case 2, (c) Case 3.

4. Describe the graphs of the functions (5–20) in (a) Case 1, (b) Case 2, (c) Case 3.

5. Show that the following equations are unstable and graph typical solutions:
(a) $(D^2 + 2D - 3)x = 0$,   (b) $(D^2 - D + 2)x = 0$.

6. Show that when a_1^2 is much larger than $4a_0$, the time constant τ is approximately a_1.

1. Max. at $t = 0.55 + n\pi$, min. at $t = 2.12 + n\pi$, contact at $t = 0.79 + n\pi/2$, $(n = 0, \pm 1, \ldots)$. 2. No crossing if $-2 \leqq v_0 \leqq -1$.

5–5 Response of second order equation to constant, ramp, and step-function inputs. We return to the stable second order equation with constant coefficients,

$$(a_0 D^2 + a_1 D + 1)x = F(t), \qquad (5\text{--}22)$$

and consider the outputs x that correspond to various types of inputs $F(t)$. The output will also be composed of a particular solution plus a transient. We are interested mainly in the appearance of the particular solution for large t, after the effects of initial conditions have become negligible.

Constant input. If $F(t) \equiv F_0$, where F_0 is constant, then $x = F_0$ is a particular solution, and hence the general output is

$$x = F_0 + \text{transient}. \qquad (5\text{--}23)$$

The form of the transient depends on the nature of the homogeneous equation related to (5–22) (Section 5–4). For example, in the underdamped case the transient is a damped oscillation $A \sin(\beta t + \alpha)$ and the outputs (5–23) have the form of Fig. 5–8. The constant F_0 is an *equilibrium value* approached by x as t increases. In the overdamped case there is a similar conclusion, except that x approaches F_0 without oscillation.

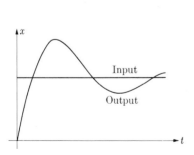

Fig. 5–8. Response to constant input.

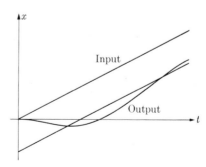

Fig. 5–9. Response to ramp input.

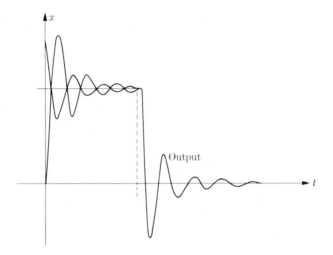

FIG. 5–10. Response to square pulse.

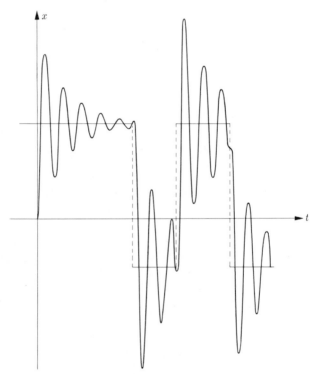

FIG. 5–11. Response to step-function input.

Ramp input. If $F(t) = kt$, a linear function, then a particular solution is $x = k(t - a_1)$. The general output is

$$x = k(t - a_1) + \text{transient}. \tag{5-24}$$

When the transient is a damped oscillation, the solution has the appearance of Fig. 5–9. Hence for large t, the output effectively coincides with the linear function $k(t - a_1)$. The slope of the output coincides with that of the input, but output lags input by time a_1.

Step-function input. When $F(t)$ is merely piecewise continuous, we can define the solutions of the differential equation as in Section 3–6. Each solution will be continuous and have a continuous derivative, but the second derivative will have a jump discontinuity at each such discontinuity of $F(t)$. Thus if $F(t)$ is continuous for $0 \leqq t < t_0$, with a jump at t_0, we can obtain the solution $x(t)$ with given initial values at $t = 0$ in the interval $0 \leqq t < t_0$; because $F(t)$ has a limit as $t \rightarrow t_0$ from the left, $x(t)$ and $x'(t)$ also have limits. The limiting values of x and x' serve as *initial values* at t_0 for the next interval $t_0 \leqq t < t_1$. Thus the solution can be prolonged in unique fashion past the discontinuities, and both $x(t)$ and $x'(t)$ remain continuous.

For example, let $F(t) = 0$ for $t \leqq 0$, $F(t) = 1$ for $0 < t \leqq 1$, and $F(t) = 0$ for $t > 1$. We consider the output for which $x = 0$, $x' = 0$ for $t = 0$. Since $x(t)$ satisfies the equation $(a_0 D^2 + a_1 D + 1)x = 0$ for $t \leqq 0$, we conclude that $x \equiv 0$ for $t \leqq 0$. For the interval $0 \leqq t \leqq 1$ the output satisfies the equation

$$(a_0 D^2 + a_1 D + 1)x = 1$$

and has initial values $x = 0$, $x' = 0$ at $t = 0$. This is simply an output which corresponds to a constant input, e.g., as in Fig. 5–8; thus $x = 1 +$ transient, and x is approaching 1. At $t = 1$ the process is interrupted, and we then solve the equation

$$(a_0 D^2 + a_1 D + 1)x = 0$$

with initial values of x, x' at $t = 1$ equal to the values obtained from the previous solution. Hence for $t \geqq 1$, x has the form of a pure transient. A typical solution is shown in Fig. 5–10.

If the initial conditions at $t = 0$ are changed, the particular solution is modified only by addition of a transient. Hence for large t the solutions effectively coincide, as shown in Fig. 5–10.

If $F(t)$ is a more complicated step function, we make a similar analysis and obtain an output such as that of Fig. 5–11, which shows the fashion in which the output imitates the input but cannot follow it precisely. The accuracy of the follow-up is governed by the time constant τ; the smaller τ is, the more rapidly the oscillations die down. However, the presence of oscillations introduces a new complication. The precise time at which the input jumps to a new value has a large influence on the initial values of the transient for the next interval, and hence on the size of the first few oscillations. In particular, if the input $F(t)$ is periodic, the precise value of the period is important in determining the size of the transient error. This relationship is analyzed quantitatively in the following section.

As in Section 3–8, we can analyze the response to a quite general input by approximating it by a step function. Hence we can predict qualitatively what the output will be like: if $F(t)$ varies very slowly, the input will follow $F(t)$ very closely; if $F(t)$ is at all jerky, transient errors, which tend to die out, are excited. It should be noted that in replacing $F(t)$ by a step function, we are in effect adding a somewhat jerky error to $F(t)$. Under certain conditions this error may be much amplified in the output, so that the output corresponding to the step function is no longer a good approximation to the output corresponding to $F(t)$; however, this defect can be remedied by making the error in approximation of $F(t)$ very small (see Prob. 6(b) below).

Throughout this section we have emphasized the case in which the transient is underdamped, hence oscillatory. In the critically damped or overdamped case the conclusions are similar, but the transient oscillations do not appear (except for the effect of *overshooting*) and the follow-up is based on exponential decay. Indeed, when a_1^2 is large relative to a_0, the equation behaves more like the first order equation $(a_1D + 1)x = F(t)$, for which exponential decay is the typical feature.

PROBLEMS

1. Graph the input and the output with initial values $x = 0$, $x' = 0$ at $t = 0$ for each of the following:

(a) $(2D^2 + 2D + 1)x = 3$,   (b) $(5D^2 + 2D + 1)x = 5t$,
(c) $(D^2 + 2D + 1)x = t(t - 1)(t - 2)$,   (d) $(D^2 + 3D + 2)x = e^{2t}$.

2. Find the time constant for each of the equations of Prob. 1.

3. Let $F(t) = 0$ for $t \leqq 0$, $F(t) = 2$ for $0 < t \leqq \pi$, $F(t) = -1$ for $t > \pi$. Find and graph the output with initial values $x = 0$, $x' = 0$ at $t = 0$ for the equation $(2D^2 + 2D + 1)x = F(t)$.

4. Let the equation $(2\gamma^2 D^2 + 2\gamma D + 1)x = t$ (where γ is a positive constant) be given.

(a) Find the solution such that $x = 0$ and $x' = 0$ when $t = 0$.

(b) In the solution of part (a) find the limiting value of x as $\gamma \to 0+$, for fixed positive t. Interpret the result in terms of the approximation: output $\sim$ input.

5. *Weighting function.* Rule 20 of Table 4–2 (Section 4–14) provides an output of the equation $(a_0 D^2 + a_1 D + 1)x = F$ of form

$$x = \int_c^t F(u) W(t - u)\, du, \tag{i}$$

where $W(t)$ is the solution of the related homogeneous equation such that $W(0) = 0$, $W'(0) = 1/a_0$. If $F(t) = 0$ for $t < c$, then (i) can be written

$$x = \int_{-\infty}^t F(u) W(t - u)\, du = \int_0^\infty F(t - v) W(v)\, dv, \tag{ii}$$

where $v = t - u$. More generally, if $F(t)$ is sufficiently small for large negative t that the improper integrals in (ii) exist, (ii) can be justified in the same way as the analogous formula (3–51) in Section 3–12. Equation (ii) expresses $x(t)$ as a weighted average of $F(t)$ over the past, as in Section 3–12, and we call $W(t)$ the *weighting function.*

(a) Find $W(t)$ for the equation $(D^2 + 2D + 2)x = 2F(t)$ (cf. Rule 15 in Table 4–2) and graph.

(b) Use the weighting function of part (a) to obtain an output graphically when $F(t) = 1$ for $0 < t < \pi$, $F(t) = 0$ otherwise. (See Fig. 3–21.)

(c) Find $W(t)$ for the equation $(D^2 + 3D + 2)x = 2F(t)$.

6. (a) Show that for a stable equation $(a_0 D^2 + a_1 D + 1)x = F(t)$, the weighting function W of Prob. 5 satisfies the inequality

$$|W(t)| < Ce^{-t/\tau} \qquad (t \geqq 0)$$

for some constant C, where τ is the time constant (Section 5–4).

(b) Apply the result of part (a) to show that if $|F(t)| < \epsilon$ for all t, the output $x(t)$ defined by Eq. (i) of Prob. 5 satisfies the inequality

$$|x(t)| \leqq C\epsilon\tau(1 - e^{-t/\tau}) < C\epsilon\tau \qquad (t > 0).$$

ANSWERS

1. Outputs: (a) $3 + e^{-t/2}(-3 \cos \frac{1}{2}t - 3 \sin \frac{1}{2}t)$,
(b) $5(t - 2) + e^{-t/5}(10 \cos \frac{2}{5}t - \frac{15}{2} \sin \frac{2}{5}t)$,
(c) $t^3 - 9t^2 + 32t - 46 + e^{-t}(46 + 14t)$,   (d) $(e^{2t} - 4e^{-t} + 3e^{-2t})/12$.

2. (a) 2, (b) 5, (c) 1, (d) 1.

3. $x = 0$ for $t \leqq 0$, $x = 2 - 2e^{-t/2}(\cos \frac{1}{2}t + \sin \frac{1}{2}t)$ for $0 \leqq t \leqq \pi$, $x = -1 + e^{-t/2}[(-2 - 3e^{\pi/2}) \cos \frac{1}{2}t + (3e^{\pi/2} - 2) \sin \frac{1}{2}t]$ for $t > \pi$.

4. (a) $t - 2\gamma + 2\gamma e^{-t/(2\gamma)} \cos [t/(2\gamma)]$, (b) limit is t.

5. (a) $2e^{-t} \sin t$, (c) $2(e^{-t} - e^{-2t})$.

5–6 Output of second order equation corresponding to sinusoidal input. With an input $A_i \sin \omega t$, $(\omega > 0)$, our equation becomes

$$(a_0 D^2 + a_1 D + 1)x = A_i \sin \omega t. \tag{5–25}$$

By the methods of Chapter 4 (Section 4–10, or Rule 8 in Table 4–2, Section 4–14) a particular output is found to be

$$x = A_i \frac{(1 - a_0\omega^2) \sin \omega t - a_1\omega \cos \omega t}{(1 - a_0\omega^2)^2 + a_1^2\omega^2} = A_o \sin (\omega t - \alpha), \tag{5–26}$$

where

$$A_o = \frac{A_i}{k}, \qquad \cos \alpha = \frac{1 - a_0\omega^2}{k}, \qquad \sin \alpha = \frac{a_1\omega}{k},$$

$$\tag{5–27}$$

$$k = [(1 - a_0\omega^2)^2 + a_1^2\omega^2]^{1/2}.$$

Thus k and α are polar coordinates of the point whose rectangular coordinates are $1 - a_0\omega^2$, $a_1\omega$, as in Fig. 5–12.

Since our equation is assumed to be stable, the general output will effectively coincide with (5–26), after a transient period has elapsed. Thus, except for a transient, the output has form similar to the input. It is again sinusoidal, with the same *frequency* ω but with a new amplitude $A_o = A_i/k$; here A_i denotes input amplitude and A_o output amplitude, and there is a phase lag α. Since a_0, a_1 are positive, Fig. 5–12 shows that $0 < \alpha < \pi$.

If we write $h = \frac{1}{2}a_1/a_0$, $\lambda = a_0^{-1/2}$, then we obtain

$$k = \left[\left(1 - \frac{\omega^2}{\lambda^2}\right)^2 + 4 \frac{h^2}{\lambda^2} \frac{\omega^2}{\lambda^2} \right]^{1/2} \tag{5–28}$$

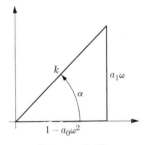

FIGURE 5–12

Thus the ratio of output amplitude to input amplitude depends only on the ratios ω/λ, h/λ:

$$\frac{A_o}{A_i} = \frac{1}{k} = \frac{1}{[(1 - \nu^2)^2 + 4\eta^2\nu^2]^{1/2}} \qquad (\nu = \omega/\lambda), \ (\eta = h/\lambda). \tag{5–29}$$

When $h = 0$, the solutions of the homogeneous equation are the undamped oscillations $c_1 \cos \lambda t + c_2 \sin \lambda t$. Hence λ is the frequency of the natural

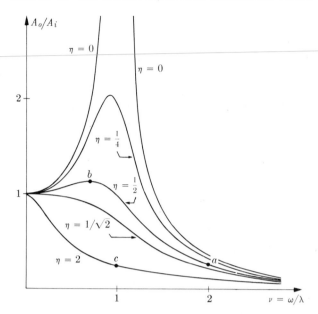

Fig. 5–13. Amplification for second order system.

undamped oscillations and $\nu = \omega/\lambda$ is the ratio of the frequency of the input to the frequency λ. The quantity η can be considered as a measure of the amount of damping, for fixed λ. The ratio A_o/A_i is called the *amplification factor* or simply the amplification. As we shall see, A_o/A_i can be both less than 1 and greater than 1.

To study the dependence of amplification on applied frequency and damping, we graph A_o/A_i against ν for different choices of η. The result is the family of curves shown in Fig. 5–13. The verification of this graph is left as an exercise (Prob. 4 following Section 5–7).

The borderline case $\eta = 0$ (that is, $h = 0$) is also shown. In this case, the amplification rises from 1 to ∞ as ν goes from 0 to 1. When $\nu = 1$, $\omega = \lambda$, and we are then considering the equation

$$(a_0 D^2 + 1)x = A_i \sin \lambda t,$$

where $a_0 = 1/\lambda^2$. The output is no longer sinusoidal, but is given by

$$x = -A_i \frac{\lambda}{2} t \cos \lambda t + c_1 \cos \lambda t + c_2 \sin \lambda t.$$

The particular solution has a factor t, which leads to an oscillation of increasing amplitude, as shown in Fig. 5–14. The addition of the com-

plementary function distorts this steady in-
crease, but cannot prevent $|x|$ from rising
to arbitrarily large values for large t. This
phenomenon of excitation of large oscillations by matching of input and natural
frequencies is known as *resonance*.

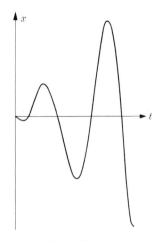

When damping is present, so that h and η
are positive, the complication disappears.
The output is always sinusoidal (plus transient) and the output amplitude simply rises
to a maximum and then falls as ν increases
or, for large η, simply decreases steadily to 0
as ν increases. The maximum occurs for
$0 < \eta < \sqrt{2}/2$, that is, for h between
0 and $\lambda/\sqrt{2}$; at each maximum point, $\nu =$
$\sqrt{1 - 2\eta^2}$ and $A_o/A_i = (2\eta\sqrt{1 - \eta^2})^{-1}$
(Prob. 4 following Section 5–7). We term

Fig. 5–14. Resonance.

the value of ω at the maximum the *resonant frequency in presence of
damping*. As the term suggests, the maximum is regarded as a form of resonance, even though for the corresponding solution $x(t)$ the amplitude does
not increase as t increases.

The dependence of phase lag α on η and ν can be analyzed similarly.
From (5–27) and (5–29),

$$\alpha = \tan^{-1}\frac{a_1\omega}{1 - a_0\omega^2} = \tan^{-1}\frac{2h\omega}{\lambda^2 - \omega^2} = \tan^{-1}\frac{2\eta\nu}{1 - \nu^2}. \quad (5\text{–}30)$$

In Fig. 5–15 the angle α is graphed against ν for various values of η (Prob. 5
following Section 5–7). We see that for fixed η, α is increased if ν is
increased.

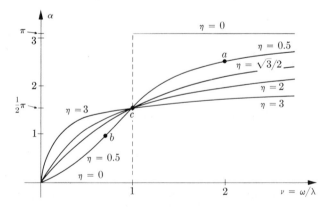

Fig. 5–15. Phase lag for second order system.

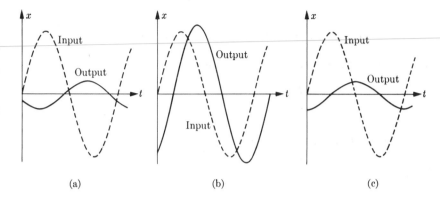

FIG. 5–16. Typical response to sinusoidal inputs.

Figure 5–16 shows graphs of input and output for three cases, corresponding to the points labeled a, b, c on Figs. 5–13 and 5–15.

Remark. If the input is taken as $A_i \cos \omega t$, instead of $A_i \sin \omega t$, then we find that the output is $A_o \cos (\omega t - \alpha)$, where A_o and α are defined as above. We can reach the same conclusion by writing

$$A_i \cos \omega t = A_i \sin \left(\omega t - \frac{3\pi}{2} \right) = A_i \sin \omega \left(t - \frac{3\pi}{2\omega} \right).$$

Thus the input $A_i \cos \omega t$ is the input $A_i \sin \omega t$ delayed by time $3\pi/(2\omega)$. Since the coefficients are constant (see Sections 3–13 and 4–14), delay of the input by a certain time interval leads to a delay of the output by the same time interval (principle of *stationarity*), and hence the output is

$$A_o \sin \left[\omega \left(t - \frac{3\pi}{2\omega} \right) - \alpha \right] = A_o \sin \left(\omega t - \alpha - \frac{3\pi}{2} \right) = A_o \cos (\omega t - \alpha).$$

An input $p \sin \omega t + q \cos \omega t$ can be written as $A_i \sin (\omega t - \beta)$, where $A_i = (p^2 + q^2)^{1/2}$, $p = A_i \cos \beta$, $-q = A_i \sin \beta$. Hence the input is $A_i \sin \omega t$ delayed by time β/ω and the output is $A_o \sin (\omega t - \beta - \alpha)$.

5–7 Output of second order equation corresponding to periodic input. As in Section 3–11, we can represent a periodic input by a Fourier series:

$$F(t) = \frac{p_0}{2} + \sum_{n=1}^{\infty} (p_n \cos n\omega t + q_n \sin n\omega t)$$

$$= \frac{p_0}{2} + \sum_{n=1}^{\infty} A_n^i \sin (n\omega t - \beta_n), \qquad (5\text{–}31)$$

$$A_n^i \cos \beta_n = q_n, \qquad -A_n^i \sin \beta_n = p_n.$$

For the differential equation

$$(a_0 D^2 + a_1 D + 1)x = F(t) \qquad (a_0 > 0,\ a_1 > 0),$$

the output is then obtained by superposition:

$$x = \frac{p_0}{2} + \sum_{n=1}^{\infty} A_n^o \sin (n\omega t - \beta_n - \alpha_n) + \text{transient}. \qquad (5\text{--}32)$$

Here we have applied the final remark of the preceding section. The constants A_n^o and α_n are given as before by the equations

$$A_n^o = \frac{A_n^i}{\left[\left(1 - \dfrac{n^2 \omega^2}{\lambda^2} \right)^2 + 4 \dfrac{h^2}{\lambda^2} \dfrac{n^2 \omega^2}{\lambda^2} \right]^{1/2}}, \qquad (5\text{--}33)$$

$$\alpha_n = \tan^{-1} \frac{2hn\omega}{\lambda^2 - n^2 \omega^2} \qquad (0 < \alpha_n < \pi).$$

In Eq. (5–32) we can write

$$\sin (n\omega t - \beta_n - \alpha_n) = \sin n\omega t \cos (\alpha_n + \beta_n) - \cos n\omega t \sin (\alpha_n + \beta_n),$$

which shows that (5–32) is a Fourier series expansion for the output. Thus the output is also periodic (plus transient).

We do not consider here the problems of convergence of the Fourier series which appear. It can be shown that if $F(t)$ is piecewise continuous, then its Fourier series (5–31) is well defined [but may not converge to $F(t)$]; the series for $x(t)$ does converge to $x(t)$ and is the Fourier series for $x(t)$. (See Reference 6.)

The above analysis and that of Section 5–6 shows that the output is obtained by modifying the terms of different frequency in the Fourier series of $F(t)$. Each term is amplified and delayed by amounts that depend on the frequency. Since h and λ are fixed, the terms correspond to a sequence of points on one of the curves of Fig. 5–13, as suggested in Fig. 5–17. Thus the first few terms may be amplified ($A_n^o / A_n^i > 1$), but the later ones will be *attenuated* (diminished in amplitude), the more so

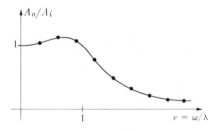

FIG. 5–17. Amplification at different frequencies.

the larger the n. This effect is also revealed by (5–33), which shows that
for large n

$$\frac{A_n^o}{A_n^i} \sim \frac{\lambda^2}{n^2\omega^2}.$$

When the maximum of Fig. 5–17 is very sharp, the few frequencies $n\omega$
for which $n\omega/\lambda$ is close to the maximum will be greatly emphasized in the
output, while the others receive moderate or negligible weight. Hence
the equation acts as a *filter*, allowing a narrow band of frequencies to pass
through, but effectively blocking the others.

EXAMPLE 1. $(3D^2 + D + 1)x = \sin t + 2 \sin 4t$.   Here $F(t)$ is given
as a Fourier series of two terms. Since $h = \frac{1}{6}$, $\lambda^2 = \frac{1}{3}$, $\omega = 1$, we find
by (5–33)

$$A_n^o = \frac{A_n^i}{[(1 - 3n^2)^2 + n^2]^{1/2}}, \qquad \alpha_n = \tan^{-1}\frac{n}{1 - 3n^2},$$

and hence (apart from a transient)

$$x = A_1^o \sin(t - \alpha_1) + A_4^o \sin(4t - \alpha_4)$$

$$= \frac{\sin[t - \tan^{-1}(-\frac{1}{2})]}{\sqrt{5}} + \frac{2\sin[4t - \tan^{-1}(-4/47)]}{5\sqrt{89}}.$$

Input and output for this example are graphed in Fig. 5–18, which shows
the smoothing effect already encountered for first order equations (Section
3–11). The high-frequency term in $F(t)$ is so greatly attenuated that its
effect almost disappears.

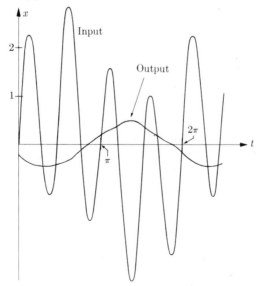

FIG. 5–18. Response to periodic input.

EXAMPLE 2. $(\frac{1}{16}D^2 + \frac{1}{8}D + 1)x = F(t)$, where $F(t)$ is a square wave: $F(t) = 1$ for $0 < t < \pi$, $F(t) = -1$ for $-\pi < t < 0$, and $F(t)$ has period 2π. As in Section 3–10, $F(t)$ can be represented by a Fourier series ($\omega = 1$):

$$F(t) = \frac{4}{\pi} \sum_{n=1}^{\infty} \frac{\sin (2n - 1)t}{2n - 1};$$

at the jumps of $F(t)$ the series converges to 0, the average of left and right limits. The homogeneous equation is $(D^2 + 2D + 16)x = 0$, so that $h = 1$, $\lambda = 4$. Hence by (5–33)

$$A_n^o = \frac{A_n^i}{\left[\left(1 - \frac{n^2}{16}\right)^2 + \frac{n^2}{64}\right]^{1/2}}, \qquad \alpha_n = \tan^{-1} \frac{2n}{16 - n^2}. \qquad (5\text{–}34)$$

Since $F(t)$ contains only the odd frequencies $1, 3, 5, \ldots$, we find as output

$$x(t) = \frac{4}{\pi} \sum_{n=1}^{\infty} \frac{\sin \left[(2n - 1)t - \tan^{-1} \overline{(4n - 2)/\{16 - (2n - 1)^2\}}\right]}{(2n - 1)\left[\left(1 - \frac{(2n - 1)^2}{16}\right)^2 + \frac{(2n - 1)^2}{64}\right]^{1/2}}$$

$$= \frac{64}{\pi} \sum_{n=1}^{\infty} \frac{\{16 - (2n - 1)^2\} \sin (2n - 1)t - (4n - 2) \cos (2n - 1)t}{(2n - 1)[\{16 - (2n - 1)^2\}^2 + (4n - 2)^2]}$$

$$= \frac{64}{\pi} \left(\frac{15 \sin t - 2 \cos t}{229} + \frac{7 \sin 3t - 6 \cos 3t}{255} \right.$$

$$\left. + \frac{-9 \sin 5t - 10 \cos 5t}{905} + \cdots \right).$$

Figure 5–19 shows $F(t)$ and the third partial sum of the output (through the term of frequency 5). It can be verified that this partial sum is a very good approximation to the output. Since $h = 1$, $\lambda = 4$, the value of η is 0.25 (Eq. 5–29). Hence the system is on the curve $\eta = \frac{1}{4}$ (Fig. 5–13), with ω/λ equal to $\frac{1}{4}$, $\frac{3}{4}$, $\frac{5}{4}$, $\ldots$ It is clear that maximum amplification occurs when $\omega = 3$. With the aid of Eq. (5–34) we can tabulate input and output amplitudes as in Table 5–1.

The exceptional weight given to the frequency 3 leads to the distortion of output relative to input in Fig. 5–19. Note that although the constants a_0, a_1 in the differential equation are quite small ($\frac{1}{16}$ and $\frac{1}{8}$), the output is not a very precise reproduction of the input. As the analysis shows, this effect is due to resonance approximately at $\omega = 3$.

TABLE 5–1

n	Frequency $2n - 1$	$A^i = 4/[(2n-1)\pi]$	A^o/A^i	A^o
1	1	1.273	1.06	1.350
2	3	0.424	1.74	0.738
3	5	0.255	1.19	0.303
4	7	0.182	0.45	0.082

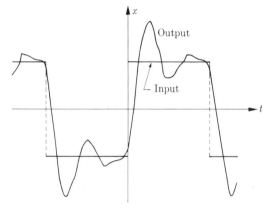

FIG. 5–19. Response to square wave input.

PROBLEMS

1. For each of the following equations find ω, h, λ, ν, η and read the approximate values of amplification and phase lag from the graphs of Figs. 5–13 and 5–15:

(a) $(\frac{1}{9}D^2 + \frac{2}{9}D + 1)x = \sin 2t$,   (b) $(4D^2 + 6D + 1)x = \sin t$,
(c) $(D^2 + 5D + 9)x = \sin 3t$.

2. For each of the following equations obtain the periodic output as a Fourier series:

(a) $(5D^2 + D + 1)x = \displaystyle\sum_{n=1}^{\infty} \frac{\sin nt}{n^2 + 1}$,

(b) $(D^2 + D + 1)x = \displaystyle\sum_{n=1}^{\infty} \frac{\sin (nt/5)}{n^3}$,

(c) $(D^2 + 3D + 2)x = |\cos t|$.

3. For each of the following equations determine the term of the Fourier series of the input which is most amplified in the output:

(a) Eq. of Prob. 2(a),   (b) Eq. of Prob. 2(b).

4. To obtain the basic features of the graphs of Fig. 5–13, let $u = [(1 - \nu^2)^2 + 4\eta^2\nu^2]^{-1/2} = A_o/A_i$ and consider u as a function of ν for fixed η, $(\nu > 0, \eta > 0)$.

(a) Show that as $\nu \to 0$, $u \to 1$.

(b) Show that as $\nu \to +\infty$, $u \to 0$.

(c) Show that for $\eta \geq \sqrt{2}/2$, $du/d\nu$ is always negative.

(d) Show that for $\eta < \sqrt{2}/2$, $du/d\nu$ is positive for $0 < \nu < \sqrt{1 - 2\eta^2}$ and negative for $\nu > \sqrt{1 - 2\eta^2}$, so that u has a maximum at $\nu = \sqrt{1 - 2\eta^2}$ and at the maximum, $u = (2\eta\sqrt{1 - \eta^2})^{-1}$.

(e) Show that for fixed ν, u decreases as η increases; as $\eta \to +\infty$, $u \to 0$; as $\eta \to 0$, $u \to |1 - \nu^2|^{-1}$.

5. To obtain the basic features of the graphs of Fig. 5–15, consider $\alpha = \tan^{-1}[2\eta\nu/(1 - \nu^2)]$ as a function of ν for fixed η, $(0 < \alpha < \pi, 0 < \eta, 0 < \nu)$.

(a) Show that $\alpha(1) = \pi/2$.

(b) Show that $\dfrac{d\alpha}{d\nu} = \dfrac{2\eta(1 + \nu^2)}{(1 - \nu^2)^2 + 4\eta^2\nu^2}$.

(c) Show that $\dfrac{d^2\alpha}{d\nu^2} = \dfrac{-4\eta\nu(\nu^4 + 2\nu^2 + 4\eta^2 - 3)}{[(1 - \nu^2)^2 + 4\eta^2\nu^2]^2}$.

(d) Show that α increases as ν increases, that as $\nu \to 0$, $\alpha \to 0$, and as $\nu \to +\infty$, $\alpha \to \pi$.

(e) Show that for $\eta^2 \geq \frac{3}{4}$ the graph of $\alpha(\nu)$ is concave down for all ν, but for $0 < \eta^2 < \frac{3}{4}$ the graph is concave up for $\nu < c_\eta$ and concave down for $\nu > c_\eta$, where $c_\eta = [-1 + 2(1 - \eta^2)^{1/2}]^{1/2}$, so that there is an inflection point at $u = c_\eta$.

(f) Show that for fixed ν, $(0 < \nu < 1)$, α increases as η increases, and $\alpha \to 0$ as $\eta \to 0$, $\alpha \to \pi/2$ as $\eta \to +\infty$; show that for fixed ν, $(\nu > 1)$, α decreases as η increases, and $\alpha \to \pi$ as $\eta \to 0$, $\alpha \to \pi/2$ as $\eta \to +\infty$.

ANSWERS

1. (a) $\omega = 2$, $h = 1$, $\lambda = 3$, $\nu = \frac{2}{3}$, $\eta = \frac{1}{3}$, $A_o/A_i = 1.4$, $\alpha = 0.67$; (b) $\omega = 1, h = \frac{3}{4}, \lambda = \frac{1}{2}, \nu = 2, \eta = \frac{3}{2}, A_o/A_i = 0.15, \alpha = 2.03$; (c) $\omega = 3$, $h = \frac{5}{2}, \lambda = 3, \nu = 1, \eta = \frac{5}{6}, A_o/A_i = \frac{3}{5}, \alpha = \pi/2$.

2. (a) $\displaystyle\sum_{n=1}^{\infty} \dfrac{(1 - 5n^2)\sin nt - n\cos nt}{(25n^4 - 9n^2 + 1)(n^2 + 1)}$,

(b) $\displaystyle\sum_{n=1}^{\infty} \dfrac{(625 - 25n^2)\sin(nt/5) - 125n\cos(nt/5)}{n^3(n^4 - 25n^2 + 625)}$,

(c) $\dfrac{1}{\pi} + \dfrac{4}{\pi}\displaystyle\sum_{n=1}^{\infty} \dfrac{(-1)^{n+1}}{4n^2 - 1} \dfrac{(2 - 4n^2)\cos 2nt + 6n\sin 2nt}{16n^4 + 20n^2 + 4}$.

3. (a) $n = 1$, (b) $n = 3$ and $n = 4$ equally.

5–8 Input-output analysis for equations of higher order. We now consider an equation with constant coefficients in the form

$$(a_0 D^n + \cdots + a_{n-1} D + 1)x = F(t), \tag{5-35}$$

so that $F(t)$ is the input. We assume the equation to be stable, so that the output is uniquely determined except for a transient. Each characteristic root is then a negative real number or a complex number with negative real part.

We could proceed to analyze the output by the methods of the preceding sections and by those of Chapter 3; however, we shall take another point of view which is very suggestive of the physical applications. It is based on *factorization* of the operator on the left side of Eq. (5–35) (see Section 4–12).

If the characteristic roots are known, then the operator can be factored as if it were an ordinary polynomial in D with these roots. The grouping of the complex roots in conjugate pairs leads to a factorization into linear and quadratic factors. By multiplying and dividing by constants, we can ensure that the constant term in each factor is 1. Hence we achieve the factorization

$$(b_1 D + 1) \cdots (b_k D + 1)(p_1 D^2 + q_1 D + 1) \cdots$$
$$(p_l D^2 + q_l D + 1)x = F(t). \tag{5-36}$$

Some factors will be equal if there are repeated roots. The roots themselves appear as the numbers

$$-\frac{1}{b_1}, \quad -\frac{1}{b_2}, \quad \cdots, \quad -\frac{1}{b_k},$$
$$\frac{-q_1 \pm i\sqrt{4p_1 - q_1^2}}{2p_1}, \quad \cdots, \quad \frac{-q_l \pm i\sqrt{4p_l - q_l^2}}{2p_l}. \tag{5-37}$$

By the stability assumption the coefficients $b_1, \ldots, b_k, p_1, q_1, \ldots, p_l, q_l$ must all be *positive*. We use quadratic factors only for the complex roots, so that the discriminants $q_1^2 - 4p_1, \ldots, q_l^2 - 4p_l$ are all assumed to be *negative*.

Following the method of Section 4–12, we now introduce auxiliary variables, and we illustrate the procedure by examples.

EXAMPLE 1. $(D + 1)(2D + 1)(D^2 + D + 1)x = F(t).$

We write

$$(D^2 + D + 1)x = u_1,$$
$$(2D + 1)u_1 = u_2, \tag{5-38}$$
$$(D + 1)u_2 = F(t).$$

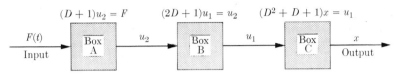

$$(D + 1)u_2 = F \qquad (2D + 1)u_1 = u_2 \qquad (D^2 + D + 1)x = u_1$$

FIG. 5-20. Block diagram for a fourth order system.

These three equations are equivalent to the given equation, for successive elimination of u_1 and u_2 leads precisely to the given equation.

We now interpret the system (5–38) as suggested in Fig. 5–20. We call u_2 the output, corresponding to input $F(t)$, for the mechanism represented by box A. Next, u_2 serves as input to box B, for which u_1 is the output. Finally, u_1 is input to box C and x is the output.

We can now describe the relationship between $F(t)$ and the final output $x(t)$ by combining what we know about the individual stages. For boxes A and B we have first order differential equations, and the analysis is as in Chapter 3. For box C the equation is of second order, and the results of this chapter apply; furthermore, since the characteristic roots are complex, the transient for box C is a damped oscillation and the curve is in the region $\eta < 1$, ($h < \lambda$), on Fig. 5–13.

Suppose, for example, that $F(t) = \sin 3t$. We then refer to Section 3–9 and find that the output u_2 is governed by Eq. (3–26) with $a = 1$, $\omega = 3$, $A_i = 1$. Ignoring the transient, we find

$$u_2 = \frac{\sin (3t - \alpha_2)}{\sqrt{10}}, \qquad \alpha_2 = \tan^{-1} 3 = 1.25.$$

We now use u_2 as input to box B. Applying Eqs. (3–32), (3–33), and (3–27), with $a = 2$, $\omega = 3$, we find

$$u_1 = \frac{1}{\sqrt{10}\sqrt{37}} \sin (3t - \alpha_2 - \alpha_1), \qquad \alpha_1 = \tan^{-1} 6 = 1.41.$$

Finally, u_1 serves as input to box C and we apply (5–29) and (5–30) (see *Remark* at end of Section 5–6) to find the output x. Since $h = \frac{1}{2}$ and $\lambda = 1$, we find $\nu = 3$, $\eta = \frac{1}{2}$, and

$$x = \frac{1}{\sqrt{73}\sqrt{10}\sqrt{37}} \sin (3t - \alpha_3 - \alpha_2 - \alpha_1),$$

$$\alpha_3 = \tan^{-1} \frac{3}{-8} = 2.78.$$

Thus the output is

$$x = 0.0061 \sin (3t - 5.44).$$

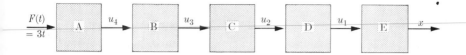

FIG. 5-21. Block diagram for a fifth order system.

Our example is typical: a sinusoidal input will be attenuated in transmission through boxes A and B, which are governed by first order equations, and will be either amplified or attenuated in passing through box C, depending on the values of ν and η; at each stage an additional phase lag is imposed.

EXAMPLE 2. $(2D + 1)^5 x = 3t$. The corresponding system is

$$(2D + 1)x = u_1, \qquad (2D + 1)u_1 = u_2, \qquad (2D + 1)u_2 = u_3,$$
$$(2D + 1)u_3 = u_4, \qquad (2D + 1)u_4 = 3t,$$

and the diagram is that of Fig. 5-21. Since all equations are of first order and the input is of ramp type, we apply the results of Section 3-4. The input $3t$ to box A is delayed by time 2 and becomes $3(t - 2) = u_4$. By stationarity (Section 3-13) this input is treated as a ramp input to box B and is simply delayed again, to become $3(t - 4) = u_3$. Similarly, $u_2 = 3(t - 6)$, $u_1 = 3(t - 8)$, $x = 3(t - 10)$. Hence, except for a transient, the final output is $3(t - 10)$.

Another description of the systems of higher order can be obtained from the *weighting function* (Section 3-12; Prob. 5 following Section 5-5). If $F(t) = 0$ for $t < c$, then by Rule 20 of Table 4-2 (Section 4-14) we obtain a corresponding output $x(t)$ as follows:

$$x(t) = \int_c^t F(u)W(t - u)\, du = \int_{-\infty}^t F(u)W(t - u)\, du$$

$$= \int_0^\infty F(t - v)W(v)\, dv, \tag{5-39}$$

where $W(t)$ is the solution of the equation $(a_0 D^n + \cdots + 1)x = 0$, such that

$$W(0) = 0, \qquad W'(0) = 0, \qquad \ldots, \qquad W^{(n-2)}(0) = 0,$$
$$W^{(n-1)}(0) = 1/a_0.$$

Equation (5-39) defines $x(t)$ as a weighted average over the past of $F(t)$.

We can verify (Prob. 4 below) that

$$\int_0^\infty W(t)\, dt = 1,$$

(5–40)

that is, the total weight is unity. However, simple examples show that $W(t)$ may take on both positive and negative values, as in Fig. 5–22 (Prob. 3 below). Thus considerable distortion of the input may be achieved by forming the weighted average (5–39). In particular, $F(t)$ may change sign with the same frequency as W, as in Fig. 5–22; for some values of t, $F(t - v)W(v)$ is then positive or zero for all v and the resulting integral will have a large positive value (this gives some insight into the phenomenon of resonance).

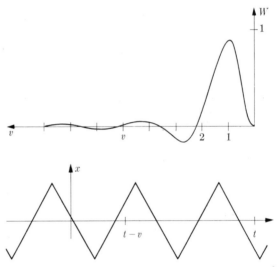

FIG. 5–22. Output as weighted average of past input.

5–9 Unstable equations. Equations with variable coefficients. In the previous sections we have assumed our basic equation to have constant coefficients and to be stable. Throughout we have neglected transients. Even for a stable equation, however, the transients can be of great importance and it may be necessary to study carefully their dependence on initial conditions.

If the equation is unstable, then the transients effectively dominate the picture. The general solution is composed of a particular solution $x^*(t)$ and a complementary function $x_c(t)$. Except for special choices of initial conditions, the complementary function does not approach 0 as $t \to +\infty$. The output is in no sense uniquely determined, but is very sensitive to initial conditions. A slight change in the input has an effect similar to that of a slight change in initial conditions and can also have considerable influence on the output.

For equations with variable coefficients there are new complications. It is difficult to obtain explicit expressions for either the complementary function or the particular solution (Chapter 9). However, if the coefficients are varying slowly, we can use an approximation in order to gain some insight on the behavior of the solutions. To illustrate, we use the second order equation

$$a_0(t)D^2x + a_1(t)Dx + a_2(t)x = F(t), \qquad (5\text{–}41)$$

and simply approximate the functions $a_0(t)$, $a_1(t)$, $a_2(t)$ by step functions (see Section 3–8). Thus for the interval 0 to t_1 we replace all three functions by suitable average values; t_1 is to be chosen so small that the functions deviate only slightly from the average values between 0 and t_1. A similar replacement is made from t_1 to t_2, and so on. Within each interval we are dealing with an equation which has constant coefficients, and the general solution can be found. If initial conditions are given at $t = 0$, we use these to determine the solution up to t_1, and hence obtain new initial conditions for the next interval, so that the solution can be prolonged indefinitely. The procedure is closely related to that of step-by-step integration (Section 1–7), but is more advantageous if the coefficients vary so slowly that the intervals 0 to t_1, t_1 to t_2, . . . can be chosen to be relatively long.

An important question now arises. If the equation with step-function coefficients is stable in each interval 0 to t_1, t_1 to t_2, . . . is it necessarily stable for all positive t? (For the second order equation (5–41), this comes down to the question of whether positiveness of the coefficients guarantee stability.) It is easily shown by examples (Prob. 5 below) that this is *not* the case. The difficulty arises because the real parts of the characteristic roots, while remaining negative, can approach zero as t increases. In general, if the coefficients remain safely within the "region of stability," never approaching the edge, then stability of the equation can be assured.

PROBLEMS

1. For each of the following equations verify stability and find an output by breaking the operation down into several stages, each governed by a first or second order equation.

(a) $(D + 1)(3D + 1)x = \sin 5t.$

(b) $(D^2 + D + 1)(2D^2 + 3D + 1)x = 7t + 5.$

(c) $(D^2 + 3D + 1)^{10}x = 4t + 2.$

(d) $(D^2 + \frac{1}{2}D + 1)^{10}x = \sin t.$

(e) $(D^2 + D + 1)^2x = \sum_{n=1}^{\infty} \frac{\sin nt}{n^2}.$

2. (a) Without finding the characteristic roots exactly, verify stability of the equation

$$(D^3 + 9D^2 + 23D + 14)x = F(t).$$

(b) In the differential equation let $F(t) = 14A_i \sin \omega t$, so that the input is $A_i \sin \omega t$. Obtain an output in the form $A_o \sin (\omega t - \alpha)$. Graph A_o/A_i as a function of ω.

3. (a) Find the weighting function for the equation

$$\tfrac{1}{10}(D + 2)(D^2 + 2D + 5)x = F(t)$$

and verify that its graph is as in Fig. 5–22.

(b) Apply the result of part (a) to obtain graphically the output for the function $F(t)$ of Fig. 5–22.

4. Prove: if Eq. (5–35) is stable and $W(t)$ is the associated weighting function, then Eq. (5–40) is valid. [*Hint:* Let $F(t) = 0$ for $t < 0$, $F(t) = 1$ for $t \geqq 0$. Show that the output $x(t)$ of Eq. (5–39) equals $\int_0^t W(v)\, dv$ for $t > 0$. Let $t \to +\infty$ and use the known form of the output as response to a step-function input.]

5. (a) Verify that for $t > 0$, the general solution of the equation $6t^2 D^2 x + tDx + x = 0$ is $x = c_1 t^{1/3} + c_2 t^{1/2}$, and hence conclude that the equation $6t^2 D^2 x + tDx + x = F(t)$ is unstable.

(b) Discuss the behavior of the characteristic roots of the equation $(6a^2 D^2 + aD + 1)x = 0$, where a is independent of t, as $a \to +\infty$.

ANSWERS

1. (a) $0.013 \sin (5t - 2.9)$,   (b) $7(t - 4) + 5$,   (c) $4(t - 30) + 2$,
(d) $-2^{10} \sin t$,

(e) $\displaystyle\sum_{n=1}^{\infty} \frac{\sin [nt - 2 \tan^{-1} \{n/(1 - n^2)\}]}{n^2(n^4 - n^2 + 1)}$.

2. (b) Output is $14A_i(\omega^6 + 35\omega^4 + 277\omega^2 + 196)^{-1/2} \sin (\omega t - \alpha)$, where α is the polar angle of $14 - 9\omega^2 + i(23\omega - \omega^3)$.

3. (a) $[2e^{-2t} + e^{-t}(\sin 2t - 2 \cos 2t)]$.

5–10 Applications of linear differential equations in mechanics. Here and in the following sections we shall give some examples of physical applications of linear differential equations.

A. *Forced vibrations of a spring-mass system.* A mass m is free to move along a line and is subject to a spring force $-k^2 x$, a frictional force $-bv$ (where $v = dx/dt$), and an external force $F(t)$, as shown in Fig. 5–23. The

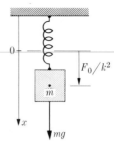

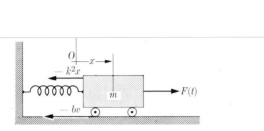

FIG. 5–23. Spring-mass system.   FIG. 5–24. Vertical
                                                spring-mass system.

coordinate x measures the displacement of the mass from its equilibrium
position under the spring force. Newton's second law gives the equation

$$m \frac{d^2 x}{dt^2} = -k^2 x - b \frac{dx}{dt} + F(t)$$

or

$$(mD^2 + bD + k^2)x = F(t). \qquad (5\text{–}42)$$

We assume m, b, and k^2 to be positive constants. Accordingly, Eq. (5–42)
is a linear differential equation of second order, to which the discussions of
Sections 5–5 and 5–6 are applicable. The *input* is $F(t)/k^2$, and

$$h = \frac{1}{2} \frac{b}{m}, \qquad \lambda^2 = \frac{k^2}{m}, \qquad \eta = \frac{1}{2} \frac{b}{k\sqrt{m}}. \qquad (5\text{–}43)$$

If, in particular, $h < \lambda$ and $F(t) = 0$, then the motions are the damped
oscillations of Fig. 5–3; in the limiting case $h = 0$, the oscillations are
purely sinusoidal and the mass moves in *simple harmonic motion*.

If $F(t)$ is constant, $F(t) \equiv F_0$, then the input is F_0/k^2 and the output is
F_0/k^2 plus a transient. This is illustrated by the oscillations of a vertical
spring (Fig. 5–24), in which the applied force F_0 is mg, the force of gravity.
The constant force simply displaces the equilibrium point about which the
oscillations take place.

If $F(t)$ is sinusoidal, $(F = B \sin \omega t)$, then the input is $A_i \sin \omega t$, where
$A_i = B/k^2$, and the output is $A_o \sin (\omega t - \alpha)$ plus a transient, as in
Section 5–6. Here the oscillations of the mass are forced; such motions
can be achieved physically in a variety of ways: for example, by forcing
the wall to which the spring is attached (Fig. 5–23) to oscillate sinusoidally.
If η is very small and ω/λ is properly chosen, the output amplitude can be
very large; that is, we can achieve *resonance*, a very common phenomenon
which is often called *sympathetic vibration*.

B. *Coupled springs.* Equations of order higher than two arise in mechanics most often by elimination in a system of simultaneous second order

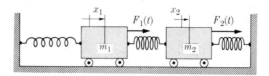

FIG. 5–25. Coupled springs.

equations. Consider the case of oscillations of two masses coupled by springs, as in Fig. 5–25. We assume that all three springs have natural length L and spring constant k^2. We let x_1, x_2 measure the displacements of m_1, m_2 from their equilibrium positions, at which the springs all have length L_1. The mass m_1 is subject to a spring force $-k^2(L_1 + x_1 - L)$ to the left and a spring force $+k^2(L_1 + x_2 - x_1 - L)$ to the right; we also assume a frictional force $-b_1\, dx_1/dt$ and an applied force $F_1(t)$. A similar analysis applies to m_2, and we obtain the differential equations

$$m_1 \frac{d^2 x_1}{dt^2} = -k^2(L_1 + x_1 - L)$$

$$+ k^2(L_1 + x_2 - x_1 - L) - b_1 \frac{dx_1}{dt} + F_1(t),$$

$$m_2 \frac{d^2 x_2}{dt^2} = -k^2(L_1 + x_2 - x_1 - L)$$

$$+ k^2(L_1 - x_2 - L) - b_2 \frac{dx_2}{dt} + F_2(t),$$

which can be written

$$(m_1 D^2 + b_1 D + 2k^2)x_1 - k^2 x_2 = F_1(t),$$
$$(m_2 D^2 + b_2 D + 2k^2)x_2 - k^2 x_1 = F_2(t). \tag{5–44}$$

Hence we have a system of two simultaneous second order differential equations.

We now obtain a single equation for x_1 by solving the first equation for x_2 and substituting in the second equation:

$$x_2 = (1/k^2)[(m_1 D^2 + b_1 D + 2k^2)x_1 - F_1(t)], \tag{5–45}$$

$$(1/k^2)(m_2 D^2 + b_2 D + 2k^2)[(m_1 D^2 + b_1 D + 2k^2)x_1 - F_1(t)]$$
$$- k^2 x_1 = F_2(t). \tag{5–46}$$

We thus obtain a fourth order equation for x_1:

$$(a_0 D^4 + a_1 D^3 + a_2 D^2 + a_3 D + a_4)x_1 = F(t), \tag{5–47}$$

$$a_0 = m_1 m_2, \qquad a_1 = m_1 b_2 + m_2 b_1,$$

$$a_2 = b_1 b_2 + 2k^2(m_1 + m_2),$$

$$a_3 = 2k^2(b_1 + b_2), \qquad a_4 = 3k^4, \tag{5-48}$$

$$F(t) = (m_2 D^2 + b_2 D + 2k^2)F_1(t) + k^2 F_2(t).$$

When $x_1(t)$ has been found from (5–47), then $x_2(t)$ can be obtained from (5–45).

It can be proved that (5–47) is always stable; we verify this for the case where $m_1 = m_2$ and $b_1 = b_2$. Then (5–46) can be written

$$[(m_1 D^2 + b_1 D + 2k^2)^2 - k^4]x_1 = F(t).$$

Hence the characteristic equation is

$$(m_1 r^2 + b_1 r + 2k^2)^2 - k^4 = 0$$

or

$$(m_1 r^2 + b_1 r + 3k^2)(m_1 r^2 + b_1 r + k^2) = 0. \tag{5-49}$$

Since m_1, b_1, and k^2 are assumed to be positive, the characteristic roots must have negative real parts.

Equation (5–47) is, then, stable, and we can discuss the relationship between inputs and outputs as in Sections 5–5, 5–6. In particular, if $F_1(t) = B \sin \omega t$ and $F_2(t) \equiv 0$, then $x_1(t)$ will follow a sinusoidal oscillation $A_o \sin \omega(t - \alpha)$, and a similar expression is obtained for $x_2(t)$.

C. *Simple pendulum.* For a simple pendulum (Fig. 5–26) we measure the position of the swinging arm by an angular coordinate θ. The angular component of acceleration is then $L d^2\theta/dt^2$, and the component of gravitational force in the same direction is $-mg \sin \theta$. If we ignore other forces, we obtain the differential equation

$$mL \frac{d^2\theta}{dt^2} = -mg \sin \theta. \tag{5-50}$$

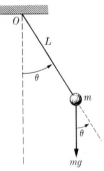

This equation is nonlinear but can be reduced to a first order equation and solved explicitly (Prob. 7 following Section 2–11). Here we are considering *small oscillations* of the pendulum, that is, motions in which θ remains small. Thus we can make the approximation $\sin \theta = \theta$ and replace (5–50) by the linear equation

$$mL \frac{d^2\theta}{dt^2} + mg\theta = 0. \tag{5-51}$$

Fig. 5–26. Simple pendulum.

Here $h = 0$ and $\lambda^2 = g/L$, so that the solutions are sinusoidal:

$$\theta = c_1 \cos \lambda t + c_2 \sin \lambda t, \qquad (5\text{-}52)$$

and have period

$$T = \frac{2\pi}{\lambda} = 2\pi \sqrt{\frac{L}{g}}. \qquad (5\text{-}53)$$

If we add frictional and driving forces, we obtain the same type of equation as for the forced vibrations of a spring-mass system.

5–11 Applications in elasticity.   The problems of elasticity lead, in general, to fourth order *partial* differential equations. In special circumstances these equations can be replaced by ordinary differential equations, as the following examples illustrate.

A. *Beam on elastic foundation.* Consider a beam whose axis is the x-axis and subject to deflections y in the y-direction (Fig. 5–27). We assume the beam to be subject to a load $F(x)$ (force per unit length) and to rest on an elastic foundation which exerts spring-type restoring forces proportional to deflection; the restoring force per unit length is $-k^2 y$. In the theory of elasticity it is shown that when the beam is in equilibrium, the deflection $y(x)$ satisfies the differential equation

$$EI \frac{d^4 y}{dx^4} + k^2 y = F(x), \qquad (5\text{-}54)$$

where E is a constant, *Young's modulus*, and I is the moment of inertia about the x-axis of a cross section of the beam perpendicular to that axis. We assume I also to be constant.

Although Eq. (5–54) suggests the input-output interpretation, there is a basic difference between this equation and those previously discussed. Here particular solutions are specified not by initial conditions but by boundary conditions. For example, if the beam is clamped at two points x_1, x_2, then at these points $y = 0$ and $dy/dx = 0$.

For derivation of Eq. (5–54) and discussion of its applications, the reader is referred to pp. 267–273 of Reference 9 at the end of this chapter.

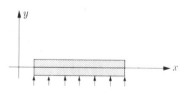

FIG. 5–27. Beam on elastic foundation.

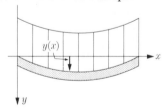

FIG. 5–28. Suspension bridge.

B. *The suspension bridge* can be considered as a beam partly supported by a cable (Fig. 5–28). Under various simplifying assumptions the following differential equation is obtained:

$$EI \frac{d^4y}{dx^4} - (H + h) \frac{d^2y}{dx^2} = F(x) - G(x) \frac{h}{H} \qquad (5\text{--}55)$$

(see pp. 277–279 of Reference 9). In Eq. (5–55) H is the horizontal tension in the cable due to dead weight, h is the extra tension caused by addition of a live load $F(x)$, and $G(x)$ is the dead load of the system. The constants E, I, H are considered known, whereas h must be chosen to satisfy an auxiliary condition. If the cable is considered to be inextensible, the auxiliary condition takes the form

$$\int_0^L G(x)y(x)\, dx = 0. \qquad (5\text{--}56)$$

In addition, boundary conditions must be imposed at the ends of the beam.

PROBLEMS

1. A 50-gm mass can stretch a spring 2 cm under its own weight. With what frequency will the spring oscillate when supporting the weight (ignore friction)?

2. If the spring-mass system of Prob. 1 is oscillating with an amplitude of 4 cm, find the maximum velocity.

3. In a frictionless spring-mass system the mass is 100 gm and the natural frequency is 2 rad/sec. The system is at rest at time $t = 0$ and a force $500 \cos 2t$ dynes (t in sec) is applied. For what value of t will the displacement first equal 1 cm?

4. In a system of coupled masses (Fig. 5–25) neglect friction and assume no outside forces are applied.

(a) Show that the characteristic roots of Eq. (5–47) are pure imaginary. [*Hint:* Show that the roots satisfy the condition

$$r^2 = \frac{k^2}{m_1 m_2}\left(-(m_1 + m_2) \pm \sqrt{(m_1 - m_2)^2 + m_1 m_2}\right)$$

and that the quantity on the right is negative.]

(b) Let $\pm\lambda_1 i$, $\pm\lambda_2 i$ be the characteristic roots obtained in part (a). Let $x_1 = \sin \lambda_1 t$ and find x_2. Describe the resulting motion.

5. The motion of a particle in the xy-plane under the influence of a central force *proportional to distance* leads to the differential equations

$$m \frac{d^2x}{dt^2} = -k^2 x, \qquad m \frac{d^2y}{dt^2} = -k^2 y.$$

Show that (except for certain degenerate cases) the particle moves in an ellipse with center at the origin.

6. Find the deflection of a beam on an elastic foundation, clamped at $x = -L$ and $x = L$ and subject to a load $F(x) = F_0 \cos (\pi x/L)$.

ANSWERS

1. 22.1 rad/sec. 2. 88.5 cm/sec. 3. 0.8 sec. 4. (b) $x_2 = \text{const} \times \sin \lambda_1 t$, so that the two masses oscillate either in phase or in opposite phase. It can be shown that for λ_1 one case arises and for λ_2 the other arises. (See Section 6–12.)

6. $y = [A/(ab + cd)]$

$\times \, [(ab + cd) \cos \alpha x + (ad + bc) \cosh \delta x \cos \delta x + (bc - ad) \sinh \delta x \sin \delta x]$,

where $A = F_0/[EI(\beta^4 + \alpha^4)]$, $\beta = (k^2/EI)^{1/4}$, $\alpha = \pi/L$, $\delta = \beta/\sqrt{2}$, $a = \sinh \epsilon$, $b = \cosh \epsilon$, $c = \sin \epsilon$, $d = \cos \epsilon$, $\epsilon = \delta L = \beta L/\sqrt{2}$.

5–12 Applications to electric circuits. We consider a simple L-R-C circuit; that is, a circuit containing an inductance L (measured in henries), resistance R (ohms), capacitance C (farads), and a driving electromotive force $\mathcal{E}$ (volts). See Fig. 5–29. The current I is measured in amperes, the charge q on the capacitor in coulombs, and $\mathcal{E}$, q, and I are all considered to be functions of time t (sec).

The differential equation which governs the circuit is based on *Kirchhoff's law:* the total potential drop around a closed circuit is zero. The drop across the inductor is $L(dI/dt)$, across the resistor is RI, across the capacitor is q/C, across the driving emf is $-\mathcal{E}$; all potential drops are measured in volts, and with respect to the direction in which I is measured positively. The equation obtained is the following:

$$L \frac{dI}{dt} + RI + \frac{q}{C} = \mathcal{E}. \qquad (5\text{--}57)$$

In addition, the current I is equal to the rate of change of q:

$$\frac{dq}{dt} = I. \qquad (5\text{--}58)$$

FIG. 5–29. L-R-C circuit.

The coefficients L, R, C are all assumed to be positive constants. We can eliminate q by differentiating (5–57) and using (5–58):

$$L \frac{d^2 I}{dt^2} + R \frac{dI}{dt} + \frac{I}{C} = \frac{d\mathcal{E}}{dt}. \qquad (5\text{--}59)$$

We can also substitute $I = dq/dt$ in (5–57) to obtain a differential equation for q:

$$L \frac{d^2 q}{dt^2} + R \frac{dq}{dt} + \frac{q}{C} = \mathcal{E}. \qquad (5\text{--}60)$$

Either (5–59) or (5–60) can be taken as the basic differential equation of the circuit.

Most commonly, initial conditions are given in the form of values of q and I at time t_0; by virtue of (5–58), this is the same as giving initial values of q and dq/dt; by (5–57) we can also find the initial value of dI/dt from the given values of q and I and knowledge of $\mathcal{E}$ as a function of t.

Equations (5–59) and (5–60) are of the form of second order equations with variable input, and hence the results of Sections 5–4 through 5–7 are applicable. Since all coefficients are positive, the equations are stable, and for both equations,

$$h = \frac{R}{2L}, \quad \lambda^2 = \frac{1}{CL}, \quad \eta = \frac{h}{\lambda} = \frac{R}{2}\sqrt{\frac{C}{L}} = \sqrt{\frac{CR^2}{4L}}. \quad (5\text{–}61)$$

Accordingly, the transients are

damped oscillations when $\eta < 1$ or $CR^2 < 4L$,

critically damped when $\eta = 1$ or $CR^2 = 4L$,

overcritically damped when $\eta > 1$ or $CR^2 > 4L$.

If the emf $\mathcal{E}$ is constant and the capacitor is removed ($1/C$ replaced by 0), then Eq. (5–57) has the steady-state solution

$$I = \mathcal{E}/R \quad \text{(Ohm's law)}. \quad (5\text{–}62)$$

When the capacitor is reinserted, we use Eq. (5–60); in the steady state, $q = C\mathcal{E} = $ const and $I = dq/dt = 0$; the driving emf is simply balanced by a corresponding drop in potential across the capacitor.

If the emf $\mathcal{E}$ equals kt (ramp input), then we apply Eq. (5–60). In the notation of Section 5–5 the input is kCt and $a_1 = CR$. Hence the steady-state output q is $kC(t - CR)$, while the steady-state current I is $dq/dt = kC$.

If $\mathcal{E} = \mathcal{E}_0 \sin \omega t$, then we apply the results of Section 5–6. The input in Eq. (5–60) is $\mathcal{E}_0 C \sin \omega t$ and the steady-state output is $q = A_o \sin(\omega t - \alpha)$, where

$$\frac{A_o}{A_i} = \frac{A_o}{\mathcal{E}_0 C} = \frac{1}{[(1 - \nu^2)^2 + 4\eta^2\nu^2]^{1/2}},$$

$$\alpha = \tan^{-1}\frac{2\eta\nu}{1 - \nu^2}, \quad \nu = \frac{\omega}{\lambda} = \omega\sqrt{CL}. \quad (5\text{–}63)$$

The steady-state current is

$$I = \frac{dq}{dt} = \omega A_o \cos(\omega t - \alpha) = \omega A_o \sin\left(\omega t - \alpha + \frac{\pi}{2}\right). \quad (5\text{–}64)$$

From (5–63) and (5–61)

$$\omega A_o \doteq \frac{\omega \mathcal{E}_0 C}{[(1 - \omega^2 CL)^2 + R^2 C^2 \omega^2]^{1/2}} \tag{5–65}$$

$$= \frac{\mathcal{E}_0}{[\{(\omega C)^{-1} - L\omega\}^2 + R^2]^{1/2}} .$$

The quantity

$$Z = \left[\left(\frac{1}{\omega C} - L\omega \right)^2 + R^2 \right]^{1/2} \tag{5–66}$$

is known as the *impedance*. From Eqs. (5–64) through (5–66)

$$I = \frac{\mathcal{E}_0}{Z} \sin \left(\omega t - \alpha + \frac{\pi}{2} \right) . \tag{5–67}$$

Thus the oscillatory emf of amplitude $\mathcal{E}_0$ leads to an oscillatory current of amplitude $\mathcal{E}_0/Z$. This can be regarded as a generalization of Ohm's law (Eq. 5–62), with the impedance Z replacing the resistance R. The current also lags the emf in phase; the lag is

$$\beta = \alpha - \frac{\pi}{2} = \tan^{-1} \frac{2\eta\nu}{1 - \nu^2} - \frac{\pi}{2} = \tan^{-1} \frac{\nu^2 - 1}{2\eta\nu} .$$

Since $0 < \alpha < \pi$, β lies between $-\pi/2$ and $\pi/2$; in particular, β can be zero (current and emf *in phase*) and negative (current *leading* the emf). From Eqs. (5–61) and (5–63) we can write

$$\beta = \tan^{-1} \frac{\omega^2 CL - 1}{RC\omega} = \tan^{-1} \frac{L\omega - (C\omega)^{-1}}{R} . \tag{5–68}$$

The fact that the spring-mass system of Section 5–10 and the *L-R-C* circuit are governed by the same type of second order differential equation enables us to study one problem by means of the other. For example, the behavior of a given spring-mass system can be studied experimentally by designing an analogous electric circuit. The principle involved here has been generalized widely in the form of the electronic analog computer, which is a very flexible electric system capable of imitating a wide variety of physical systems.

5–13 Applications to electric networks. Electric networks can generally be described by systems of simultaneous differential equations. By elimination these can be reduced to a single equation for one unknown.

We illustrate the procedure here in a simple case; a treatment of the general problem depends on the theory of simultaneous differential equations as developed in Chapter 6.

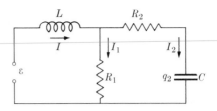

FIG. 5–30. Electric network.

Consider the network of Fig. 5–30. We can apply Kirchhoff's law concerning potential drops to each closed circuit contained in the configuration. In addition, a second law of Kirchhoff can be applied: *the total current entering each junction (node) is zero.* For each of the two junctions of the network of Fig. 5–30, this gives the relation

$$I = I_1 + I_2. \tag{5–69}$$

The first law gives

$$L \frac{dI}{dt} + R_1 I_1 = \mathcal{E}, \tag{5–70}$$

$$R_2 I_2 + \frac{q_2}{C} - R_1 I_1 = 0, \tag{5–71}$$

$$L \frac{dI}{dt} + R_2 I_2 + \frac{q_2}{C} = \mathcal{E}; \tag{5–72}$$

the first equation corresponds to the left-hand loop, the second to the right-hand loop, and the third to the exterior circuit. However, the third equation is simply the result of adding the first two, so that only two independent equations are obtained. Finally, q_2 and I_2 are related:

$$\frac{dq_2}{dt} = I_2. \tag{5–73}$$

It is the system of equations (5–69) through (5–73) that governs the behavior of the network.

We now proceed to obtain an equation for I alone by differentiating (5–69), (5–70), and (5–71) and using (5–73):

$$\frac{dI}{dt} = \frac{dI_1}{dt} + \frac{dI_2}{dt}, \tag{5–74}$$

$$L \frac{d^2 I}{dt^2} + R_1 \frac{dI_1}{dt} = \frac{d\mathcal{E}}{dt}, \tag{5–75}$$

$$R_2 \frac{dI_2}{dt} + \frac{I_2}{C} - R_1 \frac{dI_1}{dt} = 0. \tag{5–76}$$

It is now a matter of algebra to eliminate the four unknowns I_1, dI_1/dt,

I_2, dI_2/dt from the five equations (5–69), (5–70), (5–74), (5–75), and (5–76). Thus we solve (5–69) and (5–74) for I_2 and dI_2/dt and substitute in (5–76):

$$R_2 \left(\frac{dI}{dt} - \frac{dI_1}{dt} \right) + \frac{1}{C} (I - I_1) - R_1 \frac{dI_1}{dt} = 0. \tag{5–77}$$

We solve (5–70) and (5–75) for I_1 and dI_1/dt and substitute in (5–77):

$$R_2 \left[\frac{dI}{dt} - \frac{1}{R_1} \left(\frac{d\mathcal{E}}{dt} - L \frac{d^2I}{dt^2} \right) \right] + \frac{1}{C} \left[I - \frac{1}{R_1} \left(\mathcal{E} - L \frac{dI}{dt} \right) \right]$$

$$- \left(\frac{d\mathcal{E}}{dt} - L \frac{d^2I}{dt^2} \right) = 0.$$

This is the desired equation for I. It can be written as the second order equation

$$L \left(1 + \frac{R_2}{R_1} \right) \frac{d^2I}{dt^2} + \left(R_2 + \frac{L}{CR_1} \right) \frac{dI}{dt} + \frac{I}{C} = \frac{\mathcal{E}}{CR_1} + \left(1 + \frac{R_2}{R_1} \right) \frac{d\mathcal{E}}{dt},$$

$$\tag{5–78}$$

and it can be analyzed as in the previous section.

PROBLEMS

1. In the L-R-C circuit of Fig. 5–29 let $L = 10$ henries, $R = 100$ ohms, $C = (500)^{-1}$ farad, $\mathcal{E} = 1$ volt (const). If $I = 0$ and $q = 0$ for $t = 0$, find (a) I and q as functions of t, (b) the maximum values of I and q for $t > 0$.

2. In Prob. 1 we can consider the circuit to be functioning for all t, with $\mathcal{E} = 0$ for $t < 0$ and $\mathcal{E} = 1$ for $t \geqq 0$, so that $\mathcal{E}$ has a jump discontinuity at $t = 0$. Equation (5–59) is then meaningless, since $d\mathcal{E}/dt = 0$ for $t < 0$ and $t > 0$, but $d\mathcal{E}/dt = \infty$ for $t = 0$. However, Eq. (5–60) can be solved for q, with initial values $q = 0$ and $dq/dt = 0$ for $t = 0$: $q = 0$ for $t < 0$ and $q = [1 - e^{-5t} (\cos 5t + \sin 5t)]/500$ for $t \geqq 0$ (as in Prob. 1). Show that $dq/dt = I$ is well defined and continuous, even at $t = 0$, that $d^2q/dt^2 = dI/dt$ has a jump discontinuity at $t = 0$, and that $d^3q/dt^3 = d^2I/dt^2 = \infty$ at $t = 0$. (The function $I(t)$ can be interpreted as a solution of Eq. (5–59) in which the right member is $\delta(t)$, the *Dirac delta function* or *unit impulse function*.)

3. In the L-R-C circuit of Fig. 5–29 let $\mathcal{E} = \mathcal{E}_0 e^{i\omega t}$. (a) Obtain the steady-state current in the form $I_0 e^{i(\omega t - \beta)}$. The ratio $\mathcal{E}/I = Z = (\mathcal{E}_0/I_0) e^{i\beta}$ is known as the *complex impedance*. (b) Show that the impedance of Eq. (5–66) is the absolute value of the complex impedance, $|x + iy| = (x^2 + y^2)^{1/2}$.

4. For the network of Fig. 5–30 discuss the steady-state solutions for I, I_1, I_2 in each of the following cases:

(a) $\mathcal{E} = \text{const} = \mathcal{E}_0$,

(b) $\mathcal{E} = bt$ (b = const),

(c) $\mathcal{E} = \mathcal{E}_0 e^{i\omega t}$.

5. For the network of Fig. 5–31 obtain a differential equation for I alone.

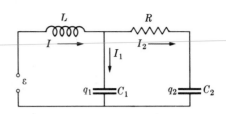

FIGURE 5–31

ANSWERS

1. (a) $q = [1 - e^{-5t}(\cos 5t + \sin 5t)]/500$, $I = (e^{-5t} \sin 5t)/50$; (b) max q is 2.08×10^{-3} coulomb, max I is 6.5×10^{-3} amp.

3. (a) $I = \mathcal{E}_0 e^{i\omega t}/\{R + [L\omega - (C\omega)^{-1}]i\} = I_0 e^{i(\omega t - \beta)}$, $I_0 = \mathcal{E}_0 \div \{R^2 + [L\omega - (C\omega)^{-1}]^2\}^{1/2}$, β = polar angle of $Z = R + [L\omega - (C\omega)^{-1}]i$.

4. (a) $I = \mathcal{E}_0/R_1 = I_1$, $I_2 = 0$; (b) $I = bR_1^{-1}t + bC - bLR_1^{-2}$, $I_1 = btR_1^{-1} - bLR_1^{-2}$, $I_2 = bC$; (c) $I = \mathcal{E}_0 e^{i\omega t}[1 + i\omega C(R_1 + R_2)]/Z_0$, $I_1 = \mathcal{E}_0 e^{i\omega t}[1 + i\omega CR_2]/Z_0$, $I_2 = \mathcal{E}_0 e^{i\omega t}[i\omega CR_1]/Z_0$, $Z_0 = R_1 - LC\omega^2(R_1 + R_2) + i\omega(L + CR_1R_2)$.

5. $[LRC_1C_2D^3 + L(C_1 + C_2)D^2 + RC_2D + 1]I = [RC_1C_2D^2 + (C_1 + C_2)D]\mathcal{E}$.

SUGGESTED REFERENCES

1. ANDRONOW, A., and CHAIKIN, C. E., *Theory of Oscillations*. Princeton, N. J.: Princeton University Press, 1949.

2. BELLMAN, RICHARD, *Stability Theory of Differential Equations*. New York: McGraw-Hill, 1953.

3. CODDINGTON, EARL A., and LEVINSON, NORMAN, *Theory of Ordinary Differential Equations*. New York: McGraw-Hill, 1955.

4. DRAPER, CHARLES S., McKAY, WALTER, and LEES, SIDNEY, *Instrument Engineering*, Vol. 2. New York: McGraw-Hill, 1953.

5. GUILLEMIN, E. A., *Mathematics of Circuit Analysis*. New York: John Wiley and Sons, Inc., 1949.

6. KAPLAN, WILFRED, *Operational Methods for Linear Systems*. Reading, Mass.: Addison-Wesley (to appear).

7. KAPLAN, WILFRED, "Stability Theory," *Proceedings of the Symposium on Non-linear Circuit Analysis*, Vol. VI, pp. 3–21. New York: Polytechnic Institute of Brooklyn, 1957.

8. LAWDEN, DEREK F., *Mathematics of Engineering Systems*. New York: John Wiley and Sons, Inc., 1954.

9. VON KÁRMÁN, T., and BIOT, M. A., *Mathematical Methods in Engineering*. New York: McGraw-Hill, 1940.

CHAPTER 6

SIMULTANEOUS LINEAR DIFFERENTIAL EQUATIONS

6-1 General principles. In this chapter we consider systems of n first order linear differential equations in n unknowns. First we consider them in a special *basic form*, which we illustrate for $n = 2$:

$$\frac{dx}{dt} = a_1 x + b_1 y + f_1(t),$$

$$\frac{dy}{dt} = a_2 x + b_2 y + f_2(t); \tag{6-1}$$

and for $n = 3$:

$$\frac{dx}{dt} = a_1 x + b_1 y + c_1 z + f_1(t),$$

$$\frac{dy}{dt} = a_2 x + b_2 y + c_2 z + f_2(t), \tag{6-2}$$

$$\frac{dz}{dt} = a_3 x + b_3 y + c_3 z + f_3(t).$$

For n equations the unknowns can be denoted by $x_1, \ldots, x_n$, and the equations have the form

$$\frac{dx_i}{dt} = a_{i1} x_1 + \cdots + a_{in} x_n + f_i(t) \qquad (i = 1, \ldots, n). \tag{6-3}$$

More general systems can also be reduced to this form.

The coefficients $a_1, b_1, \ldots, a_{i1}, \ldots a_{in}$ may depend on t, but for most of this chapter they will be *constant*.

Systems of equations are discussed in Section 1–7. Let us recall the basic facts as they relate to the systems (6–1), (6–2), or (6–3). A *solution* of (6–1) is a *pair* of functions $x(t)$, $y(t)$ which together satisfy the equations identically over some interval; if the coefficients a_1, b_1, a_2, b_2 and the functions $f_1(t)$, $f_2(t)$ are continuous over an interval $t_1 < t < t_2$, then there is a unique solution $x(t)$, $y(t)$ with prescribed initial values $x(t_0) = x_0$, $y(t_0) = y_0$ at $t = t_0$, where $t_1 < t_0 < t_2$. Similarly, a solution of (6–2) is a *triple* of functions $x(t)$, $y(t)$, $z(t)$ which satisfy the equations, and a solution of (6–3) is an n-tuple of functions $x_1(t), \ldots, x_n(t)$ which satisfy the equations; when the coefficients and the $f_i(t)$ are continuous, a unique solution satisfying prescribed initial conditions exists.

For the remainder of this section we assume that $n = 3$, and hence consider only the equations (6–2); the extension of the results to general values of n will be evident.

If the functions $f_1(t)$, $f_2(t)$, $f_3(t)$ are identically 0, the equations (6–2) are termed *homogeneous;* otherwise they are *nonhomogeneous*. If the equations

218

(6–2) are nonhomogeneous, we can replace the $f_i(t)$ by 0 and obtain the *related homogeneous system*

$$\frac{dx}{dt} = a_1 x + b_1 y + c_1 z,$$

$$\frac{dy}{dt} = a_2 x + b_2 y + c_2 z, \qquad (6\text{–}4)$$

$$\frac{dz}{dt} = a_3 x + b_3 y + c_3 z.$$

A set of k solutions of (6–4) is said to be *linearly independent* over an interval if no linear combination of the solutions, with coefficients not all 0, is identically 0 over the interval. We can denote the k solutions (which form k *triples*) by the following: $x_1(t)$, $y_1(t)$, $z_1(t)$; $x_2(t)$, $y_2(t)$, $z_2(t)$; ... ; $x_k(t)$, $y_k(t)$, $z_k(t)$. The linear combination is then itself a triple: $C_1 x_1(t) + C_2 x_2(t) + \cdots + C_k x_k(t)$, $C_1 y_1 + \cdots + C_k y_k$, $C_1 z_1 + \cdots + C_k z_k$. Linear independence of the solutions is equivalent to the condition: if

$$C_1 x_1(t) + \cdots + C_k x_k(t) \equiv 0,$$
$$C_1 y_1(t) + \cdots + C_k y_k(t) \equiv 0, \qquad (6\text{–}5)$$
$$C_1 z_1(t) + \cdots + C_k z_k(t) \equiv 0$$

over the given interval, then $C_1 = 0, C_2 = 0, \ldots, C_k = 0$.

The fact that the triples are solutions of (6–4) is not essential to the above definition; it applies equally well to any set of k triples. However, if the triples are solutions of (6–4), then each linear combination

$$x(t) = C_1 x_1(t) + \cdots + C_k x_k(t), \qquad y(t) = C_1 y_1(t) + \cdots,$$
$$z(t) = C_1 z_1(t) + \cdots \qquad (6\text{–}6)$$

forms a new triple which is itself a solution of (6–4). For example, since

$$\frac{dx_i}{dt} = a_1 x_i + b_1 y_i + c_1 z_i$$

for $i = 1, \ldots, k$, we conclude from (6–6) that

$$\frac{dx}{dt} = C_1 \frac{dx_1}{dt} + \cdots + C_k \frac{dx_k}{dt}$$
$$= C_1(a_1 x_1 + b_1 y_1 + c_1 z_1) + \cdots + C_k(a_1 x_k + b_1 y_k + c_1 z_k)$$
$$= a_1(C_1 x_1 + \cdots + C_k x_k) + b_1(C_1 y_1 + \cdots) + c_1(C_1 z_1 + \cdots)$$
$$= a_1 x + b_1 y + c_1 z.$$

Similarly, the other two equations are satisfied.

Let us suppose we have found *three* linearly independent solutions of (6–4). We then see that the corresponding linear combinations

$$x(t) = C_1 x_1(t) + C_2 x_2(t) + C_3 x_3(t),$$

$$y(t) = C_1 y_1(t) + C_2 y_2(t) + C_3 y_3(t), \qquad (6\text{--}7)$$

$$z(t) = C_1 z_1(t) + C_2 z_2(t) + C_3 z_3(t)$$

provide the *general solution* of (6–4) over the interval for which the solutions exist. If initial values x_0, y_0, z_0 at t_0 are given, then to determine the C's, we are led to the equations

$$x_0 = C_1 x_1(t_0) + C_2 x_2(t_0) + C_3 x_3(t_0),$$

$$y_0 = C_1 y_1(t_0) + C_2 y_2(t_0) + C_3 y_3(t_0),$$

$$z_0 = C_1 z_1(t_0) + C_2 z_2(t_0) + C_3 z_3(t_0).$$

These three equations in three unknowns C_1, C_2, C_3 can be solved for C_1, C_2, C_3, provided the determinant of coefficients is not zero. It turns out that linear independence implies that this determinant cannot be zero; hence the C's can always be chosen to satisfy initial conditions (Prob. 6 below).

We can now state a theorem that parallels the fundamental theorem of Section 4–3.

THEOREM 1. *Let the system (6–2) be given, with the related homogeneous system (6–4). Let the coefficients $a_1(t)$, $b_1(t)$, ..., and the functions $f_1(t), f_2(t), f_3(t)$ be continuous for $t_1 < t < t_2$.*

There exists a set of three linearly independent solutions of (6–4) over the given interval. If $x_i(t)$, $y_i(t)$, $z_i(t)$, $(i = 1, 2, 3)$, is any such set of linearly independent solutions, then (6–7) provides the general solution of (6–4) over the interval.

There exist solutions of the nonhomogeneous system (6–2) over the interval. If $x^*(t)$, $y^*(t)$, $z^*(t)$ is one such solution, and $x_i(t)$, $y_i(t)$, $z_i(t)$ are chosen as above, then

$$x = x^*(t) + C_1 x_1(t) + C_2 x_2(t) + C_3 x_3(t),$$

$$y = y^*(t) + C_1 y_1(t) + C_2 y_2(t) + C_3 y_3(t), \qquad (6\text{--}8)$$

$$z = z^*(t) + C_1 z_1(t) + C_2 z_2(t) + C_3 z_3(t)$$

is the general solution of (6–2) over the given interval.

The proof of this theorem is given in Chapter 12. The following examples illustrate its applications.

EXAMPLE 1. Consider the homogeneous system

$$\frac{dx}{dt} = 4x - 9y + 5z, \quad \frac{dy}{dt} = x - 10y + 7z, \quad \frac{dz}{dt} = x - 17y + 12z.$$

One solution is given by

$$x_1 \equiv e^t, \quad y_1 \equiv 2e^t, \quad z_1 \equiv 3e^t,$$

for substitution in the differential equations yields the identities

$$e^t = 4e^t - 18e^t + 15e^t, \quad 2e^t = e^t - 20e^t + 21e^t,$$

$$3e^t = e^t - 34e^t + 36e^t.$$

Similarly, the following triples are solutions:

$$x_2 \equiv e^{2t}, \quad y_2 \equiv 3e^{2t}, \quad z_2 \equiv 5e^{2t};$$

$$x_3 \equiv e^{3t}, \quad y_3 \equiv -e^{3t}, \quad z_3 \equiv -2e^{3t}.$$

The three triples are linearly independent for all t, for the relations (6–5) here become

$$C_1 e^t + C_2 e^{2t} + C_3 e^{3t} \equiv 0,$$

$$2C_1 e^t + 3C_2 e^{2t} - C_3 e^{3t} \equiv 0,$$

$$3C_1 e^t + 5C_2 e^{2t} - 2C_3 e^{3t} \equiv 0;$$

since e^t, e^{2t}, e^{3t} are linearly independent, each relation by itself implies that $C_1 = C_2 = C_3 = 0$. Hence the general solution is given by

$$x = C_1 e^t + C_2 e^{2t} + C_3 e^{3t},$$

$$y = 2C_1 e^t + 3C_2 e^{2t} - C_3 e^{3t}, \tag{6–9}$$

$$z = 3C_1 e^t + 5C_2 e^{2t} - 2C_3 e^{3t}.$$

EXAMPLE 2. We now consider the nonhomogeneous system

$$\frac{dx}{dt} = 4x - 9y + 5z + 1 + 13t, \quad \frac{dy}{dt} = x - 10y + 7z + 3 + 15t,$$

$$\frac{dz}{dt} = x - 17y + 12z + 2 + 26t.$$

The related homogeneous system is that of Example 1; hence the "complementary function" is given by the triple (6–9). We verify that a particular solution is given by the triple

$$x^* \equiv t, \quad y^* \equiv 3t, \quad z^* \equiv 2t.$$

Hence the general solution is given by

$$x = C_1e^t + C_2e^{2t} + C_3e^{3t} + t,$$
$$y = 2C_1e^t + 3C_2e^{2t} - C_3e^{3t} + 3t,$$
$$z = 3C_1e^t + 5C_2e^{2t} - 2C_3e^{3t} + 2t.$$

For equations with two unknowns such as (6–1) the general solution is constructed from *two* linearly independent solutions and depends on *two* arbitrary constants. For general n there are n constants and n linearly independent solutions. This is to be expected, since the system is of nth *order* (Section 1–7).

Methods for finding the linearly independent solutions of the homogeneous systems and for finding particular solutions of nonhomogeneous systems will be described later. As the above results indicate, there is a great similarity between the linear system of order n and the single linear equation of order n; this resemblance will continue to be evident in the methods described in the following sections.

PROBLEMS

1. (a) Verify that each of the pairs of functions $x = e^t$ and $y = e^t$, $x = 3e^{-t}$ and $y = 5e^{-t}$ is a solution of the system

$$\frac{dx}{dt} = 4x - 3y, \qquad \frac{dy}{dt} = 5x - 4y.$$

(b) Show that the two pairs of functions of part (a) are linearly independent for all t.

(c) With the aid of the results of parts (a) and (b), obtain the general solution of the system of part (a).

(d) Show that the pair $x = -\cos t - 4\sin t$, $y = -5\sin t$ is a solution of the system

$$\frac{dx}{dt} = 4x - 3y + 2\sin t, \qquad \frac{dy}{dt} = 5x - 4y.$$

(e) With the aid of the results of parts (c) and (d), obtain the general solution of the system of part (d).

(f) Find a solution of the system of part (d) such that $x = 1$, $y = 0$, when $t = 0$.

2. (a) Verify that each of the triples of functions

$$x = 3, \qquad y = 1, \qquad z = 2,$$
$$x = 4\sin t + 9\cos t, \qquad y = \sin t + 3\cos t, \qquad z = 2\sin t + 7\cos t,$$
$$x = 9\sin t - 4\cos t, \qquad y = 3\sin t - \cos t, \qquad z = 7\sin t - 2\cos t$$

is a solution of the homogeneous system related to the system

$$\frac{dx}{dt} = -9x + 19y + 4z + 1, \qquad \frac{dy}{dt} = -3x + 7y + z,$$

$$\frac{dz}{dt} = -7x + 17y + 2z.$$

(b) Show that the three triples of functions of part (a) are linearly independent for all t.

(c) Show that the triple $x = -3t$, $y = -t$, $z = -2t - 1$ is a solution of the system of part (a), and obtain the general solution.

(d) Find a solution of the system of part (a) such that $x = 6$, $y = 2$, $z = 3$ for $t = 0$.

3. Test each of the following sets of triples of functions for linear independence:

(a) $x = t$, $y = t - 1$, $z = 3t$; $x = t + 2$, $y = 2t$, $z = 3 + t$; $x = 3t + 2$, $y = 4t - 2$, $z = 7t + 3$;

(b) $x = e^t$, $y = e^{2t}$, $z = e^{3t}$; $x = e^t$, $y = 2e^{2t}$, $z = 5e^{3t}$; $x = e^t$, $y = 5e^{2t}$, $z = 3e^{3t}$.

4. (a) Let the general solution of Eqs. (6–4) be given by Eqs. (6–7). Let $\alpha_1, \alpha_2, \alpha_3, \beta_1, \beta_2, \beta_3, \gamma_1, \gamma_2, \gamma_3$ be constants such that

$$\begin{vmatrix} \alpha_1 & \beta_1 & \gamma_1 \\ \alpha_2 & \beta_2 & \gamma_2 \\ \alpha_3 & \beta_3 & \gamma_3 \end{vmatrix} \neq 0.$$

Let functions $x_4(t)$, $x_5(t)$, $x_6(t)$, $y_4(t)$, $\ldots$, $z_4(t)$, $\ldots$ be defined by the equations

$$x_{i+3}(t) = \alpha_i x_1(t) + \beta_i x_2(t) + \gamma_i x_3(t),$$

$$y_{i+3}(t) = \alpha_i y_1(t) + \beta_i y_2(t) + \gamma_i y_3(t),$$

$$z_{i+3}(t) = \alpha_i z_1(t) + \beta_i z_2(t) + \gamma_i z_3(t),$$

where $i = 1, 2, 3$. Show on the basis of Theorem 1 that the general solution of Eqs. (6–4) is also given by

$$x = C_1 x_4(t) + C_2 x_5(t) + C_3 x_6(t),$$

$$y = C_1 y_4(t) + C_2 y_5(t) + C_3 y_6(t),$$

$$z = C_1 z_4(t) + C_2 z_5(t) + C_3 z_6(t).$$

(b) Show on the basis of part (a) that the general solution of Example 1 in the text is given not only by (6–9) but also by

$$x = C_1(e^t + e^{3t}) + C_2 e^{2t} + C_3(e^t - e^{2t}),$$

$$y = C_1(2e^t - e^{3t}) + 3C_2 e^{2t} + C_3(2e^t - 3e^{2t}),$$

$$z = C_1(3e^t - 2e^{3t}) + 5C_2 e^{2t} + C_3(3e^t - 5e^{2t}).$$

5. Obtain for the following an equivalent system of first order equations of form (6–3):

(a) $\dfrac{d^2x}{dt^2} - 2x + y = \sin t$, $\dfrac{d^2y}{dt^2} - x - 2y = \cos t$.

[*Hint:* Set $x_1 = x$, $x_2 = dx/dt$, $x_3 = y$, $x_4 = dy/dt$.]

(b) $\dfrac{dx}{dt} - 2\dfrac{dy}{dt} + x - 3y = e^t$, $2\dfrac{dx}{dt} + 3\dfrac{dy}{dt} - x + 4y = 2e^t$.

[*Hint:* Solve for dx/dt and dy/dt.]

6. Let the three triples $x_i(t)$, $y_i(t)$, $z_i(t)$, $(i = 1, 2, 3)$, be solutions of the system (6–2) for $t_1 < t < t_2$, and let

$$W(t) = \begin{vmatrix} x_1(t) & y_1(t) & z_1(t) \\ x_2(t) & y_2(t) & z_2(t) \\ x_3(t) & y_3(t) & z_3(t) \end{vmatrix}$$

Prove: if the triples are linearly independent for $t_1 < t < t_2$, then $W(t) \neq 0$; if the triples are linearly dependent for $t_1 < t < t_2$, then $W(t) \equiv 0$. [*Hint:* Apply Theorem 1; cf. also the proof of Theorem 2 in Section 4–3.]

ANSWERS

1. (c) $x = C_1e^t + 3C_2e^{-t}$, $y = C_1e^t + 5C_2e^{-t}$;   (e) $x = C_1e^t + 3C_2e^{-t} - \cos t - 4 \sin t$, $y = C_1e^t + 5C_2e^{-t} - 5 \sin t$;   (f) $x = 5e^t - 3e^{-t} - \cos t - 4 \sin t$, $y = 5e^t - 5e^{-t} - 5 \sin t$.

2. (c) $x = 3C_1 + C_2(4 \sin t + 9 \cos t) + C_3(9 \sin t - 4 \cos t) - 3t$, $y = C_1 + C_2 (\sin t + 3 \cos t) + C_3(3 \sin t - \cos t) - t$, $z = 2C_1 + C_2(2 \sin t + 7 \cos t) + C_3(7 \sin t - 2 \cos t) - 2t - 1$; (d) $x = 6 - 3t, y = 2 - t, z = 3 - 2t$.

3. (a) dependent, (b) independent.

5. (a) $dx_1/dt = x_2$, $dx_2/dt = 2x_1 - x_3 + \sin t$, $dx_3/dt = x_4$, $dx_4/dt = x_1 + 2x_3 + \cos t$;   (b) $dx/dt = (-x + y + 7e^t)/7$, $dy/dt = (3x - 10y)/7$.

6–2 Homogeneous linear systems with constant coefficients. Here we consider the typical case of a homogeneous system:

$$\frac{dx}{dt} = a_1x + b_1y + c_1z,$$

$$\frac{dy}{dt} = a_2x + b_2y + c_2z, \qquad (6\text{--}10)$$

$$\frac{dz}{dt} = a_3x + b_3y + c_3z,$$

and assume the coefficients to be constant. We seek particular solutions from which to construct the general solution. By analogy with the single equation of nth order, we try functions $e^{\lambda t}$, for constant λ. After experimenting, we discover that we must seek solutions of form

$$x = \alpha e^{\lambda t}, \qquad y = \beta e^{\lambda t}, \qquad z = \gamma e^{\lambda t}, \tag{6–11}$$

where α, β, γ are constants. Substitution of (6–11) in (6–10) leads to the equations

$$(a_1 - \lambda)\alpha + b_1\beta + c_1\gamma = 0,$$

$$a_2\alpha + (b_2 - \lambda)\beta + c_2\gamma = 0, \tag{6–12}$$

$$a_3\alpha + b_3\beta + (c_3 - \lambda)\gamma = 0.$$

These can be regarded as homogeneous linear equations for α, β, γ; solutions other than the trivial one $\alpha = 0$, $\beta = 0$, $\gamma = 0$ are obtainable precisely when the determinant of the coefficients is 0:

$$\begin{vmatrix} a_1 - \lambda & b_1 & c_1 \\ a_2 & b_2 - \lambda & c_2 \\ a_3 & b_3 & c_3 - \lambda \end{vmatrix} = 0. \tag{6–13}$$

When (6–13) is expanded, it appears as a cubic equation in λ. It is termed the *characteristic equation* associated with the system (6–10) and its roots are called *characteristic roots*.

Let us suppose that the roots λ_1, λ_2, λ_3 of (6–13) are *real and distinct*. Corresponding to the root λ_1, we can find a set of values α_1, β_1, γ_1 (not all 0) satisfying (6–12); then the triple

$$x_1 \equiv \alpha_1 e^{\lambda_1 t}, \qquad y_1 \equiv \beta_1 e^{\lambda_1 t}, \qquad z_1 \equiv \gamma_1 e^{\lambda_1 t}$$

is a solution of Eqs. (6–10). Similarly, each of the roots λ_2, λ_3 provides a solution. *The three triples are linearly independent, and the corresponding expressions*

$$x = C_1\alpha_1 e^{\lambda_1 t} + C_2\alpha_2 e^{\lambda_2 t} + C_3\alpha_3 e^{\lambda_3 t},$$

$$y = C_1\beta_1 e^{\lambda_1 t} + C_2\beta_2 e^{\lambda_2 t} + C_3\beta_3 e^{\lambda_3 t}, \tag{6–14}$$

$$z = C_1\gamma_1 e^{\lambda_1 t} + C_2\gamma_2 e^{\lambda_2 t} + C_3\gamma_3 e^{\lambda_3 t}$$

provide the general solution of Eqs. (6–10). To verify linear independence, we need only show that the constants C_1, C_2, C_3 can be chosen so that x, y, z in (6–14) reduce identically to zero only if $C_1 = 0, C_2 = 0, C_3 = 0$.

If x, y, z do reduce to 0, then because of the linear independence of the functions $e^{\lambda_1 t}$, $e^{\lambda_2 t}$, $e^{\lambda_3 t}$, we conclude that

$$C_1\alpha_1 = 0, \quad C_2\alpha_2 = 0, \quad C_3\alpha_3 = 0, \quad C_1\beta_1 = 0, \quad \ldots, \quad C_3\gamma_3 = 0.$$

If $C_1 \neq 0$, then $\alpha_1 = 0$, $\beta_1 = 0$, $\gamma_1 = 0$, which is contrary to the way α_1, β_1, γ_1 were chosen; hence C_1 must equal 0. Similarly, $C_2 = 0, C_3 = 0$. Thus the triples are linearly independent.

EXAMPLE 1. We consider Example 1 of Section 6–1:

$$\frac{dx}{dt} = 4x - 9y + 5z, \qquad \frac{dy}{dt} = x - 10y + 7z, \qquad \frac{dz}{dt} = x - 17y + 12z.$$

The characteristic equation is

$$\begin{vmatrix} 4 - \lambda & -9 & 5 \\ 1 & -10 - \lambda & 7 \\ 1 & -17 & 12 - \lambda \end{vmatrix} = 0.$$

When expanded, this becomes

$$\lambda^3 - 6\lambda^2 + 11\lambda - 6 = 0,$$

and the roots are found to be $\lambda_1 = 1, \lambda_2 = 2, \lambda_3 = 3$. The equations for α_1, β_1, γ_1 are

$$\begin{aligned} 3\alpha_1 - \ 9\beta_1 + \ 5\gamma_1 &= 0, \\ \alpha_1 - 11\beta_1 + \ 7\gamma_1 &= 0, \\ \alpha_1 - 17\beta_1 + 11\gamma_1 &= 0. \end{aligned} \tag{6–15}$$

These equations have as determinant of coefficients the above determinant, with λ replaced by $\lambda_1 = 1$; since the determinant has value 0 (that is precisely how λ_1 is chosen), we can solve (6–15) for two of the unknowns in terms of the third, or else (6–15) merely expresses one letter in terms of the other two. Subtraction of the second equation from the third gives

$$6\beta_1 - 4\gamma_1 = 0;$$

hence $\beta_1 = \frac{2}{3}\gamma_1$. Substitution in the second equation yields $\alpha_1 = \frac{1}{3}\gamma_1$. These values check in the first and third equations. We can write the solution as

$$\alpha_1 = k, \qquad \beta_1 = 2k, \qquad \gamma_1 = 3k,$$

where k is arbitrary. We want only one set of values, other than 0, 0, 0; hence we choose $k = 1$ and $\alpha_1 = 1$, $\beta_1 = 2$, $\gamma_1 = 3$.

Determination of α_2, β_2, γ_2 and of α_3, β_3, γ_3 is similar. We are led to the sets of equations

$$2\alpha_2 - 9\beta_2 + 5\gamma_2 = 0,$$
$$\alpha_2 - 12\beta_2 + 7\gamma_2 = 0, \qquad (6\text{–}16\text{a})$$
$$\alpha_2 - 17\beta_2 + 10\gamma_2 = 0;$$

$$\alpha_3 - 9\beta_3 + 5\gamma_3 = 0,$$
$$\alpha_3 - 13\beta_3 + 7\gamma_3 = 0, \qquad (6\text{–}16\text{b})$$
$$\alpha_3 - 17\beta_3 + 9\gamma_3 = 0.$$

Solutions are found to be $\alpha_2 = 1$, $\beta_2 = 3$, $\gamma_2 = 5$, $\alpha_3 = 1$, $\beta_3 = -1$, $\gamma_3 = -2$. We thus obtain the three triples

$$x_1 = e^t, \qquad y_1 = 2e^t, \qquad z_1 = 3e^t,$$
$$x_2 = e^{2t}, \qquad y_2 = 3e^{2t}, \qquad z_2 = 5e^{2t},$$
$$x_3 = e^{3t}, \qquad y_3 = -e^{3t}, \qquad z_3 = -2e^{3t},$$

which were verified to be linearly independent solutions in Section 6–1. The general solution is given by Eqs. (6–9).

EXAMPLE 2. We consider the system of Prob. 1 following Section 6–1:

$$\frac{dx}{dt} = 4x - 3y, \qquad \frac{dy}{dt} = 5x - 4y.$$

The characteristic equation is

$$0 = \begin{vmatrix} 4 - \lambda & -3 \\ 5 & -4 - \lambda \end{vmatrix} = \lambda^2 - 1.$$

The characteristic roots are $\lambda_1 = 1$, $\lambda_2 = -1$. Corresponding to $\lambda_1 = 1$, we have the equations

$$3\alpha_1 - 3\beta_1 = 0, \qquad 5\alpha_1 - 5\beta_1 = 0,$$

for α_1, β_1; we see at once that $\alpha_1 = \beta_1 = k$, where k is arbitrary. We set $k = 1$ and obtain the solution $x_1 = e^t$, $y_1 = e^t$. Similarly, for $\lambda_2 = -1$ we have

$$5\alpha_1 - 3\beta_1 = 0, \qquad 5\alpha_1 - 3\beta_1 = 0;$$

hence $\alpha_1 = 3\beta_1/5$, or $\alpha_1 = 3k$, $\beta_1 = 5k$ (where k is arbitrary); we take $k = 1$ and obtain the solution $x_2 = 3e^{-t}$, $y_2 = 5e^{-t}$. The general solution is then

$$x = C_1 e^t + 3C_2 e^{-t}, \qquad y = C_1 e^t + 5C_2 e^{-t}.$$

6–3 Case of complex characteristic roots. If $\lambda = a \pm bi$ is a pair of complex roots of the characteristic equation, then we can write $\lambda_1 = a + bi$, $\lambda_2 = \bar{\lambda}_1 = a - bi$ (the bar denotes complex conjugate). The procedure of Section 6–2 can then be followed, but it leads to solutions $\alpha_1 e^{\lambda_1 t}$, $\beta_1 e^{\lambda_1 t}$, $\gamma_1 e^{\lambda_1 t}$ and $\alpha_2 e^{\lambda_2 t}$, $\beta_2 e^{\lambda_2 t}$, $\gamma_2 e^{\lambda_2 t}$, which are *complex* solutions of the differential equations, as in Section 4–7. However, since $\lambda_2 = \bar{\lambda}_1$, we can choose $\alpha_2 = \bar{\alpha}_1$, $\beta_2 = \bar{\beta}_1$, $\gamma_2 = \bar{\gamma}_1$. For since the coefficients a_1, b_1, c_1, a_2, ... are assumed real, the equations

$$(a_1 - \lambda_1)\alpha_1 + b_1\beta_1 + c_1\gamma_1 = 0,$$
$$a_2\alpha_1 + (b_2 - \lambda_1)\beta_1 + c_2\gamma_1 = 0, \quad \ldots,$$

on taking conjugates, imply

$$(a_1 - \bar{\lambda}_1)\bar{\alpha}_1 + b_1\bar{\beta}_1 + c_1\bar{\gamma}_1 = 0,$$
$$a_2\bar{\alpha}_1 + (b_2 - \bar{\lambda}_1)\bar{\beta}_1 + c_2\bar{\gamma}_1 = 0, \quad \ldots$$

It follows that if $C_2 = \bar{C}_1$, then

$$C_1\alpha_1 e^{\lambda_1 t} + C_2\alpha_2 e^{\lambda_2 t} \quad (\alpha_2 = \bar{\alpha}_1, \lambda_2 = \bar{\lambda}_1)$$

is real, and hence, as in Section 4–7, we obtain real solutions by writing

$$
\begin{aligned}
x &= C_1\alpha_1 e^{\lambda_1 t} + \bar{C}_1\bar{\alpha}_1 e^{\bar{\lambda}_1 t}, \\
y &= C_1\beta_1 e^{\lambda_1 t} + \bar{C}_1\bar{\beta}_1 e^{\bar{\lambda}_1 t}, \quad\quad\quad (6\text{–}17)\\
z &= C_1\gamma_1 e^{\lambda_1 t} + \bar{C}_1\bar{\gamma}_1 e^{\bar{\lambda}_1 t}.
\end{aligned}
$$

If we write $C_1 = C_1' + iC_1''$, $\bar{C}_1 = C_1' - iC_1''$, $\alpha_1 = \alpha_1' + i\alpha_1''$, $\bar{\alpha}_1 = \alpha_1' - i\alpha_1''$, $\beta_1 = \beta_1' + i\beta_1''$, ..., where C_1', C_1'', ... are real, then (6–17) can be expanded to give x, y, and z as linear combinations of $e^{at} \cos bt$, $e^{at} \sin bt$.

EXAMPLE. We take the system of Prob. 2(a) following Section 6–1:

$$\frac{dx}{dt} = -9x + 19y + 4z, \quad \frac{dy}{dt} = -3x + 7y + z,$$
$$\frac{dz}{dt} = -7x + 17y + 2z.$$

The characteristic equation is

$$0 = \begin{vmatrix} -9 - \lambda & 19 & 4 \\ -3 & 7 - \lambda & 1 \\ -7 & 17 & 2 - \lambda \end{vmatrix} = -\lambda^3 - \lambda = -\lambda(\lambda^2 + 1).$$

The roots are $\pm i$, 0. Corresponding to the root $\lambda_1 = i$, we obtain the equations for α_1, β_1, γ_1:

$$(-9 - i)\alpha_1 + 19\beta_1 + 4\gamma_1 = 0,$$
$$-3\alpha_1 + (7 - i)\beta_1 + \gamma_1 = 0,$$
$$-7\alpha_1 + 17\beta_1 + (2 - i)\gamma_1 = 0.$$

From the first and second equations we obtain

$$(3 - i)\alpha_1 + (-9 + 4i)\beta_1 = 0,$$
$$\gamma_1 = 3\alpha_1 - (7 - i)\beta_1.$$

Hence if we take $\beta_1 = (3 - i)k$, then $\alpha_1 = (9 - 4i)k$ and $\gamma_1 = (7 - 2i)k$, where k is arbitrary. We choose $k = 1$, and hence

$$\alpha_1 = 9 - 4i, \qquad \beta_1 = 3 - i, \qquad \gamma_1 = 7 - 2i.$$

The equations for α_2, β_2, γ_2 are obtained from those for α_1, β_1, γ_1 by replacing i by $-i$, and we can verify that this simply replaces i by $-i$ in the results; that is, we can choose $\alpha_2 = \bar{\alpha}_1, \beta_2 = \bar{\beta}_1, \gamma_2 = \bar{\gamma}_1$ as remarked above:

$$\alpha_2 = 9 + 4i, \qquad \beta_2 = 3 + i, \qquad \gamma_2 = 7 + 2i.$$

Corresponding to the root $\lambda_3 = 0$, we obtain the equations

$$-9\alpha_3 + 19\beta_3 + 4\gamma_3 = 0,$$
$$-3\alpha_3 + 7\beta_3 + \gamma_3 = 0,$$
$$-7\alpha_3 + 17\beta_3 + 2\gamma_3 = 0,$$

for which we find the solution

$$\alpha_3 = 3, \qquad \beta_3 = 1, \qquad \gamma_3 = 2.$$

Accordingly, the general real solution is given by

$$x = C_1(9 - 4i)e^{it} + \bar{C}_1(9 + 4i)e^{-it} + 3C_3,$$
$$y = C_1(3 - i)e^{it} + \bar{C}_1(3 + i)e^{-it} + C_3, \qquad (6\text{--}18)$$
$$z = C_1(7 - 2i)e^{it} + \bar{C}_1(7 + 2i)e^{-it} + 2C_3.$$

If we write $C_1 = \frac{1}{2}(c_1 + ic_2)$, $\bar{C}_1 = \frac{1}{2}(c_1 - ic_2)$, then

$$x = C_1(9 - 4i)e^{it} + \bar{C}_1(9 + 4i)e^{-it} + 3C_3$$
$$= 2 \operatorname{Re} [C_1(9 - 4i)e^{it}] + 3C_3$$
$$= \operatorname{Re} [(c_1 + ic_2)(9 - 4i)e^{it}] + 3C_3$$
$$= c_1(9 \cos t + 4 \sin t) - c_2(9 \sin t - 4 \cos t) + 3C_3.$$

Similarly, we find

$$y = c_1 (\sin t + 3 \cos t) - c_2(3 \sin t - \cos t) + C_3,$$

$$z = c_1(2 \sin t + 7 \cos t) + c_2(7 \sin t - 2 \cos t) + 2C_3.$$

Remark. As the example illustrates, we can obtain the solutions in real form by taking

$$x_1 = \text{Re} \, (\alpha_1 e^{\lambda_1 t}), \qquad y_1 = \text{Re} \, (\beta_1 e^{\lambda_1 t}), \qquad z_1 = \text{Re} \, (\gamma_1 e^{\lambda_1 t});$$

$$x_2 = \text{Im} \, (\alpha_1 e^{\lambda_1 t}), \qquad y_2 = \text{Im} \, (\beta_1 e^{\lambda_1 t}), \qquad z_2 = \text{Im} \, (\gamma_1 e^{\lambda_1 t});$$

$$x_3 = \alpha_3 e^{\lambda_3 t}, \qquad y_3 = \beta_3 e^{\lambda_3 t}, \qquad z_3 = \gamma_3 e^{\lambda_3 t}.$$

We can verify directly that these are three linearly independent solutions.

PROBLEMS

For each of the following systems of linear differential equations, obtain the general solution. Throughout, $D = d/dt$.

1. $Dx = 7x + 6y$, $Dy = 2x + 6y$.

2. $Dx = -x + y$, $Dy = -5x + 3y$.

3. $Dx = 16x + 14y + 38z$, $Dy = -9x - 7y - 18z$,
   $Dz = -4x - 4y - 11z$.

4. $Dx = -5x - 10y - 20z$, $Dy = 5x + 5y + 10z$, $Dz = 2x + 4y + 9z$.

5. $Dx = 3x - 5y + u$, $Dy = x - y$, $Dz = -3z - u$, $Du = 5z + u$.

6. $Dx = 2x$, $Dy = 4y$.

7. $Dx = 2x$, $Dy = 2y$.

8. $Dx = 7x + 6y$, $Dy = 2x + 6y$, $Dz = 7u + 6v$, $Du = 2u + 6v$.
[*Hint:* See Prob. 1.]

ANSWERS

1. $x = 2C_1 e^{10t} + 3C_2 e^{3t}$, $y = C_1 e^{10t} - 2C_2 e^{3t}$.

2. $x = e^t(C_1 \cos t + C_2 \sin t)$,
   $y = e^t[C_1(2 \cos t - \sin t) + C_2(2 \sin t + \cos t)]$.

3. $x = C_1 e^{2t} - 2C_2 e^{-t} - 2C_3 e^{-3t}$, $y = -C_1 e^{2t} - 3C_2 e^{-t}$, $z = 2C_2 e^{-t} + C_3 e^{-3t}$.

4. $x = -2C_1 e^{5t} + e^{2t}[C_2(20 \cos t - 10 \sin t) + C_3(15 \cos t + 5 \sin t)]$,
   $y = e^{2t}[C_2(15 \cos t + 5 \sin t) + C_3(15 \sin t - 5 \cos t)]$,
   $z = C_1 e^{5t} + e^{2t}[C_2 (-14 \cos t + 2 \sin t) + C_3 (-2 \cos t - 14 \sin t)]$.

5. $x = C_1(2 + i)e^{(1+i)t} + \overline{C}_1(2 - i)e^{(1-i)t} + C_2(3 - i)e^{(-1+i)t} + \overline{C}_2(3 + i)e^{(-1-i)t}$,

   $y = C_1e^{(1+i)t} + \overline{C}_1e^{(1-i)t} + C_2(-1 - 3i)e^{(-1+i)t} + \overline{C}_2(-1 + 3i)e^{(-1-i)t}$,

   $z = 8C_2e^{(-1+i)t} + 8\overline{C}_2e^{(-1-i)t}$,

   $u = C_2(-16 - 8i)e^{(-1+i)t} + \overline{C}_2(-16 + 8i)e^{(-1-i)t}$.

6. $x = C_1e^{2t}, y = C_2e^{4t}$.

7. $x = C_1e^{2t}, y = C_2e^{2t}$.

8. $x = 2C_1e^{10t} + 3C_2e^{3t}, y = C_1e^{10t} - 2C_2e^{3t}, z = 2C_3e^{10t} + 3C_4e^{3t}$,

   $u = C_3e^{10t} - 2C_4e^{3t}$.

6-4 Case of repeated roots. When the characteristic equation has repeated roots, the methods of the preceding sections appear to yield less than the required number of linearly independent solutions. We now show how the required additional solutions can be found. In essence, our method consists in adding terms which are linear combinations of $te^{\lambda t}$, $t^2e^{\lambda t}$, ..., $t^{k-1}e^{\lambda t}$, just as in the case of the single equation of order n. However, it can happen that for a root of multiplicity k, not all these terms are required; in fact, none of them may be needed, since the solutions are built solely from the functions $e^{\lambda t}$.

After illustrating the method by examples, we shall formulate some general rules.

EXAMPLE 1. $dx/dt = 5x, dy/dt = 5y$. The characteristic equation is

$$0 = \begin{vmatrix} 5 - \lambda & 0 \\ 0 & 5 - \lambda \end{vmatrix} = (5 - \lambda)^2.$$

The roots are 5, 5. The substitution $x = \alpha e^{5t}, y = \beta e^{5t}$ leads to the equations

$$0\alpha + 0\beta = 0, \qquad 0\alpha + 0\beta = 0,$$

which are satisfied for all values of α and β. We obtain two linearly independent solutions by choosing $\alpha_1 = 1, \beta_1 = 0$ and $\alpha_2 = 0, \beta_2 = 1$:

$$x_1 = e^{5t}, \qquad y_1 = 0; \qquad x_2 = 0, \qquad y_2 = e^{5t}.$$

Hence the general solution is

$$x = C_1e^{5t}, \qquad y = C_2e^{5t}. \tag{6-19}$$

From the form of the given differential equations it is clear that (6-19) provides all solutions.

EXAMPLE 2. $dx/dt = x + y$, $dy/dt = -x + 3y$.

The characteristic equation is

$$0 = \begin{vmatrix} 1 - \lambda & 1 \\ -1 & 3 - \lambda \end{vmatrix} = \lambda^2 - 4\lambda + 4 = (\lambda - 2)^2.$$

The roots are 2, 2. The substitution $x = \alpha e^{2t}$, $y = \beta e^{2t}$ leads to the equations

$$-\alpha + \beta = 0, \qquad -\alpha + \beta = 0. \tag{6–20}$$

The solutions have the form $\alpha = k$, $\beta = k$, where k is arbitrary. With $k = 1$, we obtain the solution

$$x_1 = e^{2t}, \qquad y_1 = e^{2t},$$

but can obtain no second linearly independent solution of form αe^{2t}, βe^{2t}. To obtain the second solution, we set

$$x = e^{2t}(\alpha_1 t + \alpha_2), \qquad y = e^{2t}(\beta_1 t + \beta_2). \tag{6–21}$$

Substitution in the differential equations and cancellation of e^{2t} leads to the equations

$$2\alpha_1 t + 2\alpha_2 + \alpha_1 = (\alpha_1 + \beta_1)t + \alpha_2 + \beta_2,$$
$$2\beta_1 t + 2\beta_2 + \beta_1 = (3\beta_1 - \alpha_1)t + 3\beta_2 - \alpha_2.$$

Since these are to be identities, we conclude that $2\alpha_1 = \alpha_1 + \beta_1, \ldots$; we thus obtain the four equations

$$-\alpha_1 + \beta_1 = 0, \qquad -\alpha_1 + \beta_1 = 0, \tag{6–22}$$
$$-\alpha_2 + \beta_2 = \alpha_1, \qquad -\alpha_2 + \beta_2 = \beta_1. \tag{6–23}$$

The first two result from comparison of terms in t, the second two from comparison of constant terms. We note that (6–22) has the same form as (6–20), so that its solutions are $\alpha_1 = k_1$, $\beta_1 = k_1$, where k_1 is arbitrary. Equations (6–23) are then satisfied when $\alpha_2 = k_2$, $\beta_2 = k_1 + k_2$, where k_2 is arbitrary. We need only one set of values α_1, β_1, α_2, β_2, with α_1 and β_1 not both 0; hence we choose $k_1 = 1$, $k_2 = 0$, so that

$$\alpha_1 = 1, \qquad \beta_1 = 1, \qquad \alpha_2 = 0, \qquad \beta_2 = 1.$$

The desired solution (6–21) is

$$x_2 = e^{2t}(t), \qquad y_2 = e^{2t}(t + 1),$$

and the general solution is

$$x = C_1 e^{2t} + C_2 t e^{2t}, \qquad y = C_1 e^{2t} + C_2 (t + 1)e^{2t}.$$

EXAMPLE 3. Consider the system of third order

$$\frac{dx}{dt} = 14x + 66y - 42z, \qquad \frac{dy}{dt} = 4x + 24y - 14z,$$

$$\frac{dz}{dt} = 10x + 55y - 33z.$$

The characteristic equation is

$$0 = \begin{vmatrix} 14 - \lambda & 66 & -42 \\ 4 & 24 - \lambda & -14 \\ 10 & 55 & -33 - \lambda \end{vmatrix} = -(\lambda - 2)^2(\lambda - 1).$$

The roots are 2, 2, 1. The substitution $x = \alpha e^{2t}$, $y = \beta e^{2t}$, $z = \gamma e^{2t}$ leads to the equations

$$12\alpha + 66\beta - 42\gamma = 0,$$

$$4\alpha + 22\beta - 14\gamma = 0, \qquad\qquad (6\text{-}24)$$

$$10\alpha + 55\beta - 35\gamma = 0.$$

These can be reduced to the single equation

$$2\alpha + 11\beta - 7\gamma = 0;$$

hence the solutions can be written

$$\alpha = -11k_1 + 7k_2, \qquad \beta = 2k_1, \qquad \gamma = 2k_2,$$

in terms of *two* arbitrary constants k_1, k_2. We obtain two linearly independent solutions by choosing $k_1 = 1$, $k_2 = 0$, and then choosing $k_1 = 0$, $k_2 = 1$:

$$x_1 = -11e^{2t}, \qquad y_1 = 2e^{2t}, \qquad z_1 = 0;$$

$$x_2 = 7e^{2t}, \qquad y_2 = 0, \qquad z_2 = 2e^{2t}.$$

The simple root $\lambda_3 = 1$ leads to the equations

$$13\alpha_3 + 66\beta_3 - 42\gamma_3 = 0,$$

$$4\alpha_3 + 23\beta_3 - 14\gamma_3 = 0,$$

$$10\alpha_3 + 55\beta_3 - 34\gamma_3 = 0.$$

These are satisfied when $\alpha_3 = 6k$, $\beta_3 = 2k$, $\gamma_3 = 5k$; with $k = 1$, we obtain the solution

$$x_3 = 6e^t, \qquad y_3 = 2e^t, \qquad z_3 = 7e^t.$$

Hence the general solution is

$$x = -11C_1e^{2t} + 7C_2e^{2t} + 6C_3e^t,$$
$$y = 2C_1e^{2t} + 2C_3e^t,$$
$$z = 2C_2e^{2t} + 5C_3e^t.$$

EXAMPLE 4. The system is

$$\frac{dx}{dt} = -8x + 47y - 8z, \qquad \frac{dy}{dt} = -4x + 18y - 2z,$$
$$\frac{dz}{dt} = -8x + 39y - 5z.$$

The characteristic equation is

$$0 = \begin{vmatrix} -8 - \lambda & 47 & -8 \\ -4 & 18 - \lambda & -2 \\ -8 & 39 & -5 - \lambda \end{vmatrix} = -(\lambda - 2)^2(\lambda - 1).$$

As in Example 3, the roots are 2, 2, 1. The substitution $x = \alpha e^{2t}$, $y = \beta e^{2t}$, $z = \gamma e^{2t}$ leads to the equations

$$-10\alpha + 47\beta - 8\gamma = 0,$$
$$-4\alpha + 16\beta - 2\gamma = 0, \qquad (6\text{--}25)$$
$$-8\alpha + 39\beta - 7\gamma = 0.$$

The solutions are $\alpha = 17k$, $\beta = 6k$, $\gamma = 14k$; hence, with $k = 1$,

$$x_1 = 17e^{2t}, \qquad y_1 = 6e^{2t}, \qquad z_1 = 14e^{2t}.$$

To obtain the second solution, we set

$$x = e^{2t}(\alpha_1 t + \alpha_2), \qquad y = e^{2t}(\beta_1 t + \beta_2), \qquad z = e^{2t}(\gamma_1 t + \gamma_2),$$

and as in Example 2 we are led to the equations

$$-10\alpha_1 + 47\beta_1 - 8\gamma_1 = 0, \qquad -4\alpha_1 + 16\beta_1 - 2\gamma_1 = 0,$$
$$-8\alpha_1 + 39\beta_1 - 7\gamma_1 = 0, \qquad (6\text{--}26)$$

and

$$-10\alpha_2 + 47\beta_2 - 8\gamma_2 = \alpha_1,$$
$$-4\alpha_2 + 16\beta_2 - 2\gamma_2 = \beta_1, \qquad (6\text{--}27)$$
$$-8\alpha_2 + 39\beta_2 - 7\gamma_2 = \gamma_1.$$

The equations (6–26) are the same as (6–25) and are satisfied when $\alpha_1 = 17k_1$, $\beta_1 = 6k_1$, $\gamma_1 = 14k_1$, where k_1 is arbitrary. If we substitute these expressions in (6–27), then (6–27) becomes a set of three equations for α_2, β_2, γ_2, in terms of k_1. However, we verify that these are *dependent* equations (in particular, the third equation is obtained by subtracting one-half the second from the first). Hence we use only the first two equations; γ_2 can be chosen arbitrarily, $\gamma_2 = k_2$, and we solve for α_2, β_2:

$$\alpha_2 = \frac{17k_2 - 5k_1}{14}, \qquad \beta_2 = \frac{2k_1 + 3k_2}{7}, \qquad \gamma_2 = k_2.$$

Thus, as in Example 2, the solutions depend on two arbitrary constants k_1, k_2; we seek a solution for which α_1, β_1, γ_1 are not all 0. We find this by setting $k_1 = 14$, $k_2 = 0$; then $\alpha_1 = 238$, $\beta_1 = 84$, $\gamma_1 = 196$, $\alpha_2 = -5$, $\beta_2 = 4$, $\gamma_2 = 0$, and the solution is

$$x_2 = e^{2t}(238t - 5), \qquad y_2 = e^{2t}(84t + 4), \qquad z = e^{2t}(196t).$$

For the third solution we set $x = \alpha e^t$, $y = \beta e^t$, $z = \gamma e^t$ and find, without difficulty, $\alpha = 6k$, $\beta = 2k$, $\gamma = 5k$, so that we can choose

$$x_3 = 6e^t, \qquad y_3 = 2e^t, \qquad z_3 = 5e^t.$$

Then the general solution is

$$x = e^{2t}[17C_1 + C_2(238t - 5)] + 6C_3 e^t,$$
$$y = e^{2t}[6C_1 + C_2(84t + 4)] + 2C_3 e^t,$$
$$z = e^{2t}[14C_1 + C_2(196t)] + 5C_3 e^t.$$

We now generalize the procedures indicated by the above examples. If λ_1 is a double root of the characteristic equation, we first seek solutions of form $x = \alpha e^{\lambda_1 t}$, $y = \beta e^{\lambda_1 t}$, ... At least one such solution can be found (with α, β, ... not all 0); it may also happen that two linearly independent solutions of this form can be found. If two independent solutions cannot be found, then the second solution is obtainable in the form

$$x = e^{\lambda_1 t}(\alpha_1 t + \alpha_2), \qquad y = e^{\lambda_1 t}(\beta_1 t + \beta_2), \dots$$

For a triple root λ_1 we proceed similarly, obtaining first as many independent solutions of form $x = \alpha e^{\lambda_1 t}$, $y = \beta e^{\lambda_1 t}$, ... as possible (at most three), then as many independent solutions of form $x = e^{\lambda_1 t}(\alpha_1 t + \alpha_2)$, $y = e^{\lambda_1 t}(\beta_1 t + \beta_2)$, ... as possible, and finally (if one more solution is needed) a solution of form

$$x = e^{\lambda_1 t}(\alpha_1 t^2 + \alpha_2 t + \alpha_3), \qquad y = e^{\lambda_1 t}(\beta_1 t^2 + \beta_2 t + \beta_3), \qquad \dots$$

For a root of multiplicity k we may have to use polynomial coefficients of degree $k - 1$ (but no higher).

Instead of gradually raising the degree of the coefficient of $e^{\lambda_1 t}$, we can obtain all the solutions associated with the characteristic root λ_1 by at once trying the polynomials of highest degree. Substitution in the differential equations leads to homogeneous linear equations for the coefficients in the polynomials. These equations can always be solved for some of the coefficients in terms of the others; when the multiplicity is k, the "others" are k in number. We can obtain k linearly independent solutions by in turn choosing one of the "others" to be 1 and the remaining ones to be 0. We illustrate this procedure by considering Examples 2 and 3 again.

For Example 2 the multiplicity is 2; we at once seek solutions of form (6–21) and are led to equations (6–22), (6–23), which can be solved for α_1, α_2 in terms of β_1, β_2:

$$\alpha_1 = \beta_1, \qquad \alpha_2 = \beta_2 - \beta_1.$$

When $\beta_1 = 1$, $\beta_2 = 0$, we obtain the solution

$$x_1 = e^{2t}(t - 1), \qquad y_1 = e^{2t}(t);$$

when $\beta_1 = 0$, $\beta_2 = 1$, we obtain the solution

$$x_2 = e^{2t}, \qquad y_2 = e^{2t}.$$

These are not the same two linearly independent solutions that we obtained in Example 2, but the general solution

$$x = C_1 e^{2t}(t - 1) + C_2 e^{2t}, \qquad y = C_1 t e^{2t} + C_2 e^{2t}$$

can be easily verified to be equivalent to the previous one.

For Example 3 the multiplicity is again 2, and we set

$$x = e^{2t}(\alpha_1 t + \alpha_2), \qquad y = e^{2t}(\beta_1 t + \beta_2), \qquad z = e^{2t}(\gamma_1 t + \gamma_2).$$

Substitution in the differential equations leads to the equations

$$12\alpha_1 + 66\beta_1 - 42\gamma_1 = 0, \qquad 4\alpha_1 + 22\beta_1 - 14\gamma_1 = 0,$$
$$10\alpha_1 + 55\beta_1 - 35\gamma_1 = 0;$$

$$12\alpha_2 + 66\beta_2 - 42\gamma_2 = \alpha_1, \qquad 4\alpha_2 + 22\beta_2 - 14\gamma_2 = \beta_1,$$
$$10\alpha_2 + 55\beta_2 - 35\gamma_2 = \gamma_1.$$

From the second group of equations we conclude that

$$2\alpha_2 + 11\beta_2 - 7\gamma_2 = \frac{\alpha_1}{6} = \frac{\beta_1}{2} = \frac{\gamma_1}{5} \, ;$$

thus $\alpha_1 = 3\beta_1$, $\gamma_1 = 5\beta_1/2$. The first equation now gives $36\beta_1 + 66\beta_1 - 105\beta_1 = 0$, so that $\beta_1 = 0$, $\alpha_1 = 0$, and $\gamma_1 = 0$. The second group can then be reduced to the equation

$$2\alpha_2 + 11\beta_2 - 7\gamma_2 = 0;$$

hence, for example, β_2 and γ_2 can be chosen as the constants in terms of which all are expressed:

$$\alpha_2 = \frac{7\gamma_2 - 11\beta_2}{2}, \qquad \alpha_1 = 0,\ \beta_1 = 0,\ \gamma_1 = 0.$$

The choices $\beta_2 = 1$, $\gamma_2 = 0$ and $\beta_2 = 0$, $\gamma_2 = 1$ give us the two solutions

$$x_1 = -\tfrac{11}{2}e^{2t}, \qquad y_1 = e^{2t}, \qquad z_1 = 0;$$
$$x_2 = \tfrac{7}{2}e^{2t}, \qquad y_2 = 0, \qquad z_2 = e^{2t}.$$

Except for a factor of 2, these are the same as those obtained in Example 3.

A proof that the method described does yield all solutions is given in Section 6–22. The whole subject can be studied more systematically with the aid of matrices; see Sections 6–15 through 6–23.

If λ_1 is a multiple complex root, we obtain complex solutions $x_1, y_1, \ldots,$ $x_2, y_2 \ldots,$ as above. We can then obtain the general real solution by forming the terms

$$C_1 x_1(t) + \overline{C}_1 \overline{x}_1(t) + C_2 x_2(t) + \overline{C}_2 \overline{x}_2(t) + \cdots,$$
$$C_1 y_1(t) + \overline{C}_1 \overline{y}_1(t) + C_2 y_2(t) + \overline{C}_2 \overline{y}_2(t) + \cdots, \qquad \cdots,$$

just as for simple roots. The justification is similar to that for the single differential equation of order n (Section 4–8). Equivalently, we can use the real solutions:

$$\text{Re}\,[x_1(t)], \qquad \text{Re}\,[y_1(t)], \qquad \ldots,$$
$$\text{Im}\,[x_1(t)], \qquad \text{Im}\,[y_1(t)], \qquad \ldots,$$
$$\text{Re}\,[x_2(t)], \qquad \text{Re}\,[y_2(t)], \qquad \ldots,$$
$$\text{Im}\,[x_2(t)], \qquad \text{Im}\,[y_2(t)], \qquad \ldots,$$
$$\ldots,$$

as a set of linearly independent solutions which correspond to the root λ_1.

PROBLEMS

Find the general solution of the systems of Probs. 1–8. Throughout, $D = d/dt$.

1. $Dx = -x$, $Dy = -y$.

2. $Dx = -4x - y$, $Dy = x - 2y$.

3. $Dx = y,\ Dy = 4x + 3y - 4z,\ Dz = x + 2y - z.$

4. $Dx = -2x + 3z,\ Dy = 4y,\ Dz = -6x + 7z.$

5. $Dx = 2x + y + 2z,\ Dy = -x - 2z,\ Dz = z.$

6. $Dx = 3x + y - z,\ Dy = -x + 2y + z,\ Dz = x + y + z.$

7. $Dx = 3x,\ Dy = 3y,\ Dz = 3z,\ Du = 3u.$

8. $Dx = -7x - 4u,\quad Dy = -13x - 2y - z - 8u,\quad Dz = 6x + y + 4u,$
$Du = 15x + y + 9u.$

9. Given a system $Dx_i = \sum_{j=1}^{n} a_{ij}x_j,\ (i = 1, \ldots, n)$, let λ_1 be a real root of multiplicity k and let Δ_1 represent the characteristic determinant, with λ replaced by λ_1, so that Δ_1 has value 0. Let Δ_1 have the property that every minor of order greater than $n - k$ is 0, but some minor of order $n - k$ is not 0 (so that Δ_1 has rank $n - k$). Show that k linearly independent solutions are obtainable in the form $x_i = \alpha_{il}e^{\lambda_1 t},\ (i = 1, \ldots, n;\ l = 1, \ldots, k)$. Can Δ_1 have rank less than $n - k$?

ANSWERS

1. $x = C_1 e^{-t},\ y = C_2 e^{-t}.$

2. $x = e^{-3t}[C_1 + C_2(t - 1)],\ y = e^{-3t}[-C_1 - C_2 t].$

3. $x = e^t[2C_1 + C_2(4t - 4)] + C_3,\ y = e^t(2C_1 + 4C_2 t),$
$z = e^t[3C_1 + (6t - 5)C_2] + C_3.$

4. $x = C_1 e^{4t} + C_2 e^t,\ y = C_3 e^{4t},\ z = 2C_1 e^{4t} + C_2 e^t.$

5. $x = e^{2t}[-C_1 - 2C_2 + C_3(t + 1)],\ y = e^t(C_1 - tC_3),\ z = C_2 e^t.$

6. $x = e^{2t}[C_1 + (t + 1)C_2 + (\tfrac{1}{2}t^2 + t)C_3],\ y = e^{2t}[C_2 + C_3(t + 2)],$
$z = e^{2t}\{C_1 + C_2(t + 1) + C_3[(t^2/2) + t + 1]\}.$

7. $x = C_1 e^{3t},\ y = C_2 e^{3t},\ z = C_3 e^{3t},\ u = C_4 e^{3t}.$

8. $x = -4C_1 e^{it} - 4\overline{C}_1 e^{-it} - 4tC_2 e^{it} - 4t\overline{C}_2 e^{-it},$
$y = (-4 - 2i)C_1 e^{it} + (-4 + 2i)\overline{C}_1 e^{-it} + C_2 e^{it}[(-4 - 2i)t - 2 + 2i]$
$\qquad\qquad\qquad\qquad\qquad + \overline{C}_2 e^{-it}[(-4 + 2i)t - 2 - 2i],$
$z = 2C_1 e^{it} + 2\overline{C}_1 e^{-it} + (2t + 2)C_2 e^{it} + (2t + 2)\overline{C}_2 e^{-it},$
$u = C_1(7 + i)e^{it} + \overline{C}_1(7 - i)e^{-it} + C_2 e^{it}[(7 + i)t + 1]$
$\qquad\qquad\qquad\qquad\qquad + \overline{C}_2 e^{-it}[(7 - i)t + 1].$

6-5 Nonhomogeneous systems. Variation of parameters. We consider the nonhomogeneous system of third order

$$Dx = a_1 x + b_1 y + c_1 z + f_1(t),$$
$$Dy = a_2 x + b_2 y + c_2 z + f_2(t),\qquad\qquad (6\text{--}28)$$
$$Dz = a_3 x + b_3 y + c_3 z + f_3(t),$$

(where $D = d/dt$), and assume that the general solution of the related homogeneous system

$$Dx = a_1x + b_1y + c_1z, \qquad Dy = a_2x + \cdots ,$$
$$Dz = a_3x + \cdots \tag{6-29}$$

is known. Let this general solution be given in the form

$$x = C_1x_1(t) + C_2x_2(t) + C_3x_3(t),$$
$$y = C_1y_1(t) + C_2y_2(t) + C_3y_3(t), \tag{6-30}$$
$$z = C_1y_1(t) + C_2y_2(t) + C_3y_3(t).$$

We now replace the constants C_1, C_2, C_3 by variables v_1, v_2, v_3:

$$x = v_1x_1(t) + v_2x_2(t) + v_3x_3(t),$$
$$y = v_1y_1(t) + v_2y_2(t) + v_3y_3(t), \tag{6-31}$$
$$z = v_1z_1(t) + v_2z_2(t) + v_3z_3(t).$$

As in Section 4–9, the equations (6–31) are considered as describing a change of variables (x, y, z are replaced by v_1, v_2, v_3). If we substitute in (6–28), we obtain differential equations for v_1, v_2, v_3 in terms of t. Thus the first equation of (6–28) becomes

$$v_1Dx_1 + v_2Dx_2 + v_3Dx_3 + x_1Dv_1 + x_2Dv_2 + x_3Dv_3$$
$$= a_1(v_1x_1 + v_2x_2 + v_3x_3) + b_1(v_1y_1 + v_2y_2 + v_3y_3)$$
$$+ c_1(v_1z_1 + v_2z_2 + v_3z_3) + f_1(t). \tag{6-32}$$

However, since the triple x_1, y_1, z_1 is a solution of the homogeneous system (6–29), we have the relation

$$v_1Dx_1 = v_1(a_1x_1 + b_1y_1 + c_1z_1).$$

We obtain similar expressions for v_2Dx_2, v_3Dx_3. Hence (6–32) reduces to the equation

$$x_1Dv_1 + x_2Dv_2 + x_3Dv_3 = f_1(t).$$

Similar reasoning applies to the remaining equations (6–28); hence the new equations for v_1, v_2, v_3 are

$$x_1Dv_1 + x_2Dv_2 + x_3Dv_3 = f_1(t),$$
$$y_1Dv_1 + y_2Dv_2 + y_3Dv_3 = f_2(t), \tag{6-33}$$
$$z_1Dv_1 + z_2Dv_2 + z_3Dv_3 = f_3(t).$$

In (6–33), $x_1, \ldots, y_1, \ldots, z_1, \ldots$ represent known functions of t, and v_1, v_2, v_3 are sought as functions of t. Since (6–33) is a set of simultaneous linear equations for Dv_1, Dv_2, Dv_3, we can solve by elimination or by determinants to obtain the expressions

$$Dv_1 = g_1(t), \qquad Dv_2 = g_2(t), \qquad Dv_3 = g_3(t).$$

Then v_1, v_2, v_3 are obtained by integration:

$$v_1 = \int g_1(t)\, dt, \qquad v_2 = \int g_2(t)\, dt, \qquad v_3 = \int g_3(t)\, dt.$$

Only one choice of the indefinite integrals is needed; this choice yields one particular solution

$$x^*(t) = v_1 x_1 + v_2 x_2 + v_3 x_3, \qquad y^*(t) = \ldots, \qquad z^*(t) = \ldots,$$

in accordance with (6–31). The general solution is then given by

$$x = x^*(t) + C_1 x_1(t) + C_2 x_2(t) + C_3 x_3(t),$$
$$y = y^*(t) + \cdots, \qquad z = z^*(t) + \cdots,$$

by Theorem 1 of Section 6–1.

EXAMPLE. The system is

$$Dx = 4x - 9y + 5z + 1 + 13t, \qquad Dy = x - 10y + 7z + 3 + 15t,$$
$$Dz = x - 17y + 12z + 2 + 26t.$$

It is the same as Example 2 of Section 6–1. The solution of the related homogeneous system is given by (6–9):

$$x = C_1 e^t + C_2 e^{2t} + C_3 e^{3t},$$
$$y = 2C_1 e^t + 3C_2 e^{2t} - C_3 e^{3t},$$
$$z = 3C_1 e^t + 5C_2 e^{2t} - 2C_3 e^{3t}.$$

We now write

$$x = v_1 e^t + v_2 e^{2t} + v_3 e^{3t},$$
$$y = 2v_1 e^t + 3v_2 e^{2t} - v_3 e^{3t},$$
$$z = 3v_1 e^t + 5v_2 e^{2t} - 2v_3 e^{3t},$$

and we try to determine v_1, v_2, v_3 as functions of t, so that the corresponding functions x, y, z satisfy the given nonhomogeneous system. The equations (6–33) become

$$e^t Dv_1 + e^{2t} Dv_2 + e^{3t} Dv_3 = 1 + 13t,$$

$$2e^t Dv_1 + 3e^{2t} Dv_2 - e^{3t} Dv_3 = 3 + 15t,$$

$$3e^t Dv_1 + 5e^{2t} Dv_2 - 2e^{3t} Dv_3 = 2 + 26t.$$

Elimination of v_3 leads to the equations

$$3e^t Dv_1 + 4e^{2t} Dv_2 = 4 + 28t, \qquad 5e^t Dv_1 + 7e^{2t} Dv_2 = 4 + 52t,$$

from which we find

$$Dv_1 = (12 - 12t)e^{-t}, \qquad Dv_2 = (-8 + 16t)e^{-2t},$$

$$v_1 = 12te^{-t}, \qquad\qquad v_2 = -8te^{-2t},$$

$$Dv_3 = (-3 + 9t)e^{-3t}, \qquad v_3 = -3te^{-3t},$$

$$x = 12t - 8t - 3t = t,$$

$$y = 24t - 24t + 3t = 3t,$$

$$z = 36t - 40t + 6t = 2t.$$

The general solution is then as given in Section 6–1:

$$x = C_1 e^t + C_2 e^{2t} + C_3 e^{3t} + t, \qquad y = \ldots, \qquad z = \ldots$$

Remark 1. Solution of Eqs. (6–33) for Dv_1, Dv_2, Dv_3 is possible, for arbitrary $f_1(t), f_2(t), f_3(t)$, only if the determinant of coefficients is never 0. As pointed out in Section 6–1, linear independence of the three solutions $x_j(t)$, $y_j(t)$, $z_j(t)$, $(j = 1, 2, 3)$, guarantees that this determinant cannot be zero. (See Prob. 6 following Section 6–1.)

Remark 2. As for the single linear equation of order n, the method of variation of parameters is applicable to equations with *variable* coefficients. However, when the coefficients are variable, solution of the related homogeneous system becomes much more difficult.

6–6 Nonhomogeneous systems. Operational methods.

Differential operators can be used to carry out an elimination process resembling that for simultaneous linear equations in algebra. However, certain precautions should be followed if we are to avoid extraneous solutions, which can be rejected only after a time-consuming check in the original differential equations. The following principle is the basis of the process: *the two linear equations*

$$f_1(D)x + g_1(D)y + \cdots = F_1(t), \tag{6–34}$$

$$f_2(D)x + g_2(D)y + \cdots = F_2(t) \tag{6–35}$$

are equivalent to the two equations

$$f_1(D)x + g_1(D)y + \cdots = F_1(t), \tag{6-36}$$

$$[\phi(D)f_1(D) + kf_2(D)]x + [\phi(D)g_1(D) + kg_2(D)]y + \cdots$$
$$= \phi(D)F_1(t) + kF_2(t); \tag{6-37}$$

that is, if $x = x(t)$, $y = y(t)$, . . . satisfy the first pair of equations, then $x = x(t)$, $y = y(t)$, . . . satisfy the second pair of equations, and conversely. The number k is a nonzero constant. The operators $f_1(D)$, $g_1(D)$, . . . , $f_2(D)$, $g_2(D)$, . . . , and $\phi(D)$ may have variable coefficients (depending on t), but we shall apply the principle only for operators with constant coefficients. We shall assume that $F_1(t)$, $F_2(t)$ have continuous derivatives of all orders required. To justify the principle, we remark that if $x(t)$, $y(t)$, . . . satisfy (6–34) and (6–35), then (6–36) holds since it is the same as (6–34), and (6–37) holds since it is obtained from (6–34), (6–35) by applying $\phi(D)$ to the first equation, multiplying the second equation by k, and adding the results. Conversely, if (6–36), (6–37) hold, then (6–34) holds since it is the same as (6–36), and (6–35) holds since it is obtained from (6–36), (6–37) by multiplying (6–36) by $\phi(D)$, subtracting (6–37), and then dividing by $-k$ (assumed different from 0).

To illustrate the procedure, we consider the two equations

$$3x - (D - 1)y = 1, \tag{6-38}$$

$$(D - 1)x + (D - 2)y = e^t. \tag{6-39}$$

We multiply Eq. (6–38) by $\phi(D) = D - 1$, Eq. (6–39) by $k = -3$, and then add, thereby eliminating x:

$$(-D^2 - D + 5)y = -1 - 3e^t. \tag{6-39'}$$

The new equation and the old equation (6–38) are equivalent to (6–38), (6–39). The numbering (6–39′) indicates that the new equation replaces (6–39). Thus *the elimination process retains the equation which has been multiplied by $\phi(D)$, and replaces the one which has been multiplied by k by a new equation.* It can happen that $\phi(D)$ reduces to a nonzero constant; in that case, the elimination process is the same as in ordinary algebra, and we can use the new equation as a replacement for either one of the given equations.

We could eliminate y from Eqs. (6–38), (6–39) by multiplying the first by $(D - 2)$, the second by $(D - 1)$, and adding the results:

$$(D^2 + D - 5)x = -2. \tag{6-40}$$

However, although Eq. (6–40) is implied by (6–38), (6–39), we cannot combine it with either of the given equations to obtain the other. It is this sort of procedure which leads to extraneous solutions. Thus from the given equations we obtain (6–39′), (6–40) and, accordingly,

$$x = \tfrac{2}{5} + c_1 e^{at} + c_2 e^{bt}, \tag{6–41}$$

$$y = -\tfrac{1}{5} - e^t + c_3 e^{at} + c_4 e^{bt}, \tag{6–42}$$

where a, b are the numbers $\tfrac{1}{2}(-1 \pm \sqrt{21})$. However, (6–41), (6–42) contain too many arbitrary constants; if we substitute in (6–38), we find

$$[3c_1 - (a - 1)c_3]e^{at} + [3c_2 - (b - 1)c_4]e^{bt} = 0;$$

hence $c_1 = \tfrac{1}{3}(a - 1)c_3$, $c_2 = \tfrac{1}{3}(b - 1)c_4$, and

$$x = \tfrac{2}{5} + \tfrac{1}{3}[(a - 1)c_3 e^{at} + (b - 1)c_4 e^{bt}]. \tag{6–43}$$

With this expression for x, and with y as given in (6–42), we find that (6–39) is also satisfied; thus the general solution contains two arbitrary constants. The same result could have been obtained without any possibility of extraneous solutions from (6–38), (6–39′); solving (6–39′) for y and substituting in (6–38) immediately gives (6–42), (6–43) as the general solution.

As a further illustration, we consider the system

$$(D - 14)x + 8y - 2z = t, \tag{I}$$

$$-41x + (D + 24)y - 7z = 0, \tag{II}$$

$$-73x + 44y + (D - 15)z = 0. \tag{III}$$

We have numbered the equations (I), (II), (III) to emphasize that at each stage there will be three equations; as primes are added, individual equations will be replaced by others. We first seek to eliminate z. Multiplication of Eq. (I) by 7, of Eq. (II) by -2, and addition yield the new equation

$$(7D - 16)x + (-2D + 8)y = 7t. \tag{II′}$$

Multiplication of Eq. (I) by $(D - 15)$, of Eq. (III) by 2, and addition yield

$$(D^2 - 29D + 64)x + (8D - 32)y = 1 - 15t. \tag{III′}$$

Multiplication of Eq. (II′) by 4, of Eq. (III′) by 1, and addition yield

$$(D^2 - D)x = 13t + 1. \tag{III″}$$

Thus our given system (I), (II), (III) has been replaced by the equivalent system (I), (II'), (III''). From (III'') we easily find x:

$$x = c_1 + c_2 e^t - \tfrac{13}{2}t^2 - 14t; \tag{IV}$$

substitution in (II') gives an equation for y:

$$(2D - 8)y = (7D - 16)x - 7t$$
$$= -16c_1 - 9c_2 + 104t^2 + 126t - 98.$$

We find

$$y = 2c_1 + \tfrac{3}{2}c_2 e^t + c_3 e^{4t} - 13t^2 - \tfrac{89}{4}t + \tfrac{107}{16}. \tag{V}$$

Substitution of the expressions (IV), (V) in (I) gives z:

$$z = c_1 - \tfrac{1}{2}c_2 e^t + 4c_3 e^{4t} - \tfrac{13}{2}t^2 + 2t + \tfrac{79}{4}. \tag{VI}$$

Equations (IV), (V), (VI) give the general solution of the given system. We note that there are three arbitrary constants, but that the expression for x contains only two constants.

It is not difficult to show that the elimination process can always be applied to successively eliminate all but one unknown. For three unknowns we obtain an equivalent system of "triangular form":

$$f_1(D)x = F_1(t),$$
$$f_2(D)x + g_2(D)y = F_2(t),$$
$$f_3(D)x + g_3(D)y + h_3(D)z = F_3(t).$$

From these equations we obtain x, y, z, in turn. For further information see p. 160 of Reference 4, pp. 148–149 of Reference 8, and pp. 138–142 of Reference 10.

Remark. The operational procedure can also be applied to homogeneous equations, as an alternative to the method of Sections 6–2 through 6–4.

PROBLEMS

1. Find a particular solution for each of the following by variation of parameters; the general solution of the related homogeneous system is given by the answer to the corresponding problem following Section 6–3:

(a) $Dx = 7x + 6y - 10e^{3t}$, $Dy = 2x + 6y - 5e^{3t}$ (see Prob. 1 following Section 6–3);

(b) $Dx = -x + y + \cos t$, $Dy = -5x + 3y$ (see Prob. 2 following Section 6–3);

(c) $Dx = 16x + 14y + 38z - 2e^{-t}$, $Dy = -9x - 7y - 18z - 3e^{-t}$, $Dz = -4x - 4y - 11z + 2e^{-t}$ (see Prob. 3 following Section 6–3).

2. Apply the operational method to obtain the general solution of each of the following parts of Prob. 1:

(a) part (a), (b) part (b), (c) part (c).

3. Obtain the general solution by the operational method:

$$Dx = x + y, \qquad Dy = y + z, \qquad Dz = z + u, \qquad Du = u + x.$$

4. Let the following nonhomogeneous linear system be given for $t > 0$:

$$t^3 Dx + (-1 - 2t^2)x + ty = t^2, \qquad t^4 Dy + (t^2 - 1)x + (t - 2t^3)y = t^3.$$

(a) Verify that the following pairs are linearly independent solutions of the related homogeneous system:

$$x_1 = t^2, \qquad y_1 = t; \qquad x_2 = t, \qquad y_2 = t^2 + 1.$$

(b) Obtain the general solution of the related homogeneous system.

(c) Apply variation of parameters to obtain the general solution of the given nonhomogeneous system.

ANSWERS

1. (a) $x = e^{3t}$, $y = e^{3t}$; (b) $x = \sin t - \cos t$, $y = 2 \sin t - \cos t$;
   (c) $x = -2te^{-t}$, $y = -3te^{-t}$, $z = 2te^{-t}$.

2. (a) $x = 2c_1 e^{10t} + e^{3t}(3c_2 + 1)$, $y = c_1 e^{10t} + e^{3t}(-2c_2 + 1)$;
   (b) $x = e^t(c_1 \cos t + c_2 \sin t) - \cos t + \sin t$,
       $y = e^t[c_1(2 \cos t - \sin t) + c_2(2 \sin t + \cos t)] + 2 \sin t - \cos t$;
   (c) $x = c_1 e^{2t} - 2e^{-t}(c_2 + t) - 2c_3 e^{-3t}$, $y = -c_1 e^{2t} - 3e^{-t}(c_2 + t)$,
       $z = 2e^{-t}(c_2 + t) + c_3 e^{-3t}$.

3. $x = c_1 + c_2 e^{2t} + e^t(c_3 \cos t + c_4 \sin t)$,
   $y = -c_1 + c_2 e^{2t} + e^t(c_4 \cos t - c_3 \sin t)$,
   $z = c_1 + c_2 e^{2t} + e^t(-c_3 \cos t - c_4 \sin t)$,
   $u = -c_1 + c_2 e^{2t} + e^t(-c_4 \cos t + c_3 \sin t)$.

4. (b) $x = c_1 t^2 + c_2 t$, $y = c_1 t + c_2(t^2 + 1)$;
   (c) $x = c_1 t^2 + c_2 t + (-6 - 2t^{-1} + t^{-2})/12$,
       $y = c_1 t + c_2(t^2 + 1) + (-6 - 2t^{-1} - 2t^{-2} + t^{-3})/12$.

6–7 Linear systems in general form. As remarked in Section 6–1, there are more general systems of simultaneous linear differential equations which are reducible to the basic form of (6–1), (6–2), or (6–3). For example, the equations

$$(D^2 - 1)x + (D + 2)y = t^2, \qquad (D^2 + D)y + 2Dx = 1 - t$$

$$(6\text{–}44)$$

are reducible to basic form by the introduction of the auxiliary variables $z = Dx$, $u = Dy$; Eqs. (6–44) are then equivalent to the four equations

$$Dx = z, \qquad Dy = u, \qquad Dz - x + u + 2y = t^2,$$
$$Du + u + 2z = 1 - t. \tag{6–45}$$

The most general case to be considered here is that of n linear equations in n unknowns, with coefficients which are polynomials in D:

$$\phi_{11}(D)x_1 + \cdots + \phi_{1n}(D)x_n = f_1(t),$$
$$\phi_{21}(D)x_1 + \cdots + \phi_{2n}(D)x_n = f_2(t),$$
$$\vdots \tag{6–46}$$
$$\phi_{n1}(D)x_1 + \cdots + \phi_{nn}(D)x_n = f_n(t).$$

When the $f_j(t)$ are all 0, the system is called *homogeneous*. For the sake of simplicity, we assume the ϕ_{ij} to have constant coefficients; in general, the coefficients would be permitted to depend on t. Equations (6–44) provide an example with two unknown functions x, y; the following is an example with three unknowns:

$$(D^2 - 2D + 1)x + (D - 3)y + Dz = 1,$$
$$(D^2 + 1)x + (2D + 1)y + (D - 2)z = t, \tag{6–47}$$
$$(4D^2 - 4)x + (26D + 20)y - (19D + 4)z = t^2.$$

For each unknown a derivative of highest order will appear. For example, in (6–47) the derivatives of highest order are D^2x, Dy, Dz. If the determinant of coefficients of the highest derivatives is zero, we call the system *degenerate;* if the determinant is not zero, we call the system *nondegenerate.* For (6–47) the determinant is

$$\begin{vmatrix} 1 & 1 & 1 \\ 1 & 2 & 1 \\ 4 & 26 & -19 \end{vmatrix} = -23;$$

hence the system is nondegenerate.

A nondegenerate system can always be reduced to the basic form of Section 6–1. For since the determinant of coefficients is nonzero, we can solve the equations for the highest derivatives in terms of the remaining quantities. Introduction of new unknowns equal to the derivatives of lower order leads at once to the basic form. Thus, for example, (6–47) can be solved for D^2x, Dy, Dz. We write the equations as follows:

$$D^2x + Dy + Dz = (2D - 1)x + 3y + 1,$$
$$D^2x + 2Dy + Dz = -x - y + 2z + t,$$
$$4D^2x + 26Dy - 19Dz = 4x - 26y + 4z + t^2.$$

Then either by elimination or by determinants we find

$$D^2x = \tfrac{1}{23}(128Dx - 15x + 211y - 86z + t^2 - 45t + 64),$$
$$Dy = -2Dx - 4y + 2z + t - 1,$$
$$Dz = \tfrac{1}{23}(-36Dx - 8x - 50y + 40z - t^2 + 22t - 18).$$

If we then set $u = Dx$, we obtain the system

$$Dx = u, \qquad Dy = -2u - 4y + 2z + t - 1,$$
$$Dz = \tfrac{1}{23}(-8x - 36u - 50y + 40z - t^2 + 22t - 18), \qquad (6\text{--}48)$$
$$Du = \tfrac{1}{23}(128u - 15x + 211y - 86z + t^2 - 45t + 64).$$

We note that if (6–47) is replaced by the related homogeneous system, then so also is (6–48).

Since a nondegenerate system can be reduced to basic form, the existence theorem of Section 6–1 is applicable. For example, we can conclude that Eqs. (6–47) have a unique solution with given initial values of x, y, z and $u = Dx$. The general solution will be formed from four linearly independent solutions of the related homogeneous system and one particular solution of the nonhomogeneous system. (The related homogeneous system is first defined with reference to the equations in basic form; however, as the above example shows, it is equivalent to the system obtained by replacing the right-hand sides of (6–46) by 0.)

If the system is degenerate, the reduction to basic form is in general impossible. The equations may in fact be contradictory, as the following example shows:

$$Dx + 2Dy = 1, \qquad Dx + 2Dy = 2.$$

The variety of other possibilities is shown by Prob. 4 following Section 6–9. For a full discussion see pp. 138–142 of Reference 10.

6–8 Homogeneous linear systems in general form. We can rewrite a nondegenerate homogeneous linear system in basic form and then follow the procedures of Sections 6–2 through 6–4. However, we can achieve the same results by substituting directly in the given equations.

We illustrate the procedure by considering the homogeneous system related to (6–47):

$$(D^2 - 2D + 1)x + (D - 3)y + Dz = 0,$$
$$(D^2 + 1)x + (2D + 1)y + (D - 2)z = 0, \qquad (6\text{--}49)$$
$$(4D^2 - 4)x + (26D + 26)y - (19D + 4)z = 0.$$

We set $x = \alpha e^{\lambda t}$, $y = \beta e^{\lambda t}$, $z = \gamma e^{\lambda t}$ and obtain

$$(\lambda^2 - 2\gamma + 1)\alpha + (\lambda - 3)\beta + \lambda\gamma = 0,$$
$$(\lambda^2 + 1)\alpha + (2\lambda + 1)\beta + (\lambda - 2)\gamma = 0, \qquad (6\text{--}50)$$
$$(4\lambda^2 - 4)\alpha + (26\lambda + 20)\beta - (19\lambda + 4)\gamma = 0.$$

These equations have a nontrivial solution α, β, γ when

$$\begin{vmatrix} \lambda^2 - 2\lambda + 1 & \lambda - 3 & \lambda \\ \lambda^2 + 1 & 2\lambda + 1 & \lambda - 2 \\ 4\lambda^2 - 4 & 26\lambda + 20 & -19\lambda - 4 \end{vmatrix} = 0; \qquad (6\text{--}51)$$

this is indeed the *characteristic equation* of the given system (or of the equivalent system in basic form). When expanded, (6–51) is a *fourth* degree equation in λ:

$$-23\lambda^4 + 76\lambda^3 + 23\lambda^2 - 76\lambda = 0.$$

The roots are found to be 0, ± 1, $76/23$, and since they are simple, we obtain four linearly independent solutions of form $\alpha e^{\lambda t}$, $\beta e^{\lambda t}$, $\gamma e^{\lambda t}$. For example, when $\lambda = 0$, Eqs. (6–50) become

$$\alpha - 3\beta = 0,$$
$$\alpha + \beta - 2\gamma = 0,$$
$$-4\alpha + 20\beta - 4\gamma = 0;$$

thus $\alpha = 3\beta$, $\gamma = 2\beta$. We can choose $\alpha = 3$, $\beta = 1$, $\gamma = 2$ and obtain the solution $x = 3$, $y = 1$, $z = 2$. Similarly, $\lambda = 1$, $\lambda = -1$, $\lambda = 76/23$ lead to solutions of the form sought, and the general solution is

$$x = 3c_1 + c_2 e^t + 11c_3 e^{-t} - 301{,}070 c_4 e^{76t/23},$$
$$y = c_1 - 2c_2 e^t + 10c_3 e^{-t} + 394{,}910 c_4 e^{76t/23}, \qquad (6\text{--}52)$$
$$z = 2c_1 - 4c_2 e^t + 4c_3 e^{-t} + 447{,}440 c_4 e^{76t/23}.$$

If we adjoin the relation

$$u = Dx = c_2 e^t - 11c_3 e^{-t} - 13{,}090 c_4 e^{76t/23}, \qquad (6\text{--}53)$$

then we have the general solution of the equations in basic form.

If λ_1 is a multiple root of the characteristic equation, then we proceed as in Section 6–4. Complex roots are also handled as in the previous case.

6–9 Nonhomogeneous linear systems in general form. Both the method of variation of parameters and the operational method can be adapted to nonhomogeneous equations in general form. We apply both methods to the following example:

$$(3D^2 + 1)x + (D^2 + 3)y = f(t),$$
$$(2D^2 + 1)x + (D^2 + 2)y = g(t).$$
$$(6\text{--}54)$$

Variation of parameters. The substitution $x = \alpha e^{\lambda t}$, $y = \beta e^{\lambda t}$ in the related homogeneous system leads to the characteristic equation $\lambda^4 - 1 = 0$. The roots are ± 1, $\pm i$. The general solution of the homogeneous system is

$$x = c_1 e^t + c_2 e^{-t} + c_3 \cos t + c_4 \sin t,$$
$$y = -c_1 e^t - c_2 e^{-t} + c_3 \cos t + c_4 \sin t. \qquad (6\text{--}55)$$

We introduce the variables $z = Dx$, $u = Dy$, so that

$$z = c_1 e^t - c_2 e^{-t} - c_3 \sin t + c_4 \cos t,$$
$$u = -c_1 e^t + c_2 e^{-t} - c_3 \sin t + c_4 \cos t. \qquad (6\text{--}56)$$

Thus (6–55), (6–56) together give the general solution of the homogeneous system in basic form.

We now replace the constants c_1, c_2, c_3, c_4 by variables v_1, v_2, v_3, v_4. Equations (6–55), (6–56) thus become equations that describe a change of variables:

$$x = v_1 e^t + v_2 e^{-t} + v_3 \cos t + v_4 \sin t,$$
$$y = -v_1 e^t - v_2 e^{-t} + v_3 \cos t + v_4 \sin t,$$
$$z = v_1 e^t - v_2 e^{-t} - v_3 \sin t + v_4 \cos t,$$
$$u = -v_1 e^t + v_2 e^{-t} - v_3 \sin t + v_4 \cos t. \qquad (6\text{--}57)$$

To obtain differential equations for $v_1, \ldots, v_4$ as functions of t, we substitute in the relations

$$3Dz + x + Du + 3y = f, \qquad 2Dz + x + Du + 2y = g,$$
$$Dx = z, \qquad Dy = u. \qquad (6\text{--}58)$$

[The first two equations in (6–58) are simply the given differential equations (6–55).] After substitution we obtain the equations

$$2v_1'e^t - 2v_2'e^{-t} - 4v_3' \sin t + 4v_4' \cos t = f,$$

$$v_1'e^t - v_2'e^{-t} - 3v_3' \sin t + 3v_4' \cos t = g,$$

$$v_1'e^t + v_2'e^{-t} + v_3' \cos t + v_4' \sin t = 0, \tag{6-59}$$

$$-v_1'e^t - v_2'e^{-t} + v_3' \cos t + v_4' \sin t = 0.$$

The terms in $v_1, \ldots, v_4$ drop out, simply because (6–55) and (6–56) are the general solution of the homogeneous system related to (6–58). From (6–59) we easily find

$$v_1' = \frac{e^t}{4}(3f - 4g), \qquad v_2' = \frac{e^t}{4}(4f - 3g), \tag{6-60}$$

$$v_3' = \tfrac{1}{2}(f \sin t - 2g \sin t), \qquad v_4' = \tfrac{1}{2}(2g \cos t - f \cos t).$$

If $f(t)$, $g(t)$ are given, $v_1, \ldots, v_4$ are easily found by integration. Substitution in (6–57) then yields a particular solution, which can be combined with (6–55) to give the general solution.

Operational method. We proceed as in Section 6–6. We obtain a system equivalent to (6–54) by replacing the second equation by the first minus the second:

$$(3D^2 + 1)x + (D^2 + 3)y = f(t), \tag{I}$$

$$D^2x + y = f(t) - g(t). \tag{II'}$$

Multiplication of Eq. (I) by -1, of Eq. (II') by $(D^2 + 3)$, and addition give

$$(D^4 - 1)x = (D^2 - 4)f - 3g. \tag{I'}$$

We solve (I') for x and then obtain y from Eq. (II'):

$$x = c_1e^t + c_2e^{-t} + c_3 \cos t + c_4 \sin t + x^*(t), \tag{6-61}$$

$$y = -c_1e^t - c_2e^{-t} + c_3 \cos t + c_4 \sin t + f(t) - g(t) - D^2x^*(t). \tag{6-62}$$

Equations (6–61) and (6–62) provide the general solution sought.

PROBLEMS

1. Verify that each of the following systems is nondegenerate, and obtain the general solution:

(a) $(D + 1)x + (D + 2)y = 0$, $(7D - 5)x + (8D - 4)y = 0$;

(b) $(2D - 3)x + (3D - 6)y + (D^2 + D + 5)z = 0$,
$\quad (7D - 12)x + (11D - 24)y + (3D^2 + 4D + 12)z = 0$,
$\quad (D - 3)x + (2D - 6)y + (D^2 + 3D - 1)z = 0$;

(c) $(D - 8)x - 4y + (D - 12)z = 0,$
   $(D + 1)x + Dy + (2D + 1)z = 0,$
   $(3D - 6)x + (2D - 4)y + (D^2 + 5D - 11)z = 0.$

2. Write each of the systems of Prob. 1 in basic form.

3. Obtain the general solution of each of the following systems. The related homogeneous systems are considered in Prob. 1.

(a) $(D + 1)x + (D + 2)y = 5,$ $(7D - 5)x + (8D - 4)y = 2.$
(b) $(2D - 3)x + (3D - 6)y + (D^2 + D + 5)z = 0,$
   $(7D - 12)x + (11D - 24)y + (3D^2 + 4D + 12)z = 0,$
   $(D - 3)x + (2D - 3)y + (D^2 + 3D - 1)z = e^{3t}.$
(c) $(D - 8)x - 4y + (D - 12)z = \cos 3t,$
   $(D + 1)x + Dy + (2D + 1)z = 0,$
   $(3D - 6)x + (2D - 4)y + (D^2 + 5D - 11)z = 0.$

4. Verify that each of the following systems is degenerate, and obtain all solutions (if there are any):

(a) $(D - 1)x + (2D + 1)y = 0,$ $(D + 1)x + (2D + 3)y = 0;$
(b) $(D - 1)x + (2D + 1)y = 2,$ $(D + 1)x + (2D + 3)y = 0;$
(c) $(D - 1)x + Dy + 2Dz = 1,$ $Dx - Dy + (D + 1)z = t,$
   $(D - 3)x + 5Dy + (4D - 2)z = 4 - 2t;$
(d) $(D^2 - 1)x + (D^2 + 1)y = 0,$ $(D + 2)x + (D + 2)y = 0;$
(e) $(D - 1)x + Dy = 0,$ $x - y = 1.$

ANSWERS

1. (a) $x = 5c_1e^{3t} + 4c_2e^{2t},$ $y = -4c_1e^{3t} - 3c_2e^{2t};$
   (b) $x = 2c_1 + c_2e^{3t} + e^{-t}[c_3(17 \cos t + 5 \sin t) + c_4(17 \sin t - 5 \cos t)],$
      $y = -c_1 - c_2e^{3t}$
         $+ e^{-t}[c_3 (-9 \cos t - 3 \sin t) + c_4 (-9 \sin t + 3 \cos t)],$
      $z = e^{-t}[c_3 (\cos t + \sin t) + c_4 (\sin t - \cos t)];$
   (c) $x = c_1e^t + c_2e^{-t} + 2c_3e^{2t} + c_4e^{2t}(2t - 1),$
      $y = c_1e^t + c_2e^{-t} - 3c_3e^{2t} + c_4e^{2t}(-3t + 2),$ $z = -c_1e^t - c_2e^{-t}.$

2. (a) $Dx = -13x - 20y,$ $Dy = 12x + 18y;$
   (b) $Dx = -3x - 6y + 5u - 15z,$ $Dy = 3x + 6y - 3u + 9z,$ $Dz = u,$
                                        $Du = - 2z - 2u;$
   (c) $Dx = 8x + 4y + 12z - u,$ $Dy = -9x - 4y - 13z - u,$ $Dz = u,$
      $Du = -6x - 4y + z.$

3. (a) $x = -4 + 5c_1e^{3t} + 4c_2e^{2t},$ $y = \frac{9}{2} - 4c_1e^{3t} - 3c_2e^{2t};$
   (b) $x = -\frac{1}{3}e^{3t} + 2c_1 + c_2e^{3t}$
         $+ e^{-t}[c_3(17 \cos t + 5 \sin t) + c_4(17 \sin t - 5 \cos t)],$
      $y = \frac{1}{3}e^{3t} - c_1 - c_2e^{3t}$
         $+ e^{-t}[c_3 (-9 \cos t - 3 \sin t) + c_4(3 \cos t - 9 \sin t)],$
      $z = e^{-t}[c_3 (\cos t + \sin t) + c_4 (\sin t - \cos t)];$

(c) $x = c_1e^t + c_2e^{-t} + 2c_3e^{2t} + c_4e^{2t}(2t - 1)$
$$- (529 \cos 3t - 150 \sin 3t)/1690,$$

$y = c_1e^t + c_2e^{-t} - 3c_3e^{2t} + c_4e^{2t}(2 - 3t)$
$$+ (241 \cos 3t - 30 \sin 3t)/1690,$$

$z = -c_1e^t - c_2e^{-t} + (\cos 3t)/10.$

4. (a) $x = ce^{-2t}$, $y = -ce^{-2t}$; (b) $x = ce^{-2t} - \frac{3}{2}$, $y = -ce^{-2t} + \frac{1}{2}$;
(c) no solution; (d) $x = 5ce^{-2t}$, $y = -3ce^{-2t}$; (e) $x = ce^{t/2}$, $y = -1 + ce^{t/2}$.

6–10 Application of Laplace transforms. The solution of a linear system, with constant coefficients, which satisfies prescribed initial conditions can be obtained efficiently with the aid of the Laplace transform. The procedures parallel those of Section 4–15. We illustrate them for the system

$$(3D^2 + 1)x + (D^2 + 3)y = f(t),$$
$$(2D^2 + 1)x + (D^2 + 2)y = g(t) \tag{6-63}$$

discussed in Section 6–9. The system is nondegenerate; hence we can prescribe initial values x_0 of x, y_0 of y, x_0' of $x' = Dx$, y_0' of $y' = Dy$, at time $t_0 = 0$. We take Laplace transforms on both sides of Eqs. (6–64) and apply the rule

$$\mathcal{L}[D^2u] = s^2\mathcal{L}[u] - [u'(0) + su(0)] \tag{6-64}$$

established in Section 4–15. We obtain the equations

$$(3s^2 + 1)\mathcal{L}[x] + (s^2 + 3)\mathcal{L}[y] = \mathcal{L}[f] + 3x_0' + 3sx_0 + y_0' + sy_0,$$
$$(2s^2 + 1)\mathcal{L}[x] + (s^2 + 2)\mathcal{L}[y] = \mathcal{L}[g] + 2x_0' + 2sx_0 + y_0' + sy_0. \tag{6-65}$$

These can be solved for $\mathcal{L}[x]$, $\mathcal{L}[y]$:

$$\mathcal{L}[x] = \frac{(s^2 + 2)\mathcal{L}[f] - (s^2 + 3)\mathcal{L}[g] + x_0s^3 + x_0's^2 - y_0s - y_0'}{s^4 - 1},$$

$$\mathcal{L}[y] = \frac{(3s^2 + 1)\mathcal{L}[g] - (2s^2 + 1)\mathcal{L}[f] + y_0s^3 + y_0's^2 - x_0s - x_0'}{s^4 - 1}.$$

If $f(t)$ and $g(t)$ are given for $t \geq 0$ (and satisfy appropriate conditions), $\mathcal{L}[f]$ and $\mathcal{L}[g]$ can be found; hence $\mathcal{L}[x]$, $\mathcal{L}[y]$ are known and x, y can be found. If a good set of tables of Laplace transforms is available, these steps usually cause little difficulty.

For example, if $f(t) = e^t$, $g(t) = e^{-t}$, $x_0 = 1$, $y_0 = 1$, $x_0' = 0$, $y_0' = 0$, we find (see Eq. 4–124)

$$\mathcal{L}[f] = \frac{1}{s - 1}, \qquad \mathcal{L}[g] = \frac{1}{s + 1},$$

$$\mathcal{L}[x] = \frac{s^5 - 2s^3 + 2s^2 + 5}{(s+1)^2(s-1)^2(s^2+1)}$$

$$= \frac{-7/8}{(s-1)^2} + \frac{1/4}{s-1} + \frac{1}{(s+1)^2} + \frac{9/8}{s+1} + \frac{(3s+3)/4}{s^2+1}.$$

$$\mathcal{L}[y] = \frac{s^3 - 5s^2 - 2}{(s+1)^2(s-1)^2(s^2+1)}$$

$$= \frac{-3/4}{(s-1)^2} + \frac{5/8}{s-1} + \frac{-1}{(s+1)^2} + \frac{-3/8}{s+1} + \frac{(3s+3)/4}{s^2+1},$$

with the aid of partial-fraction expansions. From Eqs. (4–124), (4–128), (4–129) we now conclude that

$$x = -\tfrac{7}{8}e^t + \tfrac{1}{4}te^t + \tfrac{9}{8}e^{-t} + te^{-t} + \tfrac{3}{4}(\cos t + \sin t),$$

$$y = \tfrac{5}{8}e^t - \tfrac{3}{4}te^t - \tfrac{3}{8}e^{-t} - te^{-t} + \tfrac{3}{4}(\cos t + \sin t).$$

Remark. If we do not give numerical values for the initial values but denote them as above by x_0, x_0', y_0, y_0', then we obtain the general solution, with x_0, x_0', ... as the arbitrary constants.

PROBLEMS

1. Find the solution for each of the following, with given initial values:

(a) $Dx = 7x + 6y$, $Dy = 2x + 6y$; $x = 1$, $y = 2$ for $t = 0$;

(b) $Dx = y$, $.Dy = 4x + 3y - 4z$, $Dz = x + 2y - z$; $x = 0$, $y = 0$, $z = 1$ for $t = 0$;

(c) the system of Prob. 1(b) following Section 6–9, with initial conditions $x = 0$, $y = 0$, $z = 0$, $Dz = 1$ for $t = 0$.

2. Find the general solution for each of the following, with initial values as arbitrary constants:

(a) the system of Prob. 1(a) following Section 6–9;

(b) the system of Prob. 1(c) following Section 6–9.

ANSWERS

1. (a) $x = (16e^{10t} - 9e^{3t})/7$, $y = (8e^{10t} + 6e^{3t})/7$;

(b) $x = -4 + 4e^t(1 - t)$, $y = -4te^t$, $z = e^t(5 - 6t) - 4$;

(c) $x = -6 + e^{(-1+i)t}[3 - (11/2)i] + e^{(-1-i)t}[3 + (11/2)i]$,

$y = -3 + 3e^{(-1+i)t}(-\tfrac{1}{2} + i) + 3e^{(-1-i)t}(-\tfrac{1}{2} - i)$,

$z = -\tfrac{1}{2}ie^{(-1+i)t} + \tfrac{1}{2}ie^{(-1-i)t}$.

2. (a) $x = x_0(16e^{2t} - 15e^{3t}) + 20y_0(e^{2t} - e^{3t})$,

    $y = 12x_0(e^{3t} - e^{2t}) + y_0(16e^{3t} - 15e^{2t})$;

(b) $x = x_0e^{2t}(1 + 6t) + 4y_0te^{2t} + \frac{1}{2}z_0(-e^t - e^{-t} + 2e^{2t} + 20te^{2t})$
$$+ \tfrac{1}{2}u_0(e^{-t} - e^t),$$

    $y = -9x_0te^{2t} + y_0(e^{2t} - 6te^{2t}) + \frac{1}{2}z_0(-e^t - e^{-t} + 2e^{2t} - 30te^{2t})$
$$+ \tfrac{1}{2}u_0(e^{-t} - e^t),$$

    $z = \frac{1}{2}z_0(e^t + e^{-t}) + \frac{1}{2}u_0(e^t - e^{-t})$, where $Dz = u_0$ for $t = 0$.

6-11 Input-output analysis. We show that the concepts of Chapter 5 can be extended to systems governed by simultaneous differential equations by considering a system of third order in basic form:

$$(D - a_1)x = b_1y + c_1z + F_1(t),$$

$$(D - b_2)y = a_2x + c_2z + F_2(t), \tag{6-66}$$

$$(D - c_3)z = a_3x + b_3y + F_3(t).$$

The coefficients will be assumed to be constant. We shall term the triple of functions $F_1(t)$, $F_2(t)$, $F_3(t)$ an *input;* each solution is a triple $x(t)$, $y(t)$, $z(t)$, which we call an *output.*

The system is called *stable* if for every solution $x(t)$, $y(t)$, $z(t)$ of the related homogeneous system, $x(t)$, $y(t)$, $z(t)$ all approach zero as $t \to +\infty$. As for the single equation of nth order, we verify that stability is equivalent to the condition that each characteristic root should have negative real part.

Let us consider a stable system. We seek the analog of the approximation output $\sim$ input (or output $\sim$ const $\times$ input), which is of such value for the single equation of nth order. Since we have three input functions and three output functions, it is not clear which output should approximate which input; we might try to associate F_1 with x, F_2 with y, F_3 with z, but it is easily seen that this is in general not a good choice.

More natural reasoning considers the terms in Dx, Dy, Dz as obstacles which prevent x, y, z from coinciding with the inputs. If we replace Dx, Dy, Dz by 0, we obtain

$$a_1x + b_1y + c_1z = -F_1(t),$$

$$a_2x + b_2y + c_2z = -F_2(t), \tag{6-67}$$

$$a_3x + b_3y + c_3z = -F_3(t),$$

which are simultaneous linear equations for x, y, z. Because of the stability of Eqs. (6-66), the determinant of coefficients in (6-67) is not zero (Prob. 5 below); hence we can solve for x, y, z. The results have the form

$$x = p_1 F_1(t) + p_2 F_2(t) + p_3 F_3(t),$$

$$y = q_1 F_1(t) + q_2 F_2(t) + q_3 F_3(t), \qquad (6\text{--}68)$$

$$z = r_1 F_1(t) + r_2 F_2(t) + r_3 F_3(t),$$

where $p_1, \ldots, q_1, \ldots, r_1, \ldots$ are constants (Prob. 6 below).

Equations (6–68) were obtained by neglecting the derivative terms in (6–66), and they must be regarded as an approximation to a solution of (6–66). As an approximation, they give valuable insight into the relation between output and input. Each of the three output functions depends on *all* the input functions (unless some of the constants p_j, q_j, r_j are 0); to a first approximation, *each output function is a linear combination of the three input functions.*

Particular inputs can be analyzed as in Chapter 5. If F_1, F_2, F_3 are constants, the equations (6–68) describe an exact solution; for x, y, z are then constants, their derivatives are zero, and Eqs. (6–66) are satisfied, with $Dx = 0$, $Dy = 0$, $Dz = 0$. The general solution is obtained by addition of the general solution of the homogeneous system. We assume stability, so that the terms added represent transients. Hence effectively there is only one output.

If $F_1(t)$, $F_2(t)$, $F_3(t)$ are changing very slowly, we expect the output to behave almost as if they were constant; that is, Eqs. (6–68) should again be a good approximation to the output. If $F_1(t)$, $F_2(t)$, $F_3(t)$ are step functions, and the jumps occur only at widely separated instants, the approximation (6–68) should remain accurate, except near the jumps, at which transient errors are introduced.

Sinusoidal inputs are most easily analyzed by means of their relationship to complex exponential functions. If the input is

$$F_1(t) = A_1 e^{bt}, \qquad F_2(t) = A_2 e^{bt}, \qquad F_3(t) = A_3 e^{bt}, \qquad (6\text{--}69)$$

where the A's are real constants, then there is a particular output of similar form:

$$x = \alpha e^{bt}, \qquad y = \beta e^{bt}, \qquad z = \gamma e^{bt}, \qquad (6\text{--}70)$$

provided b is not a characteristic root. For we can substitute (6–69), (6–70) in Eqs. (6–66) and solve for α, β, γ as undetermined coefficients. After canceling the factor e^{bt}, we obtain the linear equations

$$(a_1 - b)\alpha + b_1\beta + c_1\gamma = -A_1,$$

$$a_2\alpha + (b_2 - b)\beta + c_2\gamma = -A_2, \qquad (6\text{--}71)$$

$$a_3\alpha + b_3\beta + (c_3 - b)\gamma = -A_3.$$

Since b is not a characteristic root, the determinant of coefficients is not zero, and we obtain for α, β, γ a unique solution which has the form

$$\alpha = \xi_1 A_1 + \eta_1 A_2 + \zeta_1 A_3,$$
$$\beta = \xi_2 A_1 + \eta_2 A_2 + \zeta_2 A_3, \qquad (6\text{--}72)$$
$$\gamma = \xi_3 A_1 + \eta_3 A_2 + \zeta_3 A_3,$$

where the ξ_j, η_j, ζ_j are constants but do depend on the choice of b. If now $b = i\omega$, $(i = \sqrt{-1})$, then

$$F_1(t) = A_1 e^{i\omega t} = A_1 (\cos \omega t + i \sin \omega t),$$

and similar expressions hold for $F_2(t)$, $F_3(t)$. Equations (6–70) give the output in similar form. As in Section 4–13, Example 5, we verify that the real and imaginary parts correspond; that is, an output corresponding to the input triple

$$A_1 \cos \omega t, \quad A_2 \cos \omega t, \quad A_3 \cos \omega t \qquad (6\text{--}73)$$

is given by

$$x = \operatorname{Re} (\alpha e^{i\omega t}), \qquad y = \operatorname{Re} (\beta e^{i\omega t}), \qquad z = \operatorname{Re} (\gamma e^{i\omega t}); \qquad (6\text{--}74)$$

even though A_1, A_2, A_3 are real, α, β, γ will in general be complex. Thus (6–74), when expanded, gives x, y, z as sums of terms in both $\cos \omega t$ and $\sin \omega t$. If we replace the cosines by sines in (6–73), that is, use the *imaginary* parts of the inputs, then Re is replaced by Im in (6–74).

EXAMPLE. Consider the system

$$Dx = -4x + 9y - 5z + F_1(t),$$
$$Dy = -x + 10y - 7z + F_2(t),$$
$$Dz = -x + 17y - 12z + F_3(t).$$

The characteristic roots are found to be -1, -2, -3, so that the system is stable. With $F_1(t) = A_1 e^{bt}$, $F_2 = A_2 e^{bt}$, $F_3 = A_3 e^{bt}$, the equations for α, β, γ become

$$(b + 4)\alpha - 9\beta + 5\gamma = A_1,$$
$$\alpha + (b - 10)\beta + 7\gamma = A_2,$$
$$\alpha - 17\beta + (b + 12)\gamma = A_3.$$

Hence

$$\alpha = \frac{A_1(b^2 + 2b - 1) + A_2(9b + 23) + A_3(-5b - 13)}{b^3 + 6b^2 + 11b + 6},$$

$$\beta = \frac{-A_1(b + 5) + A_2(b^2 + 16b + 43) - A_3(7b + 23)}{b^3 + 6b^2 + 11b + 6},$$

$$\gamma = \frac{A_1(-b - 7) + A_2(17b + 59) + A_3(b^2 - 6b - 31)}{b^3 + 6b^2 + 11b + 6},$$

and $x = \alpha e^{bt}$, $y = \beta e^{bt}$, $z = \gamma e^{bt}$ is an output. If, in particular, $b = 2i$, and $A_1 = A_2 = A_3 = 1$, then

$$\alpha = \frac{3 - 11i}{20}, \qquad \beta = \frac{1 - 17i}{20}, \qquad \gamma = \frac{-1 - 23i}{20}.$$

If we take real parts of inputs and outputs, we obtain the expressions

$$F_1 = F_2 = F_3 = \cos 2t,$$

$$x = \frac{3 \cos 2t + 11 \sin 2t}{20}, \qquad y = \frac{\cos 2t + 17 \sin 2t}{20},$$

$$z = \frac{-\cos 2t + 23 \sin 2t}{20}.$$

Detailed studies of amplification and phase lag can be carried out, as in Section 5–6. Since the input and output are triples of functions, the description becomes much more elaborate; however, certain qualitative features are at once visible. For example, if $b = i\omega$ in the above example, then for large ω, α, β, γ have very small real and imaginary parts, approaching zero as $\omega \to +\infty$. For ω close to 0, α, β, γ are close to certain real values, and hence the phase lag will be very small.

When $F_1(t)$, $F_2(t)$, $F_3(t)$ are periodic functions of t, with common period τ, we can use Fourier series, as in Section 3–10, to obtain the output. The *superposition principle* is valid, just as for the single equation of nth order. Hence the terms of frequency ω in the Fourier series for the output functions correspond to the terms of the same frequency in the input functions. As in the example considered above, the terms of high frequency receive very little weight in the output; hence the system acts as a *smoothing device*. Some frequencies may receive exceptionally large weight; that is, there may be some form of resonance (see Prob. 3 below).

PROBLEMS

1. Determine whether each of the following systems is stable or unstable:

(a) $(D + 5)x - 2y = F_1(t)$, $(D + 2)y + x = F_2(t)$;

(b) $Dx = -x + y + z + F_1(t)$, $Dy = x - 2y + z + F_2(t)$,
$Dz = 2x + y + z + F_3(t)$;

(c) $Dx = 7x + 4u + F_1(t)$, $Dy = 13x + 2y + z + 8u + F_2(t)$,
$Dz = -6x - y - 4u + F_3(t)$, $Du = -15x - y - 9u + F_4(t)$.

2. Given the system of Prob. 1(a), determine an output for each of the following inputs:

(a) $F_1(t) \equiv 3$, $F_2(t) \equiv 7$;

(b) $F_1 = 4t$, $F_2 = 5t$;

(c) $F_1 = 2 \sin 3t$, $F_2 = 5 \sin 3t$;

(d) $F_1 = 0$, $F_2 = 0$ for $t < 0$; $F_1 = 3$, $F_2 = 3$ for $0 \leq t < 3$; $F_1 = 0$, $F_2 = 0$ for $t \geq 3$.

3. Given the system

$$Dx = -3x - y + z + F_1(t), \qquad Dy = x - 2y - z + F_2(t),$$
$$Dz = -x - y - z + F_3(t),$$

(a) test for stability;

(b) find an output when $F_1(t) = \cos \omega t$, $F_2 \equiv 0$, $F_3 \equiv 0$;

(c) for the input of part (b) find the amplitude of the sinusoidal output $x(t)$, graph the amplitude as a function of ω, and determine the maximum amplitude.

4. Given the system

$$(D + 1)x - y = F_1(t), \qquad (D - 1)y + 5x = F_2(t),$$

(a) show that the characteristic roots are $\pm 2i$;

(b) find the general solution when $F_1(t) = A \cos 2t$, $F_2(t) = \cos 2t + 2 \sin 2t$, and discuss the question of resonance. [This example illustrates the fact that for a system which can oscillate sinusoidally at frequency ω, application of forcing functions of the same frequency need not lead to resonance (i.e., oscillations of increasing amplitude). However, the forcing functions must be chosen with great care, as the slightest deviation from the proper choice will in general lead to resonance.]

5. Prove: if the system (6–66) is stable, then

$$\begin{vmatrix} a_1 & b_1 & c_1 \\ a_2 & b_2 & c_2 \\ a_3 & b_3 & c_3 \end{vmatrix} \neq 0.$$

[*Hint:* Show that if the determinant is 0, then 0 is a characteristic root.]

6. Show that if the determinant of Prob. 5 is not 0, then the solution of Eqs. (6–67) has form (6–68).

ANSWERS

1. (a) stable, (b) unstable, (c) unstable.

2. (a) $x = \frac{5}{3} + $ trans, $y = \frac{8}{3} + $ trans;   (b) $x = (36t - 13)/24 + $ trans,
$y = (84t - 29)/48 + $ trans;   (c) $x = (28 \sin 3t - 46 \cos 3t)/505 + $ trans,
$y = (64 \sin 3t - 73 \cos 3t)/75 + $ trans;   (d) $x = y = f(t) + $ trans, $f(t) = 0$
for $t \leqq 0$, $f(t) = 1 - e^{-3t}$ for $0 \leqq t \leqq 3$, $f(t) = (e^9 - 1)e^{-3t}$ for $t \geqq 3$.

3. (a) stable;   (b) $x = \operatorname{Re}\left[\{(1 - \omega^2) + 3i\omega\}(2 + i\omega)^{-3}e^{i\omega t}\right]$,
$y = \operatorname{Re}\left[(2 + i\omega)^{-2}e^{i\omega t}\right]$, $z = \operatorname{Re}\left[(-3 - i\omega)(2 + i\omega)^{-3}e^{i\omega t}\right]$;   (c) maximum
amplification of 0.3 at $\omega = 2.83$.

4. (b) $x = c_1 \cos 2t + c_2 \sin 2t + \frac{1}{4}(1 - A)(t \sin 2t - 2t \cos 2t)$,
$\quad\quad y = (c_2 - 2c_1) \sin 2t + (c_2 - 2c_2 - A) \cos 2t + \frac{1}{4}(1 - A)$
$\quad\quad\quad\quad \times (5t \sin 2t + \sin 2t - 2 \cos 2t)$;   no resonance if $A = 1$.

6–12 Applications in mechanics.   The majority of applications of
simultaneous linear differential equations in mechanics arise in the analysis
of oscillations about equilibrium. The linear nature of the equations generally results from an approximation, which is accurate only near the
equilibrium position; accordingly, the theory is often described as that
of "small vibrations."

One example, that of two coupled springs (Example B), is given in
Section 5–10. The differential equations are

$$(m_1 D^2 + b_1 D + 2k^2)x_1 - k^2 x_2 = F_1(t),$$
$$(m_2 D^2 + b_2 D + 2k^2)x_2 - k^2 x_1 = F_2(t), \tag{6–75}$$

where x_1, x_2 are coordinates measuring displacements from equilibrium,
$F_1(t)$, $F_2(t)$ are applied forces, and m_1, m_2, b_1, b_2, and k^2 are positive
constants.

The solution of (6–75) by elimination is carried out in Section 5–10.
Here we consider the exponential substitution and certain physical concepts to which it leads.

We assume that $F_1(t)$ and $F_2(t)$ are 0 and that $b_1 = b_2 = 0$, so that
there is no friction. Our equations then become

$$(m_1 D^2 + 2k^2)x_1 - k^2 x_2 = 0,$$
$$(m_2 D^2 + 2k^2)x_2 - k^2 x_1 = 0. \tag{6–76}$$

The substitution $x_1 = \alpha e^{\lambda t}$, $x_2 = \beta e^{\lambda t}$ leads to the characteristic equation

$$0 = \begin{vmatrix} m_1 \lambda^2 + 2k^2 & -k^2 \\ -k^2 & m_2 \lambda^2 + 2k^2 \end{vmatrix} = m_1 m_2 \lambda^4 + 2k^2 \lambda^2 (m_1 + m_2) + 3k^4, \tag{6–77}$$

which is a quadratic equation for λ^2, both of whose roots are negative:

$$\lambda^2 = \frac{k^2}{m_1 m_2}(-m_1 - m_2 \pm \sqrt{m_1^2 + m_2^2 - m_1 m_2});   (6\text{--}78)$$

since the quantity under the radical can be written as $(m_1 - m_2)^2 + m_1 m_2$, it must be positive and, in any case, it cannot be as large as $m_1 + m_2$, so that λ^2 must be negative. It follows that the four values for λ can be given as

$$\lambda = \pm\gamma i, \qquad \lambda = \pm\delta i,   (6\text{--}79)$$

where γ and δ are positive and unequal.

When $\lambda = \gamma i$, we obtain the equations

$$(-m_1\gamma^2 + 2k^2)\alpha - k^2\beta = 0,$$
$$-k^2\alpha + (-m_2\gamma^2 + 2k^2)\beta = 0;$$

these are satisfied if $\alpha = k^2$, $\beta = -m_1\gamma^2 + 2k^2$. Corresponding complex solutions are given by

$$x_1 = k^2 e^{i\gamma t}, \qquad x_2 = (2k^2 - m_1\gamma^2)e^{i\gamma t};   (6\text{--}80)$$

real solutions are obtained by taking real and imaginary parts:

$$x_1 = k^2 \cos \gamma t, \qquad x_2 = (2k^2 - m_1\gamma^2) \cos \gamma t;$$
$$x_1 = k^2 \sin \gamma t, \qquad x_2 = (2k^2 - m_1\gamma^2) \sin \gamma t.   (6\text{--}81)$$

Similar expressions can be obtained for the other two solutions, with γ replaced by δ. The general solution can be written

$$x_1 = A_1 k^2 \sin (\gamma t + \phi_1) + A_2 k^2 \sin (\delta t + \phi_2),$$
$$x_2 = A_1(2k^2 - m_1\gamma^2) \sin (\gamma t + \phi_1)   (6\text{--}82)$$
$$+ A_2(2k^2 - m_1\delta^2) \sin (\delta t + \phi_2),$$

where A_1, A_2, ϕ_1, ϕ_2 are arbitrary constants.

The solutions for which $A_2 = 0$ describe a sinusoidal oscillation:

$$x_1 = A_1 k^2 \sin (\gamma t + \phi),$$
$$x_2 = A_1(2k^2 - m_1\gamma^2) \sin (\gamma t + \phi_1).   (6\text{--}83)$$

The two masses oscillate with the same frequency γ, with the same phase (perfect synchronization), and with amplitudes in a fixed ratio (possibly negative; see Prob. 1 below). A similar synchronous oscillation is obtained with frequency δ. The two types of sinusoidal motion are called *normal modes* of the physical system. The general motion is a linear combination of the two types; it will in general not even be periodic, unless δ and γ are commensurable.

The absence of friction in this example leads to the conclusion that there is no dissipation of energy or that the system is *conservative* (Prob. 2 below). Many other examples of conservative systems described by a pair of simultaneous equations such as (6–75) can be given. Such systems have two degrees of freedom; that is, two coordinates can vary. These coordinates need not be rectangular (see Prob. 4 below).

6–13 Lagrangian equations. Systems with more than two degrees of freedom lead to simultaneous equations in more than two unknowns. It is shown in mechanics (see, for example, Chapter 10 of Reference 5), that the differential equations for unforced small oscillations of a conservative system about equilibrium can be written in the *Lagrangian form:*

$$\frac{d}{dt}\left(\frac{\partial L}{\partial q_j'}\right) - \frac{\partial L}{\partial q_j} = 0 \qquad (j = 1, \ldots, n). \qquad (6\text{–}84)$$

Here $q_1, \ldots, q_n$ are n coordinates which together specify the position of all parts of the system at time t, and $q_j' = Dq_j$; the coordinates are assumed to be chosen so that $q_1 = 0, \ldots, q_n = 0$ at equilibrium. The function L is the *Lagrangian* of the system and equals $T - V$, where T is the total kinetic energy and V is the total potential energy of the system. For each system, T and V are given as quadratic functions as follows:

$$T = \frac{1}{2}\sum_{i=1}^{n}\sum_{j=1}^{n} a_{ij}q_i'q_j' = \frac{1}{2}(a_{11}q_1'q_1' + a_{12}q_1'q_2' + \cdots),$$

$$\hspace{6cm} (6\text{–}85)$$

$$V = \frac{1}{2}\sum_{i=1}^{n}\sum_{j=1}^{n} b_{ij}q_iq_j = \frac{1}{2}(b_{11}q_1q_1 + b_{12}q_1q_2 + \cdots),$$

where $a_{ij} = a_{ji}$ and $b_{ij} = b_{ji}$; the a_{ij} and b_{ij} are constants. Accordingly,

$$L = \frac{1}{2}\sum\sum a_{ij}q_i'q_j' - \frac{1}{2}\sum\sum b_{ij}q_iq_j,$$

$$\frac{\partial L}{\partial q_j'} = \sum_{i=1}^{n} a_{ij}q_i', \qquad \frac{\partial L}{\partial q_j} = -\sum_{i=1}^{n} b_{ij}q_i;$$

for q_j' appears in the terms $a_{ij}q_i'q_j'$ and $a_{ji}q_j'q_i'$, and $a_{ij} = a_{ji}$. A similar remark applies to the q_j.

The differential equations (6-84) become

$$\sum_{i=1}^{n} a_{ij}D^2 q_i + \sum_{i=1}^{n} b_{ij}q_i = 0 \qquad (j = 1, \ldots, n), \qquad (6\text{-}86)$$

which is the general second order system referred to above. It clearly includes (6-76) as a special case.

The nature of the solutions of (6-86) is determined by the roots of the corresponding characteristic equation

$$\begin{vmatrix} a_{11}\lambda^2 + b_{11} & a_{21}\lambda^2 + b_{21} & \cdots \\ a_{12}\lambda^2 + b_{12} & \cdots & \cdots \\ \cdots & \cdots & a_{nn}\lambda^2 + b_{nn} \end{vmatrix} = 0, \qquad (6\text{-}87)$$

often called the *secular equation*. It can be shown (Chapter 10 of Reference 5) that for each root of (6-87), λ^2 is real, and hence the roots come in pairs of conjugate pure imaginary numbers, of two zeros, or of two real numbers of form $\pm a$. When V has its unique minimum at $q_1 = 0, \ldots, q_n = 0$, the roots λ must be pure imaginary ($\lambda = \pm i\omega$). Furthermore, the general solution then has the form

$$q_j = \sum_{k=1}^{n} (c_k \alpha_{jk} e^{i\omega_k t} + \bar{c}_k \bar{\alpha}_{jk} e^{-i\omega_k t}) \qquad (j = 1, \ldots, n), \qquad (6\text{-}88)$$

where $c_1 \ldots, c_n$ are arbitrary complex constants, the α_{jk} are complex constants, and $\pm \omega_1 i, \ldots, \pm \omega_n i$ are the characteristic roots. Some of the ω's may be equal, but terms of form $t e^{i\omega_k t}$, $t^2 e^{i\omega_k t}$, $\ldots$ do *not* appear (see Section 6-4); this conclusion follows in particular from Probs. 6, 7 below.

Equation (6-88) can be written in terms of sines and cosines and then interpreted as a linear combination of n *normal modes:*

$$q_j = A_{jk} \sin(\omega_k t + \phi_k) \qquad (j = 1, \ldots, n), \qquad (6\text{-}89)$$

with $k = 1$ for the first mode, $k = 2$ for the second, $\ldots$, $k = n$ for the nth.

To describe dissipation of energy, we can add terms in Dq_i to Eqs. (6-86). The characteristic roots can then have general form $\alpha \pm \beta i$, with $\alpha < 0$ when the system is stable. We can also add forcing terms $F_i(t)$ and consider the relation between input and output.

The expressions (6-85) for T and V should be considered as first approximations, valid near the equilibrium point $q_1 = 0, \ldots, q_n = 0$. They can be considered as the lowest degree terms of series expansions in powers of the q_i. The kinetic energy T is always a quadratic expression in the "velocities" q_j', but the coefficients a_{ij} are quite general functions of the coordinates q_j and can be expanded in power series in these variables.

The potential energy V is a general function of $q_1, \ldots q_n$. The fact that $q_1 = 0, \ldots, q_n = 0$ at the equilibrium point corresponds to the fact that the first degree terms in the power series for V are missing. The constant term is chosen as 0 for convenience; changing the constant term has no effect on the differential equations.

PROBLEMS

1. Let the characteristic roots $\pm \gamma i$, $\pm \delta i$ in Eqs. (6–79) be labeled so that γ corresponds to the plus sign in Eq. (6–78), δ to the minus sign, so that $0 < \gamma < \delta$.

(a) Show that $2k^2 - m_1 \gamma^2 > 0$, $2k^2 - m_1 \delta^2 < 0$.

(b) Show that this implies that in the normal mode (6–83) the masses oscillate in phase, but in the other normal mode ($A_1 = 0$) the masses are 180° out of phase.

2. The kinetic energy T and potential energy V associated with Eqs. (6–76) are defined as follows:

$$T = \tfrac{1}{2}m_1(Dx_1)^2 + \tfrac{1}{2}m_2(Dx_2)^2,$$
$$V = k^2(x_1^2 - x_1 x_2 + x_2^2).$$

(a) Show that for each solution of the differential equations, $T + V = \text{const}$ (conservation of energy). [Hint: Show that, by virtue of the differential equations, $D(T + V) \equiv 0$.]

(b) Show that V has only one minimum point, namely $x_1 = 0$, $x_2 = 0$.

(c) Show that Eqs. (6–76) are Lagrangian equations with the above choice of T and V.

(d) Show that the conclusion of part (a) holds for an arbitrary Lagrangian system (6–84).

3. *Ball on a surface.* Let a surface $z = f(x, y)$ be given, where $f(x, y)$ can be represented by a power series $ax^2 + 2bxy + cy^2 + dx^3 + \cdots$, for $|x|$ and $|y|$ sufficiently small. Let the z-axis be considered vertical with respect to the earth's surface, and let a tiny ball of mass m roll on the surface near the point $(0, 0, 0)$. The potential energy V is chosen as mgz, the kinetic energy T as $\tfrac{1}{2}m[(Dx)^2 + (Dy)^2 + (Dz)^2]$.

(a) Show that for small x, y, approximately $T = \tfrac{1}{2}m[(Dx)^2 + (Dy)^2]$, $V = mg(ax^2 + 2bxy + cy^2)$, provided we retain only terms in x, y of degree 0 in T and of degree 2 in V.

(b) With the approximations of part (a), set up the differential equations for the motion of the particle.

(c) Under what conditions is the motion described by the equations of part (b) sinusoidal? Interpret physically in terms of the appearance of the surface $z = f(x, y)$.

4. *Torsional vibrations.* A shaft carrying two disks, as in Fig. 6–1, is attached at each end to a wall. The distance between the disks and between disk and wall is l. Each disk can turn about the shaft, but in so doing it exerts a torque on the

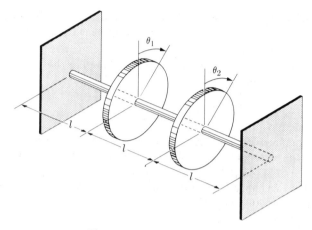

FIG. 6-1. Torsional vibrations.

shaft. The angular coordinates θ_1, θ_2 represent displacements from equilibrium, at which there is no torque. The total potential energy of the system is

$$V = \frac{c}{2}[\theta_1^2 + (\theta_2 - \theta_1)^2 + \theta_2^2],$$

where $c = C/l$, and C is the torsional stiffness of the shaft; the kinetic energy is

$$T = \frac{1}{2}[I_1(D\theta_1)^2 + I_2(D\theta_2)^2],$$

where I_1, I_2 are the two moments of inertia.

(a) Set up the Lagrangian differential equations and compare with Eqs. (6-76).

(b) Discuss the appearance of the normal modes (see Prob. 1 above).

5. *A double pendulum.* A weightless rod of length l is fixed at one end and linked at the other end to a rod of length $2a$ and of uniform density. The two rods are free to swing in the vertical xz-plane (Fig. 6-2). The position of the system at any time is specified by two angles θ, ϕ, as in the figure. The potential energy is $V = -mgz + C$, where z is the z-coordinate of the center of mass of the heavy rod, m is its mass, and C is to be chosen so that $V = 0$ when $\theta = 0$

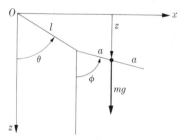

FIG. 6-2. Double pendulum.

and $\phi = 0$.   The kinetic energy is $T = \frac{1}{2}m[(Dx)^2 + (Dz)^2] + \frac{1}{2}I(D\phi)^2$, where $I = \frac{1}{3}ma^2$ is the moment of inertia of the heavy rod about its center of mass.

(a) Show that
$$V = mg[l(1 - \cos\theta) + a(1 - \cos\phi)]$$
and
$$T = \frac{1}{2}m[l^2(D\theta)^2 + \frac{4}{3}a^2(D\phi)^2 + 2al(D\theta)(D\phi)\cos(\theta - \phi)].$$

(b) For small θ, ϕ justify the approximations
$$V = \frac{1}{2}mg(l\theta^2 + a\phi^2), \quad T = \frac{1}{2}m[l^2(D\theta)^2 + (4a^2/3)(D\phi)^2 + 2al(D\theta)(D\phi)].$$

(c) Using the approximations of part (b), obtain the Lagrangian equations for the motion and show that the solutions are purely sinusoidal.

6. (a) Write out the differential equations in Lagrangian form (6–84), if
$$T = (Dq_1)^2 + \cdots + (Dq_n)^2, \quad V = k_1q_1^2 + \cdots + k_nq_n^2. \quad (*)$$

(b) Show that if $V > 0$ except when all $q_i = 0$ (we would then call V a *positive definite* quadratic expression), then the solutions are all sinusoidal. [*Remark.* It can be shown that by a transformation of coordinates we can always ensure that T and V have the form (*). See Prob. 7.]

7. Let $L = T - V$ be a given Lagrangian function, where
$$T = aq_1'^2 + bq_1'q_2' + cq_2'^2, \quad V = \alpha q_1^2 + \beta q_1q_2 + \gamma q_2^2.$$

Let a, b, c, α, β, γ be constants, and let T be greater than 0 except when $q_1' = q_2' = 0$ (so that T is *positive definite*).

(a) Show that the locus $ax^2 + bxy + cy^2 = 1$ in the xy-plane is an ellipse.

(b) Let
$$X = x\cos\phi + y\sin\phi, \quad Y = -x\sin\phi + y\cos\phi \quad (\phi = \text{const})$$

describe a rotation of axes which reduces the ellipse $ax^2 + bxy + cy^2 = 1$ to the standard form $Ax^2 + Cy^2 = 1$. Show that the corresponding change of coordinates
$$Q_1 = q_1\cos\phi + q_2\sin\phi, \quad Q_2 = -q_1\sin\phi + q_2\cos\phi$$

reduces T to the form
$$T = AQ_1'^2 + CQ_2'^2 \quad (A > 0, C > 0)$$

and changes V to a new quadratic expression in Q_1, Q_2.

(c) Let T have form $aq_1'^2 + cq_2'^2$ and let $V = \alpha q_1^2 + \beta q_1q_2 + \gamma q_2^2$. Show that a proper change of scale: $Q_1 = \eta_1q_1$, $Q_2 = \eta_2q_2$, (η_1, η_2 const), reduces T to the form $Q_1'^2 + Q_2'^2$ and changes V to a new quadratic expression in Q_1, Q_2.

(d) Let T have form $q_1'^2 + q_2'^2$ and let $V = \alpha q_1^2 + \beta q_1 q_2 + \gamma q_2^2$. Let $Q_1 = q_1 \cos \phi + q_2 \sin \phi$, $Q_2 = -q_1 \sin \phi + q_2 \cos \phi$ describe a rotation which eliminates the cross-product term $\beta q_1 q_2$ from V, so that V becomes the expression $A Q_1^2 + C Q_2^2$. Show that under this rotation, T becomes $Q_1'^2 + Q_2'^2$.

(e) Show that by a combination of the steps (b), (c), (d) we can always reduce T, V to the form

$$T = Q_1'^2 + Q_2'^2, \qquad V = A Q_1^2 + C Q_2^2.$$

[*Remark.* This process of simultaneous reduction of two quadratic expressions to the form of expressions containing only squares can be generalized to the case of n coordinates $q_1, \ldots, q_n$. It is crucial that T be positive definite, so that the locus $T = 1$ is an ellipse, an ellipsoid, or a hyperellipsoid. The geometric reasoning given above carries over to the general case. The process is equivalent to *diagonalization* of certain *matrices* as studied in higher algebra. See p. 187 of Reference 10.]

ANSWERS

3. (b) $m D^2 x + m g(2ax + 2by) = 0$, $m D^2 y + m g(2bx + 2cy) = 0$;
   (c) $a + c > 0$, $b^2 - ac < 0$; surface is then bowl-shaped.

5. (c) $l^2 D^2 \theta + a l D^2 \phi = -g l \theta$, $\frac{4}{3} a^2 D^2 \phi + a l D^2 \theta = -g a \phi$.

6. (a) $D^2 q_i + k_i q_i = 0$, $(i = 1, \ldots, n)$.

6-14 Applications to electric networks. As in Section 5–13, the analysis of a general electric network is based on Kirchhoff's laws:

I. The total voltage drop about each closed circuit (mesh) is zero.
II. The total current entering each junction (node) is zero.

A general network has a scheme such as that of Fig. 6–3. The nodes are represented by large dots, the branches by lines (possibly curved), each of which connects two nodes. Each branch has a positive direction selected, with respect to which currents and voltage drops are measured.

If there are N branches, then the N currents $I_1, \ldots, I_N$ can be taken as unknowns. By virtue of the second law, some of these can be eliminated, and n currents remain, in terms of which all others can be expressed.

An alternative procedure is to introduce n *mesh currents* $J_1, \ldots, J_n$. These are certain linear combinations of $I_1, \ldots, I_N$, in terms of which $I_1, \ldots, I_N$ can all be expressed. The term *mesh* refers to a closed cir-

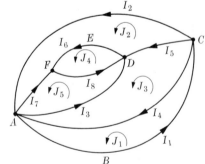

Fig. 6–3. Mesh currents in a planar network.

cuit in the network. For simplicity we restrict attention to planar networks (i.e., those that can be placed in a plane without forcing two branches to cross). For a network in a plane the boundaries of the regions into which the interior of the network is subdivided form n meshes with which $J_1, \ldots,$ J_n can be associated; each is assumed to flow in the counterclockwise direction. If the kth mesh contains a branch on the outermost boundary, we can choose the current J_k as the current I through that branch. (We can assume the positive direction on the outermost boundary to be counterclockwise, as in Fig. 6–3.) The other mesh currents are then defined by systematically applying the rule that *if the lth branch is common to two meshes, say the pth and the qth, then* $I = \pm(J_p - J_q)$*, with a plus sign when the chosen direction on the branch is the same as the counterclockwise direction in the pth mesh, and a minus sign otherwise.*

For the network of Fig. 6–3 we choose $J_1 = I_1, J_2 = I_2$, then

$$I_4 = J_1 - J_3, \qquad I_5 = J_3 - J_2,$$
$$I_6 = J_4 - J_2, \qquad I_8 = J_4 - J_5, \qquad (6\text{–}90)$$
$$I_7 = J_2 - J_5, \qquad I_3 = J_5 - J_3.$$

Hence

$$J_3 = J_1 - I_4 = I_1 - I_4, \qquad J_4 = J_2 + I_6 = I_2 + I_6,$$
$$J_5 = J_4 - I_8 = I_2 + I_6 - I_8. \qquad (6\text{–}91)$$

We can verify that by virtue of Kirchhoff's second law, all the equations (6–90) are satisfied. Thus $I_1, \ldots, I_8$ can be expressed in terms of $J_1, \ldots,$ J_5, and conversely. For the network of Fig. 6–4 we have

$$J_1 = I_1, \qquad J_2 = I_2,$$
$$I_3 = J_1 - J_2, \qquad (6\text{–}92)$$

in agreement with Kirchhoff's second law.

To obtain the differential equations for the network, we now apply Kirchhoff's first law to each mesh separately, and express the results in terms of the mesh currents. We illustrate this for the network of Fig. 6–4. For the first mesh,

$$L_1 \frac{dI_1}{dt} + R_1 I_1 + \frac{q_1}{C_1} + R_3 I_3 = \mathcal{E}_1,$$

where

$$q_1 = \int I_1 \, dt$$

is the charge on the capacitor. By (6–92) this can be rewritten

$$L_1 \frac{dJ_1}{dt} + R_1 J_1 + \frac{1}{C_1} \int J_1 \, dt + R_3(J_1 - J_2) = \mathcal{E}_1. \qquad (6\text{–}93)$$

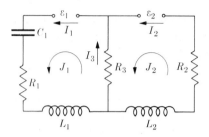

FIGURE 6-4

Similarly, for the second mesh,

$$L_2 \frac{dJ_2}{dt} + R_2 J_2 + \frac{1}{C_2} \int J_2 \, dt + R_3 (J_2 - J_1) = \mathcal{E}_2. \qquad (6\text{-}94)$$

By differentiating (6-93), (6-94), we eliminate the integral signs:

$$
\begin{aligned}
L_1 \frac{d^2 J_1}{dt^2} + R_1 \frac{dJ_1}{dt} + \frac{J_1}{C_1} + R_3 \left(\frac{dJ_1}{dt} - \frac{dJ_2}{dt} \right) = \frac{d\mathcal{E}_1}{dt}, \\
L_2 \frac{d^2 J_2}{dt^2} + R_2 \frac{dJ_2}{dt} + \frac{J_2}{C_2} + R_3 \left(\frac{dJ_2}{dt} - \frac{dJ_1}{dt} \right) = \frac{d\mathcal{E}_2}{dt}.
\end{aligned}
\qquad (6\text{-}95)
$$

We can also introduce the *mesh charges* Q_1, Q_2:

$$Q_1 = \int J_1 \, dt, \qquad Q_2 = \int J_2 \, dt, \qquad (6\text{-}96)$$

for appropriate choices of the indefinite integrals. The equations (6-95) then become

$$
\begin{aligned}
L_1 \frac{d^2 Q_1}{dt^2} + R_1 \frac{dQ_1}{dt} + \frac{Q_1}{C_1} + R_3 \left(\frac{dQ_1}{dt} - \frac{dQ_2}{dt} \right) = \mathcal{E}_1, \\
L_2 \frac{d^2 Q_2}{dt^2} + R_2 \frac{dQ_2}{dt} + \frac{Q_2}{C_2} + R_3 \left(\frac{dQ_2}{dt} - \frac{dQ_1}{dt} \right) = \mathcal{E}_2.
\end{aligned}
\qquad (6\text{-}97)
$$

For a general planar network containing inductance, resistance, capacitance, and driving emf's, we obtain a system of equations of second order

$$\sum_{\beta=1}^{n} (L_{\alpha\beta} D^2 + R_{\alpha\beta} D + \gamma_{\alpha\beta}) J_\beta = \frac{d\mathcal{E}_\alpha}{dt}, \qquad (6\text{-}98)$$

for $\alpha = 1, \ldots, n$. In terms of mesh charges $Q_\alpha = \int J_\alpha \, dt$, the system becomes

$$\sum_{\beta=1}^{n} (L_{\alpha\beta} D^2 + R_{\alpha\beta} D + \gamma_{\alpha\beta}) Q_\beta = \mathcal{E}_\alpha. \qquad (6\text{-}99)$$

The $\mathcal{E}_\alpha$ on the right-hand side is the total driving emf in the α'th mesh. From the way in which the mesh currents are defined, we can verify the *reciprocity law:*

$$L_{\alpha\beta} = L_{\beta\alpha}, \qquad R_{\alpha\beta} = R_{\beta\alpha}, \qquad \gamma_{\alpha\beta} = \gamma_{\beta\alpha}. \tag{6–100}$$

These equations follow from the fact that $L_{\alpha\beta}$ corresponds to the inductance contributed to the α'th mesh along the branch shared by the α'th mesh and β'th mesh; by the above definitions, interchanging α and β describes exactly the same inductance. A similar argument holds for the other terms.

We now introduce the two energy functions

$$T = \frac{1}{2} \sum_{\alpha,\beta=1}^{n} L_{\alpha\beta} J_\alpha J_\beta = \frac{1}{2} \sum_{\alpha,\beta=1}^{n} L_{\alpha\beta} \frac{dQ_\alpha}{dt} \frac{dQ_\beta}{dt},$$
$$V = \frac{1}{2} \sum_{\alpha,\beta=1}^{n} \gamma_{\alpha\beta} Q_\alpha Q_\beta. \tag{6–101}$$

The first is the *total electromagnetic energy;* it is analogous to the total kinetic energy in a mechanical system. The function V is the total *electrostatic energy;* it is analogous to potential energy. When R_α and $\mathcal{E}_\alpha$ are 0 for $\alpha = 1, \ldots, n$, Eqs. (6–101) can be written in the Lagrangian form

$$\frac{d}{dt}\left(\frac{\partial L}{\partial Q'_\alpha}\right) - \frac{\partial L}{\partial Q_\alpha} = 0,$$
$$L = T - V, \tag{6–102}$$

where $Q'_\alpha = dQ_\alpha/dt$. Accordingly, the results of Section 6–13 are applicable, and the network behavior is that of "small vibrations."

The terms in the $R_{\alpha\beta}$ correspond to *dissipation of energy.* The function

$$F = \tfrac{1}{2}\sum R_{\alpha\beta} Q'_\alpha Q'_\beta \tag{6–103}$$

is termed the *dissipation function;* it is one-half the rate of loss of energy (converted to heat) per unit time. It can be shown that for all values of the variables,

$$T \geqq 0, \qquad V \geqq 0, \qquad F \geqq 0. \tag{6–104}$$

From these inequalities we can conclude that the characteristic roots have real parts which are 0 or negative, so that there is some form of stability; the case of 0 real part is exceptional (see Prob. 4 below).

For further information on network theory see References 2 and 6 at the end of this chapter.

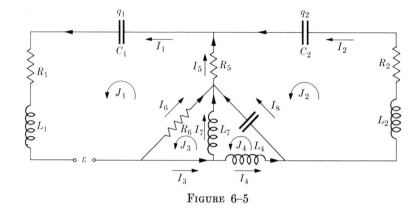

FIGURE 6–5

Finally, let us consider the network of Fig. 6–5. The mesh currents $J_1, \ldots, J_4$ are chosen as $I_1, \ldots, I_4$, respectively. Then

$$I_5 = J_1 - J_2, \qquad I_6 = J_1 - J_3,$$
$$I_7 = J_3 - J_4, \qquad I_8 = J_4 - J_2.$$

Application of Kirchhoff's first law to the four meshes gives the equations

$$L_1 \frac{dI_1}{dt} + R_1 I_1 + \frac{q_1}{C_1} + R_6 I_6 + R_5 I_5 = \varepsilon,$$

$$L_2 \frac{dI_2}{dt} + R_2 I_2 + \frac{q_2}{C_2} - R_5 I_5 - \frac{q_8}{C_8} = 0,$$

$$L_7 \frac{dI_7}{dt} - R_6 I_6 = 0,$$

$$L_4 \frac{dI_4}{dt} + \frac{q_8}{C_8} - I_7 \frac{dI_7}{dt} = 0.$$

(6–105)

We write $q_\alpha = \int I_\alpha \, dt$, $Q_\alpha = \int J_\alpha \, dt$, as above, so that $q_\alpha = Q_\alpha$ for $\alpha = 1, \ldots, 4$, and $q_5 = Q_1 - Q_2$, $q_6 = Q_1 - Q_3$, $q_7 = Q_3 - Q_4$, $q_8 = Q_4 - Q_2$. The differential equations can be written

$$\left[L_1 D^2 + (R_1 + R_5 + R_6)D + \frac{1}{C_1} \right] Q_1 - R_5 D Q_2 - R_6 D Q_3 = \varepsilon,$$

$$-R_5 D Q_1 + \left[L_2 D^2 + (R_2 + R_5)D + \left(\frac{1}{C_2} + \frac{1}{C_8} \right) \right] Q_2 - \frac{Q_4}{C_8} = 0,$$

$$-R_6 D Q_1 + (L_7 D^2 + R_6 D)Q_3 - L_7 D^2 Q_4 = 0,$$

$$-\frac{Q_2}{C_8} - L_7 D^2 Q_3 + \left[(L_4 + L_7)D^2 + \frac{1}{C_8} \right] Q_4 = 0.$$

(6–106)

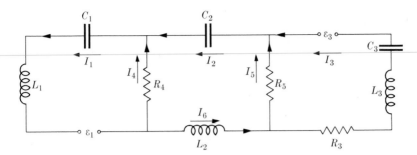

FIGURE 6-6

1. (a) For the network of Fig. 6-6 define mesh currents in terms of branch currents, and express all branch currents in terms of mesh currents. (Note that $I_2 = I_6$ by Kirchhoff's second law.)

(b) Obtain the differential equations satisfied by the mesh charges associated with the mesh currents of part (a).

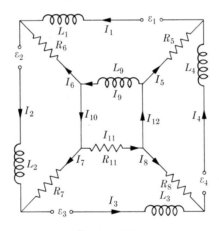

FIGURE 6-7

2. (a) Proceed as in Prob. 1(a) for the network of Fig. 6-7.
(b) Obtain the differential equations for the mesh currents of part (a).

3. (a) Show that the system (6-106) is nondegenerate (Section 6-7).
(b) Verify the reciprocity law (6-100) for the system (6-106).
(c) Find the functions T, F, V for the system (6-106).

4. *Stability of passive networks.* Let the functions T, F, V of Eqs. (6-101), (6-102) be *positive definite:* that is, let all three be ≥ 0 and let T and F reduce to 0 only when all Q'_α are 0; let V reduce to 0 only when all Q_α are 0. Show that the system (6-99) is stable. [*Hint:* Set $Q_\beta = c_\beta e^{\lambda t}$ in the related homogeneous system

to obtain the system

$$\lambda^2 \sum_{\beta=1}^{n} L_{\alpha\beta}c_\beta + \lambda \sum_{\beta=1}^{n} R_{\alpha\beta}c_\beta + \sum_{\beta=1}^{n} \gamma_{\alpha\beta}c_\beta = 0,$$

where $\alpha = 1, \ldots, n$. Let $c_\alpha = a_\beta + ib_\beta$, multiply by $\bar{c}_\alpha = a_\alpha - ib_\alpha$, and sum from $\alpha = 1$ to n to obtain

$$\lambda^2 \sum_{\alpha=1}^{n}\sum_{\beta=1}^{n} L_{\alpha\beta}c_\beta\bar{c}_\alpha + \lambda \sum_{\alpha=1}^{n}\sum_{\beta=1}^{n} R_{\alpha\beta}c_\beta\bar{c}_\alpha + \sum_{\alpha=1}^{n}\sum_{\beta=1}^{n} \gamma_{\alpha\beta}c_\beta\bar{c}_\alpha = 0.$$

Use the reciprocity law (6–100) to show that this equation reduces to

$$\lambda^2 \sum_{\alpha=1}^{n}\sum_{\beta=1}^{n} L_{\alpha\beta}(a_\alpha a_\beta + b_\alpha b_\beta) + \lambda \sum_{\alpha=1}^{n}\sum_{\beta=1}^{n} R_{\alpha\beta}(a_\alpha a_\beta + b_\alpha b_\beta)$$

$$+ \sum_{\alpha=1}^{n}\sum_{\beta=1}^{n} \gamma_{\alpha\beta}(a_\alpha a_\beta + b_\alpha b_\beta) = 0.$$

Apply the positive definiteness of T, F, V to conclude that unless $c_1 = \cdots = c_n = 0$, all coefficients in the last equation are positive. Hence show that Re $(\lambda) < 0$, so that the system is stable.]

Remark. The condition of positive definiteness is not always satisfied. For example, if many of the branches contain no inductance, T may reduce to 0 even though not all I_α are zero. We can of course reason that in a physical network each branch does contain at least a small inductance. Similar reasoning applies to the resistance and capacitance terms, and justifies the conclusion that in a physical network positive definiteness is satisfied.

ANSWERS

1. (a) $J_1 = I_1$, $J_2 = I_2 = I_6$, $J_3 = I_3$, $I_4 = J_1 - J_2$, $I_5 = J_2 - J_3$;
   (b) $[L_1D^2 + R_4D + (1/C_1)]Q_1 - R_4DQ_2 = \mathcal{E}_1$,
   $-R_4DQ_1 + [L_2D^2 + (R_4 + R_5)D + (1/C_2)]Q_2 - R_5DQ_3 = 0$,
   $-R_5DQ_2 + [L_3D^2 + (R_3 + R_5)D + (1/C_3)]Q_3 = \mathcal{E}_3$.

2. (a) $J_1 = I_1$, $J_2 = I_2$, $J_3 = I_3$, $J_4 = I_4$, $J_5 = I_1 + I_9 = I_2 + I_{10} = \cdots$, $I_5 = J_1 - J_4$, $I_6 = J_2 - J_1$, $I_7 = J_3 - J_2$, $I_8 = J_4 - J_3$, $I_9 = J_5 - J_1$, $I_{10} = J_5 - J_2$, $I_{11} = J_5 - J_3$, $I_{12} = J_5 - J_4$;
   (b) $[(L_1 + L_9)D + (R_5 + R_6)]J_1 - R_6J_2 - R_5J_4 - L_9DJ_5 = \mathcal{E}_1$,
   $-R_6J_1 + [L_2D + (R_6 + R_7)]J_2 - R_7J_3 = \mathcal{E}_2$,
   $-R_7J_2 + [L_3D + (R_7 + R_{11} + R_8)J_3] - R_8J_4 - R_{11}J_5 = \mathcal{E}_3$,
   $-R_5J_1 - R_8J_3 + [L_4D + (R_5 + R_8)J_4] = \mathcal{E}_4$,
   $-L_9DJ_1 - R_{11}J_3 + [L_9D + R_{11}]J_5 = 0$.

3. (c) $2T = L_1Q_1'^2 + L_2Q_2'^2 + L_7Q_3'^2 + (L_4 + L_7)Q_4'^2 - 2L_7Q_3'Q_4'$,
   $2V = (Q_1^2/C_1) + Q_2^2[(1/C_2) + (1/C_8)] + (Q_4^2/C_8) - 2(Q_2Q_4/C_8)$,
   $2F = (R_1 + R_5 + R_6)Q_1'^2 + (R_2 + R_5)Q_2'^2 + R_6Q_3'^2 - 2R_5Q_1'Q_2'$
   $\qquad\qquad\qquad\qquad\qquad\qquad\qquad - 2R_6Q_1'Q_3'$

APPENDIX TO CHAPTER 6

APPLICATION OF MATRICES TO SIMULTANEOUS
LINEAR DIFFERENTIAL EQUATIONS

In the following sections we give a brief introduction to matrices and their application to simultaneous linear differential equations. For a full treatment of the topic the reader is referred to References 1, 4, 7, and 10 at the end of this chapter.

6–15 Vectors. The vectors of mechanics are vectors in three-dimensional space, and can be specified in terms of their components along three coordinate axes. In terms of such components each vector $\mathbf{v}$ is then a triple (v_1, v_2, v_3) of real numbers. (Throughout, vectors will be denoted by boldface letters such as $\mathbf{u}$, $\mathbf{v}$; they can be indicated in handwriting by arrows: $\vec{u}, \vec{v}$.)

Generalizing the concept, we can think of vectors in n-dimensional space. In terms of a fixed coordinate system, each vector $\mathbf{v}$ is an n-tuple of real numbers $(v_1, v_2, \ldots, v_n)$; $v_1, \ldots, v_n$ are the components of $\mathbf{v}$ with respect to the given coordinate system. In particular, the *basis vectors*

$$\mathbf{e}_1 = (1, 0, \ldots, 0), \quad \mathbf{e}_2 = (0, 1, 0, \ldots, 0),$$
$$\ldots, \quad \mathbf{e}_n = (0, 0, \ldots, 0, 1) \tag{6-107}$$

are vectors along the coordinate axes (analogous to the vectors $\mathbf{i}$, $\mathbf{j}$, $\mathbf{k}$ of vector analysis in three-dimensional space).

Addition of n-dimensional vectors is defined as follows: if

$$\mathbf{u} = (u_1, \ldots, u_n), \quad \mathbf{v} = (v_1, \ldots, v_n),$$

then

$$\mathbf{u} + \mathbf{v} = (u_1 + v_1, \ldots, u_n + v_n). \tag{6-108}$$

We define multiplication of a vector by a real number (scalar) as follows: if k is a number and $\mathbf{v}$ is as above, then

$$k\mathbf{v} = (kv_1, \ldots, kv_n). \tag{6-109}$$

The *zero* vector $\mathbf{0}$ is defined as a vector all of whose components are 0:

$$\mathbf{0} = (0, 0, \ldots, 0). \tag{6-110}$$

From the above definitions we can verify familiar algebraic rules:

$$\mathbf{u} + \mathbf{v} = \mathbf{v} + \mathbf{u}, \quad k(\mathbf{u} + \mathbf{v}) = k\mathbf{u} + k\mathbf{v},$$
$$\mathbf{u} + (\mathbf{v} + \mathbf{w}) = (\mathbf{u} + \mathbf{v}) + \mathbf{w}, \quad k_1(k_2\mathbf{v}) = (k_1 k_2)\mathbf{v}, \tag{6-111}$$
$$\mathbf{u} + \mathbf{0} = \mathbf{0} + \mathbf{u} = \mathbf{u}, \quad k\mathbf{0} = \mathbf{0}, \quad (k_1 + k_2)\mathbf{u} = k_1\mathbf{u} + k_2\mathbf{u}.$$

Furthermore, *subtraction* is possible; there is a unique vector $\mathbf{w}$, denoted by $\mathbf{v} - \mathbf{u}$, such that

$$\mathbf{u} + \mathbf{w} = \mathbf{v};$$

in fact, $\mathbf{w} = \mathbf{v} + (-1)\mathbf{u}$ (Prob. 2 below, following Section 6–16). We can also verify that

$$\mathbf{v} = v_1\mathbf{e}_1 + v_2\mathbf{e}_2 + \cdots + v_n\mathbf{e}_n, \tag{6-112}$$

in terms of the vectors $\mathbf{e}_j$ defined by (6–107).

A set of m vectors $\mathbf{v}_1, \ldots, \mathbf{v}_m$ is said to be *linearly independent* if the only numbers $k_1, \ldots, k_m$ for which

$$k_1\mathbf{v}_1 + \cdots + k_m\mathbf{v}_m = \mathbf{0} \tag{6-113}$$

are the numbers $k_1 = 0, \ldots, k_m = 0$. Otherwise, the vectors are termed *linearly dependent*. In particular, the vectors $\mathbf{e}_1, \ldots, \mathbf{e}_n$ of Eqs. (6–107) are linearly independent. We can prove that, if $\mathbf{v}_1, \ldots, \mathbf{v}_m$ are linearly independent, then $m \leqq n$, and if $m = n$, then every vector $\mathbf{v}$ can be represented in unique fashion as a linear combination of $\mathbf{v}_1, \ldots, \mathbf{v}_n$ (Prob. 3 below, following Section 6–16):

$$\mathbf{v} = a_1\mathbf{v}_1 + \cdots + a_n\mathbf{v}_n. \tag{6-114}$$

We can then regard $\mathbf{v}_1, \ldots, \mathbf{v}_n$ as a new set of basis vectors that replaces $\mathbf{e}_1, \ldots, \mathbf{e}_n$; $a_1, \ldots, a_n$ are the components of $\mathbf{v}$ with respect to the new basis. If $\mathbf{w} = b_1\mathbf{v}_1 + \cdots + b_n\mathbf{v}_n$ and k is a scalar, then

$$\mathbf{v} + \mathbf{w} = (a_1 + b_1)\mathbf{v}_1 + \cdots + (a_n + b_n)\mathbf{v}_n,$$
$$kw = (kb_1)\mathbf{v}_1 + \cdots + (kb_n)\mathbf{v}_n; \tag{6-115}$$

that is, addition and multiplication by a scalar can be performed in terms of the new components in the same way as in terms of the old.

Accordingly, although we started with a given set of basis vectors $\mathbf{e}_1, \ldots, \mathbf{e}_n$, it is better to think of a vector space as having many possible sets of basis vectors, no one of which is preferred. For each particular application we can choose a set of basis vectors which is adapted to the problem at hand.

Throughout, we have chosen all numbers to be *real*, and hence we should speak of a *real n-dimensional vector space*. We could also allow the numbers to be complex; all the discussion remains valid, but we would then speak of a *complex n-dimensional vector space*. In the following sections we shall allow all numbers to be *complex*.

6–16 Matrices. By an $m \times n$ *matrix* is meant a rectangular array of complex numbers

$$
\begin{bmatrix}
a_{11} & a_{12} & \cdots & a_{1n} \\
a_{21} & a_{22} & & a_{2n} \\
\vdots & & & \vdots \\
a_{m1} & a_{m2} & \cdots & a_{mn}
\end{bmatrix}
\tag{6–116}
$$

which has m rows and n columns. For example,

$$
\begin{bmatrix} 3 & 2 \\ 1 & i \end{bmatrix}
\quad
\begin{bmatrix} 3 & 0 & 1 \\ 0 & i & 0 \end{bmatrix}
\quad
(1, 0, 2)
\quad
\begin{bmatrix} 1+i \\ 0 \\ 1-i \end{bmatrix}
$$

illustrate a 2×2 matrix, a 2×3 matrix, a 1×3 matrix, and a 3×1 matrix. When $m = n$, as in the first example, we speak of a *square* matrix of order n. We shall use capital letters $(A, B, \ldots)$ to denote matrices and write $A = (a_{ij})$ as abbreviations for the matrix (6–116).

If $A = (a_{ij})$ and $B = (b_{ij})$ are $m \times n$ matrices and c is a complex number, then we define $A + B$ and cA by the equations

$$
A + B = (a_{ij} + b_{ij}) =
\begin{bmatrix}
a_{11}+b_{11} & a_{12}+b_{12} & \cdots \\
a_{21}+b_{21} & \cdots & \cdots \\
& \cdots &
\end{bmatrix}
\tag{6–117}
$$

$$
cA = (ca_{ij}) =
\begin{bmatrix}
ca_{11} & ca_{12} & \cdots \\
ca_{21} & \cdots & \cdots \\
\cdots & \cdots & \cdots
\end{bmatrix}
\tag{6–118}
$$

We define 0_{mn} to be the $m \times n$ matrix whose elements all equal 0; usually we can write 0 for 0_{mn}, since the context indicates the number of rows and columns. We can then verify that the rules (6–111) hold, if we replace vectors by $m \times n$ matrices.

The $1 \times n$ matrices have form $(a_{11}, \ldots, a_{1n})$ and can be considered to be vectors in n-dimensional space; the operations of addition and multiplication by a scalar, as defined for such matrices, yield the same results as those previously defined for vectors. We call such a matrix a *row vector.* Similar remarks apply to the $m \times 1$ matrices, which are termed *column vectors.* For typographical reasons we shall denote the column vector

$$\begin{bmatrix} v_1 \\ v_2 \\ \vdots \\ v_n \end{bmatrix}$$

by $(v_1, v_2, \ldots, v_n)$, with a reminder that the symbol denotes a column vector.

Matrices arise naturally when we analyze the change of basis in an n-dimensional vector space. If $\mathbf{e}_1, \ldots, \mathbf{e}_n$ is the old basis and $\mathbf{e}'_1, \ldots, \mathbf{e}'_n$ is the new, then each vector $\mathbf{v}$ can be written in two ways:

$$\mathbf{v} = v_1\mathbf{e}_1 + \cdots + v_n\mathbf{e}_n, \qquad \mathbf{v} = v'_1\mathbf{e}'_1 + \cdots + v'_n\mathbf{e}'_n.$$

In particular, we can write

$$\mathbf{e}_j = \sum_{i=1}^{n} a_{ij}\mathbf{e}'_i \qquad (j = 1, \ldots, n)$$

for appropriate scalars a_{ij}. Therefore,

$$\mathbf{v} = \sum_{j=1}^{n} v_j\mathbf{e}_j = \sum_{j=1}^{n} \sum_{i=1}^{n} v_j a_{ij}\mathbf{e}'_i = \sum_{i=1}^{n} \left(\sum_{j=1}^{n} a_{ij}v_j \right) \mathbf{e}'_i.$$

Accordingly, new and old components are related by the equations

$$v'_i = \sum_{j=1}^{n} a_{ij}v_j \qquad (i = 1, \ldots, n). \tag{6–119}$$

Thus the square matrix $A = (a_{ij})$ describes the relation between new and old components.

Multiplication of matrices is defined as follows. If $A = (a_{ij})$ is an $m \times n$ matrix and $B = (b_{ij})$ is an $n \times p$ matrix, then $A \cdot B$ (or simply AB) is the $m \times p$ matrix $C = (c_{ij})$, where

$$c_{ij} = \sum_{k=1}^{n} a_{ik}b_{kj} \qquad (i = 1, \ldots, m; \ j = 1, \ldots, p). \tag{6–120}$$

Thus the product $A \cdot B$ is defined only when the number of columns of A equals the number of rows of B.

EXAMPLES.

$$\begin{bmatrix} 2 & 3 & 5 \\ 5 & 6 & 4 \end{bmatrix} \begin{bmatrix} 6 & 7 \\ 0 & -1 \\ -5 & 6 \end{bmatrix} = \begin{bmatrix} -13 & 41 \\ 10 & 53 \end{bmatrix}$$

$$\begin{bmatrix} 5 & 1 \\ 0 & 2 \end{bmatrix} \begin{bmatrix} 2 \\ 3 \end{bmatrix} = \begin{bmatrix} 13 \\ 6 \end{bmatrix}$$

In general, the element c_{ij} is obtained by multiplying the elements of the ith *row* of A by the corresponding elements of the jth *column* of B, and adding.

Multiplication of matrices arises in connection with successive linear substitutions such as (6–119). For example, if we have two substitutions

$$v_i' = \sum_{j=1}^{n} b_{ij} v_j, \qquad v_i'' = \sum_{k=1}^{n} a_{ik} v_k' \qquad (i = 1, \ldots, n),$$

then we have the resulting substitution

$$v_i'' = \sum_{j=1}^{n} \sum_{k=1}^{n} a_{ik} b_{kj} v_j = \sum_{j=1}^{n} c_{ij} v_j,$$

where c_{ij} is defined by (6–120). Thus $C = AB$.

From the definitions we can prove the following rules (Prob. 5 below):

$$A \cdot (B + C) = A \cdot B + A \cdot C, \qquad (A + B) \cdot C = A \cdot C + B \cdot C, \qquad (6\text{–}121)$$

$$A \cdot (kB) = (kA) \cdot B = k(A \cdot B) \qquad (k = \text{scalar}), \qquad (6\text{–}122)$$

$$A \cdot (B \cdot C) = (A \cdot B) \cdot C, \qquad (6\text{–}123)$$

$$I_m \cdot A = A \cdot I_n = A, \qquad A \cdot 0 = 0 \cdot A = 0. \qquad (6\text{–}124)$$

Here it is assumed that all operations have meaning. For example, if in (6–121) B is $n \times p$, then C must also be $n \times p$ and A must be $m \times n$ for some m. In (6–124) I_m is the square $m \times m$ matrix (δ_{ij}), where δ_{ij} is the *Kronecker delta:*

$$\delta_{ij} = 0 \quad \text{for} \quad i \neq j, \qquad \delta_{ij} = 1 \quad \text{for} \quad i = j; \qquad (6\text{–}125)$$

thus

$$I_3 = \begin{bmatrix} 1 & 0 & 0 \\ 0 & 1 & 0 \\ 0 & 0 & 1 \end{bmatrix}$$

We call I_m the *unit* matrix of order m; as with 0_{mn}, we write simply I when the order is clear.

The commutative law does not hold for multiplication. For example, if

$$A = \begin{bmatrix} 1 & 0 \\ 2 & 1 \end{bmatrix} \qquad B = \begin{bmatrix} 2 & 0 \\ 1 & 0 \end{bmatrix}$$

then

$$AB = \begin{bmatrix} 2 & 0 \\ 5 & 0 \end{bmatrix} \neq BA = \begin{bmatrix} 2 & 0 \\ 1 & 0 \end{bmatrix}$$

Multiplication of a matrix by a column vector yields a column vector:

$$\begin{bmatrix} a_{11} & a_{12} & \cdots \\ a_{21} & a_{22} & \cdots \\ \cdots & \cdots & \cdots \end{bmatrix} \begin{bmatrix} v_1 \\ v_2 \\ \vdots \end{bmatrix} = \begin{bmatrix} a_{11}v_1 + a_{12}v_2 + \cdots \\ a_{21}v_1 + a_{22}v_2 + \cdots \\ \cdots \end{bmatrix}$$

A system of linear differential equations

$$\frac{dx_i}{dt} = \sum_{j=1}^{n} a_{ij}x_j \qquad (i = 1, \ldots, n) \tag{6-126}$$

is hence equivalent to the vector equation

$$\frac{d\mathbf{x}}{dt} = A\mathbf{x}, \tag{6-127}$$

where

$$\mathbf{x} = \begin{bmatrix} x_1 \\ x_2 \\ \vdots \\ x_n \end{bmatrix} \qquad \frac{d\mathbf{x}}{dt} = \begin{bmatrix} dx_1/dt \\ dx_2/dt \\ \vdots \\ dx_n/dt \end{bmatrix} \tag{6-128}$$

PROBLEMS

1. The following vectors are in the xy-plane (real two-dimensional vector space) and are given by their components (v_x, v_y) along the x- and y-axes. In particular, $\mathbf{u}_1 = (2, 3)$, $\mathbf{u}_2 = (2, -1)$, $\mathbf{u}_3 = (0, 1)$.

(a) Graph $\mathbf{u}_1$, $\mathbf{u}_2$, and $\mathbf{u}_3$.

(b) Find $\mathbf{u}_1 + \mathbf{u}_2$, and graph.

(c) Find $\mathbf{u}_1 - \mathbf{u}_2$, and graph.

(d) Find $2\mathbf{u}_1 + 3\mathbf{u}_2$.

(e) Show that $\mathbf{u}_1$, $\mathbf{u}_2$, $\mathbf{u}_3$ are linearly dependent.

(f) Show that $\mathbf{u}_1$, $\mathbf{u}_2$ are linearly independent. If $\mathbf{v} = (v_x, v_y)$, find a_1, a_2 such that $\mathbf{v} = a_1\mathbf{u}_1 + a_2\mathbf{u}_2$.

2. (a) Prove the rules (6-111).

(b) Prove: if $\mathbf{u}$, $\mathbf{v}$, are given vectors in an n-dimensional vector space, then there is one and only one vector $\mathbf{w}$ such that $\mathbf{u} + \mathbf{w} = \mathbf{v}$.

3. (a) Prove: if $\mathbf{v}_1 = (v_{11}, \ldots, v_{1n}), \ldots, \mathbf{v}_n = (v_{n1}, \ldots, v_{nn})$ are vectors in an n-dimensional vector space and

$$V = \begin{vmatrix} v_{11} & \cdots & v_{1n} \\ v_{21} & & v_{2n} \\ \vdots & & \vdots \\ v_{n1} & \cdots & v_{nn} \end{vmatrix} = \begin{vmatrix} v_{11} & v_{21} & \cdots & v_{n1} \\ v_{12} & v_{22} & & v_{n2} \\ \vdots & & & \vdots \\ v_{1n} & v_{2n} & \cdots & v_{nn} \end{vmatrix}$$

then $V \neq 0$ if and only if $\mathbf{v}_1, \ldots, \mathbf{v}_n$ are linearly independent. [*Hint:* Consider the linear equations

$$\sum_{i=1}^{n} k_i \mathbf{v}_i = \mathbf{0};$$

that is,

$$\sum_{i=1}^{n} k_i v_{ij} = 0 \qquad (j = 1, \ldots, n).\Big]$$

(b) If the vectors $\mathbf{v}_1, \ldots, \mathbf{v}_n$ of part (a) are linearly independent, prove that every vector $\mathbf{v}$ of the space can be expressed uniquely as

$$\sum_{i=1}^{n} k_i \mathbf{v}_i.$$

[*Hint:* Consider the linear equations

$$v_j = \sum_{i=1}^{n} k_i v_{ij} \qquad (j = 1, \ldots, n),$$

and use the result of part (a).]

(c) Prove: if $\mathbf{v}_1, \ldots, \mathbf{v}_m$ are linearly independent, then so also are $\mathbf{v}_1, \ldots, \mathbf{v}_h$, for $h \leqq m$.

(d) Prove: if $\mathbf{v}_1, \ldots, \mathbf{v}_m$ are linearly independent vectors in an n-dimensional vector space, then $m \leqq n$. [*Hint:* If $m > n$, then by part (c) $\mathbf{v}_1, \ldots, \mathbf{v}_n$ are linearly independent. Apply part (b) to write

$$\mathbf{v}_{n+1} = \sum_{i=1}^{n} k_i \mathbf{v}_i,$$

and conclude that $\mathbf{v}_1, \ldots, \mathbf{v}_{n+1}$ are dependent, in contradiction of part (c).]

4 Let the matrices A, B, C, D, E be given as follows:

$$A = \begin{bmatrix} 2 & 0 & 0 \\ 1 & 0 & 1 \\ 0 & 1 & 0 \end{bmatrix} \qquad B = \begin{bmatrix} 1 & 0 \\ 0 & 3 \\ -1 & 0 \end{bmatrix} \qquad C = \begin{bmatrix} 1 & 5 \\ 2 & 3 \end{bmatrix}$$

$$D = \begin{bmatrix} -1 & 0 \\ 1 & 0 \end{bmatrix} \qquad E = \begin{bmatrix} i \\ -i \end{bmatrix}$$

Evaluate the following matrices:

(a) $3A$,   (b) $C + D$,   (c) $C - D$,   (d) AB,

(e) BC,   (f) CD,   (g) DC,   (h) DE,

(i) $B(CE)$,   (j) $(BC)E$,   (k) $B(3C + 2D)$.

5. (a) Prove (6–121). (b) Prove (6–122). (c) Prove (6–123). (d) Prove (6–124).

1. (b) $(4, 2)$, (c) $(0, 4)$, (d) $(10, 3)$, (f) $a_1 = (v_x + 2v_y)/8$, $a_2 = (3v_x - 2v_y)/8$.

4. (a) $\begin{bmatrix} 6 & 0 & 0 \\ 3 & 0 & 3 \\ 0 & 3 & 0 \end{bmatrix}$
(b) $\begin{bmatrix} 0 & 5 \\ 3 & 3 \end{bmatrix}$
(c) $\begin{bmatrix} 2 & 5 \\ 1 & 3 \end{bmatrix}$

(d) $\begin{bmatrix} 2 & 0 \\ 0 & 0 \\ 0 & 3 \end{bmatrix}$
(e) $\begin{bmatrix} 1 & 5 \\ 6 & 9 \\ -1 & -5 \end{bmatrix}$
(f) $\begin{bmatrix} 4 & 0 \\ 1 & 0 \end{bmatrix}$
(g) $\begin{bmatrix} -1 & -5 \\ 1 & 5 \end{bmatrix}$

(h) $\begin{bmatrix} -i \\ i \end{bmatrix}$
(i) and (j) $\begin{bmatrix} -4i \\ -3i \\ 4i \end{bmatrix}$
(k) $\begin{bmatrix} 1 & 15 \\ 24 & 27 \\ -1 & -15 \end{bmatrix}$

6–17 Matrices and linear transformations. Let an n-dimensional vector space, of elements $\mathbf{u}$, $\mathbf{v}$, ... be given; also let an m-dimensional vector space, of elements $\mathbf{u}'$, $\mathbf{v}'$, ... be given. Let fixed bases be chosen in both spaces. Consider the elements $\mathbf{u}$, $\mathbf{v}$, ... as $n \times 1$ column vectors and the elements $\mathbf{u}'$, $\mathbf{v}'$, ... as $m \times 1$ column vectors. If $A = (a_{ij})$ is an $m \times n$ matrix, the equations

$$v_i' = \sum_{j=1}^{n} a_{ij} v_j \qquad \text{or} \qquad \mathbf{v}' = A\mathbf{v} \qquad (6\text{–}129)$$

assign a vector $\mathbf{v}'$ to each vector $\mathbf{v}$. We write $\mathbf{v}' = T(\mathbf{v})$ and call T a *transformation* of the n-dimensional vector space into the m-dimensional vector space. The transformation T has the property (Prob. 2 below)

$$T(c_1 \mathbf{u} + c_2 \mathbf{v}) = c_1 T(\mathbf{u}) + c_2 T(\mathbf{v}), \qquad (6\text{–}130)$$

and is hence called a *linear transformation*. Thus each $m \times n$ matrix A determines a linear transformation of an n-dimensional vector space into an m-dimensional vector space.

Conversely, if T is a linear transformation of an n-dimensional vector space into an m-dimensional vector space and bases are chosen in each space, then T is given by (6–129) in terms of an appropriate matrix $A = (a_{ij})$. To find the matrix A, we let $\mathbf{v}$ be a basis vector $\mathbf{e}_j$. Then

$$T(\mathbf{e}_j) = \sum_{i=1}^{m} a_{ij}\mathbf{e}_i',$$

in terms of the basis vectors $\mathbf{e}_1', \ldots, \mathbf{e}_m'$. This defines $A = (a_{ij})$. Then for general $\mathbf{v} = \sum v_j \mathbf{e}_j$,

$$T(\mathbf{v}) = \sum_{j=1}^{n} v_j T(\mathbf{e}_j) = \sum_{j=1}^{n} \sum_{i=1}^{m} a_{ij} v_j \mathbf{e}_i' = \sum_{i=1}^{m} \left(\sum_{j=1}^{n} a_{ij} v_j \right) \mathbf{e}_i',$$

so that (6–129) holds.

6–18 Inverse matrix. Rank. Let $A = (a_{ij})$ be a *square* $n \times n$ matrix. If an $n \times n$ matrix B exists such that

$$A \cdot B = I,$$

then the matrix B is termed an inverse matrix of A. We can prove that if one such B exists, then there is only one inverse. We denote the inverse matrix B by A^{-1} and can then prove that

$$A \cdot A^{-1} = A^{-1} \cdot A = I \tag{6–131}$$

(Prob. 3 below).

We write $\det A = \det (a_{ij})$ for the determinant formed of the elements of the matrix A:

$$\det A = \begin{vmatrix} a_{11} & a_{12} & \cdots & \cdots \\ a_{21} & a_{22} & \cdots & \cdots \\ \vdots & \vdots & & \vdots \\ a_{n1} & a_{n2} & \cdots & a_{nn} \end{vmatrix} \tag{6–132}$$

By the rule for multiplication of determinants (Chapter 4 of Reference 10) we conclude that

$$\det (A \cdot C) = \det A \cdot \det C \tag{6–133}$$

for $n \times n$ matrices A and C. Since $\det I = 1$, Eq. (6–131) shows that

$$\det A \cdot \det A^{-1} = 1. \tag{6–134}$$

Hence the determinant of the inverse matrix is the reciprocal of the determinant of A. Furthermore, when A has an inverse, det $A \neq 0$. Conversely, if det $A \neq 0$, then A has an inverse (Prob. 3 below).

If we assign a linear transformation T to the matrix A, as in Section 6–17:

$$v'_i = \sum_{j=1}^{n} a_{ij} v_j \qquad (i = 1, \ldots, n)$$

and A^{-1} exists, then A^{-1} defines the inverse transformation T^{-1} of the second space into the first; that is,

$$T^{-1}[T(\mathbf{v})] = \mathbf{v}. \tag{6–135}$$

For if $A^{-1} = (b_{ij})$, then $T^{-1}[T(\mathbf{v})]$ has the components

$$v_k^* = \sum_{i=1}^{n} b_{ki} v'_i = \sum_{i=1}^{n} \sum_{j=1}^{n} a_{ij} b_{ki} v_j = \sum_{j=1}^{n} \left(\sum_{i=1}^{n} b_{ki} a_{ij} \right) v_j;$$

that is, $T^{-1}[T(\mathbf{v})] = (BA)\mathbf{v} = I\mathbf{v} = \mathbf{v}$, and (6–135) holds.

When det $A \neq 0$, A is termed *nonsingular*. When det $A = 0$, A is termed *singular*. In this case, we examine the minors of det A. A minor of order $n - k$ is obtained by deleting k rows and k columns from A. If r is the largest integer for which some minor of A of order r is not 0, then r is termed the *rank* of A. For example,

$$\begin{vmatrix} 2 & 2 & 1 \\ 1 & 2 & 0 \\ 3 & 4 & 1 \end{vmatrix} \qquad \begin{vmatrix} 1 & 2 & 2 \\ 0 & 0 & 1 \\ 1 & 0 & 1 \end{vmatrix}$$

have ranks 2 and 3, respectively.

The matrix $A = (a_{ij})$ associated with a change of basis (6–119) has *maximum rank;* that is, det $A \neq 0$. This follows in particular from the fact that since components are unique, we can solve Eqs. (6–119) uniquely for $v_1, \ldots, v_n$ in terms of $v'_1, \ldots, v'_n$; hence det $A \neq 0$.

PROBLEMS

1. Interpret each of the following pairs of equations as describing a linear transformation of vectors (x, y) in the xy-plane into vectors (u, v) in the uv-plane, and interpret geometrically:

(a) $u = 2x, \qquad v = 2y;$ \qquad\qquad (b) $u = -x, \qquad v = y;$

(c) $u = (x - y)/\sqrt{2}, \qquad v = (x + y)/\sqrt{2};$ \qquad (d) $u = y, \qquad v = x.$

2. Prove that every transformation T defined by (6–129) has the linearity property (6–130).

3. Let $A = (a_{ij})$ be a nonsingular square matrix of order n, so that det $A \neq 0$.

(a) Prove: for every column vector $\mathbf{w} = (w_1, \ldots, w_n)$, there exists a unique column vector $\mathbf{v} = (v_1, \ldots, v_n)$ such that $A\mathbf{v} = \mathbf{w}$. [*Hint:* Consider the corresponding linear equations.]

(b) Prove: if $A\mathbf{v} = \mathbf{0}$, then $\mathbf{v} = \mathbf{0}$. [*Hint:* Use the result of part (a).]

(c) Prove: there exists a unique matrix $B = A^{-1}$ such that $AB = I$. [*Hint:* Let $\mathbf{v}_j = (b_{1j}, \ldots, b_{nj})$ be the jth column vector of B. Then we must have $A\mathbf{v}_j = \mathbf{e}_j = (\delta_{1j}, \ldots, \delta_{nj})$. Now apply the result of part (a).]

(d) Prove: if $B = A^{-1}$ is chosen as in part (c), then $BA = I$. [*Hint:* Since $AB = I$, B is nonsingular; hence B has an inverse B^{-1}, and $BB^{-1} = I$. Therefore $A(BB^{-1}) = A$. Conclude that $B^{-1} = A$, so that $BA = I$.]

4. *Computation of the inverse.* Let

$$\mathbf{e}_1 = (1, 0, \ldots, 0), \quad \ldots, \quad \mathbf{e}_n = (0, 0, \ldots, 0, 1)$$

be considered as *row* vectors. The ith row of $A = (a_{ij})$ can be considered as

$$\mathbf{v}_i = \sum_{j=1}^{n} a_{ij}\mathbf{e}_j \quad (i = 1, \ldots, n).$$

If we solve these n equations for $\mathbf{e}_1, \ldots, \mathbf{e}_n$, we obtain the relations

$$\mathbf{e}_i = \sum_{j=1}^{n} b_{ij}\mathbf{v}_j \quad (i = 1, \ldots, n);$$

that is, $e_{ik} = \delta_{ik} = \sum b_{ij}v_{jk} = \sum b_{ij}a_{jk}$, so that $I = BA$ and B is A^{-1}. The following is an example.

$$A = \begin{bmatrix} 2 & 1 \\ 9 & 5 \end{bmatrix} \quad \begin{aligned} \mathbf{v}_1 &= 2\mathbf{e}_1 + \mathbf{e}_2, \\ \mathbf{v}_2 &= 9\mathbf{e}_1 + 5\mathbf{e}_2, \end{aligned}$$

$$\begin{aligned} \mathbf{e}_1 &= 5\mathbf{v}_1 - \mathbf{v}_2, \\ \mathbf{e}_2 &= -9\mathbf{v}_1 + 2\mathbf{v}_2, \end{aligned} \quad B = \begin{bmatrix} 5 & -1 \\ -9 & 2 \end{bmatrix}$$

$$AB = \begin{bmatrix} 1 & 0 \\ 0 & 1 \end{bmatrix} = BA.$$

Apply the method described to find the inverses of the following matrices:

(a) $\begin{bmatrix} 7 & 4 \\ 5 & 3 \end{bmatrix}$ (b) $\begin{bmatrix} 2 & 0 \\ 1 & 3 \end{bmatrix}$ (c) $\begin{bmatrix} 5 & 2 & 1 \\ 1 & 0 & 2 \\ 3 & 1 & 3 \end{bmatrix}$

5. Show that if the equations $\mathbf{v}_i = \sum a_{ij}\mathbf{e}_j$ of Prob. 4 are solved for $\mathbf{e}_1, \ldots, \mathbf{e}_n$ by determinants, then

$$\mathbf{e}_i = \frac{A_{1i}}{\det A}\,\mathbf{v}_1 + \cdots + \frac{A_{ni}}{\det A}\,\mathbf{v}_n,$$

so that $b_{ij} = A_{ji}/\det A$; here A_{ji} is the *cofactor* of a_{ji} in $\det A$; that is, A_{ji} is $(-1)^{i+j}$ times the determinant of the matrix obtained by removing the jth row and ith column from A.

[The matrix (c_{ji}) is called the *transpose* of the matrix $C = (c_{ij})$ and is denoted by C'; the matrix (A_{ij}) is called the *adjoint* of A and is denoted by adj A. Thus $A^{-1} = (\det A)^{-1}(\text{adj } A)'$.]

6. (a) Prove: if A and B are square matrices of order n and both have inverses, then $A \cdot B$ has an inverse, namely $B^{-1}A^{-1}$ (order *reversed*).

(b) Prove that $(A^{-1})^{-1} = A$.

ANSWERS

4. (a) $\begin{bmatrix} 3 & -4 \\ -5 & 7 \end{bmatrix}$ (b) $\dfrac{1}{6}\begin{bmatrix} 3 & 0 \\ -1 & 2 \end{bmatrix}$ (c) $-\dfrac{1}{3}\begin{bmatrix} -2 & -5 & 4 \\ 3 & 12 & -9 \\ 1 & 1 & -2 \end{bmatrix}$

6–19 Eigenvalues, eigenvectors, and similar matrices. Let T be a linear transformation which assigns to each vector $\mathbf{v}$ of a given n-dimensional space a vector $\mathbf{v}'$ of the *same* space. In particular, it may happen that $T(\mathbf{v}) = \lambda\mathbf{v}$ for some vector $\mathbf{v}$ not $\mathbf{0}$ and some scalar λ. When this case arises, we call λ an *eigenvalue* (*characteristic value*) of T, and $\mathbf{v}$ an associated *eigenvector*. There may be several eigenvectors associated with the same eigenvalue λ.

EXAMPLE. Let $\mathbf{v}$ have the components (v_x, v_y), with respect to the x-, y-axes in the xy-plane, and let $\mathbf{v}' = T(\mathbf{v})$ have the components (v'_x, v'_y), where

$$v'_x = 4v_x - 3v_y, \qquad v'_y = 5v_x - 4v_y.$$

Then $\mathbf{v}_1 = (1, 1)$ is an eigenvector associated with $\lambda = 1$, for $T(\mathbf{v}_1) = (1, 1) = \mathbf{v}_1$. Also, $\mathbf{v}_2 = (3, 5)$ is an eigenvector associated with $\lambda = -1$, for $T(\mathbf{v}_2) = (-3, -5) = -\mathbf{v}_2$.

We can represent T by a matrix $A = (a_{ij})$, necessarily square and n by n; the eigenvectors and eigenvalues are then said to be associated with matrix A. The equation $T(\mathbf{v}) = \lambda\mathbf{v}$ becomes the equation $A\mathbf{v} = \lambda\mathbf{v}$; that is,

$$(A - \lambda I)\mathbf{v} = \mathbf{0}$$

or

$$\sum_{j=1}^{n} (a_{ij} - \lambda\delta_{ij})v_j = 0 \qquad (i = 1, \ldots, n). \tag{6–136}$$

Thus $\mathbf{v} = (v_1, \ldots, v_n)$ is to be a solution, other than $(0, \ldots, 0)$, of a set of n homogeneous linear equations. Accordingly, the determinant of the system must be 0:

$$\begin{vmatrix} a_{11} - \lambda & a_{12} & \ldots \\ a_{21} & a_{22} - \lambda & \ldots \\ \ldots & \ldots & \ldots \\ \ldots & \ldots & a_{nn} - \lambda \end{vmatrix} = \det (A - \lambda I) = 0. \quad (6\text{-}137)$$

If we expand the determinant, we obtain an equation of form

$$P(\lambda) = 0, \quad (6\text{-}138)$$

where P is a polynomial of degree n; this is the *characteristic* equation. The roots of this equation are the eigenvalues; in general, there are n roots (real or complex), but some roots may coincide. When there are n distinct eigenvalues, we obtain n linearly independent eigenvectors (Prob. 3 below, following Section 6–20).

The form of the matrix A representing T depends on the choice of basis in the given vector space. If there are n linearly independent eigenvectors $\mathbf{v}_1, \ldots, \mathbf{v}_n$, then we can choose these as basis vectors. Since

$$T(\mathbf{v}_i) = \lambda_i \mathbf{v}_i \quad (i = 1, \ldots, n),$$

$$T(c_1 \mathbf{v}_1 + \cdots + c_n \mathbf{v}_n) = (c_1 \lambda_1 \mathbf{v}_1 + \cdots + c_n \lambda_n \mathbf{v}_n),$$

the matrix A representing T is the *diagonal* matrix

$$\begin{bmatrix} \lambda_1 & 0 & \ldots & 0 \\ 0 & \lambda_2 & & 0 \\ \vdots & & & \vdots \\ 0 & 0 & \ldots & \lambda_n \end{bmatrix} = (\lambda_i \delta_{ij}). \quad (6\text{-}139)$$

Let us suppose that we have chosen one basis $\mathbf{v}_1, \ldots, \mathbf{v}_n$ for the space, in terms of which the matrix representing T is $A = (a_{ij})$, and that we then choose a new basis. The relation between the old components $(v_1, \ldots, v_n)$ (column vector) and the new components $(v_1^*, \ldots, v_n^*)$ (column vector) is then described by a nonsingular matrix $B = (b_{ij})$:

$$v_i = \sum_{j=1}^{n} b_{ij} v_j^* \quad \text{or} \quad \mathbf{v} = B\mathbf{v}^*.$$

If we replace old components by new in the transformation equation $\mathbf{u} = A\mathbf{v}$, then we find

$$B\mathbf{u}^* = AB\mathbf{v}^*, \qquad \mathbf{u}^* = B^{-1}AB\mathbf{v}^*.$$

Thus A is replaced by the new matrix

$$C = B^{-1}AB, \tag{6–140}$$

where B is nonsingular. A matrix C related to A in this manner is termed *similar* to A. Thus similar matrices are different representations of the same linear transformation T of a space into itself. In particular, if A has n linearly independent eigenvectors, then A is similar to a diagonal matrix.

If A is given and A has n distinct eigenvalues $\lambda_1, \ldots, \lambda_n$, then we can find C and B such that (6–140) holds as follows. We choose C as the diagonal matrix (6–139) and B as a matrix whose kth column $(b_{1k}, \ldots, b_{nk})$ is an eigenvector $\mathbf{v}_k$ associated with λ_k. Then, since $A\mathbf{v}_k = \lambda_k\mathbf{v}_k$,

$$\sum_{j=1}^{n} a_{ij}b_{jk} = \lambda_k b_{ik} = \sum_{j=1}^{n} b_{ij}\lambda_j\delta_{jk};$$

that is, $AB = BC$ or $C = B^{-1}AB$.

EXAMPLE.

$$A = \begin{bmatrix} 3 & 5 \\ 6 & 2 \end{bmatrix}$$

The characteristic equation is

$$\begin{vmatrix} 3 - \lambda & 5 \\ 6 & 2 - \lambda \end{vmatrix} = 0;$$

the eigenvalues are found to be $\lambda_1 = 8$, $\lambda_2 = -3$. For $\lambda_1 = 8$ we find an eigenvector $\mathbf{v}_1 = (b_{11}, b_{21})$ by solving the equations

$$-5b_{11} + 5b_{21} = 0, \qquad 6b_{11} - 6b_{21} = 0.$$

Hence $b_{11} = b_{21} = c_1$, where c_1 is an arbitrary constant not 0. We choose $c_1 = 1$ and $\mathbf{v}_1 = (1, 1)$. Similarly, $\lambda_2 = -3$ leads to the equations

$$6b_{12} + 5b_{22} = 0, \qquad 6b_{12} + 5b_{22} = 0,$$

and we can choose $\mathbf{v}_2 = (5, -6) = (b_{12}, b_{22})$. Therefore,

$$B = \begin{bmatrix} 1 & 5 \\ 1 & -6 \end{bmatrix} \quad \text{and} \quad B^{-1}AB = \begin{bmatrix} 8 & 0 \\ 0 & -3 \end{bmatrix}$$

6–20 Jordan normal form. Not every matrix A has a full set of n linearly independent eigenvectors. For example, the matrix

$$\begin{bmatrix} 3 & 1 \\ -1 & 1 \end{bmatrix}$$

has as eigenvectors only the vectors $(b, -b)$, $b \neq 0$. Hence we cannot hope to find a diagonal matrix similar to A. Instead, however, we can find a similar matrix C of special form, known as the *Jordan normal form*. The matrix C is formed of zeros, except for blocks along the diagonal, each having the form $(\lambda \delta_{ij} + \delta_{i+1,j})$:

$$C = \begin{bmatrix} \boxed{C_1} & 0 & \cdots \\ 0 & \boxed{C_2} & \cdots \\ \cdots & \cdots & \cdots \\ & & \end{bmatrix} \qquad C_j = \begin{bmatrix} \lambda_j & 1 & 0 & \cdots & 0 \\ 0 & \lambda_j & 1 & & 0 \\ \vdots & & & & \vdots \\ 0 & 0 & \cdots & \cdots & \lambda_j \end{bmatrix} \qquad (6\text{–}141)$$

For example,

$$C = \begin{bmatrix} \boxed{\begin{matrix}2 & 1\\0 & 2\end{matrix}} & 0 & 0 & 0 & 0 \\ & & & & \\ 0 & 0 & \boxed{\begin{matrix}2&1&0\\0&2&1\\0&0&2\end{matrix}} & & 0 \\ & & & & \\ 0 & 0 & 0 & 0 & 0 & \boxed{3} \end{bmatrix}$$

is in the Jordan normal form. For a proof that every matrix is similar to a matrix in Jordan normal form, see Reference 10, Chapter 8, at the end of this chapter.

If the eigenvalues of A have been found, we can reduce A to normal form C as follows. We seek a matrix B such that $B^{-1}AB = C$, or

$$AB = BC, \qquad (6\text{–}142)$$

where C is in normal form (6–141). Because of the special form of C, the relation (6–142) can be interpreted as follows. Let $\mathbf{v}_j = (b_{1j}, b_{2j}, \ldots, b_{nj})$ be the jth *column* of B, considered as a column vector. Then if C_1 is $k \times k$, C_2 is $l \times l$, etc.,

$$A\mathbf{v}_1 = \lambda_1\mathbf{v}_1, \quad (A - \lambda_1 I)\mathbf{v}_2 = \mathbf{v}_1, \quad \ldots,$$

$$(A - \lambda_1 I)\mathbf{v}_k = \mathbf{v}_{k-1};$$

$$A\mathbf{v}_{k+1} = \lambda_2\mathbf{v}_{k+1}, \quad (A - \lambda_2 I)\mathbf{v}_{k+2} = \mathbf{v}_{k+1}, \quad \ldots,$$

$$(A - \lambda_2 I)\mathbf{v}_{k+l} = \mathbf{v}_{k+l-1}, \quad \ldots; \quad \ldots$$

For each block C_j there is such a group of equations. If C_j is $p \times p$, there are p equations in the group; if $p = 1$, there is one equation of form $A\mathbf{v} = \lambda\mathbf{v}$. For the block C_1 there is a chain of vectors $\mathbf{v}_1, \ldots, \mathbf{v}_k$, of which the first is an eigenvector associated with λ_1; a similar chain exists for each block C_j. Because B is nonsingular, the vectors $\mathbf{v}_1, \ldots, \mathbf{v}_n$ are linearly independent (Prob. 3 following Section 6–16); in particular, none is zero.

To reduce a particular matrix to normal form, we first compute the eigenvalues. For each simple eigenvalue λ, we choose an associated eigenvector $\mathbf{v}$. If λ is a repeated eigenvalue, of multiplicity m, we first determine how many linearly independent eigenvectors are associated with λ. If there are m of these, then we choose a set of m linearly independent eigenvectors. If there are q in number, where $q < m$, then we seek q chains as above, each starting with an eigenvector associated with λ. (This leads to simultaneous equations which are essentially those encountered in Section 6–4; see also Section 6–22.) It will be found that such chains do exist, together yielding m linearly independent vectors. In this way we obtain, in all, n linearly independent vectors $\mathbf{v}_1, \ldots, \mathbf{v}_n$, from which B can be formed. Then $B^{-1}AB = C$ is in the Jordan normal form.

EXAMPLE 1.

$$A = \begin{bmatrix} 14 & 66 & -42 \\ 4 & 24 & -14 \\ 10 & 55 & -33 \end{bmatrix}$$

The characteristic equation is

$$0 = \begin{vmatrix} 14 - \lambda & 66 & -42 \\ 4 & 24 - \lambda & -14 \\ 10 & 55 & -33 - \lambda \end{vmatrix} = -(\lambda - 2)^2(\lambda - 1).$$

We choose $\lambda_1 = 2$, and find an eigenvector $\mathbf{v}_1 = (b_{11}, b_{21}, b_{31})$ by solving the equations $(A - 2I)\mathbf{v}_1 = \mathbf{0}$; that is,

$$\begin{bmatrix} 12 & 66 & -42 \\ 4 & 22 & -14 \\ 10 & 55 & -35 \end{bmatrix} \begin{bmatrix} b_{11} \\ b_{21} \\ b_{31} \end{bmatrix} = \mathbf{0},$$

or

$$12b_{11} + 66b_{21} - 42b_{31} = 0,$$

$$4b_{11} + 22b_{21} - 14b_{31} = 0,$$

$$10b_{11} + 55b_{21} - 35b_{31} = 0.$$

All three equations are equivalent to one equation:

$$2b_{11} + 11b_{21} - 7b_{31} = 0.$$

Hence we can choose two linearly independent solutions:

$$\mathbf{v}_1 = (11, -2, 0) = (b_{11}, b_{21}, b_{31});$$

$$\mathbf{v}_2 = (7, 0, 2) = (b_{12}, b_{22}, b_{32}).$$

Since $\lambda_1 = 2$ has multiplicity 2, this takes care of all blocks of C associated with λ_1; in fact, C must have the diagonal form

$$C = \begin{bmatrix} 2 & 0 & 0 \\ 0 & 2 & 0 \\ 0 & 0 & 1 \end{bmatrix}$$

The third column of B is found from the equation $(A - I)\mathbf{v}_3 = \mathbf{0}$; we find a solution $\mathbf{v}_3 = (6, 2, 5)$. Hence

$$B = \begin{bmatrix} 11 & 7 & 6 \\ -2 & 0 & 2 \\ 0 & 2 & 5 \end{bmatrix}$$

and $B^{-1}AB = C$.

EXAMPLE 2.

$$A = \begin{bmatrix} -8 & 47 & -8 \\ -4 & 18 & -2 \\ -8 & 39 & -5 \end{bmatrix}$$

The characteristic equation is again found to have roots 2, 2, 1. The equation $(A - 2I)\mathbf{v}_1 = \mathbf{0}$ becomes

$$-10b_{11} + 47b_{21} - 8b_{31} = 0,$$

$$-4b_{11} + 16b_{21} - 2b_{31} = 0,$$

$$-8b_{11} + 39b_{21} - 7b_{31} = 0.$$

We find a solution $\mathbf{v}_1 = (17, 6, 14)$ and no other solution independent of this. Accordingly, we know that

$$C = \begin{bmatrix} 2 & 1 & 0 \\ 0 & 2 & 0 \\ 0 & 0 & 1 \end{bmatrix}$$

and we find $\mathbf{v}_2$ from the equation

$$(A - 2I)\mathbf{v}_2 = \mathbf{v}_1;$$

$$-10b_{12} + 47b_{22} - 8b_{32} = 17,$$

$$-4b_{12} + 16b_{22} - 2b_{32} = 6,$$

$$-8b_{12} + 39b_{22} - 7b_{32} = 14.$$

A solution is found to be $\mathbf{v}_2 = (-\frac{5}{14}, \frac{4}{14}, 0)$. For $\mathbf{v}_3$ we have the equation $(A - I)\mathbf{v}_3 = \mathbf{0}$, and we find a solution $\mathbf{v}_3 = (6, 2, 5)$. Hence

$$B = \begin{bmatrix} 17 & -\frac{5}{14} & 6 \\ 6 & \frac{4}{14} & 2 \\ 14 & 0 & 5 \end{bmatrix}$$

and $B^{-1}AB = C$.

For further discussion of the technique of reduction to the Jordan normal form, see References 3 and 10 at the end of this chapter.

PROBLEMS

1. Find the eigenvalues and associated eigenvectors of the following matrices:

(a) $\begin{bmatrix} 7 & 6 \\ 2 & 6 \end{bmatrix}$ (b) $\begin{bmatrix} -1 & 1 \\ -5 & 3 \end{bmatrix}$ (c) $\begin{bmatrix} -9 & 19 & 4 \\ -3 & 7 & 1 \\ -7 & 17 & 2 \end{bmatrix}$

(d) $\begin{bmatrix} 4 & 1 \\ -1 & 2 \end{bmatrix}$ (e) $\begin{bmatrix} 2 & 1 & 2 \\ -1 & 0 & -2 \\ 0 & 0 & 1 \end{bmatrix}$

2. Find matrices B, C such that $C = B^{-1}AB$ and that C is in diagonal form if A is the matrix of

(a) Prob. 1 part (a),   (b) Prob. 1 part (b),   (c) Prob. 1 part (c).

3. Let $\lambda_1, \ldots, \lambda_m$ be distinct eigenvalues of the matrix $A = (a_{ij})$ and let $\mathbf{v}_1, \ldots, \mathbf{v}_m$ be corresponding eigenvectors. Prove that $\mathbf{v}_1, \ldots, \mathbf{v}_m$ are linearly independent.

[*Hint:* Prove the result by induction. For $m = 1$ linear dependence would mean that $\mathbf{v}_1 = 0$; hence linear dependence is impossible (why?). If this theorem is true for $m = k$, then let $m = k + 1$. If $\mathbf{v}_1, \ldots, \mathbf{v}_{k+1}$ are linearly dependent, then we must have a relation $\mathbf{v}_{k+1} = c_1\mathbf{v}_1 + \cdots + c_k\mathbf{v}_k$, where $c_1, \ldots, c_k$ are uniquely determined (why?). Now $A\mathbf{v}_{k+1} = c_1 A\mathbf{v}_1 + \cdots + c_k A\mathbf{v}_k$, so that $\lambda_{k+1}\mathbf{v}_{k+1} = c_1\lambda_1\mathbf{v}_1 + \cdots + c_k\lambda_k\mathbf{v}_k$. If $\lambda_{k+1} = 0$, conclude that $c_1 = 0, \ldots,$ $c_k = 0$; if $\lambda_{k+1} \neq 0$, show that $c_1\lambda_{k+1} = c_1\lambda_1$, $c_2\lambda_{k+1} = c_2\lambda_2, \ldots,$ and again conclude that $c_1 = \cdots = c_k = 0$. Hence $\mathbf{v}_{k+1} = \mathbf{0}$, which is impossible.]

4. Find matrices B, C such that $C = B^{-1}AB$ and that C is in normal form, if A is the matrix

(a) in Prob. 1 part (d); (b) in Prob. 1 part (e);

$$(c) \begin{bmatrix} 0 & 2 & 6 & 7 \\ 2 & 3 & -6 & -10 \\ -4 & -1 & 14 & 19 \\ 2 & 1 & -6 & -8 \end{bmatrix}$$

ANSWERS

1. (a) $\lambda_1 = 10, \mathbf{v}_1 = c_1(2, 1)$; $\lambda_2 = 3, \mathbf{v}_2 = c_2(3, -2)$; (b) $\lambda_1 = 1 + i$, $\mathbf{v}_1 = c_1(1, 2 + i)$; $\lambda_2 = 1 - i, \mathbf{v}_2 = c_2(1, 2 - i)$; (c) $\lambda_1 = i, \mathbf{v}_1 = c_1(9 - 4i, 3 - i, 7 - 2i)$; $\lambda_2 = -i, \mathbf{v}_2 = c_2(9 + 4i, 3 + i, 7 + 2i)$; $\lambda_3 = 0$, $\mathbf{v}_3 = c_3(3, 1, 2)$; (d) $\lambda = 3, 3, \mathbf{v} = c(1, -1)$; (e) $\lambda = 1, 1, 1, \mathbf{v} = c_1(1, -1, 0) + c_2(2, 0, -1)$.

2. (a) $$C = \begin{bmatrix} 10 & 0 \\ 0 & 3 \end{bmatrix} \qquad B = \begin{bmatrix} 2 & 3 \\ 1 & -2 \end{bmatrix}$$

(b) $$C = \begin{bmatrix} 1 + i & 0 \\ 0 & 1 - i \end{bmatrix} \qquad B = \begin{bmatrix} 1 & 1 \\ 2 + i & 2 - i \end{bmatrix}$$

(c) $$C = \begin{bmatrix} i & 0 & 0 \\ 0 & -i & 0 \\ 0 & 0 & 0 \end{bmatrix} \qquad B = \begin{bmatrix} 9 - 4i & 9 + 4i & 3 \\ 3 - i & 3 + i & 1 \\ 7 - 2i & 7 + 2i & 2 \end{bmatrix}$$

4. (a) $$C = \begin{bmatrix} 3 & 1 \\ 0 & 3 \end{bmatrix} \qquad B = \begin{bmatrix} 1 & 1 \\ -1 & 0 \end{bmatrix}$$

(b)
$$C = \begin{bmatrix} 1 & 1 & 0 \\ 0 & 1 & 0 \\ 0 & 0 & 1 \end{bmatrix} \qquad B = \begin{bmatrix} 1 & 2 & 2 \\ -1 & -1 & 0 \\ 0 & 0 & -1 \end{bmatrix}$$

(c)
$$C = \begin{bmatrix} 2 & 1 & 0 & 0 \\ 0 & 2 & 0 & 0 \\ 0 & 0 & 2 & 0 \\ 0 & 0 & 0 & 3 \end{bmatrix} \qquad B = \begin{bmatrix} 3 & -2 & 0 & -1 \\ 0 & 2 & 2 & 1 \\ 1 & -2 & -3 & -2 \\ 0 & 1 & 2 & 1 \end{bmatrix}$$

6–21 Functions associated with matrices and vectors. If $v_1(t), \ldots, v_n(t)$ are n given scalar functions of a real variable t, all defined for $a \leqq t \leqq b$, then we can form a vector $\mathbf{v} = [v_1(t), \ldots, v_n(t)] = \mathbf{v}(t)$ which is also dependent on t. If $\mathbf{e}_j$ are fixed basis vectors, we can also write

$$\mathbf{v}(t) = v_1(t)\mathbf{e}_1 + v_2(t)\mathbf{e}_2 + \cdots + v_n(t)\mathbf{e}_n. \tag{6–143}$$

For example, $\mathbf{v} = t^2\mathbf{e}_1 + (1 - t)\mathbf{e}_2 + \cos t\,\mathbf{e}_3$ is such a vector function of t. We define the derivative and integral as follows:

$$\frac{d\mathbf{v}}{dt} = \frac{dv_1}{dt}\mathbf{e}_1 + \frac{dv_2}{dt}\mathbf{e}_2 + \cdots + \frac{dv_n}{dt}\mathbf{e}_n, \tag{6–144}$$

$$\int_a^b \mathbf{v}(t)\,dt = \int_a^b v_1(t)\,dt\,\mathbf{e}_1 + \cdots + \int_a^b v_n(t)\,dt\,\mathbf{e}_n. \tag{6–145}$$

The indefinite integral is defined in analogous fashion.

Similar definitions can be given for matrices. If $A = (a_{ij})$ and $a_{ij} = a_{ij}(t)$, then A becomes a function of t; we define dA/dt to be the matrix (da_{ij}/dt) and $\int_a^b A(t)\,dt$ to be the matrix $[\int_a^b a_{ij}(t)\,dt]$.

Various simple rules of the calculus can now be verified (Prob. 6 below). In particular,

$$\frac{d}{dt}[A(t) + B(t)] = \frac{dA}{dt} + \frac{dB}{dt}, \qquad \frac{d}{dt}(A \cdot B) = A\frac{dB}{dt} + \frac{dA}{dt}B \cdot \tag{6–146}$$

A system of homogeneous linear differential equations

$$\frac{dx_i}{dt} = \sum_{j=1}^{n} a_{ij}x_j \qquad (i = 1, \ldots, n)$$

is equivalent to the vector equation

$$\frac{d\mathbf{x}}{dt} = A\mathbf{x}, \tag{6–147}$$

where $\mathbf{x}$ is the column vector $(x_1, \ldots, x_n)$ and A is the matrix (a_{ij}). The a_{ij} may depend on t, so that $A = A(t)$.

For an arbitrary, constant, $n \times n$ *square* matrix A, we can form the powers

$$A^0 = I, \qquad A^1 = A, \qquad A^2 = A \cdot A,$$

$$A^3 = A^2 \cdot A, \qquad \ldots, \qquad A^k = A^{k-1} \cdot A. \tag{6-148}$$

These are all $n \times n$ matrices, as is a linear combination

$$B = c_0 A^k + c_1 A^{k-1} + \cdots + c_{k-1} A + c_k I,$$

where $c_0, \ldots, c_k$ are scalars. For fixed $c_0, c_1, \ldots, c_k$ we can vary A; B then varies and we can consider B as a function of A:

$$B = f(A).$$

Because of its form, we call B a *polynomial function* of A. For example, $3A^2 - 2A + I$ is a quadratic function of A.

Functions of A can be obtained also as infinite series. The most important example for us is the *exponential function*

$$e^A = I + A + \frac{1}{2!} A^2 + \cdots + \frac{1}{k!} A^k + \cdots \tag{6-149}$$

For fixed A, the kth partial sum of this series is a well-defined matrix B_k; it can be shown that as $k \to \infty$, the elements $b_{ij}^{(k)}$ of B_k approach limits b_{ij} which form the matrix $B = e^A$ (Prob. 7 below). If A has diagonal form, $A = (\lambda_i \delta_{ij})$, then we find

$$e^A = \begin{bmatrix} e^{\lambda_1} & 0 & \cdots \\ 0 & e^{\lambda_2} & \cdots \\ \cdots & \cdots & \cdots \end{bmatrix} \tag{6-150}$$

(Prob. 7 below). The exponential function satisfies familiar identities:

$$e^{c_1 A + c_2 A} = e^{c_1 A} e^{c_2 A}, \tag{6-151}$$

$$(e^A)^{-1} = e^{-A}, \tag{6-152}$$

$$(e^A)^m = e^{mA} \qquad (m = 2, 3, 4, \ldots), \tag{6-153}$$

$$e^0 = I \tag{6-154}$$

(Prob. 8 below).

We can go one step further and consider a function of A and t:

$$e^{tA} = I + tA + \frac{t^2}{2!} A^2 + \cdots, \tag{6-155}$$

which for fixed A defines a matrix function of t. We can then differentiate
with respect to t:

$$\frac{d}{dt} e^{tA} = A + tA^2 + \frac{t^2}{2!} A^3 + \cdots$$

$$= A \left(I + tA + \frac{t^2}{2!} A^2 + \cdots \right) = Ae^{tA}. \qquad (6\text{-}156)$$

The formal steps can be justified by standard theorems on the term-by-term differentiation of power series.

6–22 Linear differential equations. Let A be an $n \times n$ square matrix
of constant elements a_{ij}. Then, as remarked in Section 6–21, the equation

$$\frac{d\mathbf{x}}{dt} = A\mathbf{x}, \qquad (6\text{-}157)$$

where $\mathbf{x} = (x_1, \ldots, x_n)$ is considered as a column vector, is equivalent
to a system of homogeneous linear differential equations for $x_1, \ldots, x_n$.
We seek a particular solution

$$\mathbf{x} = e^{\lambda t}\mathbf{v}, \qquad (6\text{-}158)$$

where λ is a constant scalar and $\mathbf{v}$ is a constant vector, not $\mathbf{0}$. By (6–146)

$$\frac{d\mathbf{x}}{dt} = \lambda e^{\lambda t}\mathbf{v}, \qquad (6\text{-}159)$$

and substitution in the differential equation gives

$$\lambda e^{\lambda t}\mathbf{v} = e^{\lambda t}A\mathbf{v} \qquad \text{or} \qquad (A - \lambda I)\mathbf{v} = \mathbf{0}. \qquad (6\text{-}160)$$

Hence (6–158) is a solution of (6–157) precisely when $\mathbf{v}$ is an eigenvector
of A, associated with the eigenvalue λ. If A has n distinct eigenvalues
$\lambda_1, \ldots, \lambda_n$, then we can choose corresponding eigenvectors $\mathbf{v}_1, \ldots, \mathbf{v}_n$;
these vectors are linearly independent (Section 6–19). If $c_1, \ldots, c_n$ are
arbitrary (complex) constants, the linear combinations

$$\mathbf{x} = c_1 e^{\lambda_1 t}\mathbf{v}_1 + \cdots + c_n e^{\lambda_n t}\mathbf{v}_n \qquad (6\text{-}161)$$

are solutions of (6–157). Because $\mathbf{v}_1, \ldots, \mathbf{v}_n$ are linearly independent,
we can choose $c_1, \ldots, c_n$ to satisfy the arbitrary initial conditions

$$\mathbf{x} = \mathbf{x}_0 = (x_1^0, \ldots, x_n^0) \qquad \text{for } t = 0. \qquad (6\text{-}162)$$

Hence Eq. (6–161) gives the general solution. If A has some multiple eigenvalues, it may nevertheless be possible to find n linearly independent eigenvectors $\mathbf{v}_1, \ldots, \mathbf{v}_n$, so that (6–161) will again provide the general solution. If A has less than n linearly independent eigenvectors, we must seek solutions of form

$$\mathbf{x} = e^{\lambda t}(\alpha\mathbf{v}_1 + \beta t\mathbf{v}_2), \qquad \mathbf{x} = e^{\lambda t}(\alpha\mathbf{v}_1 + \beta t\mathbf{v}_2 + \gamma t^2\mathbf{v}_3), \qquad \ldots,$$

as in Section 6–4. We return to this substitution below.

Another approach to the problem is to introduce new variables $y_1, \ldots, y_n$ by a nonsingular linear transformation:

$$\mathbf{x} = B\mathbf{y}, \qquad B = (b_{ij}), \tag{6–163}$$

where the b_{ij} are constants and $\mathbf{y} = (y_1, \ldots y_n)$ is considered as a column vector. Hence, by (6–146),

$$\frac{d\mathbf{x}}{dt} = B\frac{d\mathbf{y}}{dt} = A\mathbf{x} = AB\mathbf{y}.$$

Multiplication by B^{-1} yields the new differential equation

$$\frac{d\mathbf{y}}{dt} = C\mathbf{y}, \qquad C = B^{-1}AB. \tag{6–164}$$

Here C is similar to A (Section 6–19); in particular, for proper choice of B, C is in Jordan normal form (Section 6–20). Accordingly, by an appropriate change of variables we can reduce our differential equations to the following form:

$$\frac{dy_1}{dt} = \lambda_1 y_1, \qquad \frac{dy_2}{dt} = \lambda_1 y_2 + y_1, \qquad \ldots, \qquad \frac{dy_k}{dt} = \lambda_1 y_k + y_{k-1},$$
$$\tag{6–165}$$
$$\frac{dy_{k+1}}{dt} = \lambda_2 y_{k+1}, \qquad \frac{dy_{k+2}}{dt} = \lambda_2 y_{k+2} + y_{k+1}, \qquad \ldots$$

The general solution can be obtained by solving each equation in turn (Prob. 9 below):

$$y_1 = c_1 e^{\lambda_1 t}, \qquad y_2 = e^{\lambda_1 t}(c_1 t + c_2), \qquad \ldots,$$

$$y_k = e^{\lambda_1 t}\left(c_1\frac{t^{k-1}}{(k-1)!} + c_2\frac{t^{k-2}}{(k-2)!} + \cdots + c_{k-1}t + c_k\right), \tag{6–166}$$

$$y_{k+1} = c_{k+1}e^{\lambda_2 t}, \qquad y_{k+2} = e^{\lambda_2 t}(c_{k+1}t + c_{k+2}), \qquad \ldots$$

Finally, $\mathbf{x} = B\mathbf{y}$; that is, $x_i = \sum b_{ij}y_j$, and

$$
\begin{aligned}
x_1 &= \sum_{j=1}^{n} b_{1j}y_j = b_{11}c_1 e^{\lambda_1 t} + b_{12}e^{\lambda_1 t}(c_1 t + c_2) + \cdots \\
&= c_1 e^{\lambda_1 t}\left(b_{11} + b_{12}t + \cdots + b_{1k}\frac{t^{k-1}}{(k-1)!}\right) \\
&\quad + c_2 e^{\lambda_1 t}\left(b_{12} + b_{13}t + \cdots + b_{1k}\frac{t^{k-2}}{(k-2)!}\right) + \cdots \qquad (6\text{-}167) \\
&\quad + c_k b_{1k}e^{\lambda_1 t} + c_{k+1}e^{\lambda_2 t}(b_{1,k+1} + \cdots) + \cdots , \\
x_2 &= c_1 e^{\lambda_1 t}(b_{21} + b_{22}t + \cdots) + \cdots , \\
&= \cdots
\end{aligned}
$$

This gives the solution, as in Section 6–4.

We can seek the solutions of the preceding paragraph by setting

$$
x = e^{\lambda t}\left(\mathbf{v}_k + t\mathbf{v}_{k-1} + \cdots + \frac{t^{k-1}}{(k-1)!}\mathbf{v}_1\right), \qquad (6\text{-}168)
$$

in the differential equation (6–157). Here, λ is an eigenvalue of multiplicity $\geq k$, and $\mathbf{v}_1, \ldots, \mathbf{v}_k$ are constant vectors to be found. Substitution in Eq. (6–157) yields

$$
\begin{aligned}
\frac{d\mathbf{x}}{dt} &= e^{\lambda t}\left[\lambda\mathbf{v}_k + (\lambda t + 1)\mathbf{v}_{k-1} + \cdots + \left(\frac{\lambda t^{k-1}}{(k-1)!} + \frac{t^{k-2}}{(k-2)!}\right)\mathbf{v}_1\right] \\
&= A\mathbf{x} = e^{\lambda t}\left[A\mathbf{v}_k + tA\mathbf{v}_{k-1} + \cdots + \frac{t^{k-1}}{(k-1)!}A\mathbf{v}_1\right].
\end{aligned}
$$

If we cancel $e^{\lambda t}$ and compare terms of the same degree in t, we find

$$
A\mathbf{v}_1 = \lambda\mathbf{v}_1, \quad (A - \lambda I)\mathbf{v}_2 = \mathbf{v}_1, \quad \ldots, \quad (A - \lambda I)\mathbf{v}_k = \mathbf{v}_{k-1}.
$$
$$(6\text{-}169)$$

Thus $\mathbf{v}_1, \ldots \mathbf{v}_k$ form a chain of vectors, as required in reduction of A to Jordan normal form (Section 6–20). As pointed out in Section 6–20, we can find such chains, formed of the column vectors of matrix B; if p linearly independent eigenvectors are associated with λ; we can in fact find p chains (formed of linearly independent vectors), each starting with an eigenvector and together yielding l vectors, where l is the multiplicity of λ. If $\mathbf{v}_1, \ldots, \mathbf{v}_k$ is one such chain, then there are k corresponding linearly independent solutions of the differential equation, namely

$$e^{\lambda t}\left(\mathbf{v}_k + t\mathbf{v}_{k-1} + \cdots + \frac{t^{k-1}}{(k-1)!}\mathbf{v}_1\right),$$

$$e^{\lambda t}\left(\mathbf{v}_{k-1} + \mathbf{v}_{k-2}t + \cdots + \frac{t^{k-2}}{(k-2)!}\mathbf{v}_1\right), \quad \ldots, \quad (6\text{–}170)$$

$$e^{\lambda t}(\mathbf{v}_2 + \mathbf{v}_1 t), \qquad e^{\lambda t}\mathbf{v}_1.$$

Indeed, substitution in the differential equation of each of these functions yields an identity, by virtue of Eqs. (6–169). Linear independence follows from the fact that the coefficients of $e^{\lambda t}$ are of differing degree in t.

6–23 Application of e^{tA}. We can use the exponential function e^{tA} defined in Section 6–21 to obtain the solution of the vector differential equation

$$\frac{d\mathbf{x}}{dt} = A\mathbf{x} \qquad (6\text{–}171)$$

in very concise form. Indeed, by (6–146) and (6–156), if $\mathbf{c}$ is a constant column vector, then

$$\frac{d}{dt}(e^{tA}\mathbf{c}) = Ae^{tA}\mathbf{c},$$

so that

$$\mathbf{x} = e^{tA}\mathbf{c} \qquad (6\text{–}172)$$

is a solution for every choice of the constant vector $\mathbf{c}$. When $t = 0$, $x = \mathbf{c}$, so that arbitrary initial conditions can be satisfied; hence (6–172) provides the general solution of (6–171). When A is in diagonal form, (6–172) is equivalent to the equations

$$x_1 = c_1 e^{\lambda_1 t}, \qquad \ldots, \qquad x_n = c_n e^{\lambda_n t}.$$

For the nonhomogeneous equation

$$\frac{d\mathbf{x}}{dt} - A\mathbf{x} = \mathbf{u}(t), \qquad (6\text{–}173)$$

we have the integrating factor e^{-tA}. For after multiplication by this factor, the equation can be written

$$\frac{d}{dt}(e^{-tA}\mathbf{x}) = e^{-tA}\mathbf{u}(t),$$

so that the solutions are given by

$$e^{-tA}\mathbf{x} = \int e^{-tA}\mathbf{u}(t)\,dt + \mathbf{c},$$

where **c** is an arbitrary constant vector; that is,

$$\mathbf{x} = e^{tA} \int e^{-tA} \mathbf{u}(t) \, dt + e^{tA} \mathbf{c}. \qquad (6\text{--}174)$$

The formulas of this section are admirably concise and are very useful in theoretical work. However, to solve particular systems of equations and obtain the results in explicit form, we must effectively reduce A to diagonal form or Jordan normal form, so that concise formulas here save no steps.

PROBLEMS

1. Let A be the matrix $\begin{bmatrix} 7 & 6 \\ 2 & 6 \end{bmatrix}$. Write the system of linear differential equations represented by the equation $d\mathbf{x}/dt = A\mathbf{x}$, and verify that

$$\mathbf{x} = e^{3t}(3, -2)$$

is a solution.

2. Let A be the matrix of Prob. 1.
   (a) Evaluate A^2, A^3.
   (b) Show that $A^2 - 13A + 30I = 0$. This is a special case of the *Hamilton-Cayley theorem*, which asserts that every square matrix satisfies its own characteristic equation. See p. 136 of Reference 10.

3. (a) Prove the identity $A^2 - I = (A + I)(A - I)$.
   (b) Is the identity $(A + B)^2 = A^2 + 2AB + B^2$ generally valid for $n \times n$ matrices A, B?

4. Let A be the matrix of Prob. 1(c) following Section 6–20. Obtain the general solution of the differential equation $d\mathbf{x}/dt = A\mathbf{x}$, with the aid of the eigenvectors of A, as found in the solution to that problem.

5. Obtain the general solution of the differential equation $d\mathbf{x}/dt = A\mathbf{x}$, where A is the matrix of Prob. 4(c) following Section 6–20, with the aid of the matrices B, C determined in the solution to that problem.

6. Prove the rules (6–146).

7. (a) Prove that if $A = (a_{ij})$ is a square matrix of order n, and $|a_{ij}| \leqq c$ for all i, j, then each element of the matrix A^2 is at most nc^2 in absolute value.
   (b) Use induction to extend the result of part (a) by proving that each element of A^k is at most $n^{k-1}c^k$, $(k = 1, 2, \ldots)$, in absolute value.
   (c) Show from (6–149) that each element of e^A is given by a series, which converges absolutely by comparison with the series

$$1 + c + \frac{nc^2}{2!} + \cdots + \frac{n^{k-1}c^k}{k!} + \cdots$$

(d) If A has diagonal form, show that A^k has diagonal form and is obtained from A by raising each element to the kth power. Conclude from (6–149) that e^A is given by (6–150).

8. (a) Prove the rule (6–151). [*Hint:* Since the series (6–149) defines each element of e^A by an absolutely convergent series, as in Prob. 7(c), we can multiply the series for $e^{c_1 A}$, $e^{c_2 A}$ and then arrange the terms in any order desired.]

(b) Prove the rule (6–152). [*Hint:* Use (6–151).]
(c) Prove the rule (6–153). [*Hint:* Use (6–151).]
(d) Prove the rule (6–154).

9. Show that the general solution of the system (6–165) is given by Eqs. (6–166).

ANSWERS

2. (a) $\begin{bmatrix} 61 & 78 \\ 26 & 48 \end{bmatrix}$ $\begin{bmatrix} 583 & 834 \\ 278 & 444 \end{bmatrix}$ 3. (b) It is valid when and only when $AB = BA$.

4. $c_1 e^{it}(9 - 4i, 3 - i, 7 - 2i) + c_2 e^{-it}(9 + 4i, 3 + i, 7 + 2i) + c_3(3, 1, 2)$.

5. $\mathbf{x} = B\mathbf{y}$, where $\mathbf{y}$ is the column vector $e^{2t}(c_1 + c_2 t, c_2, c_3, 0) + e^{3t}(0, 0, 0, c_4)$.

SUGGESTED REFERENCES

1. Birkhoff, Garrett, and MacLane, Saunders, *A Survey of Modern Algebra*, Rev. ed. New York: MacMillan, 1953.
2. Bode, Henrik W., *Network Analysis and Feedback Amplifier Design*. New York: Van Nostrand, 1945.
3. Browne, E. T., *Introduction to the Theory of Determinants and Matrices*, Chap. 17. Chapel Hill: The University of North Carolina Press, 1958.
4. Frazer, R. A., Duncan, W. J., and Collar, A. R., *Elementary Matrices*. Cambridge, Eng.: Cambridge University Press, 1938.
5. Goldstein, Herbert, *Classical Mechanics*. Reading, Mass.: Addison-Wesley, 1950.
6. Guillemin, E. H., *Communication Networks*, Vol. 1. New York: John Wiley and Sons, Inc., 1931.
7. Guillemin, E. H., *Mathematics of Circuit Analysis*. New York: John Wiley and Sons, Inc., 1949.
8. Ince, E. L., *Ordinary Differential Equations*. New York: Dover, 1956.
9. Kaplan, Wilfred, *Operational Methods for Linear Systems*. Reading, Mass.: Addison-Wesley (to appear).
10. Perlis, Sam, *Theory of Matrices*. Reading, Mass.: Addison-Wesley, 1952.
11. Wade, T. L., *The Algebra of Vectors and Matrices*. Reading, Mass.: Addison-Wesley, 1951.

CHAPTER 7

EXACT DIFFERENTIAL EQUATIONS

7-1 Exact equations of second order. We recall (Section 2–2) that a first order differential equation

$$P(x, y)\, dx + Q(x, y)\, dy = 0$$

is termed *exact* if there exists a function $u(x, y)$ such that

$$\frac{\partial u}{\partial x} = P, \qquad \frac{\partial u}{\partial y} = Q,$$

so that $du = P\, dx + Q\, dy$. The general solution is then given by the equation

$$u(x, y) = c.$$

If the differential equation is written in the form

$$Q(x, y)y' + P(x, y) = 0,$$

then the left-hand side can be interpreted as the derivative of $u(x, y)$ with respect to x, when y is considered as a function of x:

$$\frac{du}{dx} = \frac{\partial u}{\partial x} \cdot 1 + \frac{\partial u}{\partial y}\, y' \equiv P(x, y) + Q(x, y)y'.$$

Along each solution curve of the differential equation, $P + Qy' = 0$, so that $du/dx = 0$ and u reduces to a constant.

The definition can be extended to equations of higher order. A second order equation

$$p(x, y, y')y'' + q(x, y, y') = 0 \tag{7-1}$$

is termed *exact* if there exists a function $F(x, y, y')$ such that when y is considered as a function of x,

$$\frac{dF}{dx} = F_x + F_y y' + F_{y'} y'' \equiv p(x, y, y')y'' + q(x, y, y'). \tag{7-2}$$

For example, the differential equation

$$yy'' + y'^2 + 2x = 0$$

is exact, since

$$yy'' + y'^2 + 2x \equiv \frac{d}{dx}\,(yy' + x^2).$$

If the differential equation (7–1) is exact and (7–2) holds, then along each solution curve F remains constant; that is, the solutions satisfy the equation

$$F(x, y, y') = c_1, \qquad (7\text{–}3)$$

where c_1 is an arbitrary constant. We can think of (7–3) as the result of integrating the relation

$$\frac{dF}{dx} = 0$$

with respect to x. The equation $F = c_1$ (or often the function F itself) is called an *integral* (or *first integral*) of the differential equation.

In the example given above, the equation can be written

$$\frac{d}{dx}\,(yy' + x^2) = 0,$$

and we obtain the integral

$$yy' + x^2 = c_1.$$

Here the variables are separable and we find

$$\tfrac{1}{2}y^2 + \tfrac{1}{3}x^3 = c_1 x + c_2$$

as the general solution of the differential equation.

In general, if an equation of second order is exact, then integration leads to an equation of first order (containing an arbitrary constant); the general solution of this first order equation is the general solution of the given second order equation.

For the first order equation $P\,dx + Q\,dy = 0$, a simple test for exactness is given in Section 2–5. The equation is exact if and only if $P_y \equiv Q_x$. There is an analogous test for the second order equation (7–1): *the equation (7–1) is exact if and only if the following two relations hold:*

$$p_{xx} + 2z p_{xy} + z^2 p_{yy} \equiv q_{xz} + z q_{yz} - q_y,$$

$$p_{xz} + z p_{yz} + 2p_y \equiv q_{zz}, \qquad (7\text{–}4)$$

where z represents y'. These relations are considered in Prob. 5 below, following Section 7–3.

When Eqs. (7–4) hold, the integral F is obtainable from the equations

$$F_x = q - z(q_z - p_x - z p_y),$$

$$F_y = q_z - p_x - z p_y, \qquad (7\text{–}5)$$

$$F_z = p.$$

(See Prob. 5(c) below.) For the equation $yy'' + y'^2 + 2x = 0$ we have $p = y$, $q = z^2 + 2x$, so that (7-4) holds:

$$p_{xx} + 2zp_{xy} + z^2 p_{yy} = 0 = q_{xz} + zq_{yz} - q_y,$$

$$p_{xz} + zp_{yz} + 2p_y = 2 = q_{zz}.$$

Equations (7-5) then become

$$F_x = 2x, \qquad F_y = z, \qquad F_z = y.$$

From the first of these, $F = G(y, z) + x^2$, where G is some function of y and z. The second and third relations give

$$G_y = z, \qquad G_z = y.$$

The first equation implies that $G = yz + \phi(y)$. Substitution in the second shows that $\phi(y)$ must be a constant. Thus F can be chosen as $yz + x^2$, that is, as $yy' + x^2$, as above.

Equations (7-5) give an expression for dF:

$$dF = F_x\, dx + F_y\, dy + F_z\, dz$$

and, when Eqs. (7-4) hold, this expression is exact; that is, F can be found. In the example,

$$dF = 2x\, dx + z\, dy + y\, dz,$$

and we see at once that $F = x^2 + yz + \text{const}$. As in Section 2-6, F can also be given by a line integral:

$$F = \int_{(x_0, y_0, z_0)}^{(x, y, z)} F_x\, dx + F_y\, dy + F_z\, dz, \qquad (7\text{-}6)$$

where the path is any convenient curve from (x_0, y_0, z_0) to (x, y, z).

Remark. Throughout this chapter, as in Chapter 2, the emphasis will be on formal properties, and continuity conditions will in general be omitted; the results obtained will be valid if we assume that the functions concerned (for example, the functions p and q for Eq. 7-1) have continuous partial derivatives through the second order in some open region of the space of the variables.

7-2 Integrating factors. If the differential equation (7-1) is not exact, we can seek an *integrating factor:* that is, a function $v(x, y, y')$ (not identically 0) such that the differential equation

$$vpy'' + vq = 0 \qquad (7\text{-}7)$$

is exact.

If v is an integrating factor for (7–1), then the left-hand side of Eq. (7–7) is of the form dF/dx for some function $F(x, y, y')$. Accordingly, the equation remains exact if we multiply by any function of F. For example, if we multiply by F^2, the left-hand side becomes the derivative of $F^3/3$:

$$F^2 \frac{dF}{dx} \equiv \frac{d}{dx} \left(\frac{F^3}{3} \right).$$

Thus once we have found one integrating factor, we can find an infinity of integrating factors. (See Section 2–7.)

It may happen that two integrating factors are found, neither of which is obtainable from the other as in the preceding paragraph. For example, the equation $yy'' + y'^2 + 2x = 0$ has the two integrating factors

$$v_1 = 1, \qquad v_2 = x.$$

The first is a consequence of the exactness of the equation as given; the second is a consequence of the exactness of the equation after multiplication by x:

$$xyy'' + xy'^2 + 2x^2 \equiv \frac{d}{dx} \left(xyy' - \tfrac{1}{2}y^2 + \tfrac{2}{3}x^3 \right).$$

Hence we have the two integrals

$$F_1 \equiv yy' + x^2 = c_1,$$
$$F_2 \equiv xyy' - \tfrac{1}{2}y^2 + \tfrac{2}{3}x^3 = c_2;$$

and it is clear that the integrating factor v_1 is not obtainable from a function of F_2, nor is v_2 obtainable from a function of F_1. The integrating factors are independent, in a certain sense, and the two corresponding integrals provide the general solution. Indeed, elimination of y' between the equations $F_1 = c_1$, $F_2 = c_2$ leads, as before, to the equation

$$\tfrac{1}{2}y^2 + \tfrac{1}{3}x^3 = c_1 x + c_2.$$

In general, we call the integrating factors v_1, v_2 of Eq. (7–1) *independent* if the corresponding integrals $F_1(x, y, y')$, $F_2(x, y, y')$ are *functionally independent;* that is, if neither one is expressible as a function of the other. (For a detailed discussion of functional independence, the reader is referred to pp. 132–136 of Reference 4.) This can be interpreted geometrically as follows: $F_1(x, y, y')$, $F_2(x, y, y')$ are functionally independent if the equations $F_1(x, y, z) = \text{const}$, $F_2(x, y, z) = \text{const}$ represent surfaces in xyz-space which are never tangent, so that each surface $F_1 = c_1$ meets each surface $F_2 = c_2$ (if at all) along a *curve.*

If now $v_1(x, y, y')$ and $v_2(x, y, y')$ are independent integrating factors of Eq. (7–1), then the corresponding integrals

$$F_1(x, y, y') = c_1, \qquad F_2(x, y, y') = c_2 \qquad (7\text{–}8)$$

do describe the general solution. We do not attempt to prove this here, but give the following geometric description, which can be made the basis of a proof. The solutions of the second order equation (7–1) can be interpreted as curves in xyy'-space, with exactly one curve through each point (under appropriate continuity conditions); see Section 1–7. An integral, $F_1 = c_1$, describes a family of surfaces in xyy'-space; each surface is constructed of complete solution curves. Thus the family of surfaces $F_1 = $ const provides a *stratification* of the family of solution curves (Fig. 7–1). A similar statement applies to the family of surfaces $F = c_2$. Each surface of the second family $F_2 = $ const cuts each surface $F_1 = $ const in a curve. Since the curve lies in both surfaces, it must be a solution curve, and hence is defined by the two equations $F_1 = c_1$, $F_2 = c_2$.

From this geometric picture, it is clear why we need functionally independent integrals. If the two integrals are functionally dependent, the surfaces do not in general meet in curves. For example, if $F_2 = F_1^3$, then the families of surfaces $F_1 \doteq$ const, $F_2 = $ const are the *same;* together they provide one stratification in surfaces, but no intersection that is a solution curve.

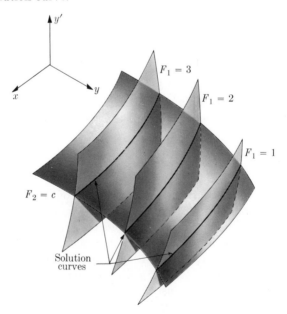

FIG. 7–1. Level surfaces of functionally independent integrals.

7-3 Exact equations of higher order.

An equation of nth order

$$p(x, y, y', \ldots, y^{(n-1)})y^{(n)} + q(x, y, y', \ldots, y^{(n-1)}) = 0 \qquad (7\text{-}9)$$

is called *exact* if there exists a function $F(x, y, \ldots, y^{(n-1)})$ such that when y is considered as a function of x,

$$\frac{dF}{dx} = F_x + F_y y' + \cdots \equiv p y^{(n)} + q. \qquad (7\text{-}10)$$

The function F is then an integral of the equation (7-9), and $F = $ const along each solution curve. Hence knowledge of an integral permits us to lower the order of the equation by one.

The conditions (7-4) for exactness and the expressions (7-5) can be generalized to the equation of nth order. The formulas are rather complicated; they are given on p. 67 of Reference 3.

Integrating factors are defined as for the equation of first or second order. If two integrating factors are known and the corresponding integrals F_1, F_2 are functionally independent, then the equations

$$F_1(x, y, \ldots, y^{(n-1)}) = c_1, \qquad F_2(x, y, \ldots, y^{(n-1)}) = c_2$$

do not provide the general solution for $n > 2$. However, elimination of $y^{(n-1)}$ between the two equations leads to an equation of order $n - 2$. In general, knowledge of k integrating factors provides k integrals and, if they are functionally independent, permits lowering the order of the equation by k. If $k = n$, the integrals provide the general solution.

PROBLEMS

1. Verify that the following equations are exact and find the general solution:

(a) $y^2 y'' + 2yy'^2 = 0$,

(b) $(x + 2xy)y'' + 2xy'^2 + 2y' + 4yy' = 0$,

(c) $yy'' + y'' + y'^2 + 1 = 0$,

(d) $(x + y)y'' + y'^2 - 1 = 0$.

2. Find the general solution for each of the following, with the aid of the integrating factors given:

(a) $xyy'' + xy'^2 + 2yy' + 2 = 0, v = x$;

(b) $x^2 y'' - xy' - 3y = 0, v = x^{-4}$;

(c) $yy'' + yy' + y'^2 + x + 1 = 0, v_1 = 1, v_2 = e^x$;

(d) $xyy'' - 2xy'^2 + 2yy' = 0, v_1 = x/(xy' - y)^2, v_2 = y/(xy' - y)^2$.

3. For each of the following equations demonstrate the properties stated.

(a) $y'' + q(y) = 0$. Show that $v = y'$ is an integrating factor.

(b) $y'' + q(y') = 0$. Show that $v = y'/q(y')$ is an integrating factor.

(c) $y'' + q(x, y') = 0$. Show that this becomes a first order equation if we write $y' = z$, $y'' = dz/dx$. If $F_1(x, z) = c_1$ is the general solution of the first order equation, show that $F_1(x, y') = c_1$ is an integral of the given equation. The general solution of the differential equation $F_1(x, y') = c_1$ is the general solution of $y'' + q(x, y') = 0$.

(d) $y'' + q(y, y') = 0$. Show that this becomes a first order equation if we write $z = y'$ and

$$y'' = \frac{dz}{dx} = \frac{dz}{dy}\frac{dy}{dx} = z\frac{dz}{dy}.$$

If $F_1(y, z) = c_1$ is the general solution of the first order equation, show that $F_1(y, y') = c_1$ is an integral of the given equation. The general solution of the differential equation $F_1(y, y') = c_1$ is the general solution of the equation $y'' + q(y, y') = 0$.

4. Find the general solution for each of the following, with the aid of the properties given in the corresponding part of Prob. 3:

(a) $y'' + e^y = 0$,   (b) $y'' + y'^2 = 0$,

(c) $y'' + (x + y')/(x - y') = 0$,   (d) $y'' + y'e^y = 0$.

5. It can be shown (part (c) below) that the second order equation

$$p(x, y, y')y'' + q(x, y, y') = 0$$

is exact: $F_x + F_y y' + F_{y'}y'' \equiv py'' + q$ for some $F(x, y, y')$ if and only if, under appropriate continuity assumptions,

$$F_x = q - z(q_z - p_x - zp_y), \qquad F_y = q_z - p_x - zp_y, \qquad F_z = p, \qquad \text{(i)}$$

where $z = y'$.

(a) Show that the conditions (i) imply the two conditions

$$
\begin{aligned}
p_{xx} + 2zp_{xy} + z^2 p_{yy} &= q_{xz} + zq_{yz} - q_y, \\
p_{xz} + zp_{yz} + 2p_y &= q_{zz}.
\end{aligned}
\qquad \text{(ii)}
$$

[*Hint:* Apply the relations $F_{xy} = F_{yx}$, $F_{xz} = F_{zx}$, $F_{yz} = F_{zy}$. It can be shown with the aid of line integrals, as in Section 2–6, that the conditions (ii) imply existence of a function F for which relations (i) hold. See p. 280 of Reference 4.]

(b) Show that conditions (ii) reduce to the condition $p_0'' - p_1' + p_2 = 0$ for the linear equation

$$p_0(x)y'' + p_1(x)y' + p_2(x)y = 0.$$

(c) Prove that equations (i) are a test for exactness. [*Hint:* From the identity

$$F_x + F_y z + F_z z' \equiv pz' + q,$$

we at once conclude that

$$F_x + F_y z = q, \qquad F_z = p. \qquad \text{(iii)}$$

Differentiate the first equation with respect to z, the second with respect to x and y; then combine the results to obtain (i). Conversely, show that Eqs. (i) imply Eqs. (iii).]

ANSWERS

1. (a) $y^3 = c_1 x + c_2$,   (b) $xy^2 + xy + c_1 x + x_2 = 0$,   (c) $x^2 + y^2 + c_1 x + 2y + c_2 = 0$, (d) $2y - 2x - c_1 \log |2x + 2y + c_1| = c_2$.

2. (a) $xy^2 + 2x^2 + 2c_1 x = c_2$,   (b) $xy + c_1 x^4 = c_2$,   (c) $x^2 + y^2 = c_1 + c_2 e^{-x}$, (d) $c_1 x + c_2 y - xy = 0$.

4. (a) $x \pm \int (dy/\sqrt{c_1 - 2e^y}) = c_2$,   (b) $e^y = c_1 x + c_2$,
    (c) $y = \frac{1}{2} x^2 \pm \frac{1}{2} \sqrt{2} [x \sqrt{x^2 + c_1} + c_1 \log (x + \sqrt{x^2 + c_1})] + c_2$,
    (d) $e^{-y} = c_1 + c_2 e^{-x/c_1}$ and $e^{-y} = x + c_2$.

7–4 Special methods for linear equations. Let the second order linear equation

$$p_0(x)y'' + p_1(x)y' + p_2(x)y = 0 \qquad (7\text{--}11)$$

be given, where p_0, p_1, p_2 have continuous first and second derivatives in an interval of the x-axis, and p_0 is not identically 0. If this equation is exact and $F(x, y, y')$ is an integral, then

$$F_x + F_y y' + F_{y'} y'' \equiv p_0(x)y'' + p_1(x)y' + p_2(x)y. \qquad (7\text{--}12)$$

We assume that F has continuous first and second derivatives for x in the given interval and that y, y' are unrestricted. Since (7–12) is an identity, we must have

$$F_{y'} \equiv p_0(x).$$

Hence, by integration,

$$F = p_0(x)y' + G(x, y).$$

If this is substituted in (7–12), we find

$$p_0' y' + G_x + G_y y' \equiv p_1(x)y' + p_2(x)y;$$

hence

$$G_x = p_2(x)y, \qquad G_y = p_1 - p_0'.$$

The second equation gives

$$G = (p_1 - p_0')y + H(x), \qquad G_x = (p_1' - p_0'')y + H'(x).$$

Hence

$$(p_1' - p_0'')y + H'(x) \equiv p_2(x)y.$$

This is possible only if $H(x)$ is constant and

$$p_1' - p_0'' \equiv p_2. \tag{7-13}$$

The constant can be disregarded and we find that

$$F = p_0 y' + (p_1 - p_0')y. \tag{7-14}$$

Conversely, if (7–13) holds, then we verify (Prob. 2 below) that F satisfies the relation (7–12), so that Eq. (7–11) is exact. We conclude:

The linear equation $p_0 y'' + p_1 y' + p_2 y = 0$ is exact if and only if

$$p_0'' - p_1' + p_2 \equiv 0. \tag{7-15}$$

If (7–15) holds, then (7–14) defines an integral $F(x, y, y')$.

Now let the equation (7–11) be given, not necessarily exact. We seek an integrating factor $v(x)$. Thus

$$v p_0 y'' + v p_1 y' + v p_2 y = 0$$

is to be exact, and the condition (7–15) becomes

$$(v p_0)'' - (v p_1)' + v p_2 = 0.$$

Hence the integrating factor v must satisfy the differential equation

$$p_0 v'' + (2 p_0' - p_1) v' + (p_0'' - p_1' + p_2) v = 0. \tag{7-16}$$

This equation is known as the *adjoint* of Eq. (7–11). Each solution v (not identically 0) of Eq. (7–16) provides an integrating factor of Eq. (7–11). It can be shown (Prob. 4 below) that two linearly independent solutions of Eq. (7–16) determine two functionally independent integrals of (7–11).

EXAMPLE 1. $x^2 y'' + 2xy' - 2y = 0$. The equation is not exact, since

$$p_0'' - p_1' + p_2 = 2 - 2 - 2 \neq 0.$$

The adjoint equation is

$$x^2 v'' + 2xv' - 2v = 0,$$

which has the same form as the original equation (which is then *self-adjoint*). By experiment we find that $v = x$ is a solution, so that

$$x^3 y'' + 2x^2 y' - 2xy = 0$$

must be exact and have the integral

$$x^3 y' - x^2 y = c_1.$$

This is a first order linear equation, for which x^{-4} is an integrating factor. The general solution is easily found to be

$$y = c_1 x^{-2} + c_2 x \qquad (x \neq 0).$$

The results obtained are also useful for a nonhomogeneous equation

$$p_0 y'' + p_1 y' + p_2 y = Q(x). \tag{7–17}$$

Indeed, an integrating factor $v(x)$ for the related homogeneous equation is also an integrating factor for the nonhomogeneous equation. The results can be extended to linear equations of arbitrary order (Prob. 6 below).

If one solution $y_1(x)$ of Eq. (7–11) is known, then it can be proved that

$$v(x) = \frac{y_1(x)}{p_0(x)} q(x), \qquad q(x) = \exp \int \frac{p_1(x)}{p_0(x)} \, dx \tag{7–18}$$

defines an integrating factor v, and that

$$y = c_1 y_1(x) \int \frac{dx}{q(x) y_1^2(x)} + c_2 y_1(x) \tag{7–19}$$

is the general solution (Prob. 4 below).

We can also take advantage of the known solution $y_1(x)$ of the homogeneous equation to obtain the general solution of the nonhomogeneous equation. This can be done by means of the integrating factor $v(x)$ or as follows. We introduce a new dependent variable u by the equation

$$y = y_1(x) u; \tag{7–20}$$

then

$$y' = u y_1' + u' y_1, \qquad y'' = u y_1'' + 2 u' y_1' + u'' y_1,$$

$$p_0 y'' + p_1 y' + p_2 y = u(p_0 y_1'' + p_1 y_1' + p_2 y_1)$$
$$+ u'(p_1 y_1 + 2 p_0 y_1') + u'' p_0 y_1.$$

The first term on the right equals 0, since y_1 satisfies Eq. (7–11); hence

$$y_1 p_0 u'' + (p_1 y_1 + 2 p_0 y_1') u' = Q(x). \tag{7–21}$$

We now set

$$w = \frac{du}{dx}. \tag{7–22}$$

Then Eq. (7–21) becomes a first order linear equation for w:

$$y_1 p_0 w' + (p_1 y_1 + 2 p_0 y_1') w = Q(x). \tag{7–23}$$

When w has been found, Eq. (7–22) can be integrated to give u and Eq. (7–20) gives y.

EXAMPLE 2. $x^2y'' + xy' - 4y = x + 2$. Here $y = x^2$ is found to be a solution of the homogeneous equation. The substitutions $y = x^2u$, $w = u'$ lead to the equations

$$x^4u'' + 5x^3u' = x + 2,$$

$$x^4w' + 5x^3w = x + 2,$$

$$w = \frac{1}{3x^2} + \frac{1}{x^3} + \frac{c_1}{x^5} \qquad (x \neq 0),$$

$$u = -\frac{1}{3x} - \frac{1}{2x^2} - \frac{c_1}{4x^4} + c_2,$$

$$y = -\frac{x}{3} - \frac{1}{2} - \frac{c_1}{4x^2} + c_2 x^2.$$

This method can be applied to a linear equation of arbitrary order n. If $y_1(x)$ is a solution of the related homogeneous equation, then the substitutions $y = y_1u$, $w = u'$ lead to a linear equation of order $n - 1$ for w. If *two* linearly independent solutions $y_1(x)$, $y_2(x)$ of the related homogeneous equation are known, then we can reduce the order of the equation to $n - 2$. For by means of $y_1(x)$ we obtain a reduction to order $n - 1$, as above. When $y = y_2(x)$, then $u = y_2(x)/y_1(x) = u_2(x)$ and $w = u_2'(x) = w_2(x)$; the fact that $y_2(x)$ satisfies the homogeneous equation related to the given equation implies that $w_2(x)$ satisfies the homogeneous equation related to the equation for w. Hence we can apply the previous method to obtain an equation of one order lower: we set $w = w_2(x) \cdot U(x)$, $W(x) = U'(x)$, and then $W(x)$ satisfies an equation of order $n - 2$. The reasoning can be extended to the case of k known linearly independent solutions; thus the particular solutions serve the same purpose as integrals. It should be remarked, however, that the method often leads to complicated integration problems. Another method of using two known solutions is indicated in Prob. 8 below.

PROBLEMS

1. Show that the following linear differential equations are exact and find the general solution of each:
   (a) $x^2y'' + (3x + 1)y' + y = 0$,
   (b) $xe^{2x}y'' + (2 + 4x)e^{2x}y' + 4(1 + x)e^{2x}y = 0$.

2. Prove that if $F(x, y, y')$ satisfies Eq. (7–14), and Eq. (7–13) holds, then F satisfies Eq. (7–12).

3. Obtain the adjoint of each of the following linear equations, obtain a solution of the adjoint of the form suggested, and use this solution as an integrating factor for the given equation to obtain the general solution.

(a) $x^2y'' + 5xy' + 4y = 0$, solution of adjoint of form x^m;

(b) $xy'' + (6x + 2)y' + (9x + 6)y = 0$, solution of adjoint of form e^{ax}.

4. (a) Show that the adjoint of the adjoint of the equation $p_0(x)y'' + p_1(x)y' + p_2(x)y = 0$ has the same form as the given equation.

(b) An equation $p_0y'' + p_1y' + p_2y = 0$ is termed *self-adjoint* if its adjoint has the same coefficients as the given equation. Show that the equation is self-adjoint if and only if $p_0' = p_1$.

(c) Show that a self-adjoint equation can be written in the form $(p_0y')' + p_2y = 0$.

(d) Show that a general equation $p_0y'' + p_1y' + p_2y = 0$ becomes self-adjoint when multiplied by

$$\frac{1}{p_0} e^{\int (p_1/p_0)\, dx}.$$

(e) Show that a particular solution $y_1(x)$ of $p_0y'' + p_1y' + p_2y = 0$ serves as an integrating factor for the adjoint equation.

(f) Show that if $y_1(x)$ is a particular solution of $p_0y'' + p_1y' + p_2y = 0$, then Eq. (7–18) defines an integrating factor of the equation and Eq. (7–19) gives the general solution. [*Hint:* Apply the result of part (d).]

(g) Show that if $p_0(x) \neq 0$ and $v_1(x)$ and $v_2(x)$ are linearly independent solutions of the adjoint equation of $p_0y'' + p_1y' + p_2y = 0$, then the corresponding integrals F_1 and F_2 are functionally independent. [*Hint:* By Eq. (7–14), $F_1 = v_1p_0y' + [v_1p_1 - (v_1p_0)']y$, $F_2 = v_2p_0y' + [v_2p_1 - (v_2p_0)']y$. For fixed x, the loci $F_1 = c_1$, $F_2 = c_2$ are straight lines in the yy'-plane. Show that these lines are parallel only if $v_1v_2' - v_2v_1' = 0$. But $v_1v_2' - v_2v_1'$ is the Wronskian determinant of v_1, v_2 and cannot be zero (Section 4–3); hence the lines meet at just one point, and the surfaces $F_1 = c_1$, $F_2 = c_2$ meet along a curve.]

5. Find the general solution for each of the following, with the aid of the indicated solution $y_1(x)$ of the related homogeneous equation:

(a) $y'' - 2y' + y = e^{-x}$, $y_1 = e^x$;

(b) $(1 - x)y'' + xy' - y = (2 - x)e^x$, $y_1 = x$;

(c) $xy'' + y' + xy = e^x$, $y_1 = J_0(x)$ [$J_0(x)$ = Bessel function of order 0; see Prob. 11 following Section 9–8];

(d) $x^2y'' + xy' - 4y = x + 2$, $y_1 = x^2$.

6. Let a linear equation of order n be given:

$$p_0(x)y^{(n)} + p_1(x)y^{(n-1)} + \cdots + p_n(x)y = 0.$$

(a) Prove that if the equation is exact: that is,

$$p_0(x)y^{(n)} + p_1(x)y^{(n-1)} + \cdots + p_n(x)y \equiv F_x + F_yy' + \cdots + F_{y^{(n-1)}}y^{(n)}, \quad \text{(i)}$$

where $F = F(x, y, \ldots, y^{(n-1)})$, then F is linear in $y, y', \ldots, y^{(n-1)}$:

$$F = q_0(x)y^{(n-1)} + q_1(x)y^{(n-2)} + \cdots + q_{n-1}(x)y.$$

[*Hint:* Use mathematical induction. Verify the assertion for $n = 1$. Assume that it is true for order $n - 1$ and consider the case of order n. Show that the identity (i) implies that $F = p_0(x)y^{(n-1)} + G(x, y, \ldots, y^{(n-2)})$ and that G is an integral of an equation of order $n - 1$.]

(b) Show that the functions $q_0(x), \ldots, q_{n-1}(x)$ of part (a) must satisfy the relations

$$p_0 = q_0, \qquad p_1 = q_0' + q_1, \qquad \ldots, \qquad p_{n-1} = q_{n-2}' + q_{n-1}, \qquad p_n = q_{n-1}'$$

and that

$$p_n - p_{n-1}' + p_{n-2}'' - \cdots + (-1)^n p_0^{(n)} = 0. \tag{ii}$$

Show that, conversely, if (ii) holds, then the equation is exact, with integral $F = q_0 y^{(n-1)} + q_1 y^{(n-2)} + \cdots$, where

$$q_0 = p_0, \qquad q_1 = p_1 - p_0', \qquad q_2 = p_2 - p_1' + p_0'', \qquad \ldots,$$

$$q_{n-1} = p_{n-1} - p_{n-2}' + p_{n-3}'' - \cdots + (-1)^{n-1} p_0^{(n-1)}.$$

(c) Show that if $v(x)$ is an integrating factor of the given equation, then v satisfies the linear equation (*adjoint equation*)

$$r_0 v^{(n)} - r_1 v^{(n-1)} + \cdots + (-1)^n r_n v = 0,$$

where

$$r_0 = p_0, \qquad r_1 = p_1 - \binom{n}{1} p_0',$$

$$r_2 = p_2 - \binom{n-1}{1} p_1' + \binom{n}{2} p_0'', \qquad \ldots,$$

$$r_k = p_k - \binom{n-k+1}{1} p_{k-1}' + \binom{n-k+2}{2} p_{k-2}'' - \cdots,$$

$$r_n = p_n - p_{n-1}' + p_{n-2}'' - \cdots,$$

and $\binom{n}{k}$ is the binomial coefficient:

$$\binom{n}{k} = \frac{n!}{k!(n-k)!}.$$

7. Test for exactness and solve by repeated integration (see Prob. 6):
(a) $x^3 y''' + 9x^2 y'' + 18xy' + 6y = 0$,
(b) $xy''' + (3 + x^2)y'' + 4xy' + 2y = 0$.

8. *Linear equation with two solutions of related homogeneous equation known.* Consider the third order equation

$$p_0(x)y''' + p_1(x)y'' + p_2(x)y' + p_3(x)y = Q(x)$$

and let $y_1(x)$, $y_2(x)$ be linearly independent solutions of the related homogeneous equation. Let $y = y_1u + y_2v$, where u and v are functions of x such that $y_1u' + y_2v' = 0$, so that we can write $u' = wy_2$, $v' = -wy_1$, in terms of a new variable w. Replace y by $y_1u + y_2v$ in the differential equation and show that this reduces to a *first* order equation for w:

$$p_0(x)r(x)w' + [2p_0(x)r'(x) + p_1(x)r(x)]w = Q(x),$$

where $r(x) = y_1'(x)y_2(x) - y_2'(x)y_1(x)$. Show that

$$w = \frac{1}{r^2q} \int \frac{rQq}{p_0} \, dx + \frac{c_3}{r^2q}, \qquad q = \exp \int \frac{p_1}{p_0} \, dx,$$

$$y = y_1 \left[\int y_2w \, dx + c_1 \right] + y_2 \left[\int y_1w \, dx + c_2 \right].$$

9. Find the general solution, with the aid of the given particular solutions:
(a) $(2x - 3)y''' + (7 - 6x)y'' + 4xy' - 4y = 0$, $y_1 = e^x$, $y_2 = e^{2x}$;
(b) $x^3y''' - 3x^2y'' + (6x - x^3)y' + (x^2 - 6)y = 0$, $y_1 = xe^x$, $y_2 = xe^{-x}$.
[*Hint:* Apply the method described in the text or the method of Prob. 8.]

10. *Cauchy's linear equation.* This term is used for an equation of form

$$a_0x^ny^{(n)} + a_1x^{n-1}y^{(n-1)} + \cdots + a_{n-1}xy' + a_ny = Q(x),$$

where $a_0, a_1, \ldots, a_n$ are constants.
(a) Show that $y = x^r$ is a solution of the related homogeneous equation provided r satisfies the equation

$$a_0r(r - 1) \cdots (r - n + 1) + a_1r(r - 1) \cdots (r - n + 2) + \cdots$$
$$+ a_{n-1}r + a_n = 0.$$

If this equation has distinct real roots $r_1, \ldots, r_n$, obtain the general solution.
(b) Prove that the substitution $x = e^t$ reduces the equation to one with constant coefficients for y in terms of t. [*Hint:* Show first that if D stands for the operator d/dt, then

$$x\frac{dy}{dx} = Dy,$$

$$x^2\frac{d^2y}{dx^2} = D(D - 1)y,$$

$$\vdots$$

$$x^k\frac{d^ky}{dx^k} = D(D - 1) \ldots (D - k + 1)y;$$

the general case can be established by induction.]

11. Obtain the general solution of each of the following (see Prob. 10):

(a) $x^2 (d^2y/dx^2) - 2x (dy/dx) + 2y = 0$,

(b) $x^2 (d^2y/dx^2) + x (dy/dx) + 4y = 0$,

(c) $x^2 (d^2y/dx^2) - x (dy/dx) + y = x$.

ANSWERS

1. (a) $y = c_1 x^{-1} \exp (x^{-1}) \int x^{-1} \exp (-x^{-1}) \, dx + c_2 x^{-1} \exp (x^{-1})$,

(b) $xe^{2x} y = c_1 x + c_2$.

3. (a) Adjoint is $x^2 v'' - xv' + v = 0$, integrating factor is $v = x$, $x^2 y = c_1 \log x + c_2$; (b) adjoint is $v'' - 6v' + 9v = 0$, integrating factor is $v = e^{3x}$, $xe^{3x} y = c_1 x + c_2$.

5. (a) $y = \frac{1}{4} e^{-x} + c_1 x e^x + c_2 e^x$, (b) $y = c_1 e^x + c_2 x + xe^x$,

(c) $y = J_0 \int x^{-1} J_0^{-2} [\int \int J_0 e^x \, dx] \, dx + c_1 J_0 \int x^{-1} J_0^{-2} \, dx + c_2 J_0$,

(d) $y = c_1 x^2 + c_2 x^{-2} - \frac{1}{3} x - \frac{1}{2}$.

7. (a) $x^3 y = c_1 + c_2 x + c_3 x^2$,

(b) $y = c_1 x^{-1} + c_2 x^{-1} \exp (-x^2/2) + c_3 x^{-1} \exp (-x^2/2) \int \exp (x^2/2) \, dx$.

9. (a) $y = c_1 x + c_2 e^{2x} + c_3 e^x$, (b) $y = x^2 + c_1 x + c_2 x e^{-x} + c_3 x e^x$.

10. (a) $y = c_1 x^{r_1} + \cdots + c_n x^{r_n}$.

11. (a) $y = c_1 x + c_2 x^2$, (b) $y = c_1 \cos \{2 \log |x|\} + c_2 \sin \{2 \log |x|\}$,

(c) $y = c_1 x + c_2 x \log |x| + \frac{1}{2} x \log^2 |x|$.

7–5 Integrals of systems of differential equations. Consider a system of first order equations:

$$\frac{dy_1}{dx} = f_1(x, y_1, \ldots, y_n),$$

$$\frac{dy_2}{dx} = f_2(x, y_1, \ldots, y_n), \qquad\qquad (7\text{–}24)$$

$$\vdots$$

$$\frac{dy_n}{dx} = f_n(x, y_1, \ldots, y_n).$$

As pointed out in Section 1–7, all differential equations and systems of differential equations can in general be reduced to this form. The solutions of Eq. (7–24) are curves

$$y_1 = y_1(x), \qquad \ldots, \qquad y_n = y_n(x) \qquad\qquad (7\text{–}25)$$

in the space of the $n + 1$ variables $x, y_1, \ldots, y_n$; under appropriate assumptions on $f_1, f_2, \ldots, f_n$, one and only one such curve passes through each point of an open region D of the space.

An *integral* (or *first integral*) of (7–24) is defined as a function

$$F(x, y_1, \ldots, y_n)$$

which is not identically constant, but is constant along each solution curve (7–25). We shall consider only integrals having continuous first partial derivatives in D. If F is such an integral, then along each solution curve,

$$\frac{dF}{dx} = F_x + F_{y_1} \frac{dy_1}{dx} + \cdots + F_{y_n} \frac{dy_n}{dx}$$
$$= F_x + F_{y_1} f_1 + \cdots + F_{y_n} f_n = 0. \tag{7–26}$$

Since each point of D lies on a solution curve, the expression $F_x + F_{y_1} f_1 + \cdots + F_{y_n} f_n$ must be identically 0 in D. In other words, *if a nonconstant function F is an integral of the system (7–24), then F satisfies the partial differential equation*

$$\frac{\partial F}{\partial x} + f_1(x, y_1, \ldots, y_n) \frac{\partial F}{\partial y_1} + \cdots + f_n(x, y_1, \ldots, y_n) \frac{\partial F}{\partial y_n} = 0 \tag{7–27}$$

in D. Conversely, if F satisfies (7–27), then along each solution curve (7–26) holds, so that F remains constant and is an integral of (7–24).

Let F be an integral of (7–24); then

$$dF = F_x \, dx + F_{y_1} \, dy_1 + \cdots + F_{y_n} \, dy_n.$$

Equation (7–27) can be written

$$0 = F_x \, dx + F_{y_1} f_1 \, dx + \cdots + F_{y_n} f_n \, dx.$$

If the last two equations are subtracted, we find

$$dF = F_{y_1} (dy_1 - f_1 \, dx) + \cdots + F_{y_n} (dy_n - f_n \, dx). \tag{7–28}$$

Hence we can regard $F_{y_1}, \ldots, F_{y_n}$ as *a set of integrating factors*. If the equations (7–24) in the form

$$dy_1 - f_1 \, dx = 0, \qquad \ldots, \qquad dy_n - f_n \, dx = 0$$

are multiplied by

$$v_1 = F_{y_1}, \qquad \ldots, \qquad v_n = F_{y_n}$$

and added, we obtain precisely the equation $dF = 0$. Integration yields

$$F(x, y_1, \ldots, y_n) = c_1; \tag{7–29}$$

as for the single equation of nth order, we call such a relation, $F = $ const, an integral.

If (7–24) is given and functions $v_1(x, y_1, \ldots, y_n), \ldots, v_n(x, y_1, \ldots, y_n)$ (not all identically 0) can be found such that

$$v_1 (dy_1 - f_1 \, dx) + \cdots + v_n (dy_n - f_n \, dx) \equiv dF, \qquad (7\text{--}30)$$

for some function $F(x, y_1, \ldots, y_n)$, then F is an integral of (7–24), the v's form a set of integrating factors, and necessarily,

$$v_1 = \frac{\partial F}{\partial y_1}, \qquad \ldots, \qquad v_n = \frac{\partial F}{\partial y_n}. \qquad (7\text{--}31)$$

For since (7–30) holds, and $dy_1 \equiv f_1 \, dx, \ldots, dy_n \equiv f_n \, dx$ along each solution curve, we must have $dF \equiv 0$ on each solution curve, that is, $F = $ const. Thus F is an integral. Equation (7–30) can be written

$$dF = v_1 \, dy_1 + \cdots + v_n \, dy_n - (v_1 f_1 + v_2 f_2 + \cdots v_n f_n) \, dx,$$

from which it follows that (7–31) holds, so that the v's are the integrating factors as defined above. Since not all v's are identically 0, F cannot be identically constant.

Knowledge of a single integral $F = c_1$ permits reduction of the order of (7–24) by one, for we can (in principle) solve (7–29) for one of the variables $y_1, \ldots, y_n$ in terms of the remaining variables and x, and thereby eliminate it from the differential equations.

Each integral (7–29) determines a family of hypersurfaces in the region D. Each solution curve lies wholly in such a surface. Hence the integral provides a stratification of the $(n + 1)$-dimensional region D (Fig. 7–1). If two integrals F_1, F_2 are known, the loci

$$F_1 = c_1, \qquad F_2 = c_2$$

are hypersurfaces of dimension $n - 1$, provided F_1, F_2 are functionally independent. If n functionally independent integrals $F_1, \ldots, F_n$ are known, the loci

$$F_1 = c_1, \qquad F_2 = c_2, \qquad \ldots, \qquad F_n = c_n$$

are curves, precisely the solution curves (7–25).

It is natural to ask whether integrals can always be found for an equation or system of nth order. The answer is that n functionally dependent integrals always do exist, provided appropriate continuity assumptions are made and provided attention is restricted to a sufficiently small part of the region of the variables (see Section 12–11).

The determination of integrals and the related problem of finding integrating factors are very difficult. Even for an equation of first order, the problems are far from simple (Section 2–7). In many cases, however, special symmetries give clues to the form of the integrals. (Several basic physical laws: conservation of energy, conservation of linear momentum, and conservation of angular momentum, are described mathematically by means of integrals of specific systems of equations. These are illustrated in Section 7–6.)

EXAMPLE 1. The system $dy/dx = g(x, y, z)$, $dz/dx = 0$ has the integral $F \equiv z = $ const, for $z = $ const along each solution curve. The level surfaces of F are simply the planes $z = $ const.

EXAMPLE 2. The system

$$\frac{dx}{dt} = 3yz, \qquad \frac{dy}{dt} = -xz, \qquad \frac{dz}{dt} = -2xy$$

has the two integrals

$$F_1(x, y, z) = x^2 + y^2 + z^2, \qquad F_2(x, y, z) = x^2 - y^2 + 2z^2.$$

For, along each solution curve,

$$\frac{dF_1}{dt} = 2x \frac{dx}{dt} + 2y \frac{dy}{dt} + 2z \frac{dz}{dt} = 6xyz - 2xyz - 4xyz = 0,$$

$$\frac{dF_2}{dt} = 2x \frac{dx}{dt} - 2y \frac{dy}{dt} + 4z \frac{dz}{dt} = 6xyz + 2xyz - 8xyz = 0.$$

The solutions are curves

$$x = x(t), \qquad y = y(t), \qquad z = z(t)$$

which can be plotted in xyz-space, with t as parameter. Since t does not appear on the right-hand side of the differential equations, it can be eliminated and the equations written, e.g., as follows:

$$\frac{dy}{dx} = -\frac{x}{3y}, \qquad \frac{dz}{dx} = -\frac{2x}{3z};$$

in this form, the equations have one solution through each point of xyz-space (except perhaps where y or z is 0). The solution curves are given by the two integrals

$$x^2 + y^2 + z^2 = c_1, \qquad x^2 - y^2 + 2z^2 = c_2;$$

they represent a family of spheres and a family of hyperboloids which intersect to form the solution curves. We can express y and z in terms of x on each curve:

$$y^2 = \tfrac{1}{3}(2c_1 - c_2 - x^2), \qquad z^2 = \tfrac{1}{3}(c_1 + c_2 - 2x^2).$$

The dependence on the parameter t can be analyzed as follows:

$$\frac{dx}{dt} = 3yz = \pm\,[(c_1 + c_2 - 2x^2)(2c_1 - c_2 - x^2)]^{1/2},$$

$$t = \pm \int \frac{dx}{[(c_1 + c_2 - 2x^2)(2c_1 - c_2 - x^2)]^{1/2}} + c_3.$$

The last integral is, in general, elliptic (p. 179 of Reference 4). If x is known in terms of t, y and z can be found from their expressions in terms of x.

Remark. For general simultaneous differential equations which are reducible to form (7–24), we can obtain a definition of integral by referring to the equivalent system (7–24). For example, an integral of equations

$$y'' = f(x, y, y', z, z'), \qquad z'' = g(x, y, y', z, z')$$

is a function $F(x, y, y', z, z')$, not identically constant, which remains constant on each solution $y = y(x)$, $z = z(x)$.

7–6 Physical applications. Here we consider several examples, of increasing generality and complexity.

EXAMPLE 1. *Motion of a particle along a line.* This is discussed in Section 2–11. When force is dependent only on position, the differential equation has the form

$$m\,\frac{d^2x}{dt^2} = f(x) \tag{7–32}$$

and there is an *energy integral*

$$\frac{1}{2}m\left(\frac{dx}{dt}\right)^2 + U(x) = c, \tag{7–33}$$

where U, the potential energy, is defined as

$$U = -\int f(x)\,dx. \tag{7–34}$$

EXAMPLE 2. *Central forces.* A particle of mass m moves in the xy-plane, subject to a force directed along the line that passes through the particle and the origin O. The differential equations are

$$m\,\frac{d^2x}{dt^2} = \frac{x}{r}\,f(r), \qquad m\,\frac{d^2y}{dt^2} = \frac{y}{r}\,f(r), \tag{7–35}$$

where $|f(r)|$ is the magnitude of the force and r is the distance of the particle from O. We can write Eqs. (7–35) as a system of four first order equations:

$$\frac{dx}{dt} = v_x, \qquad \frac{dy}{dt} = v_y, \qquad m\frac{dv_x}{dt} = x\frac{f(r)}{r}, \qquad m\frac{dv_y}{dt} = y\frac{f(r)}{r}. \quad (7\text{–}36)$$

The system has an *integral of energy*

$$F_1 = \tfrac{1}{2}m(v_x^2 + v_y^2) + U(x, y) = c_1, \qquad (7\text{–}37)$$

where $U = -\int f(r)\, dr,\ r = (x^2 + y^2)^{1/2}$, for

$$\frac{dF_1}{dt} = mv_x\frac{dv_x}{dt} + mv_y\frac{dv_y}{dt} + \frac{dU}{dr}\frac{dr}{dt}$$

$$= \frac{f(r)}{r}(xv_x + yv_y) - f(r)\left(\frac{x}{r}\frac{dx}{dt} + \frac{y}{r}\frac{dy}{dt}\right) = 0.$$

The two terms in F_1 are the kinetic energy and the potential energy of the particle. The system also has an *integral of angular momentum*

$$F_2 = m(xv_y - yv_x) = c_2, \qquad (7\text{–}38)$$

for

$$\frac{dF_2}{dt} = m\left(x\frac{dv_y}{dt} - y\frac{dv_x}{dt} + \frac{dx}{dt}v_y - \frac{dv}{dt}v_x\right)$$

$$= \frac{xy\, f(r)}{r} - \frac{xy\, f(r)}{r} + v_x v_y - v_y v_x = 0.$$

The function F_2 is the *angular momentum* of the particle with respect to O (or with respect to a z-axis perpendicular to the xy-plane at O); it can be written in polar coordinates as mr^2 $d\theta/dt$ (Prob. 5 below).

When $f(r) = -k/r^2$, we are led to the problem of motion of the planets, as discussed in Section 2–11.

EXAMPLE 3. *The problem of three bodies.* Let three particles of masses m_1, m_2, m_3 move in space, subject to mutual attractions in accordance with Newton's law of gravitation (Fig. 7–2). If the positions with respect to a "fixed" xyz-coordinate system are given by coordinates

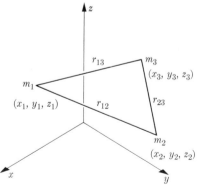

FIG. 7–2. Problem of three bodies.

(x_1, y_1, z_1), (x_2, y_2, z_2), (x_3, y_3, z_3), then the differential equations are

$$m_i \frac{d^2 x_i}{dt^2} = -km_i \sum_j{}' m_j \frac{x_i - x_j}{r_{ij}^3},$$

$$m_i \frac{d^2 y_i}{dt^2} = -km_i \sum_j{}' m_j \frac{y_i - y_j}{r_{ij}^3}, \qquad (i = 1, 2, 3), \qquad (7\text{–}39)$$

$$m_i \frac{d^2 z_i}{dt^2} = -km_i \sum_j{}' m_j \frac{z_i - z_j}{r_{ij}^3},$$

where $\sum_j{}'$ indicates a summation over $j = 1, 2, 3$, omitting the value i. The distance from the ith particle to the jth is denoted by r_{ij}; k is a positive constant.

We define the *potential energy* to be

$$V = -k \left(\frac{m_1 m_2}{r_{12}} + \frac{m_2 m_3}{r_{23}} + \frac{m_3 m_1}{r_{31}} \right). \qquad (7\text{–}40)$$

The differential equations can then be written (Prob. 6 below)

$$m_i \frac{d^2 x_i}{dt^2} = -\frac{\partial V}{\partial x_i}, \qquad m_i \frac{d^2 y_i}{dt^2} = -\frac{\partial V}{\partial y_i}, \qquad m_i \frac{d^2 z_i}{dt^2} = -\frac{\partial V}{\partial z_i}. \qquad (7\text{–}41)$$

This can be written as a system of 18 first order equations in the variables x_i, y_i, z_i, $v_{x_i} = dx_i/dt$, $v_{y_i} = dy_i/dt$, $v_{z_i} = dz_i/dt$; however, for the sake of simplicity, we shall leave the equations in second order form.

The system (7–39) has an *integral of energy* (Prob. 7 below)

$$\frac{1}{2} \sum_{i=1}^{3} m_i \left[\left(\frac{dx_i}{dt} \right)^2 + \left(\frac{dy_i}{dt} \right)^2 + \left(\frac{dz_i}{dt} \right)^2 \right] + V = c_1. \qquad (7\text{–}42)$$

The first term represents the *total* kinetic energy.

The system (7–39) also has three *integrals of linear momentum* (Prob. 8 below)

$$\sum_{i=1}^{3} m_i \frac{dx_i}{dt} = c_2,$$

$$\sum_{i=1}^{3} m_i \frac{dy_i}{dt} = c_3, \qquad (7\text{–}43)$$

$$\sum_{i=1}^{3} m_i \frac{dz_i}{dt} = c_4.$$

The three left-hand sides represent the x, y, and z components of the total linear momentum of the system. Equations (7–43) can be integrated to yield the equations of motion of the center of mass:

$$\sum_{i=1}^{3} m_i x_i = c_2 t + c_5,$$

$$\sum_{i=1}^{3} m_i y_i = c_3 t + c_6, \qquad\qquad (7\text{–}44)$$

$$\sum_{i=1}^{3} m_i z_i = c_4 t + c_7.$$

If $m = m_1 + m_2 + m_3$ denotes the total mass, then $\sum m_i x_i$ is simply $m\bar{x}$, where $\bar{x}$ is the x-coordinate of the center of mass; similarly, $\sum m_i y_i = m\bar{y}$, $\sum m_i z_i = m\bar{z}$. Hence Eqs. (7–44) can be written

$$m\bar{x} = c_2 t + c_5, \qquad m\bar{y} = c_3 t + c_6, \qquad m\bar{z} = c_4 t + c_7; \qquad (7\text{–}45)$$

thus the center of mass moves on a straight line at constant speed. Equations (7–44) can be combined with Eqs. (7–43) to yield three integrals of the original differential equations (7–39):

$$\sum_{i=1}^{3} m_i x_i - t \sum_{i=1}^{3} m_i \frac{dx_i}{dt} = c_5, \qquad \ldots, \qquad (7\text{–}46)$$

We also know three other integrals, namely, those of *angular momentum* (Prob. 9 below)

$$\sum_{i=1}^{3} m_i \left(y_i \frac{dz_i}{dt} - z_i \frac{dy_i}{dt} \right) = c_8,$$

$$\sum_{i=1}^{3} m_i \left(z_i \frac{dx_i}{dt} - x_i \frac{dz_i}{dt} \right) = c_9, \qquad\qquad (7\text{–}47)$$

$$\sum_{i=1}^{3} m_i \left(x_i \frac{dy_i}{dt} - y_i \frac{dx_i}{dt} \right) = c_{10}.$$

The three left-hand sides are the components on the coordinate axes of the total angular momentum of the system.

The original system was of 18th order, but only ten integrals have been shown. No others are known. However, by means of the ten integrals, the system can be reduced to eighth order, and by further substitutions it can be reduced to sixth order. (See Chapter 9 of Reference 1 and Chapter 13 of Reference 5.)

The problem of three bodies is of central importance in celestial mechanics. The planets and sun form a system of n bodies, governed by differential equations similar to (7–39); indeed, the ten integrals listed above have counterparts for the n-body problem. To a first approximation, the motion of the earth about the sun can be considered as a *two*-body problem (Prob. 12 below). A next approximation takes into account the influence of the moon or of Jupiter (the largest of the planets), and this introduces a three-body problem.

EXAMPLE 4. *Lagrangian equations.* A special case of these is considered in Section 6–13. In general, the equations have the form

$$\frac{d}{dt}\left(\frac{\partial L}{\partial q_j'}\right) - \frac{\partial L}{\partial q_j} = 0 \qquad (j = 1, \ldots, n), \qquad (7\text{–}48)$$

where L is a function of $q_1, \ldots, q_n, q_1', \ldots, q_n', t$, and $q_j' = dq_j/dt$. We consider the case in which L does not depend on t; the equations are then called *conservative*. In the applications, L appears as $T - V$, where

$$T \doteq \sum_{j=1}^{n} \sum_{i=1}^{n} a_{ij} q_i' q_j', \qquad V = V(q_1, \ldots, q_n), \qquad (7\text{–}49)$$

and the a_{ij} may depend on $q_1, \ldots, q_n$.

Now by the chain rule of calculus (p. 86 of Reference 4),

$$\frac{d}{dt}\left(\frac{\partial L}{\partial q_j'}\right) = \frac{\partial^2 L}{\partial q_1' \partial q_j'}\frac{dq_1'}{dt} + \cdots + \frac{\partial^2 L}{\partial q_n' \partial q_j'}\frac{dq_n'}{dt}$$

$$+ \frac{\partial^2 L}{\partial q_1 \partial q_j'}\frac{dq_1}{dt} + \cdots + \frac{\partial^2 L}{\partial q_n \partial q_j'}\frac{dq_n}{dt}. \qquad (7\text{–}50)$$

Since $dq_j'/dt = d^2 q_j/dt^2$, it appears that (7–48) can be considered as a set of n simultaneous differential equations of second order for $q_1, \ldots, q_n$ in terms of t. These equations have an *integral of energy*

$$\sum_{j=1}^{n} q_j' \frac{\partial L}{\partial q_j'} - L = c_1, \qquad (7\text{–}51)$$

for

$$\frac{d}{dt}\left(\sum_{j=1}^{n} q_j' \frac{\partial L}{\partial q_j'} - L\right) = \sum_{j=1}^{n}\left(\frac{dq_j'}{dt}\frac{\partial L}{\partial q_j'} + q_j'\frac{d}{dt}\frac{\partial L}{\partial q_j'}\right)$$

$$- \sum_{j=1}^{n}\left(\frac{\partial L}{\partial q_j} q_j' + \frac{\partial L}{\partial q_j'}\frac{dq_j'}{dt}\right) = \sum_{j=1}^{n} q_j'\left(\frac{d}{dt}\frac{\partial L}{\partial q_j'} - \frac{\partial L}{\partial q_j}\right) = 0,$$

by virtue of (7–48).

In the special case when $L = T - V$ and (7–49) holds, we verify (Prob. 10 below) that the integral (7–51) takes the form

$$T + V = c_1. \tag{7–52}$$

In the applications the function T is the total kinetic energy and V is the total potential energy.

Other integrals are often obtained because of *ignorable coordinates*. With respect to Eqs. (7–48), we call q_j ignorable if L is independent of q_j, though it may depend on q_j'. The jth equation is then

$$\frac{d}{dt}\left(\frac{\partial L}{\partial q_j'}\right) = 0.$$

Hence we have the integral

$$\frac{\partial L}{\partial q_j'} = \text{const.}$$

It can be shown that integrals of linear and angular momentum can be interpreted as integrals corresponding to ignorable coordinates (see Chapter 3 of Reference 5).

PROBLEMS

1. Verify that the following systems have the integrals given:
   (a) $dy/dx = (2xz - y + 2z^3)/x$, $dz/dx = x + z^2$, $F = xy - z^2$;
   (b) $dx/dt = xy^2 + xz^2$, $dy/dt = xy - y^3$, $dz/dt = -xz - z^3$, $F = xyz$.

2. Let the system
   $$dy/dx = f(x, y, z), \qquad dz/dx = g(x, y, z)$$

   be given. (a) Show that $F(x, y, z)$ is an integral of the system,

   $$dF \equiv u(x, y, z)\,(dy - f\,dx) + v(x, y, z)\,(dz - g\,dx),$$

   if and only if, under appropriate differentiability assumptions,

   $$F_x = -fu - gv, \qquad F_y = u, \qquad F_z = v.$$

   (b) Show that u, v is a set of integrating factors for the system if and only if

   $$fu_y + f_y u + u_x + gv_y + g_y v = 0,$$
   $$fu_z + f_z u + v_x + gv_z + g_z v = 0,$$
   $$u_x - v_y = 0.$$

3. Show that the functions u, v which are given serve as integrating factors for the following systems (cf. Prob. 2), and find the general solutions:

(a) $dy/dx = (x + z)/(y + z)$, $dz/dx = (x - y)/(y + z)$; $u = 1$, $v = -1$; $u = y$, $v = z$;

(b) $dy/dx = (2x^2y + yz)/(-xz - 2xy^2)$, $dz/dx = (2x^2z - 2y^2z)/(xz + 2xy^2)$; $u = xz$, $v = xy$; $u = 2y$, $v = -1$.

4. Show that $F(x_1, \ldots, x_n, t) = b_1(t)x_1 + \cdots + b_n(t)x_n$ is an integral of the system of linear differential equations

$$\frac{dx_i}{dt} = a_{i1}(t)x_1 + \cdots + a_{in}(t)x_n \qquad (i = 1, \ldots, n)$$

if and only if $y_1 = b_1(t), \ldots, y_n = b_n(t)$ is a solution of the system

$$\frac{dy_i}{dt} = -a_{1i}(t)y_1 - a_{2i}(t)y_2 - \cdots - a_{ni}(t)y_n \qquad (i = 1, \ldots, n).$$

This system is termed the *adjoint system* of the given system.

5. Show that the angular momentum integral (7–38) can be written as $mr^2 \, d\theta/dt = $ const. [*Hint:* Differentiate the relations $x = r \cos \theta$, $y = r \sin \theta$ with respect to t.]

6. Show that in view of Eq. (7–40), Eqs. (7–41) are the same as Eq. (7–39). [*Hint:* $r_{ij} = [(x_i - x_j)^2 + (y_i - y_j)^2 + (z_i - z_j)^2]^{1/2}$.]

7. Show that Eqs. (7–39) have the integral (7–42). [*Hint:* Use the form (7–41) of the equations.]

8. Show that Eqs. (7–39) have the integrals (7–43).

9. Show that Eqs. (7–39) have the integrals (7–47).

10. Show that when $L = T - V$ and (7–49) holds, the energy integral (7–51) becomes (7–52).

11. *The two-body problem.* Two particles of masses m_1, m_2 move in space, subject to gravitational attraction. Let (x_1, y_1, z_1), (x_2, y_2, z_2) be the coordinates of the particles; let r be the distance between them.

(a) Write the differential equations for the motion and state the order.

(b) Show that the equations have Lagrangian form, with $L = T - V$,

$$T = \frac{1}{2} m_1 \left[\left(\frac{dx_1}{dt}\right)^2 + \left(\frac{dy_1}{dt}\right)^2 + \left(\frac{dz_1}{dt}\right)^2 \right]$$

$$+ \frac{1}{2} m_2 \left[\left(\frac{dx_2}{dt}\right)^2 + \left(\frac{dy_2}{dt}\right)^2 + \left(\frac{dz_2}{dt}\right)^2 \right],$$

$$V = -k \frac{m_1 m_2}{r}.$$

(c) Obtain the energy integral.

(d) Obtain the integrals of linear momentum and the equations of motion of the center of mass.

(e) Obtain the integrals of angular momentum.

(f) Show that if the center of mass is at rest at the origin and the x and y components of total angular momentum are 0, then the motion takes place in the xy-plane and the particle m_1 moves as if attracted by a mass $m_2^3(m_1 + m_2)^{-2}$ at the origin.

ANSWERS

3. (a) $x - y + z = c_1,\ x^2 - y^2 - z^2 = c_2;$
   (b) $xyz = c_1,\ x^2 + y^2 - z = c_2.$

11. (a) $m_1 D^2 x_1 = -km_1 m_2(x_1 - x_2)r^{-3},$
    $m_1 D^2 y_1 = -km_1 m_2(y_1 - y_2)r^{-3},$
    $m_1 D^2 z_1 = -km_1 m_2(z_1 - z_2)r^{-3},$
    $m_2 D^2 x_2 = -km_1 m_2(x_2 - x_1)r^{-3},$
    $m_2 D^2 y_2 = -km_1 m_2(y_2 - y_1)r^{-3},$
    $m_2 D^2 z_2 = -km_1 m_2(z_2 - z_1)r^{-3};$ order is 12;   (c) $T + V = c_1;$
    (d) $m_1 D x_1 + m_2 D x_2 = c_2,\ m_1 D y_1 + m_2 D y_2 = c_3,$
    $m_1 D z_1 + m_2 D z_2 = c_4,\ (m_1 + m_2)\bar{x} = c_2 t + c_5,$
    $(m_1 + m_2)\bar{y} = c_3 t + c_6,\ (m_1 + m_2)\bar{z} = c_4 t + c_7;$
    (e) $m_1(y_1 D z_1 - z_1 D y_1) + m_2(y_2 D z_2 - z_2 D y_2) = c_8,$
    $m_1(z_1 D x_1 - x_1 D z_1) + m_2(z_2 D x_2 - x_2 D z_2) = c_9,$
    $m_1(x_1 D y_1 - y_1 D x_1) + m_2(x_2 D y_2 - y_2 D x_2) = c_{10}.$

SUGGESTED REFERENCES

1. BIRKHOFF, GEORGE D., *Dynamical Systems*. American Mathematical Society Colloquium Publications, Vol. 9. New York: American Mathematical Society, 1927.

2. INCE, E. L., *Ordinary Differential Equations*. New York: Dover Publications, Inc., 1956.

3. KAMKE, E., *Differentialgleichungen, Lösungsmethoden und Lösungen*, Vol. 1, 2nd ed. Leipzig: Akademische Verlagsgesellschaft, 1943.

4. KAPLAN, WILFRED, *Advanced Calculus*. Reading, Mass.: Addison-Wesley, 1952.

5. WHITTAKER, E. T., *Analytical Dynamics*, 4th ed. Cambridge, Eng.: Cambridge University Press, 1937.

CHAPTER 8

EQUATIONS NOT OF FIRST DEGREE

8–1 Reduction to first degree equation. A differential equation of nth order has the general form

$$F(x, y, y', \ldots, y^{(n)}) = 0. \tag{8-1}$$

The equation is said to be of first degree (Section 1–2) when it has the form

$$f(x, y, \ldots, y^{(n-1)})y^{(n)} + g(x, y, \ldots, y^{(n-1)}) = 0. \tag{8-2}$$

Thus the differential equation is in the form of a first degree equation for the highest derivative; hence we can solve for the highest derivative:

$$y^{(n)} = -g/f = G(x, y, \ldots, y^{(n-1)}), \tag{8-3}$$

except where f is 0.

The differential equation is said to be of kth degree if it is of degree k in the highest derivative. For example,

$$y'^2 + 2x^2y^3y' + x^4 + y^4 = 0 \tag{8-4}$$

is a quadratic equation for y', and hence has degree two. We can replace Eq. (8–4) by two first degree equations by solving for y':

$$y' = -x^2y^3 \pm (x^4y^6 - x^4 - y^4)^{1/2}. \tag{8-5}$$

Each of the two first degree equations

$$y' = -x^2y^3 + (x^4y^6 - x^4 - y^4)^{1/2},$$
$$y' = -x^2y^3 - (x^4y^6 - x^4 - y^4)^{1/2} \tag{8-6}$$

can be analyzed as in Chapter 2. The solutions form two families of curves in the xy-plane. In general, there will be *two* solution curves through each point (one with each of the slopes 8–6). Similarly, an equation of degree k can (in principle) be replaced by k first order equations.

The general equation (8–1) need not have the form of an algebraic equation in $y^{(n)}$; for example, the equation

$$e^{y''} - y'^2 + xy = 0 \tag{8-7}$$

is not algebraic in y''. However, we can solve for y'':

$$y'' = \log (y'^2 - xy),\qquad\qquad (8\text{–}8)$$

and obtain one equivalent first degree equation. Other nonalgebraic equations may be very difficult to solve for the highest derivative and may have infinitely many solutions.

Equation (8–4) illustrates another complication. The equation was replaced by the two first order equations (8–6). These are meaningful, however, only when the quantity under the radical is positive; thus for some points (x, y) the equation may prescribe imaginary slopes, for others two equal slopes, and for still others unequal slopes. For the equation of degree k there is a corresponding variety of possibilities; for given values of $x, y, \ldots, y^{(n-1)}$ some roots may be real and others imaginary.

EXAMPLE 1.

$$y'^2 + y'(x^2y - xy) - x^3y^2 = 0.$$

We can solve for y' by factoring:

$$(y' - xy)(y' + x^2y) = 0;$$

$$y' = xy, \qquad y' = -x^2y.$$

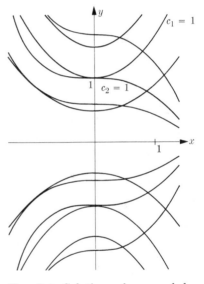

The two first degree equations can then be solved by separation of variables. We find the solutions to be

$$y = c_1 e^{x^2/2}, \qquad y = c_2 e^{-x^3/3}.$$

They are graphed in Fig. 8–1. As the figure shows, there are, in general, two solutions through each point. The x-axis is an exception; along this line both slopes are zero, and

FIG. 8–1. Solutions of a second degree equation.

hence there is just one solution through each point, the axis itself. Along the y-axis both slopes are zero, and along the line $x = -1$ both slopes equal $-y$; however there are still two (tangent) solutions through each point. These comprise all the exceptional points, for equality of the two slopes occurs where

$$xy = -x^2y,$$

and hence where $x = 0$, $y = 0$, or $x = -1$.

EXAMPLE 2. $y'^2 - 4y = 0$. We find at once

$$y' = 2\sqrt{y}, \quad y' = -2\sqrt{y} \quad (y \geqq 0);$$

$$\sqrt{y} = x + c_1, \quad \sqrt{y} = -x + c_2.$$

The solutions are shown in Fig. 8-2. The slopes are imaginary when $y < 0$; hence there are no solutions in the lower half-plane. The slopes are equal and both are zero along the x-axis; the line $y = 0$ is a solution common to both families. Through each point of the upper half-plane pass two solutions. Each solution is half of a parabola. For example, $\sqrt{y} = x$ is half of the parabola $y = x^2$; it contains only the points for which $x \geqq 0$. The missing half of the parabola is contained in the other family; it is the curve $\sqrt{y} = -x$. In general, we can combine the two families into one by squaring:

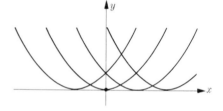

$$y = (x + c)^2;$$

the solutions now appear as complete parabolas.

FIG. 8-2. Solutions of $y'^2 - 4y = 0$.

The line $y = 0$ is not a special case of the "general solution" $y = (x + c)^2$; however, we verify that it is a solution of the differential equation. (We would naturally be led to examine the locus $y = 0$, since separation of variables in the differential equations $y' = \pm 2\sqrt{y}$ requires division by y.) Since each parabola is tangent to the x-axis, there are still two solutions through each point, even though there is only one slope. The line $y = 0$ is an example of an *envelope* of a family of curves (the parabolas), and it is called a *singular solution* of the differential equation. (See Section 8-2 below.)

Remark. When the family of solutions has an envelope, as in the example, we can construct further solutions by piecing together parts of solutions. For example, the curve formed of the three portions

$$y = x^2, x \leqq 0, \quad y = 0, 0 \leqq x \leqq 1, \quad y = (x - 1)^2, x \geqq 1,$$

defines a function $y = f(x)$ having a continuous derivative and satisfying the differential equation for all x.

EXAMPLE 3. $y'^4 - 2y'^2 + 1 - y^4 = 0$. We find four slopes:

$$y' = \pm \sqrt{1 \pm y^2}.$$

Two are imaginary when $y^2 > 1$; two are equal (both zero) when $y^2 = 1$; there are two equal pairs ($+1$ and -1) when $y = 0$; otherwise all four slopes are real and distinct. The solutions are found to be (Prob. 1(k) below)

$$y = \sin(x + c_1), \qquad y = \pm\sinh(x + c_2), \qquad y = \pm 1.$$

They are graphed in Fig. 8–3. The solutions $y = \pm 1$ are envelopes of the solutions $y = \sin(x + c_1)$ and are singular solutions (Section 8–2). The line $y = 0$ (along which there are two equal pairs of slopes) is not a solution.

EXAMPLE 4. $y''^2 - y^2 = 0$. There are two first degree equations: $y'' = \pm y$, for which the solutions are

$$y = c_1 \cos x + c_2 \sin x, \qquad y = c_3 e^x + c_4 e^{-x}.$$

Since the two equations are of second order, a solution of each passes through each point (x, y) with prescribed slope y' (that is, a unique solution passes through each point (x, y, y') of the *three*-dimensional space of these variables; see Section 1–8). The two solutions have different values of y'' (hence different *curvatures*) except where $y = 0$. The points $y = 0$ form a solution which is included in both families of general solutions (whether or not it is a singular solution is a matter of terminology; Section 8–2).

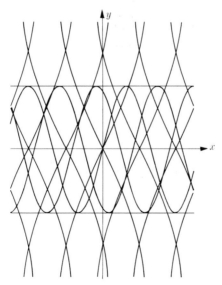

FIG. 8–3. Solutions of a fourth degree equation.

PROBLEMS

1. Reduce to first degree equations and find all solutions for the following:

(a) $y'^2 + xy' - 2x^2 = 0$,   (b) $y'^2 - y^2 + 2e^x y - e^{2x} = 0$,

(c) $y'^2 - xy = 0$,   (d) $xy'^2 - 2yy' + x = 0$,

(e) $x^2 y'^2 + 2xyy' + x^2 + y^2 - 1 = 0$,

(f) $x^2 y'^2 + 2xyy' + y^2 - y^4 - x^2 y^4 = 0$,

(g) $y''^2 + (x^2 - x)y'' - x^3 = 0$,

(h) $y''^2 + y''(y' + 2y) + 4yy' - 8y^2 = 0$,

(i) $y'^3 + yy'^2 - x^2 y^2 y' - x^2 y^3 = 0$,

(j) $xy'^3 - (2y + xy)y'^2 + (2y^2 + 4x)y' - 4xy = 0$,

(k) $y'^4 - 2y'^2 + 1 - y^4 = 0$ (Example 3 above),

(l) $\sin y' - x = 0$.

ANSWERS

1. (a) $y = c_1 - x^2$, $y = c_2 + \frac{1}{2}x^2$; (b) $y = e^x(c_1 - x)$, $y = \frac{1}{2}e^x + c_2 e^{-x}$;

(c) $9y = (x^{3/2} + c)^2$ for $x \geqq 0$, $-9y \doteq [(-x)^{3/2} + c]^2$ for $x \leqq 0$ and the line $y = 0$;   (d) $y = (1 + c^2 x^2)/2c$, $y = \pm x$;

(e) $2xy = \pm [x\sqrt{1 - x^2} + \sin^{-1} x] + c$;

(f) $y = [cx \pm \sqrt{1 + x^2} \pm x \log (x + \sqrt{1 + x^2})]^{-1}$, $y = 0$;

(g) $y = c_1 x + c_2 + \frac{1}{6}x^3$, $y = c_3 x + c_4 - \frac{1}{12}x^4$;

(h) $y = c_1 e^{-2x} + c_2 e^x$, $y = c_3 \cos 2x + c_4 \sin 2x$;

(i) $y = c_1 e^{-x}$, $y = c_2 e^{x^2/2}$, $y = c_3 e^{-x^2/2}$;

(j) $y = c_1 e^x$, $y = (c_2^2 x^2 + 4)/2c_2$;

(k) $y = \sin (x + c_1)$, $y = \pm \sinh (x + c_2)$, $y = \pm 1$;

(l) $y = x (\sin^{-1} x + 2n\pi) + \sqrt{1 - x^2} + c_1$,

$y = x[(2n + 1)\pi - \sin^{-1} x] - \sqrt{1 - x^2} + c_2$; $\sin^{-1} x$ is principal value, $n = 0, \pm 1, \pm 2, \ldots$

8-2 Singular solutions. In Section 8-1 it has been shown that an equation

$$F(x, y, y') = 0 \qquad (8\text{-}9)$$

can be reduced to first degree form by *solving for* y'; in general, there are several solutions and, accordingly, several first degree equations.

It is not obvious that every equation (8-9) can be solved for y' in terms of x and y; indeed, we can easily give examples for which this is not possible:

$$x^2 + y^2 + y'^2 = 0; \qquad e^y + y'^2 = 0. \qquad (8\text{-}10)$$

It is shown in advanced calculus (pp. 117-121 of Reference 1) that solution for y' is possible *except where* $\partial F/\partial y' = 0$. More precisely, let $F(x, y, y')$ be continuous and have continuous first partial derivatives with respect to x, y, and y' in some open region D of xyy'-space. If x_0, y_0, y_0' is a triple in D such that $F(x_0, y_0, y_0') = 0$, and $\partial F/\partial y' \neq 0$ for $x = x_0, y = y_0, y' = y_0'$, then there is a *unique* function $y' = f(x, y)$, defined for x and y close enough to x_0, y_0 (respectively) such that $y_0' = f(x_0, y_0)$, and satisfying the equation $F(x, y, y') = 0$; that is,

$$F[x, y, f(x, y)] \equiv 0.$$

Moreover, $f(x, y)$ has continuous first partial derivatives, so that the existence theorem of Section 1–8 applies to the differential equation $y' = f(x, y)$.

The theorem just formulated is known as the *implicit function theorem* for the equation $F(x, y, y') = 0$. Its significance can be seen geometrically. The equation $F(x, y, y') = 0$ describes a *surface* in xyy'-space (Fig. 8–4). The tangent plane to this surface at (x_0, y_0, y_0') has the equation

$$\frac{\partial F}{\partial x} (x - x_0) + \frac{\partial F}{\partial y} (y - y_0) + \frac{\partial F}{\partial y'} (y' - y_0') = 0, \qquad (8\text{–}11)$$

where the partial derivatives are evaluated at (x_0, y_0, y_0'). If $\partial F/\partial y' \neq 0$ at that point, the tangent plane is *not* vertical, and y' is a well-defined function of x, y near (x_0, y_0). If $\partial F/\partial y' = 0$ at the point, the tangent plane is vertical and y' is in general not a "well-behaved" function of x, y near (x_0, y_0). The two cases are shown in Fig. 8–4.

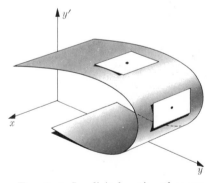

FIG. 8–4. Implicit function theorem.

In the first example in (8–10), the equation is satisfied at $x = 0$, $y = 0$, $y' = 0$; however, $\partial F/\partial y' = 2y' = 0$ at the point, and no function $y' = f(x, y)$ is obtained. In the second example the equation is not satisfied for any values of x, y, y'. These examples are extreme, in that no function $y' = f(x, y)$ is obtained. More typical is the following, which has some exceptional points:

$$x^2 + y^2 + y'^2 - 1 = 0. \qquad (8\text{–}12)$$

The surface in xyy'-space is a sphere, whose tangent plane is vertical where $x^2 + y^2 = 1$. For $x^2 + y^2 < 1$ we obtain two continuous functions:

$$y' = \pm\sqrt{1 - x^2 - y^2}. \tag{8-13}$$

The exceptional points $x^2 + y^2 = 1$ are precisely those at which $\partial F/\partial y' = 0$, for $\partial F/\partial y' = 2y' = 0$ when $y' = 0$, and hence, by (8-12), when $x^2 + y^2 = 1$.

We call the locus of all points (x, y) such that for some y',

$$F(x, y, y') = 0, \qquad \frac{\partial F}{\partial y'}(x, y, y') = 0 \tag{8-14}$$

the *singular locus* relative to the differential equation $F(x, y, y') = 0$. For given F it is in general possible to eliminate y' between the two equations (8-14) and to obtain a single equation

$$Q(x, y) = 0 \tag{8-15}$$

which describes the singular locus. [The singular locus is also termed the *p-discriminant* locus; the term arises as a result of the notation p for y'; the equations (8-14) become $F(x, y, p) = 0$, $F_p(x, y, p) = 0$.]

The singular locus may contain a solution of the given differential equation; that is, a solution curve $y = y(x)$, for which every point (x, y) lies in the singular locus. Such a solution is called a *singular solution* of the differential equation. A singular solution must thus be a function $y(x)$ such that, for each x, both equations (8-14) are satisfied when $y = y(x)$, $y' = y'(x)$. Not every curve $y = y(x)$ in the singular locus need be a singular solution.

EXAMPLE 1. $x^2 + y^2 + y'^2 - 1 = 0$. This is the Eq. (8-12) discussed above. The equations (8-14) become

$$x^2 + y^2 + y'^2 - 1 = 0, \qquad 2y' = 0,$$

and the singular locus (8-15) is the curve

$$x^2 + y^2 = 1,$$

which can be represented by two functions of x:

$$y = \pm\sqrt{1 - x^2}.$$

One or both of these may be singular solutions.

We find
$$y' = \frac{\pm x}{\sqrt{1 - x^2}},$$

$$x^2 + y^2 + y'^2 - 1 \equiv x^2 + 1 - x^2 + \frac{x^2}{1 - x^2} - 1 \equiv \frac{x^2}{1 - x^2} \not\equiv 0.$$

The differential equation is satisfied only at $x = 0$. Hence there are no singular solutions.

EXAMPLE 2. $y'^2 - 4y = 0$. This is Example 2 of Section 8–1 (Fig. 8–2). Here the equations (8–14) are

$$y'^2 - 4y = 0, \qquad 2y' = 0.$$

The singular locus (8–15) is the line $y = 0$. It is a solution of the differential equation; hence $y = 0$ is a singular solution.

EXAMPLE 3. $y'^4 - 2y'^2 + 1 - y^4 = 0$. This is Example 3 of Section 8–1 (Fig. 8–3). The equations (8–14) are

$$y'^4 - 2y'^2 + 1 - y^4 = 0, \qquad 4y'^3 - 4y' = 0.$$

Elimination of y' leads to the equations

$$y^4 = 0, \qquad y^4 = 1;$$

the singular locus is the combination of these two loci; we can represent both by one equation:
$$y^4(y^4 - 1) = 0,$$

but both parts must be examined separately. The line $y = 0$ does not satisfy the differential equation; the locus $y^4 = 1$ consists of two lines, $y = \pm 1$, both of which satisfy the differential equation. Hence the lines $y = \pm 1$ are singular solutions.

EXAMPLE 4. $e^{y'} - y' + xy - x - 1 = 0$. The equations (8–14) and the singular locus (8–15) are

$$e^{y'} - y' + xy - x - 1 = 0, \qquad e^{y'} - 1 = 0;$$

$$xy - x = 0.$$

The locus contains the lines $y = 1$, $x = 0$. Only the former has the form $y = g(x)$, and it does satisfy the differential equation. Hence $y = 1$ is the singular solution. In this example, finding the general solution would be very difficult, but the singular solution is found with ease.

Singular solutions of higher order equations. For a second order equation

$$F(x, y, y', y'') = 0 \qquad (8\text{-}16)$$

we reduce to first degree form by solving for y''. This is possible (under appropriate continuity assumptions) except where $\partial F / \partial y'' = 0$. A solution $y = g(x)$ of Eq. (8-16) such that $\partial F / \partial y'' = 0$ at each point is called a singular solution.

Similarly, for an nth order equation

$$F(x, y, y', \ldots, y^{(n)}) = 0 \qquad (8\text{-}17)$$

a singular solution is a solution $y = g(x)$ such that $\partial F / \partial y^{(n)} = 0$ at each point.

8-3 Envelopes. Let a family of curves be given in implicit form:

$$G(x, y, c) = 0; \qquad (8\text{-}18)$$

for each constant value of c, Eq. (8-18) is to represent one or more curves in the xy-plane. For example,

$$(x - c)^2 - y = 0 \qquad (8\text{-}19)$$

represents such a family; the curves are parabolas.

A curve C in the xy-plane is called an *envelope* of the family (8-18) if at each point of C at least one member of the family is tangent to C, different members being tangent at differing points of C. This is illustrated by the family (8-19); the curve C is the x-axis. As shown in Fig. 8-2, the x-axis has one tangent parabola at each point.

We note that for each point (x, y) of the upper half-plane there are *two* parabolas through the point; that is, Eq. (8-19) has two solutions for c in terms of x, y, when $y > 0$:

$$c = x + \sqrt{y}, \quad c = x - \sqrt{y} \quad (y > 0). \qquad (8\text{-}20)$$

These solutions are two continuous functions of x, y. As in Section 8-2, the implicit function theorem ensures that an equation (8-18) can be solved for c in this manner, except where $\partial G / \partial c = 0$. The locus in the xy-plane described by the equations

$$G(x, y, c) = 0, \qquad G_c(x, y, c) = 0 \qquad (8\text{-}21)$$

is termed the *c-discriminant locus*. For Eq. (8-19) this is the locus

$$(x - c)^2 - y = 0, \qquad 2(x - c) = 0;$$

hence it is the line $y = 0$, which is precisely the locus of points where the two functions (8–20) become discontinuous (since there are no values at all for $y < 0$). Furthermore, in this example the c-discriminant locus is an envelope of the family.

It can be shown (see Prob. 5 below) that a curve C in the c-discriminant locus will represent such an envelope of the family (8–18), provided $\partial G/\partial x$ and $\partial G/\partial y$ are never both zero at a point of C, and provided C is representable in the form $x = x(c)$, $y = y(c)$ in terms of the parameter c.

Equation (8–19) can be considered as a family of solutions of a first order differential equation. Differentiation and elimination of c lead to the equation

$$y'^2 = 4y. \tag{8–22}$$

There are two slopes $\pm 2\sqrt{y}$ at each point (x, y) for $y > 0$. On the line $y = 0$ the two slopes coincide, and we verify that the line $y = 0$ is a singular solution of the differential equation. Thus, in this example, the singular solution and envelope are the same. This is normally the case; however, there are exceptions (see Prob. 4 below). For a full discussion the reader is referred to pp. 83–93 of Reference 2.

PROBLEMS

1. Find all singular solutions for each of the following:
(a) $y'^2 + 2xy' + x^2 - y^2 = 0$,   (b) $y'^2 - 4xy' + 2x^2 + 2y = 0$,
(c) $y'^3 - 3x^2y' + 4xy = 0$,
(d) $3y'^5 + 5xy'^3 - 30x^2y' + 33xy = 0$.

2. For each of the following differential equations find all solutions; also graph and discuss the singular locus:
(a) $y'^2 = x$,   (b) $(3y - 1)^2y'^2 - 4y = 0$,
(c) $(x^2 - 1)y'^2 + x^2 = 0$,   (d) $y'^3 - y = 0$.

3. Graph each of the following families of curves; also, find the c-discriminant locus for each, and determine whether it is an envelope of the family:
(a) $y - 2cx + c^2 = 0$,   (b) $y^3 = (x - c)^2$,
(c) $x^2 + y^2 = c^2$,   (d) $(x - c)^2 + y^2 = 1$,
(e) $y(y - 1)^2 = (x - c)^2$, [cf. Prob. 2 (b)].

4. (a) Show that the line $y = 0$ is a singular solution of the equation $y'^3 - y^3 = 0$ but is not an envelope of the remaining solutions.

(b) Show that the line $y = 0$ is an envelope of the family of solutions of the equation $y' - y^{2/3} = 0$ but is not a singular solution of the equation. [*Remark.* In this example, $F \equiv y' - y^{2/3}$ does not have continuous first partial derivatives along the line $y = 0$, for $F_y = -\frac{2}{3}y^{-1/3}$. Hence the existence theorem of Section 1–8 does not apply and, in fact, there is not a *unique* solution through each point on the x-axis. When F has continuous partial derivatives along the envelope, then $F_{y'}$ must be 0, so that the curve is a singular solution. For otherwise we could solve for y', apply the existence theorem, and obtain a unique solution through each point; but this is impossible along an envelope.]

5. Prove: if $x = x(c)$, $y = y(c)$ represents a curve C in the c-discriminant locus (8–21) and G_x, G_y are not both zero along C, then the curve C is an envelope of the family $G(x, y, c) = 0$. [*Hint:* The slope of the tangent to each curve of the family is determined by the equation $G_x \, dx + G_y \, dy = 0$. Now

$$G[x(c), y(c), c] = 0.$$

Differentiate this equation with respect to c, and use the equation $G_c = 0$ to conclude that $G_x \, dx + G_y \, dy = 0$ along C also.]

6. Determine all singular solutions of the second order equation

$$y''^2 - 2y'y'' + y^2 = 0.$$

[*Hint:* By the definition each singular solution must satisfy both of the equations

$$y''^2 - 2y'y'' + y^2 = 0, \qquad 2y'' - 2y' = 0.$$

Find all solutions of the second equation and determine which of these satisfy the first equation.]

7. Show that a linear equation

$$p_0(x)y'' + p_1(x)y' + p_2(x)y = Q(x) \qquad [p_0(x) \neq 0]$$

has no singular solution.

1. (a) None, (b) $y = x^2$, (c) $y = \pm \frac{1}{2}x^2$, (d) $y = \pm \frac{2}{3}x^{3/2}$.

2. (a) $y = \pm \frac{2}{3}x^{3/2} + c$; singular locus $x = 0$, a locus of cusps, not a singular solution; (b) $y(y - 1)^2 = (x - c)^2$, $y > 0$; singular locus $y = 0$ (envelope, a singular solution), line $y = \frac{1}{3}$ is a locus of vertical tangents; (c) $x^2 + (y - c)^2 = 1$; singular locus $x = 0$, a tac locus (or locus along which two solutions touch), not a singular solution; lines $x = \pm 1$ can also be considered as part of the singular locus; they are loci of vertical tangents and are envelopes; (d) $27y^2 = 8(x - c)^3$; singular locus $y = 0$, a locus of cusps, a singular solution.

3. (a) $y = x^2$, an envelope; (b) $y = 0$, not an envelope (locus of cusps); (c) point $(0, 0)$, not an envelope; (d) $y = \pm 1$, an envelope; (e) $y = 0$, an envelope; $y = 1$, not an envelope (locus of nodes, i.e., self-intersections).

6. $y = ce^x$.

8–4 First order equations solvable for y, or for x. If a first order equation is solvable for y in terms of x, y', or for x in terms of y, y', then it may be possible to obtain the solutions by *differentiating* the equation. We illustrate the method by examples and give a discussion below.

EXAMPLE 1. $y'^2 - 4y = 0$. This is Example 2 of Section 8–1. We write

$$p = y', \tag{8–23}$$

so that on solving for y, we obtain

$$y = \tfrac{1}{4}p^2. \tag{8–24}$$

Differentiation with respect to x gives, in view of Eq. (8–23),

$$p = \frac{1}{4}\, 2p\, \frac{dp}{dx},$$

or

$$p\left(1 - \frac{1}{2}\, \frac{dp}{dx}\right) = 0.$$

Hence we obtain the two equations

$$p = 0 \qquad \text{or} \qquad \frac{dp}{dx} = 2.$$

The first, when combined with (8–24), gives

$$y = 0,$$

which we recognize as the *singular solution*. The second can be integrated:

$$p = 2(x + c);$$

when this relation is combined with Eq. (8–24), we find

$$y = (x + c)^2,$$

which we recognize as the remaining solutions.

The fact that the equation obtained by differentiation has a factor which, when equated to zero, leads to the singular solution is no accident. (An explanation is given below.) This factor may provide only a singular *locus*, and hence we must check in each case to see whether a solution is obtained.

EXAMPLE 2. $xy'^2 - 2yy' + x = 0$. We can solve for y:

$$y = \frac{xp}{2} + \frac{x}{2p} \qquad \left(p = \frac{dy}{dx}\right). \tag{8–25}$$

Differentiation gives

$$p = \left(\frac{x}{2} - \frac{x}{2p^2}\right)\frac{dp}{dx} + \frac{p}{2} + \frac{1}{2p},$$

$$p^3 - p = (xp^2 - x)\frac{dp}{dx},$$

$$(p^2 - 1)\left(x\frac{dp}{dx} - p\right) = 0.$$

Again we obtain two equations:

$$p^2 = 1, \qquad x\frac{dp}{dx} - p = 0.$$

The first, when combined with (8–25), gives

$$y = \pm x;$$

both functions are singular solutions. The second is a first order equation, whose general solution we find to be

$$p = cx;$$

this relation, when combined with (8–25), yields

$$y = \frac{cx^2}{2} + \frac{1}{2c} \qquad (c \neq 0),$$

which gives all other solutions of the given differential equation.

EXAMPLE 3. $y'^3 + xy' - y = 0$. We can solve either for y or for x. We solve for x and then differentiate *with respect to y*:

$$x = \frac{y}{p} - p^2,$$

$$\frac{dx}{dy} = \frac{1}{p} = \left(\frac{-y}{p^2} - 2p\right)\frac{dp}{dy} + \frac{1}{p},$$

$$(y + 2p^3)\frac{dp}{dy} = 0,$$

$$y + 2p^3 = 0, \qquad \frac{dp}{dy} = 0.$$

The first equation gives a singular solution

$$x = \frac{y}{-y^{1/3}2^{-1/3}} - \frac{y^{2/3}}{2^{2/3}}, \qquad 4x^3 = -27y^2;$$

the second equation can be integrated to give the other solutions:

$$p = c, \qquad x = \frac{y}{c} - c^2, \qquad y = cx + c^3.$$

Discussion. The basis of the method can be found in the notion of *integral* of a differential equation (Section 7–1). We consider an equation

$$F(x, y, y') = 0. \tag{8–26}$$

Differentiation with respect to x leads to the second order equation

$$F_x + F_y y' + F_{y'} y'' = 0. \tag{8–27}$$

This equation is *exact* and $F(x, y, y')$ is an integral; however, the relation

$$F(x, y, y') = c_1 = \text{const} \tag{8–28}$$

tells us nothing new. In fact, by Eq. (8–26), we are interested only in those solutions for which $c_1 = 0$. We may be able to find another integral of Eq. (8–27); this in general requires multiplication by an integrating factor. Let

$$G(x, y, y') = c \tag{8–29}$$

be such an integral, with G functionally independent of F. Then the two equations (8–28), (8–29) together provide all solutions of Eq. (8–27). But we want only those solutions for which $c_1 = 0$; that is, we want the functions $y(x)$ defined by the two equations

$$F(x, y, y') = 0, \qquad G(x, y, y') = c. \tag{8–30}$$

This is precisely the way in which the solutions are obtained in Examples 1–3 above.

We can describe the process geometrically. Our solutions correspond to the curves

$$y = y(x), \qquad y' = y'(x)$$

which lie in the integral surface $F(x, y, y') = 0$. They also satisfy the second order equation (8–27); all the solutions of Eq. (8–27) form a family of curves in xyy'-space (one curve through each point, except where $F_{y'} = 0$). The second integral (8–29) stratifies these solution curves; for each fixed value of c the intersection of the two integral surfaces (represented by Eqs. 8–30) is one of the solution curves sought. See Fig. 8–5. Equation (8–27) may not be solvable for y'' wherever $F_{y'} = 0$; the triples (x, y, y') for which this holds and for which (8–26) holds describe the singular locus. It may happen that $F_{y'}$ can be factored out of the left-hand side of Eq. (8–27); in that case, equating the factor to zero, together with Eq. (8–26), gives the singular locus.

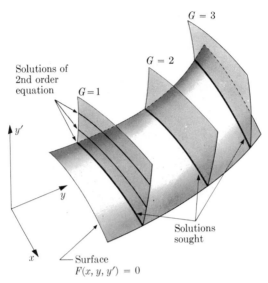

$$G = 3$$

$$G = 2$$

Solutions of
2nd order
equation $$G = 1$$

$$y'$$

$$y$$

Solutions
sought

$$x$$ Surface
$$F(x, y, y') = 0$$

FIG. 8–5. Solution of a differential equation by differentiation.

When Eq. (8–26) is solved for y:

$$y = f(x, p) \qquad (p = y'),$$

the differentiation process leads to the second order equation

$$p = f_x + f_p \frac{dp}{dx} \qquad \left(\frac{dp}{dx} = y'' \right);$$

since only x and p appear, we can consider this as a first order equation.
An integral

$$G(x, p) = c$$

must be functionally independent of the integral $y - f(x, p)$, since y does
not appear in G. Thus the two equations

$$y = f(x, p), \qquad G(x, p) = c$$

give the solutions, except those in the singular locus.

Equations solvable for x can be analyzed in the same way. The differentiation with respect to y is the same as differentiation with respect to x
followed by division by p:

$$\frac{du}{dy} = \frac{du}{dx} \frac{dx}{dy} = \frac{1}{p} \frac{du}{dx}.$$

Remark. Elimination of y' between the two equations (8–30) may not always be feasible. In that case, Eqs. (8–30) themselves define the solutions in implicit form. Occasionally, we can solve Eqs. (8–30) for x and y:

$$x = \phi(p, c), \qquad y = \psi(p, c);$$

this gives the solutions in parametric form, with p as parameter.

8–5 Clairaut equation. This name is given (after its discoverer) to the equation of form

$$y = xy' + f(y') \equiv xp + f(p). \tag{8–31}$$

Application of the procedure of differentiation (Section 8–4) leads to the equations

$$p = p + [x + f'(p)]\frac{dp}{dx},$$

$$x + f'(p) = 0, \qquad \frac{dp}{dx} = 0. \tag{8–32}$$

The first relation, when combined with Eq. (8–31), leads to the singular locus. It may not be possible to eliminate p, but the equations

$$x = -f'(p), \qquad y = xp + f(p) = -pf'(p) + f(p) \tag{8–33}$$

serve as a parametrization of the singular locus, in terms of the parameter p. It can be shown (Prob. 5(a) below) that (8–33) represents a singular solution, except where $f''(p) = 0$. The second part of (8–32) can be integrated and combined with (8–31) to yield the other solutions:

$$p = c,$$

$$y = cx + f(c); \tag{8–34}$$

they are straight lines. It can be shown (Prob. 5(b) below) that they are tangent to the curve (8–33), which is thus an envelope of the family (8–34). Indeed, *the solutions of the Clairaut equation consist of a curve and the lines tangent to it* (Fig. 8–6).

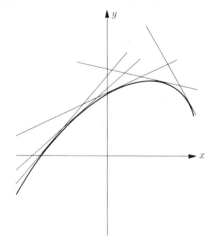

FIG. 8–6. Solutions of a Clairaut equation.

We have tacitly assumed continuity of $f(p)$, $f'(p)$, $f''(p)$ in some interval of p. It is noteworthy that, without any such hypothesis, the lines (8–34) are solutions of Eq. (8–31), wherever $f(c)$ is defined (Prob. 5(c) below).

PROBLEMS

1. Obtain all solutions of each of the following by solving for y and differentiating:

(a) $2y'^2 + x^3y' - 2x^2y = 0,$ (b) $y'^2 + 2xy' - y = 0,$

(c) $(1 + y)y'^2 - 2xy' + 1 + y = 0,$ (d) $xy'^2 - yy' - xy = 0.$

2. Obtain all solutions of the following by solving for x:

(a) $4yy'^2 + 2xy' - y = 0,$

(b) part (b) of Prob. 1,

(c) part (d) of Prob. 1.

3. Write the following as Clairaut equations, obtain all solutions, and graph:

(a) $y'^3 + 3xy' - 3y = 0,$

(b) $xy'^3 - yy'^2 + xy' - y = 1,$

(c) $e^{xy'}(y'^2 + 1) - e^y = 0.$

4. (a) Extend the result of Section 8–5 to the *Clairaut equation in implicit form*:

$$F(y - xy', y') = 0.$$

[*Hint:* Differentiate with respect to x and obtain the integral: $p = c$.]

(b) Find all solutions:

$$(xy' - y)^3 + xy' - y + y'^2 + 1 = 0.$$

5. (a) Show that Eqs. (8–33) represent the singular locus of Eq. (8–31) and that if $f''(p) \neq 0$, the locus is a singular solution. [*Hint:* For a curve given parametrically in terms of the parameter p: $x = x(p)$, $y = y(p)$, we have $dy/dx = (dy/dp)/(dx/dp)$. Show here that $dy/dx = p$.]

(b) Show that the line (8–34) is tangent to the curve (8–33) at the point for which $p = c$.

(c) Show that if $f(p)$ is defined for $p = p_0$ (and perhaps for no other value), then the line $y = p_0x + f(p_0)$ is a solution of Eq. (8–31).

6. Find the differential equation satisfied by each of the following curves and its tangent lines:

(a) the parabola $y = x^2$,   (b) the circle $x^2 + y^2 = 1$,

(c) the curve $y = e^x$,   (d) the curve $y = f(x)$.

ANSWERS

1. (a) $y = cx^2 + 4c^2$, $y = -x^4/6$; (b) $3xp^2 + 2p^3 = c$, where $p = -x \pm \sqrt{x^2 + y}$;   (c) $(y + 1)^2 = 2cx - c^2$, $x - y - 1 = 0$, $x - y + 1 = 0$; (d) $xp^2 - yp - xy = 0$, $p + x = c(p - x)^3$ and $y = 0$.

2. (a) $y^2 = cx + c^2$, $y = -\frac{1}{4}x^2$.

3. (a) $c^3 + 3cx - 3y = 0$; $4x^3 + 9y^2 = 0$; (b) $xc^3 - yc^2 + xc - y = 1$; $x = -2p(1 + p^2)^{-2}$, $y = (-1 - 3p^2)(1 + p^2)^{-2}$; (c) $y = cx + \log(1 + c^2)$, $x = -2p(1 + p^2)^{-1}$, $y = \log(1 + p^2) - 2p^2(1 + p^2)^{-1}$.

4. (b) $(xc - y)^3 + xc - y + c^2 + 1 = 0$; singular solution: $3x(xp - y)^2 + x + 2p = 0$, $(xp - y)^3 + xp - y + p^2 + 1 = 0$.

6. (a) $y = xp - \frac{1}{4}p^2$, (b) $(y - xp)^2 = 1 + p^2$, (c) $y = xp + p(1 - \log p)$, (d) $y = xp - pg(p) + f[g(p)]$, where $g(p)$ is the solution of $p = f'(x)$ for x in terms of p.

MISCELLANEOUS EXERCISES

Find all solutions.

1. $y'^2 + y'(y - x^2y) - x^2y^2 = 0$.

2. $xy'^2 + 1 - y = 0$.

3. $xyy'^2 - y'(x^2 + y^2) + xy = 0$.

4. $y'^3 - xy' - 2y = 0$.

5. $xy'e^{y'} - ye^{y'} + y' = 0$.

6. $y'^3 + 2y'^2 - y' - 2 = 0$.

7. $y'^3 - y'^2(x + y + x^2) + y'(x^3 + xy + x^2y) - x^3y = 0$.

8. $y'^2(1 + 4x^2 + 8y) + 2xy' + x^2 = 0$.

9. $y'^2 + x^2 - 5y = 0$.

10. $x^2y'^2 - 2xyy' + y^2 - 1 = 0$.

11. $x^2y'^2 + 2x(x - y)y' + 2y^2 - 2xy = 0$.

12. $y''^2 - y''(x + y') + xy' = 0$.

ANSWERS

1. $y = c_1e^{x^3/3}$, $y = c_2e^{-x}$. 2. $(y - 1 - x - c)^2 = 4cx$, $y = 1$. 3. $y = c_1x$, $y^2 - x^2 = c_2$. 4. $x = 2cp^{-1/3} + (3p^2/7)$, $y = -cp^{2/3} + (2p^3/7)$, and $y = 0$. 5. $y = cx + ce^{-c}$ and $x = (p - 1)e^{-p}$, $y = p^2e^{-p}$. 6. $y = x + c_1$, $y = -x + c_2$, $y = -2x + c_3$. 7. $y = \frac{1}{2}x^2 + c_1$, $y = \frac{1}{3}x^3 + c_2$, $y = c_3e^x$. 8. $x^2 + 4(y - c)^2 + 2y = 0$, $y = -\frac{1}{2}x^2$. 9. $(p - 2x)^4 = c(2p - x)$, where $p^2 + x^2 - 5y = 0$. 10. $y = cx \pm 1$. 11. $x \pm \sqrt{x^2 - y^2} \pm y\sin^{-1}(y/x) = y\log|x| + c$, where $\pm$ signs agree for $x > 0$, are opposite for $x < 0$. 12. $y = c_1 + c_2e^x$, $y = c_3 + c_4x + (x^3/6)$.

SUGGESTED REFERENCES

1. COURANT, RICHARD J., *Differential and Integral Calculus*, Vol. 2, transl. by E. J. McShane. New York: Interscience, 1947.

2. INCE, E. L., *Ordinary Differential Equations*. New York: Dover Publications, Inc., 1956.

CHAPTER 9

SOLUTIONS IN TERMS OF POWER SERIES

9-1 Application of Taylor's series. Let $f(x)$ be defined in some interval $a < x < b$ and let x_0 be a point of this interval. If all derivatives of $f(x)$ exist at x_0, we can form the Taylor series of $f(x)$:

$$f(x_0) + \frac{f'(x_0)}{1!}(x - x_0) + \frac{f''(x_0)}{2!}(x - x_0)^2 + \cdots$$

$$+ \frac{f^{(n)}(x_0)}{n!}(x - x_0)^n + \cdots \tag{9-1}$$

Under appropriate hypotheses, this power series will converge to $f(x)$ for all x or else in an interval with x_0 as midpoint (p. 359 of Reference 4). For example,

$$e^x = 1 + x + \frac{x^2}{2!} + \cdots + \frac{x^n}{n!} + \cdots \qquad \text{(all } x\text{)},$$

$$\tan x = x + \frac{x^3}{3} + \cdots \qquad \left(|x| < \frac{\pi}{2}\right).$$

We illustrate the application of the Taylor series to differential equations by an example.

EXAMPLE 1. $y' = x^2 - y^2$, $\quad y = 1$ when $x = 1$. $\tag{9-2}$

If we assume that the solution $y = f(x)$ can be represented by its Taylor series, then we can obtain $f(x)$ in the form of a series as follows. We take $x_0 = 1$ in Eq. (9-1). Then $f(1)$ is given as 1. By Eq. (9-2) we obtain expressions for the higher derivatives:

$$y'' = 2x - 2yy', \qquad y''' = 2 - 2y'^2 - 2yy'',$$

$$y^{(iv)} = -6y'y'' - 2yy''', \qquad \cdots \tag{9-3}$$

Since $y = 1, y' = 0$ when $x = 1$, we find successively $y'' = 2, y''' = -2$, $y^{(iv)} = 4, \ldots$; that is, $f''(1) = 2$, $f'''(1) = -2$, $f^{(iv)}(1) = 4, \ldots$ Accordingly,

$$f(x) = 1 + \frac{2}{2!}(x - 1)^2 - \frac{2(x - 1)^3}{3!} + \frac{4(x - 1)^4}{4!} + \cdots \tag{9-4}$$

This is the desired solution in the form of a power series.

There are several important questions which now arise. Does the series (9–4) converge for some x? For which x? Does the sum $f(x)$ satisfy the differential equation? Can we find the general term of the series? How rapidly does the series converge? If we take only the first k terms of the series as an approximation to $f(x)$, how great can the error be?

A detailed discussion of convergence is given in Section 9–4. Here we state merely that there is an interval $|x - 1| < a$ within which the series (9–4) converges to $f(x)$, that $f(x)$ is the desired solution of the differential equation, and that the rapidity of convergence can be estimated. Finding the general term of the series is a difficult problem; only for certain special equations can a simple expression be found for the general term. It should be noted that although it is convenient to have such an expression, the general term is well defined by the procedure described above, and that for each numerical application of the series, only a finite number of terms will be used.

The method described above applies equally well to equations of higher order, as in the following example.

EXAMPLE 2.

$$y'' = e^{-x}y' + e^{-x}y^2 - 1, \qquad y = 1, \quad y' = 1 \text{ when } x = 0. \qquad (9\text{–}5)$$

We find

$$y''' = e^{-x}(y'' + 2yy' - y' - y^2),$$
$$y^{(iv)} = e^{-x}(y''' + 2y'^2 + 2yy'' - 2y'' - 4yy' + y^2 + y').$$

Hence, when $x = 1$, $y'' = 1$, $y''' = 1$, $y^{(iv)} = 1$, and

$$y = 1 + x + \frac{x^2}{2!} + \frac{x^3}{3!} + \frac{x^4}{4!} + \cdots$$

The form of the series suggests that $y = e^x$, and it can be easily verified that $y = e^x$ is indeed a solution of Eq. (9–5). In this example the series converges for all x and the solution is valid for all x.

The method can be applied in similar fashion to systems of equations (Prob. 4 below). If the initial values are left general, the method provides a general solution. It is more common to fix the initial value of x and leave the initial values of y, y', ... general; we then obtain a general solution only for an interval of x (an interval which may depend on the initial values of y, y', ...).

EXAMPLE 3. $y' = x^2 - y^2, \qquad y = c \text{ for } x = 0. \qquad (9\text{–}6)$

The differential equation is the same as that of Example 1.   Hence

Eqs. (9–3) apply and we find $y' = -c^2$, $y'' = 2c^3$, $y''' = 2 - 6c^4$, $y^{(iv)} = 24c^5 - 4c, \ldots$, for $x = 0$, so that the solution sought is

$$y = c - c^2 x + c^3 x^2 - (c^4 - \tfrac{1}{3})x^3 + \left(c^5 - \frac{c}{6}\right)x^4 + \cdots$$

It can be shown that this series provides the general solution for $|x| < 1/|c|$.

PROBLEMS

1. Obtain the first four nonzero terms of the power series solution for each of the following differential equations with initial conditions:

(a) $y' = x^2 y^2 + 1$, $y = 1$ for $x = 1$;

(b) $y' = \sin(xy) + x^2$, $y = 3$ for $x = 0$;

(c) $y'' = x^2 - y^2$, $y = 1$ and $y' = 0$ for $x = 0$;

(d) $y''' = xy + yy'$, $y = 0$, $y' = 1$, $y'' = 2$ for $x = 0$;

(e) $y' = x + e^y$, $y = c$ for $x = 0$;

(f) $y'' + y = 0$, $y = c_1$, $y' = c_2$ for $x = 0$.

2. Given the differential equation $3y^2 y' = y^3 - x$,

(a) find the first four terms of the power series solution such that $y = 1$ when $x = 0$;

(b) graph the result of part (a) and compare with the graph of $y = (1 + x)^{1/3}$, which represents the same solution.

3. To obtain the power series solution of a general first order equation $y' = F(x, y)$, it is necessary to differentiate the equation successively to obtain y'', $y''', \ldots$ Show that $y'' = F_x + F_y y'$ and find expressions for y''', $y^{(iv)}$.

4. Obtain a solution in series form for each of the following, up to terms in $(x - x_0)^3$:

(a) $\dfrac{dy}{dx} = y + 2xz$, $\dfrac{dz}{dx} = xy - z$, $y = 0$ and $z = 1$ for $x = 1$;

(b) $\dfrac{dy}{dx} = yz$, $\dfrac{dz}{dx} = xz + y$, $y = 1$ and $z = 0$ for $x = 0$.

[*Hint*: Differentiate both equations to obtain expressions for higher derivatives in terms of lower ones.]

5. Obtain a solution in series form for the following, up to terms in x^3:

$$y'^3 + 3xy'^2 + x - y = 0, \qquad y = 1 \text{ for } x = 0.$$

[*Hint*: Differentiate the equation as it stands to find the higher derivatives.]

ANSWERS

1. (a) $1 + 2(x - 1) + 3(x - 1)^2 + \frac{19}{3}(x - 1)^3 + \cdots$,   (b) $3 + \frac{3}{2}x^2 + \frac{1}{3}x^3 - \frac{3}{4}x^4 + \cdots$,   (c) $1 - \frac{1}{2}x^2 + \frac{1}{6}x^4 - (7/360)x^6 + \cdots$,   (d) $x + x^2 + \frac{1}{24}x^4 + \frac{1}{15}x^5 + \cdots$,   (e) $c + e^c x + (1 + e^{2c})(x^2/2) + (2e^{3c} + e^c)(x^3/6) + \cdots$,
(f) $c_1[1 - (x^2/2!) + \cdots] + c_2[x - (x^3/3!) + \cdots]$.

2. (a) $y = \frac{1}{81}(81 + 27x - 9x^2 + 5x^3 + \cdots)$.

3. $y''' = F_{xx} + 2F_{xy}y' + F_{yy}y'^2 + F_y y''$, $y^{(iv)} = F_{xxx} + 3F_{xxy}y' + 3F_{xyy}y'^2 + F_{yyy}y'^3 + 3F_{xy}y'' + 3F_{yy}y'y'' + F_y y'''$.

4. (a) $y = 2(x - 1) + (x - 1)^2 + \frac{2}{3}(x - 1)^3 + \cdots$,
$z = 1 - (x - 1) + \frac{3}{2}(x - 1)^2 + \frac{1}{2}(x - 1)^3 + \cdots$;
(b) $y = 1 + 0 + \frac{1}{2}x^2 + 0 + \cdots$, $z = 0 + x + 0 + \frac{1}{2}x^3 + \cdots$

5. $1 + x - \frac{1}{2}x^2 + \frac{5}{18}x^3 + \cdots$

9–2 Method of undetermined coefficients.

The method to be described resembles that of Section 4–10. We seek a solution of a differential equation in the form of a power series:

$$y = c_0 + c_1(x - x_0) + c_2(x - x_0)^2 + \cdots \qquad (9\text{-}7)$$

Substitution in the differential equation then yields relations between the c's from which they can be determined (or expressed in terms of a certain set of c's which serve as arbitrary constants). The method is most effective for linear equations.

The steps depend on the basic rules for operation with convergent series and on the following two principles for power series:

A. if a power series (9–7) converges for $|x - x_0| < a$, then the series obtained by differentiating term-by-term also converges for $|x - x_0| < a$ and represents y';

B. if a power series (9–7) has a sum which is identically 0 for

$$|x - x_0| < a,$$

then each coefficient is 0.
(See pp. 353–355 of Reference 4.)

EXAMPLE 1. $y'' + xy' + y = 0$. We seek a solution of form (9–7) with $x_0 = 0$. By principle A, the function and its derivatives are given as follows:

$$y = c_0 + c_1 x + c_2 x^2 + \cdots + c_n x^n + \cdots,$$

$$y' = c_1 + 2c_2 x + \cdots + nc_n x^{n-1} + \cdots,$$

$$y'' = 2c_2 + 6c_3 x + \cdots + n(n - 1)c_n x^{n-2} + \cdots$$

If the three expressions are multiplied by 1, x, 1, and added, we obtain a series for $y'' + xy' + y$. If y satisfies the differential equation, we can equate this series to 0:

$$(c_0 + 2c_2) + x(2c_1 + 6c_3) + x^2(3c_2 + 12c_4)$$
$$+ \cdots + x^n[(n + 1)c_n + (n + 1)(n + 2)c_{n+2}] + \cdots = 0.$$

By principle B, each coefficient must equal 0. Hence we obtain the equations

$$c_0 + 2c_2 = 0, \qquad 2c_1 + 6c_3 = 0, \qquad 3c_2 + 12c_4 = 0,$$
$$\ldots, \qquad c_n + (n + 2)c_{n+2} = 0, \qquad \ldots$$

Therefore there is a *recursion formula* for the coefficients:

$$c_{n+2} = -\frac{c_n}{n + 2},$$

and we find

$$c_2 = -\frac{c_0}{2}, \qquad c_4 = -\frac{c_2}{4} = \frac{c_0}{2 \cdot 4}, \qquad c_6 = -\frac{c_0}{2 \cdot 4 \cdot 6}, \qquad \cdots,$$

$$c_3 = -\frac{c_1}{3}, \qquad c_5 = \frac{c_1}{3 \cdot 5}, \qquad c_7 = -\frac{c_1}{3 \cdot 5 \cdot 7}, \qquad \cdots$$

It appears that c_0, c_1 are arbitrary constants; they are, in fact, simply the initial values of y and y' at $x = 0$. The solutions can now be written in the form

$$y = c_0 \left(1 - \frac{1}{2} x^2 + \frac{1}{2 \cdot 4} x^4 - \frac{1}{2 \cdot 4 \cdot 6} x^6 + \cdots \right.$$
$$\left. + \frac{(-1)^n}{2 \cdot 4 \cdot 6 \cdots 2n} x^{2n} + \cdots \right)$$
$$+ c_1 \left(x - \frac{1}{3} x^3 + \frac{1}{3 \cdot 5} x^5 + \cdots \right.$$
$$\left. + \frac{(-1)^{n+1}}{3 \cdot 5 \cdot 7 \cdots (2n - 1)} x^{2n-1} + \cdots \right). \quad (9\text{-}8)$$

Here an application of the ratio test (Prob. 7 below) shows that both series converge for all x. Furthermore, each series satisfies the differential equation for all x. Since the corresponding functions

$$y_1(x) = 1 + \sum_{n=1}^{\infty} \frac{(-1)^n x^{2n}}{2 \cdot 4 \cdots 2n}, \qquad y_2(x) = \sum_{n=1}^{\infty} \frac{(-1)^{n+1} x^{2n-1}}{1 \cdot 3 \cdots (2n - 1)}$$

are linearly independent, the function

$$y = c_0 y_1(x) + c_1 y_2(x)$$

gives the general solution of the differential equation.

It can be verified that the method of undetermined coefficients is equivalent to the Taylor series method described in Section 9–1; that is, both yield the same series. For linear equations the method of undetermined coefficients is somewhat more convenient, but for nonlinear equations the method of undetermined coefficients becomes awkward (see Prob. 8 below).

In the linear equation of Example 1 the coefficients are powers of x and the equation is homogeneous. When more general coefficients appear or the right-hand side is not a polynomial, we are forced to expand the coefficients or right-hand side in Taylor's series.

EXAMPLE 2. $y'' + (\sin x)y = e^{x^2}$. If our solution is again to have form

$$y = \sum_{n=0}^{\infty} c_n x^n,$$

then we must expand $\sin x$ and e^{x^2} in their Taylor series about $x = 0$:

$$\sin x = x - \frac{x^3}{3!} + \frac{x^5}{5!} - \cdots, \qquad e^{x^2} = 1 + x^2 + \frac{x^4}{2!} + \frac{x^6}{3!} + \cdots$$

Accordingly,

$$(\sin x)y = \left(x - \frac{x^3}{3!} + \frac{x^5}{5!} + \cdots\right)(c_0 + c_1 x + c_2 x^2 + \cdots)$$

$$= c_0 x + c_1 x^2 + x^3 \left(c_2 - \frac{c_0}{3!}\right) + \cdots$$

Substitution in the differential equation leads to the relation

$$(2c_2 + 6c_3 x + 12c_4 x^2 + 20c_5 x^3 + \cdots) + c_0 x + c_1 x^2 + x^3 \left(c_2 - \frac{c_0}{3!}\right)$$

$$+ \cdots = 1 + x^2 + \frac{x^4}{2!} + \cdots$$

By principle B, we can now compare coefficients of like powers of x:

$$2c_2 = 1, \qquad 6c_3 + c_0 = 0, \qquad 12c_4 + c_1 = 1,$$

$$20c_5 + c_2 - \frac{c_0}{3!} = 0, \qquad \cdots$$

Accordingly,

$$c_3 = -\frac{1}{6}c_0, \qquad c_2 = \frac{1}{2}, \qquad c_5 = \frac{1}{20}\left(\frac{c_0}{6} - c_2\right) = \frac{c_0}{120} - \frac{1}{40},$$

$$\cdots, \qquad c_4 = \frac{1}{12} - \frac{c_1}{12}, \qquad \cdots$$

Again c_0, c_1 are arbitrary:

$$y = c_0 + c_1 x + \frac{1}{2}x^2 - \frac{1}{6}c_0 x^3$$

$$+ \left(\frac{1}{12} - \frac{c_1}{12}\right)x^4 + \left(\frac{c_0}{120} - \frac{1}{40}\right)x^5 + \cdots$$

$$= c_0\left(1 - \frac{1}{6}x^3 + \frac{1}{120}x^5 + \cdots\right)$$

$$+ c_1\left(x - \frac{x^4}{12} + \cdots\right) + \frac{1}{2}x^2 + \frac{1}{12}x^4 + \cdots$$

The solution appears in the form

$$y = c_0 y_1(x) + c_1 y_2(x) + y^*(x),$$

where $y^*(x)$ is a particular solution. The general terms of the series can be found with some effort and the convergence of the series analyzed. However, a general convergence theorem (Section 9–6) is applicable to show that the series converge for all x.

At first sight, the solutions in series form appear to be rather unsatisfactory. It is difficult to obtain the general term of the series, there are awkward questions of convergence, and the solutions are generally not related to familiar functions. However, we can take the point of view stressed in Chapter 1 that a differential equation is a way of describing new functions. A number of new functions (e.g., Bessel functions, hypergeometric functions) do arise naturally in this way. From the differential equations which define them, various properties of the functions can be deduced; the series expansion is simply one aspect of each function and it can be helpful in forming a table of function values. Once the new functions have been studied and tabulated, they become "familiar functions."

9–3 Techniques for forming series solutions. We can shorten the process of finding series solutions by judicious application of the $\sum$ notation. The following relation is important:

$$\sum_{n=k}^{\infty} a_n = \sum_{n=k-l}^{\infty} a_{n+l} = a_k + a_{k+1} + \cdots \qquad (9\text{–}9)$$

We consider again the equation of Example 1 in Section 9–2:

$$y'' + xy' + y = 0.$$

We write our series solution as

$$y = \sum_{n=0}^{\infty} c_n x^n.$$

Then

$$y' = \sum_{n=1}^{\infty} n c_n x^{n-1}, \qquad xy' = \sum_{n=1}^{\infty} n c_n x^n,$$

$$y'' = \sum_{n=2}^{\infty} n(n-1) c_n x^{n-2}.$$

Hence substitution in the differential equation gives

$$\sum_{n=2}^{\infty} n(n-1) c_n x^{n-2} + \sum_{n=1}^{\infty} n c_n x^n + \sum_{n=0}^{\infty} c_n x^n = 0.$$

We apply the rule (9–9) to the first term, replacing n by $n + 2$:

$$\sum_{n=0}^{\infty} (n+2)(n+1) c_{n+2} x^n + \sum_{n=1}^{\infty} n c_n x^n + \sum_{n=0}^{\infty} c_n x^n = 0.$$

The general term in all three series now contains the same nth power of x, so that we can combine the series into one power series. However, since there is no term $n = 0$ in the middle series, we write out the terms $n = 0$ separately:

$$2c_2 + c_0 + \sum_{n=1}^{\infty} [(n+2)(n+1) c_{n+2} + n c_n + c_n] x^n = 0.$$

Hence, as before,

$$2c_2 + c_0 = 0, \quad (n+2)(n+1) c_{n+2} + (n+1) c_n = 0$$

$$(n = 1, 2, \ldots)$$

and we obtain the recursion formula

$$c_{n+2} = -\frac{c_n}{n+2}.$$

The general solution is then obtained as in Section 9–2.

As a second example, consider the equation

$$(x^3 + 1)y'' + x^2 y' - 4xy = 0.$$

The work then proceeds as follows:

$$y = \sum_{n=0}^{\infty} c_n x^n, \qquad y' = \sum_{n=1}^{\infty} n c_n x^{n-1}, \qquad y'' = \sum_{n=2}^{\infty} n(n-1) c_n x^{n-2},$$

$$xy = \sum_{n=0}^{\infty} c_n x^{n+1}, \qquad x^2 y' = \sum_{n=1}^{\infty} n c_n x^{n+1},$$

$$x^3 y'' = \sum_{n=2}^{\infty} n(n-1) c_n x^{n+1},$$

$$\sum_{n=2}^{\infty} n(n-1) c_n x^{n+1} + \sum_{n=2}^{\infty} n(n-1) c_n x^{n-2}$$

$$+ \sum_{n=1}^{\infty} n c_n x^{n+1} - 4 \sum_{n=0}^{\infty} c_n x^{n+1} = 0,$$

$$\sum_{n=3}^{\infty} (n-1)(n-2) c_{n-1} x^n + \sum_{n=0}^{\infty} (n+2)(n+1) c_{n+2} x^n$$

$$+ \sum_{n=2}^{\infty} (n-1) c_{n-1} x^n - 4 \sum_{n=1}^{\infty} c_{n-1} x^n = 0,$$

$$2c_2 + (6c_3 - 4c_0)x + (12c_4 + c_1 - 4c_1)x^2$$

$$+ \sum_{n=3}^{\infty} [(n-1)(n-2) c_{n-1} + (n+2)(n+1) c_{n+2}$$

$$+ (n-1) c_{n-1} - 4 c_{n-1}] x^n = 0,$$

$$2c_2 = 0, \qquad 6c_3 - 4c_0 = 0, \qquad 12c_4 - 3c_1 = 0,$$

$$(n^2 - 2n - 3) c_{n-1} + (n+2)(n+1) c_{n+2} = 0$$
$$(n = 3, 4, \ldots),$$

$$c_2 = 0, \qquad c_3 = \tfrac{2}{3} c_0, \qquad c_4 = \tfrac{1}{4} c_1,$$

$$c_{n+2} = - \frac{n^2 - 2n - 3}{(n+2)(n+1)} c_{n-1} = - \frac{n-3}{n+2} c_{n-1},$$

$$c_5 = 0, \qquad c_8 = 0, \qquad c_{11} = 0, \qquad \ldots,$$

$$c_6 = -\tfrac{1}{6} c_3 = -\tfrac{1}{6} \cdot \tfrac{2}{3} c_0, \qquad c_9 = -\tfrac{4}{9} c_6 = \tfrac{4}{9} \cdot \tfrac{1}{6} \cdot \tfrac{2}{3} c_0,$$

$$c_{3k} = (-1)^k \frac{3k-5}{3k} \cdot \frac{3k-8}{3k-3} \cdots \frac{1}{6} \frac{-2}{3} c_0,$$

$$c_7 = -\tfrac{2}{7} c_4 = -\tfrac{2}{7} \cdot \tfrac{1}{4} c_1, \qquad c_{10} = -\tfrac{5}{10} c_7 = \tfrac{5}{10} \cdot \tfrac{2}{7} \cdot \tfrac{1}{4} c_1,$$

$$c_{3k+1} = (-1)^k \frac{3k - 4}{3k + 1} \cdot \frac{3k - 7}{3k - 2} \cdots \frac{2}{7} \cdot \frac{-1}{4} c_1,$$

$$y = c_0 \left(1 + \sum_{k=1}^{\infty} (-1)^k \frac{3k - 5}{3k} \frac{3k - 8}{3k - 3} \cdots \frac{1}{6} \cdot \frac{-2}{3} x^{3k} \right)$$

$$+ c_1 \left(x + \sum_{k=1}^{\infty} (-1)^k \frac{3k - 4}{3k + 1} \frac{3k - 7}{3k - 2} \cdots \frac{2}{7} \cdot \frac{-1}{4} x^{3k+1} \right).$$

The procedures shown are applicable to linear equations with *polynomial* coefficients. Substitution of $y = \sum c_n x^n$ in the differential equation and equating the coefficient of x^n to zero leads to a recursion formula which connects the coefficients c_n. In each of the two examples above, the recursion formula relates *two* coefficients: c_n, c_{n+2} in the first; c_{n-1}, c_{n+2} in the second. Hence we say there is a *two-term recursion formula*. When only two coefficients are related, the general term of the solution can be found as in the example. When more than two terms are related, it becomes difficult to find the general term; in such cases the recursion formula defines c_n as a new function of n, just as the differential equation defines y as a new function of x.

The following example has a three-term recursion formula:

$$y'' + y' + xy = 0,$$

$$\sum_{n=2}^{\infty} n(n - 1)c_n x^{n-2} + \sum_{n=1}^{\infty} nc_n x^{n-1} + \sum_{n=0}^{\infty} c_n x^{n+1} = 0,$$

$$\sum_{n=0}^{\infty} (n + 2)(n + 1)c_{n+2} x^n + \sum_{n=0}^{\infty} (n + 1)c_{n+1} x^n + \sum_{n=1}^{\infty} c_{n-1} x^n = 0,$$

$$2c_2 + c_1 = 0, \qquad (n + 2)(n + 1)c_{n+2} + (n + 1)c_{n+1} + c_{n-1} = 0,$$

$$c_{n+2} = -\frac{(n + 1)c_{n+1} + c_{n-1}}{(n + 2)(n + 1)} \qquad (n = 1, 2, \ldots),$$

$$c_2 = -\tfrac{1}{2}c_1, \qquad c_3 = -\frac{2c_2 + c_0}{6} = \frac{c_1 - c_0}{6},$$

$$c_4 = -\frac{3c_3 + c_1}{12} = \frac{c_0 - 3c_1}{24}, \qquad \ldots,$$

$$y = c_0 + c_1 x - \frac{1}{2}c_1 x^2 + \frac{c_1 - c_0}{6} x^3 + \frac{c_0 - 3c_1}{24} x^4 + \cdots$$

$$= c_0 \left(1 - \frac{x^3}{6} + \frac{x^4}{24} + \cdots \right) + c_1 \left(x - \frac{x^2}{2} + \frac{x^3}{6} - \frac{x^4}{8} + \cdots \right).$$

We can consider the equations for the c_n as simultaneous linear equations, to be solved for c_2, c_3, ... in terms of c_0, c_1. The first n equations involve only the n unknowns c_2, c_3, ..., c_{n+2}; hence they can be solved by determinants. We find $c_{n+2} = P_n/Q_n$, where Q_n is the determinant

$$
\begin{vmatrix}
2 & 0 & 0 & 0 & 0 & \cdots & 0 & 0 & 0 & 0 \\
2 & 6 & 0 & 0 & 0 & & 0 & 0 & 0 & 0 \\
0 & 3 & 12 & 0 & 0 & & 0 & 0 & 0 & 0 \\
1 & 0 & 4 & 20 & 0 & & 0 & 0 & 0 & 0 \\
0 & 1 & 0 & & & & 0 & 0 & 0 & 0 \\
\vdots & & & & & & & & & \vdots \\
0 & 0 & 0 & 0 & 0 & \cdots & 1 & 0 & n+1 & n^2 + 3n + 2
\end{vmatrix}
$$

and P_n is obtained from Q_n by replacing the elements of the last column by $-c_1$, $-c_0$, 0, ..., 0. It should also be noted that the recursion formula is a linear finite-difference equation and can be studied by special methods for such equations (Reference 3 at the end of this chapter).

In all the examples the power series have had the form $\sum c_n x^n$; that is, they are Taylor series about $x_0 = 0$. To obtain series solutions in powers of $x - x_0$ for $x_0 \neq 0$, it is simpler to make the substitution $u = x - x_0$ and obtain a differential equation for y in terms of u. A series solution $\sum c_n u^n$ then becomes the desired series $\sum c_n (x - x_0)^n$.

PROBLEMS

1. For each of the following obtain the general solution in the form of a power series $\sum c_n x^n$ and show that the series converges for all x:

(a) $y'' + 2xy' + 4y = 0$,   (b) $y'' - x^2 y = 0$.

2. For each of the following obtain the general solution in the form of a power series $\sum c_n (x - a)^n$ and show that the series converges for all x:

(a) $y'' + (x - 1)^2 y' + (x - 1)y = 0$, $a = 1$;
(b) $y'' + 2(x - 2)y' + y = 0$, $a = 2$.

3. (a) Find a particular solution of the equation $y'' + 2xy' + 4y = e^x$. [*Hint:* Assume $y = 0$, $y' = 0$ at $x = 0$, and hence take

$$
y = \sum_{n=2}^{\infty} c_n x^n.
$$
]

(b) Find the general solution of the differential equation of part (a). [See Prob. 1(a).]

4. Find the general solution of the differential equation $y'' - x^2 y = \sin x$. [*Hint:* Proceed as in Prob. 3(a) to find a particular solution. Then use the result of Prob. 1(b).]

5. Find a solution $y = \sum c_n x^n$ up to terms in x^6 for each of the following:

(a) $y'' + y' + xy = 0$, $y = 1$, $y' = 0$ for $x = 0$;

(b) $y''' + xy'' + y' + xy = 1 + x$, $y = 0$, $y' = 0$, $y'' = 0$ for $x = 0$.

6. Obtain solutions as series in powers of x, up to terms in x^3:

(a) $\dfrac{dy}{dx} = y + xz$, $\dfrac{dz}{dx} = y + (x - 1)z$, $y = 1$ and $z = 2$ for $x = 0$;

(b) $\dfrac{dy}{dx} = y - z$, $\dfrac{dz}{dx} = xz + y$, $y = 0$ and $z = 1$ for $x = 0$.

7. (a) Show that in Eq. (9–8) the series multiplied by c_0 is equal to $e^{-x^2/2}$.

(b) Show that both series in Eq. (9–8) converge for all x.

(c) Use the method of Section 7–4 to obtain the general solution of the equation $y'' + xy' + y = 0$ from the known solution $y_1 = e^{-x^2/2}$.

8. For each of the following find the solution as a power series by the method of undetermined coefficients:

(a) $y' = x^2 - y^2$, $y = 1$ when $x = 1$ (Example 1 of Section 9–1); obtain terms up to $(x - 1)^4$;

(b) $y' = x + e^y$, $y = c$ when $x = 0$ (Prob. 1(e) following Section 9–1); obtain terms up to x^3. [*Hint:* $y = c + c_1 x + c_2 x^2 + \cdots$,

$$e^y = e^c e^{y-c} = e^c\{1 + (y - c) + \tfrac{1}{2}(y - c)^2 + \cdots\}$$
$$= e^c\{1 + (c_1 x + c_2 x^2 + \cdots) + \tfrac{1}{2}(c_1 x + c_2 x^2 + \cdots)^2 + \cdots\}$$
$$= e^c\{1 + c_1 x + (c_2 + \tfrac{1}{2}c_1^2)x^2 + \cdots\}.]$$

9. (a) Show that the equation $x^2 y'' + xy' + y = 0$ has no solution of form $\sum_{n=0}^{\infty} c_n x^n$ except $y \equiv 0$. Show also that the existence theorem of Section 1–8 does not apply at $x = 0$.

(b) Show that the existence theorem of Section 1–8 does not apply at $x = 0$ to the equation $x^2 y'' - 3xy' + 3y = 0$, but that solutions of form $\sum_{n=0}^{\infty} c_n x^n$ are obtainable.

10. Let the differential equation

$$a_0(x)y'' + a_1(x)y' + a_2(x)y = 0$$

be given, where $a_0(x)$, $a_1(x)$, $a_2(x)$ are polynomials and $a_0(0) \neq 0$. Let L be the corresponding operator $a_0(x)D^2 + a_1(x)D + a_2(x)$. Assume that the substitution $y = \sum c_n x^n$ leads to a two-term recursion formula for the c's.

(a) Show that the differential equation must have the form

$$(\alpha x^k + \beta)y'' + \gamma x^{k-1} y' + \delta x^{k-2} y = 0,$$

where α, β, γ, δ are constants and $\beta \neq 0$, $k \geq 2$. Show that we then have

$$L(x^n) = f(n)x^{n-2} + g(n)x^{n+k-2} \qquad (n = 0, 1, 2, \ldots),$$

and find $f(n)$, $g(n)$.

  (b) Let $y(x) = x^m(\sum_{s=0}^{\infty} c_s x^{sk})$, where $c_0 = 1$. Show that

$$L(y) = x^{m-2}\left(f(m) + \sum_{s=1}^{\infty} \{c_s f(m + sk) + c_{s-1}g[m + (s - 1)k]\}x^{sk}\right).$$

  (c) Conclude from the result of part (b) that $y(x)$ is a solution of the differential equation if and only if $f(m) = 0$, so that $m = 0$ or $m = 1$, and

$$c_s f(m + sk) + c_{s-1}g[m + (s - 1)k] = 0 \qquad (s = 1, 2, \ldots).$$

Take $c_0 = 1$ and obtain an expression for c_s, $(s = 1, 2, \ldots)$. Show that the choices $m = 0$, $m = 1$ lead to the two solutions

$$y_1 = 1 + \sum_{s=1}^{\infty} (-1)^s x^{sk} \frac{g(0)g(k) \cdots g[(s - 1)k]}{f(k)f(2k) \cdots f(sk)},$$

$$y_2 = x + x \sum_{s=1}^{\infty} (-1)^s x^{sk} \frac{g(1)g(k + 1) \cdots g[1 + (s - 1)k]}{f(1 + k)f(1 + 2k) \cdots f(1 + sk)}.$$

  (d) With the aid of the expressions for $f(n)$, $g(n)$ obtained in part (a), apply the ratio test to show that the power series for y_1, y_2 of part (c) converge for $|x| < |\beta/\alpha|^{1/k}$ if $\alpha \neq 0$ and for all x if $\alpha = 0$.

ANSWERS

1. (a) $c_0\left(1 + \sum_{n=1}^{\infty} \frac{(-1)^n 2^n x^{2n}}{1 \cdot 3 \cdots (2n - 1)}\right) + c_1 \sum_{n=1}^{\infty} \frac{(-1)^n x^{2n-1}}{(n - 1)!}$,

  (b) $c_0\left(1 + \sum_{n=1}^{\infty} \frac{x^{4n}}{3 \cdot 4 \cdot 7 \cdot 8 \cdots (4n - 1)(4n)}\right)$

$$+ c_1\left(x + \sum_{n=1}^{\infty} \frac{x^{4n+1}}{4 \cdot 5 \cdot 8 \cdot 9 \cdots (4n)(4n + 1)}\right).$$

2. (a) $c_0\left(1 + \sum_{n=1}^{\infty} \frac{1 \cdot 4 \cdots (3n - 2)\,(-1)^n(x - 1)^{3n}}{2 \cdot 3 \cdot 5 \cdot 6 \cdots (3n - 1)(3n)}\right)$

$$+ c_1(x - 1)\left(1 + \sum_{n=1}^{\infty} \frac{2 \cdot 5 \cdots (3n - 1)\,(-1)^n(x - 1)^{3n}}{3 \cdot 4 \cdot 6 \cdot 7 \cdots (3n)(3n + 1)}\right),$$

(b) $c_0\left(1 + \sum_{n=1}^{\infty} \dfrac{1 \cdot 5 \cdots (4n-3)\,(-1)^n(x-2)^{2n}}{(2n)!}\right)$

$+ c_1(x-2)\left(1 + \sum_{n=1}^{\infty} \dfrac{3 \cdot 7 \cdots (4n-1)\,(-1)^n(x-2)^{2n}}{(2n+1)!}\right).$

3. (a) $\frac{1}{2}x^2 + \frac{1}{6}x^3 - \frac{7}{24}x^4 - \frac{3}{40}x^5 + (85/720)x^6 + \cdots = y^*$, (b) $y^* + y_c$, where y_c is the answer to Prob. 1(a).

4. $\dfrac{1}{3!}x^3 - \dfrac{1}{5!}x^5 + \dfrac{21}{7!}x^7 - \dfrac{41}{9!}x^9 + \cdots + y_c$, where y_c is the answer to Prob. 1(b).

5. (a) $1 - \frac{1}{6}x^3 + \frac{1}{24}x^4 - (1/120)x^5 + (1/144)x^6 + \cdots$,
   (b) $\frac{1}{6}x^3 + \frac{1}{24}x^4 - \frac{1}{40}x^5 - (1/180)x^6 + \cdots$

6. (a) $y = 1 + x + \frac{3}{2}x^2 + \frac{1}{6}x^3 + \cdots$, $z = 2 - x + 2x^2 - \frac{1}{2}x^3 + \cdots$;
   (b) $y = -x - \frac{1}{2}x^2 - \frac{1}{6}x^3 + \cdots$, $z = 1 - \frac{1}{6}x^3 + \cdots$

7. (c) $y = c_1 e^{-x^2/2} + c_2 e^{-x^2/2}\int e^{x^2/2}\,dx.$

9. (b) $c_1 x + c_2 x^3$ is the general solution.

10. (a) $f(n) = \beta n(n-1)$, $g(n) = \alpha n^2 + (\gamma - \alpha)n + \delta$;

   (c) $c_s = (-1)^s \dfrac{g(m)g(m+k) \cdots g[m+(s-1)k]}{f(m+k)f(m+2k) \cdots f(m+sk)}.$

9-4 Convergence of power series solutions of $y' = F(x, y)$. We now consider a differential equation of first order:

$$y' = F(x, y) \tag{9-10}$$

and assume $F(x, y)$ can be represented by a power series in $x - x_0$, $y - y_0$ for $|x - x_0| \le a$, $|y - y_0| \le b$, where $a > 0$, $b > 0$. By translation of axes we can restrict attention to the case $x_0 = 0$, $y_0 = 0$. Then our assumption is that

$$F(x, y) = c_{0,0} + (c_{1,0}x + c_{0,1}y) + (c_{2,0}x^2 + c_{1,1}xy + c_{0,2}y^2)$$
$$+ \cdots + (c_{n,0}x^n + \cdots + c_{0,n}y^n) + \cdots \tag{9-11}$$

for $|x| \le a$, $|y| \le b$. Such a function F is termed an *analytic* function of x and y (for $|x| < a$, $|y| < b$). It can be shown that the coefficients $c_{n,k}$ are obtained by partial differentiation:

$$c_{n-k,k} = \frac{1}{k!(n-k)!} \frac{\partial^n F}{\partial x^{n-k}\,\partial y^k}\bigg|_{x=0,\ y=0} \tag{9-12}$$

In particular, F may be a *polynomial* in x and y:

$$F = c_{0,0} + c_{1,0}x + c_{0,1}y + \cdots + c_{n,0}x^n + \cdots + c_{0,n}y^n. \qquad (9\text{--}13)$$

(See pp. 369–371 of Reference 4.)

In addition, we make a *boundedness assumption* concerning $F(x, y)$, which, for simplicity, we formulate as follows. We assume that the series of positive terms

$$|c_{0,0}| + |c_{1,0}|a + |c_{0,1}|b + |c_{2,0}|a^2 + |c_{1,1}|ab + |c_{0,2}|b^2$$
$$+ \cdots + |c_{n,0}|a^n + \cdots + |c_{0,n}|b^n + \cdots \qquad (9\text{--}14)$$

converges; let M be chosen as a number greater than or equal to the sum. From the comparison test for series, we conclude that

$$|F(x, y)| \leqq M \qquad \text{for} \qquad |x| \leqq. a, \ |y| \leqq b. \qquad (9\text{--}15)$$

We can now formulate a theorem on the power series solution of Eq. (9–10).

THEOREM 1. *Let $F(x, y)$ satisfy the·hypotheses described above. Then there is a unique solution $y = f(x)$ of Eq. (9–10) satisfying the initial condition $y = 0$ when $x = 0$. The solution is valid and can be represented by its Taylor series for $|x| < \rho$, where*

$$\rho = a(1 - e^{-1/c}) \qquad (c = 2Ma/b). \qquad (9\text{--}16)$$

If $f(x) = a_1 x + a_2 x^2 + \cdots + a_n x^n + \cdots$ and

$$R_n = f(x) - (a_1 x + a_2 x^2 + \cdots + a_n x^n) \qquad (9\text{--}17)$$

is the corresponding remainder after n terms, then

$$|R_n| \leqq \phi(r) \frac{|x|^{n+1}}{r^n(r - |x|)} \qquad (|x| < r < \rho), \qquad (9\text{--}18)$$

where

$$\phi(r) = b\left[2 + \left\{ c \left| \log\left(1 - \frac{r}{a}\right) \right| \right\}^{1/2} \right]. \qquad (9\text{--}19)$$

A proof of the theorem is given in Chapter 12 (Sections 12–12 and 12–13). Here we discuss its significance and apply it to examples.

First, the theorem guarantees that when $F(x, y)$ satisfies the conditions stated, the methods described earlier in this chapter do provide meaningful solutions in series form. Equation (9–16) gives a *lower* estimate ρ for the *radius of convergence;* the series may converge for much larger values of $|x|$. (If it does converge for such values, then so long as $|x| < a$, $|y| < b$

we can be sure that the series continues to satisfy the differential equation. For $|x| > a$ or $|y| > b$, $F(x, y)$ may be undefined.)

The inequality (9–18) shows how rapidly the series converges and how large an error is made in stopping at n terms. Only an upper estimate is given for the error. The particular estimate also depends on the choice of r. In any case, for fixed x,

$$\frac{|x|^{n+1}}{r^n(r - |x|)} = \frac{|x|}{r - |x|} \cdot \left(\frac{|x|}{r}\right)^n \to 0 \text{ as } n \to \infty,$$

so that $R_n \to 0$ as $n \to \infty$. From inequality (9–18) we can determine how many terms of the series are sufficient to evaluate $f(x)$ with error less than a prescribed ϵ.

Remark. If r is fixed, the expression on the right side of inequality (9–18) decreases as $|x|$ decreases. Hence, if we have found from (9–18) that $|R_n| \leq k$ for a particular choice of $|x|$, then we can be sure that $|R_n| \leq k$ for all smaller $|x|$.

EXAMPLE 1. $y' = 1 + y^2$. The solution with initial value 0 when $x = 0$ is found to be

$$y = f(x) = x + \tfrac{1}{3}x^3 + \tfrac{2}{15}x^5 + (17/315)x^7 + \cdots$$

The function $F(x, y) = 1 + y^2$ is analytic for all x, y, so that a, b can be chosen as large as we wish. If we choose $a = 10$, $b = 10$, then we can choose $M = 1 + 10^2 = 101$. According to Eq. (9–16), the series converges for $|x| < \rho$, where

$$\rho = 10(1 - e^{-10/2020}) = 0.05.$$

If we choose $a = 1$, $b = 1$, we obtain $M = 2$ and the larger estimate $\rho = 0.22$. With the latter values, we can try to compute $f(x)$ with an error of at most 0.01 for $x = 0.2$. We apply (9–18), (9–19), with $a = 1$, $r = 0.21$, $b = 1$, $M = 2$. Hence

$$\phi(r) = 2 + 2 \, |\log 0.79|^{1/2} = 2.97,$$

$$|R_n| \leq 2.97 \frac{|x|^{n+1}}{(0.21)^n(0.21 - |x|)} = 2.70 \left(\frac{10}{21}\right)^n.$$

The equation $2.70 \, (10/21)^n = 0.01$ is easily solved (by taking logarithms); we find $n = 4.5$. Hence for $n \geq 5$, $|R_n| < 0.01$; that is, the terms of the series through x^5 give $f(x)$ for $x = 0.1$ with an error of less than 0.01. Accordingly,

$$f(0.1) = 0.1 + \frac{0.001}{3} + \frac{0.00002}{15} + \cdots = 0.10033,$$

with an error of less than 0.01.

In this example we verify that the solution in question is simply $y = \tan x$, for which the power series is known to converge for $|x| < \frac{1}{2}\pi = 1.57\ldots$ For $x = 0.1$, $y = \tan (0.1) = 0.10033$.   Thus the radius of convergence is much larger than the 0.22 found above, and five terms of the series give $f(x)$ accurate to five significant figures when $x = 0.1$. We see, then, that the formulas of Theorem 1 provide a large *safety margin*.

EXAMPLE 2. $y' = \sin (xy) + 1$.   The solution with initial value 0 when $x = 0$ is found to be

$$y = x + \tfrac{1}{3}x^3 + \cdots$$

Now from the series for $\sin x$ we find $1 + \sin xy = 1 + xy - (x^3y^3/3!) + (x^5y^5/5!) - \cdots$ Hence $F(x, y)$ is analytic for all x, y.   Let us choose $a = 1$, $b = 1$; to find M we need to estimate

$$1 + 1 + \frac{1}{3!} + \frac{1}{5!} + \cdots = 1 + \sinh 1 = 2.175.$$

Then Eq. (9–16) gives $\rho = 0.215$.   If we use only the two terms of the series exhibited and choose $r = 0.2$, then $\phi(r) = 2.99$ and

$$|R_3| \leqq 2.99 \, \frac{|x|^4}{0.008(0.2 - |x|)}.$$

For $x = 0.05$, $|R_3| \leqq 0.16$; for $x = 0.1$, $|R_3| \leqq 0.37$; for $x = 0.01$, $|R_3| \leqq 0.000020$.

9–5 Convergence conditions for systems and equations of higher order.
We consider a system of N first order equations

$$y_i' = F_i(x, y_1, \ldots, y_N) \qquad (i = 1, \ldots, N). \tag{9–20}$$

Each solution is determined by initial values $y_{1,0}, \ldots, y_{N,0}$ of $y_1, \ldots, y_N$ at $x = x_0$.   By translating axes we can restrict attention to the case $x_0 = 0$, $y_{1,0} = 0$, $\ldots$, $y_{N,0} = 0$.   It is assumed that each F_i can be expanded in power series in $x, y_1, \ldots, y_N$ for $|x| \leqq a$, $|y_i| \leqq b$, $(i = 1, \ldots, N)$, where $a > 0$, $b > 0$.   The F_i are then termed analytic in the variables.   It is assumed, furthermore, that each power series converges when $x = a$, $y_i = b$ and every coefficient is replaced by its absolute value; for each i, M_i denotes the sum of the series of constants so obtained.   We choose a number M greater than or equal to all M_i, $(i = 1, \ldots, N)$.

THEOREM 2.   *Under the assumptions stated, the system* (9–20) *has a unique solution* $y_1 = f_1(x), \ldots, y_N = f_N(x)$, *such that* $y_1 = 0, \ldots, y_N = 0$

for $x = 0$. The functions $f_i(x)$ are represented by their Taylor series for $|x| < \rho$, where

$$\rho = a(1 - e^{-1/c}), \qquad c = \frac{M(N+1)a}{b}. \qquad (9\text{--}21)$$

For each series the remainder R_n after n terms satisfies the inequality

$$|R_n| \leqq \phi(r) \frac{|x|^{n+1}}{r^n(r - |x|)} \qquad (|x| < r < \rho), \qquad (9\text{--}22)$$

where

$$\phi(r) = b\left[2 + \left\{c\left|\log\left(1 - \frac{r}{a}\right)\right|\right\}^{1/(N+1)}\right]. \qquad (9\text{--}23)$$

A proof is given in Chapter 12. Here we consider an example:

$$\frac{dy}{dx} = xy^2 - z, \qquad \frac{dz}{dx} = yz - y,$$

so that $N = 2$. We seek the solution for which $y = 0$, $z = 0$ for $x = 0$. Since the F_i are polynomials, a and b can be chosen as large as desired. For $a = 1$, $b = 1$, we find $M_1 = 2$, $M_2 = 2$, $M = 2$, $c = 6$, $\rho = 1 - e^{-1/6} = 0.16$. If $r = 0.15$, then $\phi(r) = 3.0$ and

$$|R_n| \leqq 3.0 \frac{|x|^{n+1}}{(1.5)^n(1.5 - |x|)} \qquad (|x| < 0.15).$$

Hence if an error of less than 0.1 is desired for $|x| \leqq 0.1$, we find that ten terms are sufficient.

The equation of Nth order

$$y^{(N)} = F(x, y, \ldots, y^{(N-1)}) \qquad (9\text{--}24)$$

can be reduced to a system of form (9–20) and then analyzed similarly. If initial values $y_0, y_0', \ldots, y_0^{(N-1)}$ are prescribed at $x = x_0$, we assume that F can be expanded in a power series in all N variables $x, y, \ldots, y^{(N-1)}$ for $|x - x_0|, |y - y_0|, \ldots, |y^{(N-1)} - y_0^{(N-1)}|$ sufficiently small. Rather than carrying out a translation to obtain 0 initial values, it is simpler to translate the x only and then to introduce a new dependent variable

$$u = y - p(x), \qquad (9\text{--}25)$$

$$p(x) = y_0 + y_0'x + \cdots + y_0^{(N-1)}\frac{x^{N-1}}{(N-1)!}. \qquad (9\text{--}26)$$

Then F becomes a function of $u, u', \ldots$:

$$F(x, y, \ldots, y^{(N-1)})$$
$$\equiv F[x, u + p(x), u' + p'(x), \ldots, u^{(N-1)} + p^{(N-1)}(x)]$$
$$= G(x, u, u', \ldots, u^{(N-1)}). \qquad (9\text{--}27)$$

Since $u^{(N)}(x) = y^{(N)}(x)$, we obtain the new differential equation

$$u^{(N)} = G(x, u, u', \ldots, u^{(N-1)}), \qquad (9\text{--}28)$$

and u has initial values

$$\text{for } x = 0: \quad u = 0, \quad u' = 0, \quad \ldots, \quad u^{(N-1)} = 0. \quad (9\text{--}29)$$

If we obtain the series solution of Eq. (9–28) with initial values (9–29), it will start with terms of degree N:

$$u = c_N x^N + c_{N+1} x^{N+1} + \cdots$$

and from (9–25),

$$y = p(x) + u = y_0 + y_0' x + \cdots + y_0^{(N-1)} \frac{x^{N-1}}{(N-1)!}$$
$$+ c_N x^N + c_{N+1} x^{N+1} + \cdots$$

Thus u provides the missing "tail" of the series for y. The omission of all terms after the term of degree n in the series for u results in the same error as that committed by the omission of all terms after the term of degree n in the series for y. Hence it is sufficient to study an equation of form (9–28) with initial conditions (9–29).

Now let Eqs. (9–28), (9–29) be given. The assumption made above that F is expressible in a power series in its variables implies that G satisfies a similar condition. In any case we assume that G can be so expanded for $|x| \leqq a$, $|u| \leqq b$, $\ldots$, $|u^{(N-1)}| \leqq b$. We also assume that this power series converges when $x = a$, $u = b$, $\ldots$, $u^{(N-1)} = b$ and every coefficient is replaced by its absolute value; the sum of this series of constants is denoted by M_0. We choose a number M greater than or equal to the larger of M_0 and b.

THEOREM 3. *Under the assumptions described, Eq. (9–28) has a unique solution $u = f(x)$ satisfying conditions (9–29). The function $f(x)$ is represented by its Taylor series for $|x| < \rho$, where ρ is defined by Eq. (9–21). The remainder R_n after terms of degree n satisfies the inequality (9–22), where $\phi(r)$ is defined by Eq. (9–23) and $n \geqq N$.*

This theorem is a consequence of Theorem 2 (Prob. 6 below).

EXAMPLE. $y'' = x^2y^2 + y' - 2$, $y = 1$ and $y' = 2$ for $x = 0$. We set $u = y - (1 + 2x)$. Then

$$u' = y' - 2, \qquad u'' = y'' = x^2(u + 1 + 2x)^2 + u'$$

or

$$u'' = u' + x^2 + 4x^3 + 2x^2u + 4x^4 + 4x^3u + x^2u^2.$$

We can choose $a = 0.1$, $b = 0.5$. Then $M_0 = 0.53$, $M = 0.53$, $c = 0.32$, $\rho = 0.096$. If r is chosen as 0.08, then $\phi(r) = 1.4$ and

$$|R_n| \leqq 1.4 \frac{|x|^{n+1}}{(0.08)^n(0.08 - |x|)} \qquad (|x| < 0.08).$$

Thus for $|x| = 0.04$, $|R_n| \leqq (1.4)2^{-n} < 2^{1-n}$. If five terms are retained, the error is less than $2^{-4} = 0.063$. From the original differential equation we find the solution to be

$$y = 1 + 2x + \frac{x^4}{12} + \frac{13}{60}x^5 + \cdots;$$

hence retention of the terms through x^5 provides an approximation to the solution with an error of less than 0.063 for $|x| = 0.04$ and, in fact, for $|x| \leqq 0.04$.

9–6 Convergence conditions for linear differential equations. For linear differential equations and linear systems we can give additional information on the radius of convergence of the series solution.

THEOREM 4. *Let the linear differential equation*

$$y^{(N)} + a_1(x)y^{(N-1)} + \cdots + a_N(x)y = Q(x) \qquad (9\text{--}30)$$

be given, where $a_1(x)$, ..., $a_N(x)$, $Q(x)$ can be expanded in Taylor series about $x = 0$, converging to these functions for $|x| < a$. Then, for arbitrary initial conditions at $x = 0$, each solution of Eq. (9–30) is also representable by its Taylor series for $|x| < a$.

Let the system of linear differential equations

$$\frac{dy_i}{dx} = \sum_{j=1}^{N} a_{ij}(x)y_j + Q_i(x) \qquad (i = 1, \ldots, N) \qquad (9\text{--}31)$$

be given, where the $a_{ij}(x)$ and $Q_i(x)$ can be expanded in Taylor series about $x = 0$, converging to these functions for $|x| < a$. Then, for arbitrary initial values at $x = 0$, each solution of Eq. (9–31) is of form $y_1 = f_1(x)$, ..., $y_N = f_N(x)$, where each of the functions is representable by its Taylor series for $|x| < a$.

This theorem is proved in Chapter 12. For an equation such as

$$y'' + (x^2 - 1)y' + x^3 y = e^x,$$

we can now conclude that the Taylor series solutions converge for *all* x. For the equation

$$y'' + \frac{y'}{1 - x} + y = x^3,$$

we can conclude that the solution with given initial values at $x = 0$ can be represented by the series solution for $|x| < 1$; indeed,

$$\frac{1}{1 - x} = 1 + x + x^2 + \cdots \qquad (|x| < 1).$$

For the equation

$$a_0(x)y'' + a_1(x)y' + a_2(x)y = Q(x), \qquad (9\text{–}32)$$

we must first divide by $a_0(x)$:

$$y'' + \frac{a_1(x)}{a_0(x)}\, y' + \frac{a_2(x)}{a_0(x)}\, y = \frac{Q(x)}{a_0(x)}\,. \qquad (9\text{–}33)$$

We must now consider the expansions of $a_1(x)/a_0(x)$, $a_2(x)/a_0(x)$, and $Q(x)/a_0(x)$ in powers of x. If $a_0(0) = 0$, no such series can be obtained and the theorem is inapplicable. If $a_0(0) \neq 0$ and all $a_j(x)$ and $Q(x)$ are expandable in Taylor series about $x = 0$, then it can be shown that $a_1(x)/a_0(x)$, ... can also be expanded in such series for $|x| < a$ and proper choice of a. In particular, these expansions are possible if the $a_j(x)$ and $Q(x)$ are *polynomials* and $a_0(0) \neq 0$. In this case, the number a can be chosen as $|\alpha + \beta i| = (\alpha^2 + \beta^2)^{1/2}$, where $\alpha + \beta i$ is the root (real or complex) of $a_0(x)$ nearest to the origin in the complex plane. The results described depend on the theory of analytic functions of a complex variable (see Chapter 9 of Reference 4 and Chapters 5 and 10 of Reference 8).

EXAMPLE. $(x^2 + 2x + 2)y'' + (x + 2)y' + y = 1 - x^2$. Here $a_0 = x^2 + 2x + 2$ and $a_0(x)$ has roots $-1 \pm i$, so that $a = (1 + 1)^{1/2} = \sqrt{2}$, for the series about $x = 0$. To expand about $x = -1$, we take a new origin on the x-axis at $x = -1$ and introduce a new independent variable $t = x + 1$. It can be verified that the equation in t has a coefficient a_0 with roots at $t = \pm i$; hence the number a is 1, which is precisely the distance from the new origin $x = -1$ to the old roots $x = -1 \pm i$.

When $a_0(0) = 0$, the equation (9–33) is meaningless at $x = 0$. We say that the equation (9–32) has a *singular point* at $x = 0$. Similarly, if $a_0(x_0) = 0$, the equation has a singular point at x_0. *Under certain conditions we can obtain power series solutions $\sum a_n(x - x_0)^n$, even though x_0 is a singular point.* This topic, important for applications, is considered in Sections 9–7 to 9–11.

When $a_0(x_0) \neq 0$, so that we can divide by $a_0(x)$ and apply Theorem 4, we say that x_0 is an *ordinary point* of the differential equation (9–32).

PROBLEMS

1. For the following first order differential equations verify that the conditions of Theorem 1 are satisfied for some a, b. Show, with the aid of the choices of a, b given, that the remainder satisfies the inequality stated.

(a) $y' = xy + 1$, $y = 0$ for $x = 0$; $a = 1$, $b = 1$, $|R_n| \leqq 3 \cdot 2^{-n}$ for $|x| \leqq 0.1$.

(b) $y' = xe^y + y$, $y = 0$ for $x = 0$; $a = 0.1$, $b = 0.5$, $|R_n| \leqq 3(2/3)^n$ for $|x| \leqq 0.06$.

(c) $y' = \dfrac{y}{10(1 - x)} + \dfrac{x^2}{10}$, $y = 0$ for $x = 0$; $a = 0.5$, $b = 0.5$, $|R_n| \leqq 4(3/4)^n$ for $|x| \leqq 0.3$.

2. Let the differential equation $y' = [(1 - x)(1 - y)]^{-1}$ be given, with $y = 0$ for $x = 0$.

(a) Find the solution explicitly and show that it is defined only for

$$|x| < 1 - e^{-1/2} = 0.39.$$

(b) Find the series solution, obtaining terms through x^4.

(c) Find R_4 for $x = 0.35$.

3. For the following second order equations obtain the series solution with given initial values, exhibiting terms through x^4, and (with the aid of Theorem 3) find how many terms are sufficient to obtain an error of less than 0.1 for $|x| \leqq 0.1$.

(a) $y'' = x + xy^2$, $y = 0$, $y' = 0$ for $x = 0$.

(b) $y'' = 0.01(xy' + y^3 - 1)$, $y = 1$, $y' = 3$ for $x = 0$.

4. For the following systems of first order equations show (with the aid of Theorem 2) that the series solutions, with given initial values, converge in the intervals stated.

(a) $dy/dx = x^2 + y^2 - z^2$, $dz/dx = xyz$, $y = 0$, $z = 0$ for $x = 0$, convergence at least for $|x| \leqq 0.15$.

(b) $dy/dx = xe^y - z$, $dz/dx = x - z$, $y = 1$, $z = 2$ for $x = 3$, convergence at least for $|x - 3| \leqq 0.002$.

5. For each of the following linear equations state whether or not the given initial value x_0 of x is a singular point. If x_0 is not a singular point, find, on the basis of Theorem 4, an interval in which the series solutions in powers of $x - x_0$ are convergent.

(a) $x^2y'' + (x - 1)y' + y = e^x$, $x_0 = 0$.
(b) $(1 + x^2)y'' + x^2y' + (x^4 - 1)y = x^3$, $x_0 = 0$.
(c) $(x^2 - 2x - 3)y'' + y' + xy = 1 - x$, $x_0 = 2$.
(d) $(x^2 - 2x - 3)y''' + x^2y'' + y = e^x$, $x_0 = -1$.
(e) $(x^2 + 4x + 5)y'' - (x + 1)y' + y = \sin x$, $x_0 = 0$.

6. Show that Theorem 3 is a consequence of Theorem 2. [*Hint:* Replace Eq. (9–28) by a system of first order equations, as in Section 1–7.]

ANSWERS

2. (a) $y = 1 - [1 + 2\log(1 - x)]^{1/2}$,
    (b) $x + x^2 + \frac{4}{3}x^3 + \frac{25}{12}x^4$,   (c) $R_4 = 0.067$.

3. (a) $\frac{1}{6}x^3 + 0 + \cdots$, eleven terms suffice;
    (b) $1 + 3x + 0.02x^3 + 0.0225x^4 + \cdots$, six terms suffice.

5. (a) Singular; (b) not singular, $|x| < 1$; (c) not singular, $|x - 2| < 1$;
(d) singular; (e) not singular, $|x| < \sqrt{5}$.

9–7 Series solutions for linear equations at a singular point. We now consider a homogeneous linear equation of second order:

$$a_0(x)y'' + a_1(x)y' + a_2(x)y = 0. \tag{9–34}$$

The results to be described can be extended to equations of higher order, to systems of linear differential equations, and to nonhomogeneous equations. For the general theory see References 2, 6, and 8 at the end of this chapter.

It will be assumed that the coefficients $a_0(x)$, $a_1(x)$, $a_2(x)$ in Eq. (9–34) are analytic in an interval $|x - x_0| < a$ and that $a_0(x)$ is not identically 0 for $|x - x_0| < a$. For simplicity, x_0 will be taken to be 0; a translation reduces the general case to this one. Thus it is assumed that

$$a_0(x) = \sum_{n=0}^{\infty} \alpha_n x^n, \qquad a_1(x) = \sum_{n=0}^{\infty} \beta_n x^n, \qquad a_2 = \sum_{n=0}^{\infty} \gamma_n x^n \tag{9–35}$$

for $|x| < a$. Finally it is assumed that $a_0(0) = 0$; that is, that the constant term α_0 is 0. Accordingly, $x = 0$ is a singular point of the equation.

Under the assumptions made, the existence theorem of Section 1–8 is not applicable at the initial value $x_0 = 0$ and we cannot expect to obtain solutions with arbitrary initial values of y and y' at $x = 0$. Indeed, there may be no solution defined at $x = 0$ other than the trivial solution $y \equiv 0$.

Instead of seeking solutions which satisfy conditions at $x = 0$, we seek solutions represented by series for $0 < |x| < b$, that is, solutions defined *near* 0 but not at 0. Occasionally, the solutions turn out to be valid also at $x = 0$. It will be seen that under certain conditions solutions can be found having the form

$$x^m \sum_{n=0}^{\infty} c_n x^n \qquad (0 < |x| < b). \tag{9-36}$$

If $m = 1$, the solution continues to be valid for $x = 0$. But if $m = -3$, for example, the solution is meaningless at $x = 0$.

Regular singular points. An equation such as

$$xy'' + (x^2 - x)y = 0 \tag{9-37}$$

has a singular point at $x = 0$, according to the above definition. However, the singular point can be removed by dividing by x:

$$y'' + (x - 1)y = 0. \tag{9-38}$$

(Equation (9–37) is equivalent to the *differential* equation (9–38) and the *algebraic* relation $x = 0$.) Such a singular point is called *removable.* In the subsequent discussion only singular points which are *not* removable will be considered.

Equation (9–34) is said to have a *regular singular point* at $x = 0$ if, after cancellation of the highest power of x which is a common factor of the coefficients, the equation takes one of the two forms

$$(\alpha_1 x + \alpha_2 x^2 + \cdots)y'' + (\beta_0 + \cdots)y'$$
$$+ (\gamma_0 + \cdots)y = 0 \qquad (\alpha_1 \neq 0), \tag{9-39a}$$

$$(\alpha_2 x^2 + \alpha_3 x^3 + \cdots)y'' + (\beta_1 x + \cdots)y'$$
$$+ (\gamma_0 + \cdots)y = 0 \qquad (\alpha_2 \neq 0). \tag{9-39b}$$

An equation of form (9–39a) or (9–39b) has at least one solution and possibly two linearly independent solutions of form (9–36).

EXAMPLE. $x^2 y'' + xy' + (x^3 - 1)y = 0$. Here $a_0 = x^2$, $a_1 = x$, so that (9–39b) applies and we have a regular singular point at $x = 0$. We set

$$y = x^m(c_0 + c_1 x + \cdots + c_n x^n + \cdots)$$

and try to choose m and the c's so that the differential equation is satisfied (with y not identically 0). On substituting in the equation and collecting terms, we find the relation

$$x^m[c_0(m^2 - 1) + c_1(m^2 + 2m)x + c_2(m^2 + 4m + 3)x^2$$
$$+ x^3\{c_0 + c_3(m^2 + 6m + 5)\} + \cdots$$
$$+ x^n\{c_{n-3} + c_n(\overline{m + n}^2 - 1)\} + \cdots] = 0.$$

For $x \neq 0$ we can ignore the factor x^m. Equating the coefficient of each power of x to 0, we obtain the relations

$$c_0(m^2 - 1) = 0,$$

$$c_1(m^2 + 2m) = 0,$$

$$c_2(m^2 + 4m + 3) = 0,$$

$$c_0 + c_3(m^2 + 6m + 8) = 0,$$

$$\vdots$$

$$c_{n-3} + c_n(\overline{m + n}^2 - 1) = 0 \qquad (n = 4, 5, \ldots).$$

The first equation gives $c_0 = 0$ or $m^2 - 1 = 0$. We can always assume $c_0 \neq 0$, for any other case can be reduced to this one by factoring out a suitable power of x and hence modifying the value of m. If $c_0 \neq 0$, then $m^2 - 1 = 0$; hence $m = -1$ or $m = 1$. For $m = -1$ we have

$$-c_1 = 0, \qquad 0 \cdot c_2 = 0, \qquad c_0 + 3c_3 = 0,$$

$$c_{n-3} + (n^2 - 2n)c_n = 0,$$

for $n = 4, 5, \ldots$; hence we have the recursion formula

$$c_n = -\frac{c_{n-3}}{n^2 - 2n} = -\frac{c_{n-3}}{n(n - 2)}.$$

From this we conclude that $c_1 = 0, c_4 = 0, c_7 = 0, \ldots$ If n is a multiple of 3, then

$$c_n = \frac{-1}{n(n - 2)} \cdot \frac{-1}{(n - 3)(n - 5)} \cdots \frac{-1}{3 \cdot 1} \cdot c_0.$$

If n is of form $3s + 2$, then

$$c_n = \frac{-1}{n(n - 2)} \frac{-1}{(n - 3)(n - 5)} \cdots \frac{-1}{5 \cdot 3} c_2.$$

Thus c_0 and c_2 are arbitrary and the solutions found can be written

$$y = x^{-1} \left[c_0 \left(1 + \sum_{s=1}^{\infty} (-1)^s x^{3s} \frac{1}{(1 \cdot 3)(4 \cdot 6) \cdots (3s - 2)(3s)} \right) \right.$$

$$\left. + c_2 \left(x^2 + x^2 \sum_{s=1}^{\infty} (-1)^s x^{3s} \frac{1}{(3 \cdot 5)(6 \cdot 8) \cdots (3s)(3s + 2)} \right) \right].$$

This equation is clearly of the form $y = c_0 y_1(x) + c_2 y_2(x)$, and we can verify that y_1, y_2 are linearly independent, so that y is the general solution.

It follows from the ratio test that the series converge for all x, so that the solutions are valid for all x except 0 (because of the factor x^{-1}). We note that

$$y_2(x) = x + x \sum_{s=1}^{\infty} (-1)^s x^{3s} \frac{1}{(3 \cdot 5) \cdot (6 \cdot 8) \cdots (3s)(3s + 2)},$$

and this solution is valid for all x.

In our analysis we were led to the equation $m^2 - 1 = 0$ and we chose only the root $m = -1$. If we choose the other root, $m = 1$, we obtain a solution $x \sum c_n x^n$, which is found to be a constant times $y_2(x)$.

9–8 Indicial equation. Equations with a two-term recursion formula. The method of the example of Section 9–7 can be applied to an arbitrary equation (9–34) with a regular singular point at $x = 0$. The substitution $y = x^m \sum c_n x^n$ leads to relations between the c's, of which the first has form

$$c_0 f(m) = 0,$$

where $f(m)$ is a polynomial of *second* degree in m. The equation $f(m) = 0$ is called the *indicial* equation. If m_1, m_2 are the roots of this equation, then by choosing $m = m_1$ or $m = m_2$ we obtain two sets of constants c_n and hence two series solutions $y_1(x)$, $y_2(x)$. The later constants are determined from the earlier ones by a recursion formula, which may involve two or more of the c's.

If $m_2 = m_1$, the method provides only one solution (up to a constant factor); if $m_2 \neq m_1$, but $m_2 - m_1$ is a positive integer, then the constants c_n that correspond to $m = m_1$ may fail to be defined. Thus complications can arise in applying the method. It may also happen that m_1, m_2 are complex; we then obtain complex solutions, from which the real solutions can be obtained as in Section 4–8.

The difficulties mentioned all arise for the equations with polynomial coefficients and with a two-term recursion formula. These equations include many which are very important for applications. Hence we shall concentrate attention on these equations, for which we can formulate precise results.

THEOREM 5. *Let L be the differential operator*

$$a_0(x)D^2 + a_1(x)D + a_2(x),$$

where $a_0(x)$, $a_1(x)$, $a_2(x)$ are polynomials without common factor. Let the differential equation $Ly = 0$ have a regular singular point at $x = 0$. Let

$$L(x^n) = f(n)x^{n+h} + g(n)x^{n+h+k} \qquad (k > 0), \qquad (9\text{–}40)$$

where $f(n) \not\equiv 0$. Then $h = 0$ or -1, $f(n)$ is a polynomial in n of second degree, and $g(n)$ is a polynomial in n of degree at most 2. The substitution

$$y = x^m \sum_{n=0}^{\infty} c_n x^n \qquad (9\text{-}41)$$

in the differential equation leads to the indicial equation $f(m) = 0$ and to a two-term recursion formula for the c_n. If m_1, m_2 are the roots of the indicial equation and $(m_1 - m_2)/k$ is not an integer or zero, then two linearly independent solutions $y_1(x)$, $y_2(x)$ are given by

$$y_1(x) = \phi(x, m_1), \qquad y_2(x) = \phi(x, m_2), \qquad (9\text{-}42)$$

where

$$\phi(x, m) = x^m \left\{ 1 + \sum_{s=1}^{\infty} (-1)^s x^{ks} \frac{g(m)g(m + k) \cdots g[m + (s - 1)k]}{f(m + k)f(m + 2k) \cdots f(m + sk)} \right\}.$$

$$(9\text{-}43)$$

If $g(n)$ is of degree 0 or 1, the solutions are valid for all x except perhaps $x = 0$. If $g(n)$ is of degree 2, then the solutions are valid for $0 < |x| < a$, where

$$a = \left| \frac{f_0}{g_0} \right|^{1/k}, \qquad f(n) = f_0 n^2 + \cdots, \qquad g(n) = g_0 n^2 + \cdots$$

Proof. Since $x = 0$ is a regular singular point, either (9–39a) or (9–39b) holds. If the former holds, then

$$L(x^n) = n(n - 1)(\alpha_1 x + \alpha_2 x^2 + \cdots)x^{n-2} + n(\beta_0 + \cdots)x^{n-1}$$
$$+ (\gamma_0 + \cdots)x^n$$
$$= x^{n-1}[\alpha_1 n^2 + (\beta_0 - \alpha_1)n]$$
$$+ x^n[\alpha_2 n^2 + (\beta_1 - \alpha_2)n + \gamma_0] + \cdots$$

Since $\alpha_1 \neq 0$, the first bracket is $f(n)$, not identical with 0, a polynomial of second degree. By assumption only one other power of x appears. This must be x^{n-1+k}, where $k > 0$, and the corresponding term is

$$x^{n-1+k}[\alpha_{k+1} n^2 + (\beta_k - \alpha_{k+1})n + \gamma_{k-1}] = g(n)x^{n-1+k}.$$

Thus $g(n)$ has the form described and $h = -1$. When (9–39b) holds, we find in the same way that $h = 0$, and

$$f(n) = \alpha_2 n^2 + (\beta_1 - \alpha_2)n + \gamma_0,$$
$$g(n) = \alpha_{k+2} n^2 + (\beta_{k+1} - \alpha_{k+2})n + \gamma_k.$$

If y is given by (9–41), then

$$Ly = L\left(\sum_{n=0}^{\infty} c_n x^{m+n}\right) = \sum_{n=0}^{\infty} c_n L(x^{m+n})$$

$$= \sum_{n=0}^{\infty} c_n[f(m+n)x^{m+n+h} + g(m+n)x^{m+n+h+k}]$$

$$= x^{m+h}\left\{\sum_{n=0}^{\infty} c_n f(m+n)x^n + \sum_{n=0}^{\infty} c_n g(m+n)x^{n+k}\right\}$$

$$= x^{m+h}\left\{c_0 f(m) + c_1 f(m+1)x + \cdots + c_{k-1}f(m+k-1)x^{k-1}\right.$$

$$\left. + \sum_{n=k}^{\infty} [c_n f(m+n) + c_{n-k}g(m+n-k)]x^n\right\}.$$

Hence $f(m) = 0$ is the indicial equation and there is a two-term recursion formula which relates c_n and c_{n-k}. From the form of the recursion formula it appears that we will obtain independent solutions containing only the powers x^{m+sk} $(s = 0, 1, 2, \ldots)$. It is simpler, however, to take y in a corresponding form to start with:

$$y = x^m \sum_{s=0}^{\infty} c_s x^{ks}. \tag{9–44}$$

We can also take $c_0 = 1$ and later insert an arbitrary constant as factor. We now find as before

$$Ly = L\left(\sum_{s=0}^{\infty} c_s x^{ks+m}\right)$$

$$= \sum_{s=0}^{\infty} c_s[f(m+ks)x^{m+ks+h} + g(m+ks)x^{m+k(s+1)+h}]$$

$$= x^{m+h}\left(f(m) + \sum_{s=1}^{\infty} \{c_s f(m+ks) + c_{s-1}g[m+k(s-1)]\}x^{ks}\right).$$

Now the recursion formula is

$$c_s = -\frac{g[m+k(s-1)]}{f(m+ks)}c_{s-1}. \tag{9–45}$$

From this relation and the condition $c_0 = 1$, we find

$$c_s = (-1)^s \frac{g[m+k(s-1)] \cdots g(m+k)g(m)}{f(m+sk) \cdots f(m+2k)f(m+k)} \qquad (s = 1, 2, \ldots).$$

$$\tag{9–46}$$

If the c_s are so chosen, then y becomes the function $\phi(x, m)$ defined by Eq. (9–43), and

$$Ly = x^{m+h}f(m). \tag{9–47}$$

Now if $m = m_1$, where $f(m_1) = 0$, then y becomes $y_1 = \phi(x, m_1)$:

$$y_1 = x^{m_1} \left\{ 1 + \sum_{s=1}^{\infty} (-1)^s x^{ks} \frac{g(m_1) \cdots g[m_1 + (s-1)k]}{f(m_1 + k) \cdots f(m_1 + sk)} \right\}.$$

Equation (9–47) shows that $L(y_1) = 0$. Similarly, if $m = m_2$, then y becomes $y_2 = \phi(x, m_2)$ and $L(y_2) = 0$.

We have tacitly assumed that the constants c_s are well defined by Eq. (9–46) when $m = m_1$ or $m = m_2$. Difficulty can arise only if there is a zero in the denominator, that is, if $f(m + sk) = 0$ for some choice of s and $m = m_1$ or $m = m_2$. But $f(m_1 + sk) = 0$ can only mean $m_1 + sk = m_2$, that is, that $(m_2 - m_1)/k$ is an integer s. This is ruled out by assumption; similarly, $f(m_2 + sk) = 0$ is ruled out. Hence $\phi(x, m_1)$ and $\phi(x, m_2)$ are well defined.

Application of the ratio test shows that the series converge as stated in Theorem 5, so that $y_1(x)$ and $y_2(x)$ are solutions in the intervals given. The two solutions are linearly independent, for otherwise one would be a constant times the other; this is impossible since $m_1 \neq m_2$ (Prob. 13 below). Hence $c_1 y_1(x) + c_2 y_2(x)$ provides the general solution in the interval stated.

EXAMPLE. $2x^2 y'' + (x^2 - x)y' + y = 0$. There is a regular singular point at $x = 0$. We find

$$L(x^n) = x^n(2n^2 - 3n + 1) + nx^{n+1}.$$

Hence

$$f(n) = 2n^2 - 3n + 1 = (2n - 1)(n - 1), \quad g(n) = n, \quad k = 1, \quad h = 0,$$

$$\phi(x, m)$$

$$= x^m \left[1 + \sum_{s=1}^{\infty} (-1)^s x^s \frac{m(m+1) \cdots (m+s-1)}{(2m+1)(m) \cdots (2m+2s-1)(m+s-1)} \right].$$

The indicial equation has roots $\frac{1}{2}$, 1; their difference is not an integral multiple of k. Hence we obtain two solutions:

$$y_1 = \phi(x, \tfrac{1}{2}) = x^{1/2} \left[1 + \sum_{s=1}^{\infty} (-1)^s x^s \frac{\frac{1}{2} \cdot \frac{3}{2} \cdots [(2s-1)/2]}{(2 \cdot \frac{1}{2})(4 \cdot \frac{3}{2}) \cdots (2s)[(2s-1)/2]} \right]$$

$$= x^{1/2} \left[1 + \sum_{s=1}^{\infty} (-1)^s x^s \frac{1}{2^s s!} \right],$$

$$y_2 = \phi(x, 1) = x\left[1 + \sum_{s=1}^{\infty} (-1)^s x^s \frac{1 \cdot 2 \cdots s}{(3 \cdot 1)(5 \cdot 2) \cdots (2s + 1)s}\right]$$

$$= x\left[1 + \sum_{s=1}^{\infty} (-1)^s x^s \frac{1}{3 \cdot 5 \cdots (2s + 1)}\right].$$

Since $g(n)$ is of degree 1, the series converge for all x and the general solution is

$$y = c_1 y_1(x) + c_2 y_2(x)$$

for all x except 0. We note that $y_2(x)$ is also valid as a solution at $x = 0$, and that $y_1(x)$ is imaginary for $x < 0$, because of the square root. A corresponding real solution for $x < 0$ is a constant multiple of $y_1(x)$,

$$y(x) = i y_1(x) = i x^{1/2}(1 + \cdots)$$
$$= \sqrt{-1}\,\sqrt{x}\,(1 + \cdots) = \sqrt{-x}\,(1 + \cdots).$$

Remark 1. If $L(x^n)$ involves more than two powers of x, then the reasoning of the above proof shows that the recursion formula will relate more than two of the c's. Hence the assumption of a two-term recursion formula forces $L(x^n)$ to have the form shown. This in turn implies that the coefficients $a_0(x)$, $a_1(x)$, $a_2(x)$ are of special form (Prob. 6 below). It can happen that $L(x^n)$ involves only *one* power of x. In this case, the result of Prob. 6 implies that the differential equation can be written

$$\alpha_2 x^2 y'' + \beta_1 xy + \gamma_0 y = 0,$$

where α_2, β_1, γ_0 are constants. This is a Cauchy equation, discussed in Prob. 9 following Section 7–5. As shown there, $L(x^n) = f(n)x^n$ and, if $f(n)$ has distinct roots m_1, m_2, $y_1 = x^{m_1}$ and $y_2 = x^{m_2}$ are linearly independent solutions.

Remark 2. If the differential equation has an *ordinary* point at $x = 0$ and $L(x^n)$ involves only two powers of x, then the results of Theorem 5 are still applicable. Since $a_0(0) \neq 0$, we find that $f(n)$ is $\alpha_0 n(n - 1)$, so that the indicial equation has roots 0, 1. A detailed analysis is given in Prob. 10 following Section 9–3.

PROBLEMS

1. For each of the following differential equations determine all singular points and determine whether each is removable, regular, or irregular:
   (a) $x^2 y'' + y' + x^2 y = 0$,
   (b) $x^2 y'' + x^3 y' + (x^4 - x^2)y = 0$,
   (c) $(x^2 - x)y'' + (x + 1)y' + y = 0$,
   (d) $(x^2 - 2x - 3)^2 y'' + (x - 3)y' + y = 0$.

2. Verify that each of the following equations has a regular singular point at $x = 0$, show that the conditions of Theorem 5 are satisfied, and find two linearly independent solutions near $x = 0$:

  (a) $xy'' + (x^3 - 1)y' + 3x^2y = 0,$
  (b) $x^2y'' + x^3y' - 2y = 0,$
  (c) $xy'' - y' + x^2y = 0,$
  (d) $(3x^2 + 3x^3)y'' - (x + 6x^2)y' + y = 0,$
  (e) $2xy'' + (3 + 2x)y' - 2y = 0.$

3. Let $a + bi$ be a complex constant. We define the function $y = x^{a+bi}$ for $x > 0$ as follows:

$$y = x^{a+bi} = x^a x^{bi} = x^a e^{bi \log x} = x^a [\cos (b \log x) + i \sin (b \log x)].$$

Show that $y' = (a + bi)x^{a+bi-1}$ (see Section 4–8). The definition and conclusion can be extended to negative x by the rule

$$y = x^{a+bi} = (-|x|)^{a+bi} = (-1)^{a+bi}|x|^{a+bi} = \text{const} \cdot |x|^{a+bi}$$

$$= \text{const} \cdot |x|^a [\cos (b \log |x|) + i \sin (b \log |x|)] \qquad (x < 0).$$

The constant has, in general, one of an infinite set of values (Reference 4, pp. 529–530), but we can choose the particular value

$$(-1)^{a+bi} = e^{-b\pi} (\cos a\pi + i \sin a\pi).$$

4. Let the following differential equation be given:

$$x^2y'' + (x + x^2)y' + y = 0.$$

  (a) Show that there is a regular singular point at $x = 0$ and that the roots of the indicial equation are $\pm i$.
  (b) Use the result of Prob. 3 to obtain complex solutions valid for $|x| > 0$.
  (c) Write out the series for $m = i$ of part (b), as far as the term in x^3. Then take real and imaginary parts to obtain linearly independent real solutions for $|x| > 0$.

5. Show that each of the following equations has an irregular singular point at $x = 0$. Attempt to find a solution of form $x^m\sum_{n=0}^{\infty} c_n x^n$ and explain the difficulties which arise.

  (a) $x^3y'' + (x + 2x^2)y' - (1 + x)y = 0.$
  (b) $x^3y'' + x^2y' + (1 + x)y = 0.$

6. Let the equation $L(y) = (\alpha_0 + \alpha_1 x + \cdots)y'' + (\beta_0 + \beta_1 x + \cdots)y' + (\gamma_0 + \gamma_1 x + \cdots)y = 0$ be given. Let $L(x^n) = f(n)x^{n+h} + g(n)x^{n+k+h}$, $k \geq 0$. Show that $\alpha_{j+2}, \beta_{j+1}, \gamma_j$ are 0 for $j \neq h$, $j \neq k + h$.

7. An equation $a_0(x)y'' + a_1(x)y' + a_2(x)y = 0$, with polynomial coefficients, is said to have a *regular singular point at $x = \infty$* if the substitution

$t = 1/x$ leads to an equation with a regular singular point at $t = 0$. Show that each of the following differential equations has a regular singular point at $x = \infty$:

  (a) $x^5y'' + x^4y' + (1 - x^3)y = 0$,   (b) $2x^3y'' + (2x + x^2)y' + 2y = 0$.

8. For an equation $L(y) = 0$ with a regular singular point at infinity (Prob. 7), we can obtain series solutions by setting $t = 1/x$ and obtaining series in terms of t. We can also proceed directly as follows (in case of a two-term recursion formula). Let

$$L(x^n) = f(n)x^{n+h-k} + g(n)x^{n+h}   (k > 0).$$

The fact that the singular point at infinity is regular forces $g(n)$ to be of second degree, and the equation $g(m) = 0$ becomes the indicial equation, with roots m_1, m_2. We then seek solutions

$$y = x^m\left[1 + \sum_{s=1}^{\infty} (-1)^s c_s x^{-ks}\right].$$

If $(m_2 - m_1)/k$ is not an integer or 0, show that two linearly independent solutions for $|x| > a$ are given by

$$y_1 = \Phi(x, m_1),   y_2 = \Phi(x, m_2),$$

where

$$\Phi(x, m) = x^m\left\{1 + \sum_{s=1}^{\infty} (-1)^s x^{-ks} \frac{f(m) \cdots f[m - (s-1)k]}{g(m-k) \cdots g(m-sk)}\right\}.$$

9. Apply the method of Prob. 8 to obtain linearly independent solutions for large $|x|$ of the equations (a) and (b) of Prob. 7.

10. *The Legendre equation* is given by

$$(1 - x^2)y'' - 2xy' + N(N+1)y = 0,$$

where N is a constant.

(a) Show that $x = 0$ is an ordinary point, so that the indicial equation has roots 0, 1. Apply the method of Theorem 5 to obtain the solutions for $|x| < 1$:

$$y_1 = 1 + \sum_{s=1}^{\infty} (-1)^s x^{2s}$$

$$\times \frac{\{N(N-2) \cdots (N-2s+2)\}\{(N+1)(N+3) \cdots (N+2s-1)\}}{(2s)!},$$

$$y_2 = x\left(1 + \sum_{s=1}^{\infty} (-1)^s x^{2s}\right.$$

$$\left. \times \frac{\{(N-1)(N-3) \cdots (N-2s+1)\}\{(N+2)(N+4) \cdots (N+2s)\}}{(2s+1)!}\right).$$

(b) Show that when N is a positive even integer or 0, the solution y_1 of part (a) reduces to a polynomial of degree N. [For $N > 0$ this polynomial is

$$(-1)^{N/2} \frac{2 \cdot 4 \cdots N}{1 \cdot 3 \cdots (N-1)} P_N(x),$$

where $P_N(x)$ is the *Legendre polynomial* of even degree N. For $N = 0$, $y_1(x) \equiv P_0(x) \equiv 1$.]

(c) Show that when N is a positive odd integer, the solution y_2 of part (a) reduces to a polynomial of degree N. [For $N > 0$ this polynomial is

$$(-1)^{(N-1)/2} \frac{2 \cdot 4 \cdots (N-1)}{1 \cdot 3 \cdots N} P_N(x),$$

where $P_N(x)$ is the *Legendre polynomial* of odd degree N. For $N = 1$, $y_1(x) \equiv P_1(x) \equiv x$.]

(d) Show that the differential equation has a regular singular point at infinity (Probs. 7, 8, 9) and that the indicial equation has roots N, $-N - 1$. Assume that $N + \frac{1}{2}$ is not an integer and obtain the linearly independent solutions for $|x| > 1$:

$$y_1 = x^N \left(1 + \sum_{s=1}^{\infty} (-1)^s x^{-2s} \frac{N(N-1) \cdots (N-2s+2)(N-2s+1)}{2^s s! (2N-1) \cdots (2N+1-2s)}\right),$$

$$y_2 = x^{-N-1} \left(1 + \sum_{s=1}^{\infty} x^{-2s} \frac{(N+1)(N+2) \cdots (N+2s-1)(N+2s)}{2^s s! (2N+3) \cdots (2N+2s+1)}\right).$$

(e) Show that when N is a positive integer or 0, the solution $y_1(x)$ of part (d) remains valid and is a polynomial. [For $N > 0$ this polynomial is

$$\frac{N!}{1 \cdot 3 \cdots (2N-1)} P_N(x),$$

where $P_N(x)$ is the Legendre polynomial of degree N. For $N = 0, y_1(x) \equiv P_0(x) \equiv 1$.]

11. *The hypergeometric equation* is given by

$$x(1-x)y'' + [c - (a+b+1)x]y' - aby = 0,$$

where a, b, c are constants.

(a) Show that the equation has a regular singular point at $x = 0$ and that the indicial equation has roots 0, $1 - c$.

(b) Show that if c is not a negative integer or 0, then one solution is given by

$$y_1(x) = 1 + \frac{a \cdot b}{1 \cdot c} x + \frac{a(a+1)}{2!} \frac{b(b+1)}{c(c+1)} x^2 + \cdots \qquad (|x| < 1).$$

This function is denoted by $F(a, b, c; x)$ and its series is termed the *hypergeometric series*. (When $a = 1$ and $b = c$, the series reduces to the *geometric* series $\sum x^n$).

(c) Show that if c is not an integer, then two linearly independent solutions for $0 < |x| < 1$ are given by $y_1 = F(a, b, c; x)$ and

$$y_2 = x^{1-c}F(a - c + 1, b - c + 1, 2 - c; x).$$

12. *The Bessel equation* is given by

$$x^2 y'' + xy' + (x^2 - N^2)y = 0,$$

where N is a constant.

(a) Show that there is a regular singular point at $x = 0$ and, for N not an integer, obtain the linearly independent solutions which are valid for $|x| > 0$,

$$y_1(x) = x^N \left(1 + \sum_{s=1}^{\infty} (-1)^s \frac{x^{2s}}{2^{2s}s!(N + 1) \cdots (N + s)}\right),$$

$$y_2(x) = x^{-N} \left(1 + \sum_{s=1}^{\infty} (-1)^s \frac{x^{2s}}{2^{2s}s!(1 - N) \cdots (s - N)}\right).$$

(b) Show that the solution $y_1(x)$ of part (a) is valid for $|x| > 0$ provided N is not a negative integer. [This function is

$$2^N \Gamma(N + 1)J_N(x),$$

where $J_N(x)$ is the *Bessel function of first kind* of order N and $\Gamma(x)$ is the Gamma function (p. 381 of Reference 4). When N is a positive integer or 0, $y_1(x)$ is $2^N N! J_N(x)$.]

(c) Show that the Bessel equation has an irregular singular point at $x = \infty$.

13. Prove that the two solutions $y_1(x)$, $y_2(x)$ of Theorem 5 are linearly independent. [*Hint:* If they were linearly dependent, we would have an identity of form

$$x^{m_1}[1 + p(x)] = cx^{m_2}[1 + q(x)],$$

where $p(x)$ and $q(x)$ are continuous at $x = 0$ and $c \neq 0$. Divide by the function on the right-hand side and let $x \to 0$ to conclude that $x^{m_1-m_2} \to c$ as $x \to 0$; show that this implies $m_1 = m_2$.]

ANSWERS

1. (a) $x = 0$, irregular; (b) $x = 0$, removable; (c) $x = 0$, regular and $x = 1$, regular; (d) $x = 3$, regular, $x = -1$, irregular.

2. (All summations, except in part (e), are with respect to s, from 1 to ∞).
(a) $1 + \sum [(-1)^s x^{3s}/\{1 \cdot 4 \cdots (3s - 2)\}]$,
$x^2(1 + \sum [(-1)^s x^{3s}/\{3^s s!\}])$, all x;

(b) for $x \neq 0$, $x^{-1}(1 + \sum[\,(-1)^s x^{2s}/\{2^s s!\}])$;
for all x, $x^2(1 + \sum[\,(-1)^s x^{2s}/\{5 \cdot 7 \cdots (2s + 3)\}])$;

(c) $1 + \sum[\,(-1)^s x^{3s}/\{3^s s!1 \cdot 4 \cdots (3s - 2)\}]$, all x,
and $x^2(1 + \sum[\,(-1)^s x^{3s}/\{3^s s!5 \cdot 8 \cdots (3s + 2)\}])$, all x;

(d) $x + \frac{6}{5}x^2 + \frac{9}{20}x^3$, all x, and $x^{1/3}(1 + \sum[\,(-1)^s x^s\,(-8)\,(-5) \cdots$
$$(3s - 11)/\{3^s s!\}]), \quad 0 < |x| < 1;$$

(e) $1 + \frac{2}{3}x$, all x, and $x^{-1/2}(1 + 3x + \frac{1}{2}x^2$
$$+ 3\sum_{s=3}^{\infty}[\,(-1)^s x^s/\{s!(2s - 3)(2s - 1)\}]), \quad |x| > 0.$$

4. (b) $(\cos \log |x| \pm i \sin \log |x|)(1 + \sum_{s=1}^{\infty}(-1)^s x^s\,(\pm i)(1 \pm i) \cdots$
$$(s - 1 \pm i)]/\{(1 \pm 2i) \cdots (s^2 \pm 2is)\}]), \quad x \neq 0;$$

(c) $\cos(\log |x|)[1 - \frac{2}{5}x + \frac{1}{10}x^2 - (17/780)x^3 + \cdots]$
$$+ \sin(\log |x|)[\frac{1}{5}x - \frac{1}{20}x^2 + (6/780)x^3 + \cdots],$$
$\sin(\log |x|)[1 - \frac{2}{5}x + \frac{1}{10}x^2 - (17/780)x^3 + \cdots]$
$$- \cos(\log |x|)[\frac{1}{5}x - \frac{1}{20}x^2 + (6/780)x^3 + \cdots].$$

5. (a) Indicial equation has only one root and series diverges for $x \neq 0$,
(b) indicial equation has no roots.

9. (Summations are over s, from 1 to ∞).

(a) $x(1 + \sum[\,(-1)^s x^{-3s}/\{3^s s!1 \cdot 4 \cdots (3s - 2)\}])$,
$x^{-1}(1 + \sum[\,(-1)^s x^{-3s}/\{3^s s!5 \cdot 8 \cdots (3s + 2)\}])$, $x \neq 0$;

(b) $1 - [2/(3x)]$, $x^{1/2}(1 + \sum[\,(-1)^s x^{-s}3 \cdot 1 \cdots (5 - 2s)/$
$$\{s!1 \cdot 3 \cdots (2s - 1)\}]), \quad x \neq 0.$$

9–9 Indicial equation with equal roots. When the indicial equation has equal roots $m_1 = m_2$, the formulas of Theorem 5 provide only one solution $y_1 = \phi(x, m_1)$. This solution is well defined, since $f(m_1 + ks)$ is never 0 for $s = 1, 2, \ldots$ We now seek a second solution.

THEOREM 6. *Let all the hypotheses of Theorem 5 hold, except that m_2 is to equal m_1. Then two linearly independent solutions of the differential equation $Ly = 0$ are given by*

$$y_1 = \phi(x, m_1), \qquad y_2 = \phi_1(x, m_1), \qquad (9\text{--}48)$$

where

$$\phi_1(x, m) = \frac{\partial \phi}{\partial m} = \phi(x, m) \log x$$

$$+ x^m \sum_{s=1}^{\infty} (-1)^s x^{ks} \frac{g(m) \cdots g[m + (s - 1)k]}{f(m + k) \cdots f(m + sk)} \left\{ \frac{g'(m)}{g(m)} + \cdots \right.$$

$$+ \frac{g'[m + (s - 1)k]}{g[m + (s - 1)k]} - \frac{f'(m + k)}{f(m + k)} - \cdots - \frac{f'(m + sk)}{f(m + sk)} \right\}. \quad (9\text{--}49)$$

The solutions are valid over the same intervals as in Theorem 5.

Proof. From the definition of $\phi(x, m)$ and the proof of Theorem 5 (Eq. 9–47) we know that

$$L[\phi(x, m)] = x^{m+h}f(m), \qquad (9\text{–}50)$$

so that $Ly_1 = x^{m_1+h}f(m_1) = 0$. Since m_1 is a double root of $f(m)$, $f'(m_1) = 0$. Now

$$L\left[\frac{\partial \phi}{\partial m}\right] = \frac{\partial}{\partial m} L[\phi]. \qquad (9\text{–}51)$$

This is equivalent to asserting that the order of differentiation with respect to x and m can be interchanged; such an interchange is permitted (see p. 112 of Reference 4) if all derivatives concerned are continuous. Since ϕ and its derivatives are defined by infinite series, we must verify uniform convergence of these series (pp. 338–348 of Reference 4); this can be carried out by the M-test. From Eqs. (9–50) and (9–51) we now conclude that

$$L[\phi_1] = L\left[\frac{\partial \phi}{\partial m}\right] = \frac{\partial}{\partial m}\left[x^{m+h}f(m)\right] = x^{m+h}\log x\, f(m) + x^{m+h}f'(m).$$

Now if $m = m_1$, then $f(m_1) = 0$, $f'(m_1) = 0$, so that $L[\phi_1] = 0$ and $\phi_1(x, m_1)$ is a solution. The series converge over the same intervals as in Theorem 5, so that the solutions are valid over these intervals. The two solutions $y_1(x) = \phi(x, m_1)$, $y_2(x) = \phi_1(x, m_1)$ are clearly linearly independent, so that $y = c_1y_1(x) + c_2y_2(x)$ provides the general solution over the intervals considered. The differentiation of $\phi(x, m)$ in Eq. (9–49) is carried out with the aid of the following rule of calculus (pp. 19, 28 of Reference 4):

$$\left(\frac{u_1 \ldots u_n}{v_1 \ldots v_n}\right)' = \left(\frac{u_1 \ldots u_n}{v_1 \ldots v_n}\right)\left[\frac{u_1'}{u_1} + \cdots + \frac{u_n'}{u_n} - \frac{v_1'}{v_1} - \cdots - \frac{v_n'}{v_n}\right].$$

Remark. The function $\log x$ is imaginary when x is negative. We can verify that replacement of $\log x$ by $\log(-x)$ in Eq. (9–49) leads to a solution valid for negative x. We can include both cases by replacing $\log x$ by $\log |x|$ in Eq. (9–49).

EXAMPLE. $x^2y'' + (2x^3 - x)y' + (1 + x^2)y = 0$.   Here $L(x^n) = (n - 1)^2x^n + (2n + 1)x^{n+2}$, so that $f(n) = (n - 1)^2$, $g(n) = 2n + 1$, $k = 2$. Accordingly,

$$\phi(x, m) = x^m\left[1 + \sum_{s=1}^{\infty} (-1)^s x^{2s}\right.$$

$$\left. \times \frac{(2m + 1)(2m + 5) \cdots (2m + 4s - 3)}{(m + 1)^2(m + 3)^2 \cdots (m + 2s - 1)^2}\right],$$

$$\phi_1(x, m) = \frac{\partial \phi}{\partial m} = \phi(x, m) \log x$$

$$+ x^m \sum_{s=1}^{\infty} (-1)^s x^{2s} \frac{(2m + 1) \cdots (2m + 4s - 3)}{(m + 1)^2 \cdots (m + 2s - 1)^2}$$

$$\cdot \left(\frac{2}{2m + 1} + \cdots + \frac{2}{2m + 4s - 3} - \frac{2}{m + 1} - \cdots \right.$$

$$\left. - \frac{2}{m + 2s - 1} \right),$$

$$y_1 = \phi(x, 1) = x \left[1 + \sum_{s=1}^{\infty} (-1)^s x^{2s} \frac{3 \cdot 7 \cdots (4s - 1)}{2^2 4^2 \cdots (2s)^2} \right],$$

$$y_2 = \phi_1(x, 1) = y_1(x) \log x + x \left[\sum_{s=1}^{\infty} (-1)^s x^{2s} \frac{3 \cdot 7 \cdots (4s - 1)}{2^2 4^2 \cdots (2s)^2} \right.$$

$$\left. \cdot \left(\frac{2}{3} + \cdots + \frac{2}{4s - 1} - \frac{2}{2} - \cdots - \frac{2}{2s} \right) \right].$$

Since $g(n)$ has degree 1, the solutions are valid for all x except 0; for $x < 0$, $\log x$ should be replaced by $\log |x|$. We can verify convergence of the series for y_1 by the ratio test. For the second series in the expression for y_2 we can use a combination of the comparison and ratio tests. First the terms are increased in absolute value by replacing the quantity in the final parenthesis by

$$b_s = 2 \left(\frac{1}{2} + \frac{1}{3} + \cdots + \frac{1}{s} \right).$$

Then application of the ratio test proceeds as for the series for y_1, since $b_s \to \infty$ as $s \to \infty$, so that

$$\frac{b_{s+1}}{b_s} = 1 + \frac{2}{(s + 1)b_s} \to 1.$$

9–10 Indicial equation with roots differing by multiple of k. If $m_2 = m_1 + Nk$, where N is a positive integer, then the series for $\phi(x, m_1)$ (Section 9–8) has a zero in the denominator and the solution $y_1(x) = \phi(x, m_1)$ fails to be defined. However, the series for $y_2(x) = \phi(x, m_2)$ will then be well defined, and we obtain at least the solution $y_2(x)$.

The difficulty with regard to $\phi(x, m_1)$ may dispose of itself. This will be the case *if $g(n)$ has a root of form $m_1 + \nu k$*, where ν is 0 or a positive integer and $0 \leq \nu \leq N - 1$, for then

$$\phi(x, m_1) = x^{m_1} \left\{ 1 + \sum_{s=1}^{\infty} (-1)^s x^{ks} \right.$$

$$\left. \times \frac{g(m_1) \cdots g(m_1 + \nu k) \cdots g[m_1 + (s - 1)k]}{f(m_1 + k) \cdots f(m_1 + Nk) \cdots f(m_1 + sk)} \right\}.$$

Thus a zero appears in the numerator at $s = \nu + 1$ to cancel the zero in the denominator at $s = N$. More precisely, $g(n)$ has a factor $n - (m_1 + \nu k)$, so that $g(m + \nu k)$ has a factor $m - m_1$; similarly, $f(n)$ has a factor $n - (m_1 + Nk)$, so that $f(m + Nk)$ has a factor $m - m_1$. In the series for $\phi(x, m)$, the factor $m - m_1$ appears in both numerator and denominator in all terms for which $s \geq N$; we cancel these factors and then set $m = m_1$. The resulting series is simply the limit of $\phi(x, m)$ as $m \to m_1$. Since $L[\phi(x, m)] = f(m)x^{m+h}$, we can let $m \to m_1$ on both sides to conclude that as $m \to m_1$, $\phi(x, m) \to y_1$, and y_1 is a solution.

EXAMPLE. $x^2y'' + (x^3 - 3x)y' - 2x^2y = 0$.   We find $L(x^n) = (n^2 - 4n)x^n + (n - 2)x^{n+2}$ so that $f(n) = n(n - 4)$, $m_1 = 0$, $m_2 = 4$, $k = 2$. Hence $m_2 - m_1 = 2k$, so that the zero appears in the denominator. However, $g(n) = n - 2$, so that g has a root $2 = 1 \cdot k$. Thus the conditions described above are satisfied with $N = 2$, $\nu = 1$. Now

$$\phi(x, m) = x^m \left[1 + \sum_{s=1}^{\infty} (-1)^s x^{2s} \right.$$

$$\left. \times \frac{(m - 2)(m) \cdots (m + 2s - 4)}{(m + 2)(m - 2)(m + 4)m \cdots (m + 2s)(m + 2s - 4)} \right].$$

Since the factor m in the denominator would cause trouble for $m = 0 = m_1$, we cancel this factor (and other common factors) before setting $m = 0$:

$$\phi(x, m) = x^m \left[1 + \sum_{s=1}^{\infty} (-1)^s x^{2s} \frac{1}{(m + 2)(m + 4) \cdots (m + 2s)} \right].$$

Now

$$y_1 = \phi(x, 0) = 1 + \sum_{s=1}^{\infty} (-1)^s x^{2s} \frac{1}{2 \cdot 4 \cdots 2s},$$

$$y_2 = \phi(x, 4) = x^4 \left[1 + \sum_{s=1}^{\infty} (-1)^s x^{2s} \frac{1}{6 \cdot 8 \cdots (2s + 4)} \right].$$

Here $y = c_1y_1(x) + c_2y_2(x)$ is the general solution for all x.

When the factor $m - m_1$ does not appear in the numerator so that the solution is saved, it is necessary to modify the expression for $\phi(x, m)$. We introduce the factor $m - m_1$ artificially and consider the function

$$\psi(x, m) = (m - m_1)\phi(x, m). \tag{9–52}$$

Since the factor $m - m_1$ appears in the denominator of the terms of the series for ϕ, a cancellation can be made in all terms for which $s \geq N$. It will be assumed that this cancellation has been carried out, so that $\psi(x, m_1)$ is well defined. Now from Eq. (9–52)

$$L[\psi(x, m)] = (m - m_1)L[\phi(x, m)] = (m - m_1)x^{m+h}f(m). \tag{9–53}$$

Hence $L[\psi(x, m_1)] = 0$, so that $\psi(x, m_1)$ is a solution; however, $\psi(x, m_1)$ *is simply a constant times* $y_2 = \phi(x, m_2)$ (Prob. 5 below). Thus we have not yet found the desired second solution. To obtain it, we note that since $f(m)$ has a root m_1, $(m - m_1)f(m)$ has a double root m_1. Therefore we can apply the reasoning of the preceding section to conclude that $\psi_1(x, m_1)$ is a solution, where $\psi_1(x, m) = \partial\psi/\partial m$. Since $\psi_1(x, m_1)$ will contain a term in $\log x$, $y_1(x) = \psi_1(x, m_1)$ and $y_2(x) = \phi(x, m_2)$ must be linearly independent and provide the general solution over the interval for which the series converge.

EXAMPLE. $(x + x^3)y'' + (x^2 - 3)y' + xy = 0$. We find $L(x^n) = x^{n-1}(n^2 - 4n) + x^{n+1}(n^2 + 1)$. Here $f(n) = n(n - 4)$, $m_1 = 0$, $m_2 = 4$, $k = 2$, so that $m_2 - m_1 = 2k$; $g(n) = n^2 + 1$ has no real roots, and there is no cancellation. We find

$$\phi(x, m) = x^m \left\{ 1 + \sum_{s=1}^{\infty} (-1)^s x^{2s} \right.$$

$$\left. \times \frac{(m^2 + 1)(m^2 + 4m + s) \cdots [(m + 2s - 2)^2 + 1]}{(m + 2)(m - 2)(m + 4)m \cdots (m + 2s)(m + 2s - 4)} \right\}.$$

Thus $\phi(x, 0)$ is undefined, but

$$y_2 = \phi(x, 4) = x^4 \left\{ 1 + \sum_{s=1}^{\infty} (-1)^s x^{2s} \frac{17 \cdot 37 \cdots [(2s + 2)^2 + 1]}{6 \cdot 2 \cdot 8 \cdot 4 \cdots (2s + 4)(2s)} \right\}.$$

To obtain the second solution, we set

$$\psi(x, m) = m\phi(x, m)$$

$$= x^m \left[m + \sum_{s=1}^{\infty} (-1)^s x^{2s} \frac{m(m^2 + 1) \cdots}{(m + 2)(m - 2)(m + 4)m \cdots} \right].$$

A cancellation of the factor m can now be made. However, the factor m appears in the denominator only for $s \geq 2$. Hence we write

$$\psi(x, m) = x^m \left\{ m + (-1)x^2 \frac{m(m^2 + 1)}{(m + 2)(m - 2)} \right.$$

$$+ (-1)^2 x^4 \frac{(m^2 + 1)(m^2 + 4m + 5)}{(m + 2)(m - 2)(m + 4)} + \sum_{s=3}^{\infty} (-1)^s x^{2s}$$

$$\left. \times \frac{(m^2 + 1) \cdots [(m + 2s - 2)^2 + 1]}{(m + 2)(m - 2)(m + 4) \cdot 1 \cdots (m + 2s)(m + 2s - 4)} \right\}.$$

If we set $m = 0$, we find

$$\psi(x, 0) = x^4 \frac{1 \cdot 5}{2 \cdot (-2) \cdot 4} + \sum_{s=3}^{\infty} (-1)^s x^{2s} \frac{1 \cdot 5 \cdots [(2s-2)^2 + 1]}{2 \cdot (-2) \cdot 4 \cdots (2s)(2s-4)}$$

$$= -\frac{5}{16} x^4 \left\{ 1 + \sum_{s=3}^{\infty} (-1)^s x^{2s-4} \frac{17 \cdot 37 \cdots [(2s-2)^2 + 1]}{6 \cdot 2 \cdot 8 \cdot 4 \cdots (2s)(2s-4)} \right\}.$$

This is simply $-\frac{5}{16} y_2(x)$. However,

$$\psi_1(x, m) = \frac{\partial \psi}{\partial m} = \log x \cdot x^m (m + \cdots)$$

$$+ x^m \left\{ 1 - x^2 \frac{m(m^2+1)}{(m+2)(m-2)} \left(\frac{1}{m} + \frac{2m}{m^2+1} - \frac{1}{m+2} - \frac{1}{m-2} \right) \right.$$

$$+ x^4 \frac{(m^2+1)(m^2+4m+5)}{(m+2)(m-2)(m+4)}$$

$$\times \left(\frac{2m}{m^2+1} + \frac{2m+4}{m^2+4m+5} - \frac{1}{m+2} - \frac{1}{m-2} - \frac{1}{m+4} \right)$$

$$+ \sum_{s=3}^{\infty} (-1)^s x^{2s} \frac{(m^2+1) \cdots [(m+2s-2)^2 + 1]}{(m+2)(m-2)(m+4) \cdot 1 \cdots (m+2s)(m+2s-4)}$$

$$\left(\frac{2m}{m^2+1} + \cdots + \frac{2(m+2s-2)}{(m+2s-2)^2+1} - \frac{1}{m+2} - \cdots \right.$$

$$\left. \left. - \frac{1}{m+2s-4} \right) \right\}.$$

If we set $m = 0$, the coefficient of $\log x$ is simply $\psi(x, 0) = -\frac{5}{16} y_2(x)$. The coefficient of x^2 inside the brace has a term with m as factor in both numerator and denominator; this should be canceled before setting $m = 0$. We thus find

$$y_1(x) = \log x \, \psi(x, 0) + 1 + \frac{x^2}{4} - \frac{5}{16} x^4 \left(\frac{4}{5} - \frac{1}{2} + \frac{1}{2} - \frac{1}{4} \right)$$

$$+ \sum_{s=3}^{\infty} (-1)^s x^{2s} \frac{1 \cdot 5 \cdots [(2s-2)^2 + 1]}{2 \cdot (-2) \cdot 4 \cdot 1 \cdots (2s)(2s-4)}$$

$$\times \left(\frac{4}{5} + \cdots + \frac{4s-4}{(2s-2)^2+1} - \frac{1}{2} + \frac{1}{2} - \cdots - \frac{1}{2s} - \frac{1}{2s-4} \right).$$

Since $g(n)$ has degree 2, the series converge for $|x| < |f_0/g_0| = 1$.

9–11 Summary. We present in Table 9–1 a brief review of the results achieved.

TABLE 9–1

Solutions of $\overset{\prime}{L}(y) = 0$ at a regular singular point at $x = 0$:

$$L = a_0(x)D^2 + a_1(x)D + a_2(x),$$

$$L(x^n) = f(n)x^{n+h} + g(n)x^{n+h+k},$$

$f(n)$ of degree 2, $k > 0$, $f(m) = 0$ for $m = m_1, m_2$,

$$\phi(x, m) = x^m\left(1 + \sum_{s=1}^{\infty} (-1)^s x^{ks} \frac{g(m) \cdots g[m + (s-1)k]}{f(m+k) \cdots f(m+sk)}\right),$$

$$\phi_1(x, m) = \frac{\partial \phi}{\partial m},$$

$$\psi(x, m) = (m - m_1)\phi(x, m), \quad \psi_1(x, m) = \frac{\partial \psi}{\partial m}.$$

Case	Nature of m_1, m_2	$y_1(x)$	$y_2(x)$	Reference to text
I	$(m_2 - m_1)/k$, not an integer	$\phi(x, m_1)$	$\phi(x, m_2)$	Theorem 5, Section 9–8
II	$m_1 = m_2$	$\phi(x, m_1)$	$\phi_1(x, m_1)$	Theorem 6, Section 9–9
III	$m_2 = m_1 + Nk$, $g(m_1 + \nu k) = 0$, N, ν integers, $0 \leqq \nu \leqq N - 1$	$\phi(x, m_1)$	$\phi(x, m_2)$	Section 9–10, second paragraph
IV	$m_2 = m_1 + Nk$, N a positive integer, $g(m_1 + \nu k) \neq 0$ for $\nu = 0, \ldots, N - 1$	$\psi_1(x, m_1)$	$\phi(x, m_2)$	Section 9–10, last two paragraphs

PROBLEMS

1. Verify that there is a regular singular point at $x = 0$ and that the indicial equation has equal roots. Obtain linearly independent solutions valid near $x = 0$:

(a) $x^2 y'' - xy' + (1 + x^2)y = 0$.

(b) $x^2 y'' + (x^2 - 3x)y' + (4 - 2x)y = 0$.

2. Verify that there is a regular singular point at $x = 0$ and that the indicial equation has roots differing by a multiple of k, but that Case III of Table 9–1 applies. Obtain linearly independent solutions valid near $x = 0$.

(a) $xy'' + (x - 1)y' - y = 0$.
(b) $(x^2 + x^5)y'' - (x^4 + 3x)y' - 5y = 0$.

3. Verify that there is a regular singular point at $x = 0$ and that the indicial equation has roots differing by a multiple of k, and that Case IV of Table 9–1 applies. Obtain linearly independent solutions valid near $x = 0$.

(a) $xy'' + (x - 1)y' - 2y = 0$.   (b) $x^2y'' - 3xy' + (x^3 - 5)y = 0$.

4. For each of the following differential equations determine which of the four cases of Table 9–1 is applicable, but do not find the solutions:

(a) $(x^2 + x^5)y'' + (x^4 - x)y' + (4x^3 - 3)y = 0$,
(b) $(x^2 + x^4)y'' + (x^3 - x)y' + (4x^2 - 3)y = 0$,
(c) $x^2y'' + (3x + x^2)y' + (1 - x)y = 0$,
(d) $(4x^2 + x^4)y'' + (x^3 - 4x)y' + 3y = 0$,
(e) $x^2y'' + (x^2 + x)y' + (x - 1)y = 0$,
(f) $x^2y'' + (x^2 + x)y' - (1 + x)y = 0$,
(g) $(4x^2 + x^4)y'' + (8x + x^3)y' + (1 - x^2)y = 0$.

5. Show that if $\psi(x, m)$ is defined by Eq. (9–52) and $f(m)$ has roots m_1, $m_2 = m_1 + Nk$, where N is a positive integer, then

$$\psi(x, m_1) = \frac{(-1)^N g(m_1) \cdots g[m_1 + (N - 1)k]}{(Nkf_0)f(m_1 + k) \cdots f[m_1 + (N - 1)k]} \phi(x, m_2),$$

where $f(n) = f_0 n^2 + \cdots$

6. *The Legendre equation* is given by

$$(1 - x^2)y'' - 2xy' + N(N + 1)y = 0$$

(see Prob. 10 following Section 9–8).

(a) Show that the equation has regular singular points at $x = \pm 1$.
(b) To study the singular point at $x = 1$, let $t = x - 1$ to obtain the equation in t:

$$(t^2 + 2t)\frac{d^2y}{dt^2} + 2(1 + t)\frac{dy}{dt} - N(N + 1)y = 0,$$

with a singular point at $t = 0$. Show that the indicial equation has roots 0, 0, so that Case II of Table 9–1 applies.

(c) Obtain the solution

$$y_1 = \phi(t, 0)$$

$$= 1 + \sum_{s=1}^{\infty} t^s \frac{N(N - 1) \cdots (N - s + 1)(N + 1) \cdots (N + s)}{2^s(s!)^2} \quad (|t| < 2)$$

and show that it reduces to a polynomial when N is a positive integer. (This is $P_N(x)$, the Legendre polynomial of degree N, with $x = 1 + t$.)

(d) Assume that N is not an integer and obtain the solution

$$y_2 = \phi_1(t, 0) = \log t\, \phi(t, 0)$$

$$+ \left\{ \sum_{s=1}^{\infty} t^s\, \frac{N(N-1)\cdots(N-s+1)(N+1)\cdots(N+s)}{2^s(s!)^2} \right.$$

$$\times \left[-\frac{1}{N} - \frac{1}{N-1} - \cdots - \frac{1}{N-s+1} + \frac{1}{N+1} + \frac{1}{N+2} + \cdots \right.$$

$$\left. \left. + \frac{1}{N+s} - 2\left(1 + \frac{1}{2} + \cdots + \frac{1}{s}\right) \right] \right\}.$$

7. *The hypergeometric equation* is given by

$$x(1-x)y'' + [c - (a+b+1)x]y' - aby = 0$$

(see Prob. 11 following Section 9–8).

Show that if c is a negative integer, and neither a nor b is 0 or a negative integer between 0 and c (inclusive), then Case IV applies.

8. *The Bessel equation* is given by

$$x^2 y'' + xy' + (x^2 - N^2)y = 0$$

(see Prob. 12 following Section 9–8).

(a) Let N be 0. Show that Case II applies.

(b) Let N be a positive integer. Show that Case IV applies.

ANSWERS

1. (a) $\sum_{s=0}^{\infty} (-1)^s x^{2s+1} 2^{-2s}(s!)^{-2} = y_1(x)$, all x; $y_2(x) = y_1(x) \log |x| -$
    $\sum_{s=1}^{\infty} (-1)^s x^{2s+1} 2^{-2s}(s!)^{-2}(1 + \frac{1}{2} + \cdots + s^{-1})$, $|x| > 0$;
   (b) $y_1(x) = x^2$, $y_2(x) = x^2 \log |x| + \sum_{s=1}^{\infty} (-1)^s x^{s+2} s^{-1}(s!)^{-1}$, $|x| > 0$.

2. (a) $y_1(x) = \sum_{s=0}^{\infty} (-1)^s x^s/s! = e^{-x}$,
    $y_2(x) = x^2 + 2\sum_{s=1}^{\infty} (-1)^s x^{s+2}/(s+2)! = 2(e^{-x} - 1 + x)$;
   (b) $y_1(x) = x^{-1} + \sum_{s=1}^{\infty} (-1)^s x^{3s-1} (-1) \cdot 2 \cdots (3s - 4)3^{-s}/s!$,
    $|x| > 0$;
    $y_2(x) = x^5 + 2\sum_{s=1}^{\infty} (-1)^s x^{3s+5}\, 5 \cdot 8 \cdots (3s + 2)3^{-s}/(s + 2)!$, all x.

3. (a) $y_2(x) = x^2$, $y_1(x) = -x^2 \log |x| + 1 - 2x + 2x^2 -$
    $2\sum_{s=3}^{\infty}(-1)^s x^s(s - 2)^{-1}(s!)^{-1}$, $|x| > 0$;
   (b) $y_2(x) = 2\sum_{s=0}^{\infty} (-1)^s x^{3s} 3^{-2s}(s!)^{-1}[(s + 2)!]^{-1}$, all x;
    $y_1(x) = -y_2(x) \log |x|/54 + x^{-1} + x^2/9 + x^5/324 -$
    $\sum_{s=3}^{\infty} (-1)^s x^{3s-1} 3^{-2s}(s!)^{-1}[(s - 2)!]^{-1}$
    $\times (1 + (s - 1)^{-1} + s^{-1} - 2\{1 + \frac{1}{2} + \cdots + s^{-1}\})$, $|x| > 0$.

4. (a) I, (b) IV, (c) II, (d) I, (e) III, (f) IV, (g) II.

9–12 Series solutions for nonlinear equations near isolated singular points. For an equation of first order

$$y' = F(x, y), \qquad (9\text{-}54)$$

a singular point is a point (x, y) at which the existence theorem is inapplicable. For example, the equation

$$y' = \frac{x + xy}{x^2 - y^2}$$

has a singular point at $(0, 0)$, since the right-hand member is meaningless at this point. For an equation in differential form

$$P(x, y) \, dx + Q(x, y) \, dy = 0, \qquad (9\text{-}55)$$

where P and Q have continuous first partial derivatives in a domain D, we consider as singular points in D only those points at which *both P and Q* are 0 (Sections 1–9, 2–1); indeed, if $Q(x_0, y_0) \neq 0$, we can divide by Q and write the equation in form (9-54), so that the existence theorem is applicable at (x_0, y_0); if $P(x_0, y_0) \neq 0$, we can divide by P and obtain a similar equation for dx/dy. Where both $P(x, y)$ and $Q(x, y)$ are 0, there is no way of applying the existence theorem.

When an equation (9-54) is given in which $F(x, y)$ is expressed as the ratio of two functions

$$\frac{dy}{dx} = \frac{M(x, y)}{N(x, y)}, \qquad (9\text{-}56)$$

it is common practice to consider the equation as equivalent to one in differential form

$$M(x, y) \, dx - N(x, y) \, dy = 0; \qquad (9\text{-}57)$$

thus again (if M and N have continuous first partial derivatives) the singular points are only those points at which both $M(x, y)$ and $N(x, y)$ are 0.

For a system of order n:

$$\frac{dy_i}{dx} = F_i(x, y_1, \ldots, y_n) \qquad (i = 1, \ldots, n) \qquad (9\text{-}58)$$

a corresponding differential form is

$$dx = \frac{dy_1}{F_1} = \frac{dy_2}{F_2} = \cdots = \frac{dy_n}{F_n}$$

or, more generally,

$$\frac{dx}{P_0} = \frac{dy_1}{P_1} = \frac{dy_2}{P_2} = \cdots = \frac{dy_n}{P_n}, \qquad (9\text{-}59)$$

where $F_1 = P_1/P_0, \ldots, F_n = P_n/P_0$ and $P_i = P_i(x, y_1, \ldots, y_n)$. If the P_i have continuous first partial derivatives in a domain D of $(x, y_1, \ldots, y_n)$-space, then we can apply the existence theorem at each point of that space for which not all of the P_i are 0. For example, if $P_1 \neq 0$, we can take y_1 as independent variable:

$$\frac{dx}{dy_1} = \frac{P_0}{P_1}, \qquad \frac{dy_2}{dy_1} = \frac{P_2}{P_1}, \qquad \cdots$$

Thus the singular points in D are only those points at which $P_0, P_1, \ldots, P_n$ are all 0.

A single equation of order n can be replaced by an equivalent system (9–58) and singular points are defined accordingly. However, for a linear equation

$$a_0(x)y^{(n)}(x) + \cdots + a_n(x)y = Q(x),$$

it is customary to consider only y as dependent variable; hence the existence theorem is applicable wherever $a_0(x), \ldots, a_n(x), Q(x)$ are continuous and $a_0(x) \neq 0$. The "singular points" are then defined as the values of x for which $a_0(x) = 0$. For a first order linear equation

$$a_0(x)y' + a_1(x)y = Q(x),$$

a "singular point" $x = x_0$ actually describes a *straight line* in the xy-plane, along which the existence theorem is inapplicable, with y as dependent variable.

Isolated singular points. Let $M(x, y)$ and $N(x, y)$ have continuous first partial derivatives in a domain D. A singular point (x_0, y_0) of Eq. (9–56) in D is termed *isolated* if there are no singular points other than (x_0, y_0) within a sufficiently small circle with center (x_0, y_0). Thus $M(x_0, y_0) = 0$ and $N(x_0, y_0) = 0$, but the equations

$$M(x, y) = 0, \qquad N(x, y) = 0 \qquad (9\text{--}60)$$

will have no other common solution inside such a circle about (x_0, y_0). The Eqs. (9–60) in general describe two curves in the xy-plane; most commonly, these two curves will meet only at isolated points; exceptionally, the points of intersection, which are the singular points of Eq. (9–56), will be infinite in number and will have a *point of condensation*, which is then not an isolated singular point. A similar definition of isolated singular point can be given for the system (9–59).

Parametric form for first order equation. For the remainder of this section and throughout the following section, we confine attention to the first order equation (9–56) and assume there is an isolated singular point at the origin $(0, 0)$. It will further be assumed that $M(x, y)$ and $N(x, y)$ can be expanded in power series for $|x| < a, |y| < b$:

$$M(x, y) = a_2x + b_2y + c_2x^2 + d_2xy + e_2y^2 + \cdots,$$
$$N(x, y) = a_1x + b_1y + c_1x^2 + d_1xy + e_1y^2 + \cdots; \tag{9-61}$$

[the constant terms are missing, since $M(0, 0) = 0, N(0, 0) = 0$]. We now replace the equation (9–56) by a corresponding set of *parametric equations*

$$\frac{dx}{dt} = N(x, y) = a_1x + b_1y + c_1x^2 + \cdots,$$

$$\frac{dy}{dt} = M(x, y) = a_2x + b_2y + c_2x^2 + \cdots \tag{9-62}$$

If $x = x(t), y = y(t)$ is a solution of Eqs. (9–62), then

$$\frac{dy}{dx} = \frac{dy/dt}{dx/dt} = \frac{M(x, y)}{N(x, y)};$$

that is, elimination of t provides a solution of Eq. (9–56). Where $N = 0$, we must consider dx/dy instead of dy/dx; if the solution does not pass through a singular point, one of the two procedures is applicable. It can be verified that all solutions of Eq. (9–56) can be obtained in this way from the solutions of Eqs. (9–62).

We now consider the solutions of Eqs. (9–62). Theorem 2 of Section 9–5 is applicable and guarantees existence of a series solution $x(t), y(t)$, with $x(t_0) = 0, y(t_0) = 0$. But this solution we recognize at once; it is the trivial solution $x = 0, y = 0$! Hence the only solution through the singular point $(0, 0)$ is the solution $x = 0, y = 0$ for all t. In mechanics Eqs. (9–62) arise commonly; the point $x = 0, y = 0$ corresponds to an *equilibrium point*. A system in equilibrium will remain there indefinitely, unless some disturbance causes a change of initial conditions.

Equations (9–62) will have other solutions $x = x(t), y(t)$ with initial values x_0, y_0 not both 0. In particular, solutions may exist for which x and y *approach* 0 as $t \to +\infty$ or as $t \to -\infty$. When t is eliminated, these solutions appear as paths in the xy-plane "ending" at the singular point. It is such solutions that we seek.

EXAMPLE 1. $y' = y/x$. The parametric equations are

$$\frac{dx}{dt} = x, \qquad \frac{dy}{dt} = y;$$

the solutions are $x = c_1e^t, y = c_2e^t$. As $t \to -\infty$, x and $y \to 0$. The solutions are simply straight lines through the origin, as shown in Fig. 9–1.

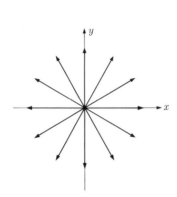

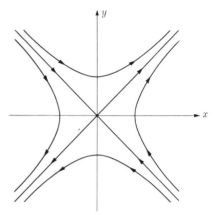

FIG. 9-1. Solutions of $dx/dt = x$, $dy/dt = y$.

FIG. 9-2. Solutions of $dx/dt = y$, $dy/dt = x$.

EXAMPLE 2. $y' = x/y$. The parametric equations are

$$\frac{dx}{dt} = y, \qquad \frac{dy}{dt} = x;$$

the solutions are

$$x = c_1 e^t + c_2 e^{-t}, \qquad y = c_1 e^t - c_2 e^{-t}.$$

As $t \to +\infty$, x and $y \to \pm\infty$ unless $c_1 = 0$; as $t \to -\infty$, x and $y \to \pm\infty$ and $\mp\infty$ unless $c_2 = 0$. Thus the solutions leading to the singular point are $x = c_1 e^t$, $y = c_1 e^t$ and $x = c_2 e^{-t}$, $y = -c_2 e^{-t}$; these provide two lines $y = \pm x$, as shown in Fig. 9-2.

EXAMPLE 3. $y' = -x/y$. The parametric equations are

$$\frac{dx}{dt} = y, \qquad \frac{dy}{dt} = -x;$$

the solutions are

$$x = c_1 \cos t + c_2 \sin t,$$

$$y = -c_1 \sin t + c_2 \cos t.$$

Hence no solution leads to the singular point $(0, 0)$ as $t \to \pm\infty$. The paths in the xy-plane are the circles $x^2 + y^2 = \text{const}$, as Fig. 9-3 shows.

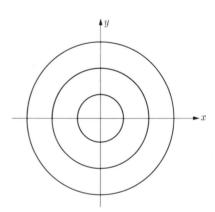

FIG. 9-3. Solutions of $dx/dt = y$, $dy/dt = -x$.

In the examples chosen, infinite series were not necessary, and the solutions ending at the singular point were easily recognized. However, we now ask how, for Eqs. (9–62), we can ascertain whether there exist solutions $x = x(t)$, $y = y(t)$ which lead to the singular point as $t \to \pm\infty$. How do we find these solutions explicitly? These questions are answered in Section 9–13. The same questions can be asked for equations of higher order and for systems of equations.

9–13 Method of Picard. We now rewrite Eqs. (9–62) in a slightly different form:

$$u\frac{dx}{du} = a_1 x + b_1 y + c_1 x^2 + d_1 xy + e_1 y^2 + \cdots ,$$

$$u\frac{dy}{du} = a_2 x + b_2 y + c_2 x^2 + d_2 xy + e_2 y^2 + \cdots ; \tag{9-63}$$

thus dt has been replaced by du/u; this is equivalent to using $u = e^t$ as parameter. We now seek solutions of Eqs. (9–63) of form

$$x = \sum_{n=1}^{\infty} \alpha_n u^{n\lambda}, \qquad y = \sum_{n=1}^{\infty} \beta_n u^{n\lambda}, \tag{9-64}$$

where λ, α_1, α_2, $\ldots$, β_1, β_2, $\ldots$ are constants to be determined. We proceed formally and discuss convergence below. Substitution in Eqs. (9–63) leads to the equations

$$\sum_{n=1}^{\infty} n\alpha_n \lambda u^{n\lambda} = a_1 \sum_{n=1}^{\infty} \alpha_n u^{n\lambda} + b_1 \sum_{n=1}^{\infty} \beta_n u^{n\lambda} + c_1 \left(\sum_{n=1}^{\infty} \alpha_n u^{n\lambda} \right)^2 + \cdots ,$$

$$\sum_{n=1}^{\infty} n\beta_n \lambda u^{n\lambda} = a_2 \sum_{n=1}^{\infty} \alpha_n u^{n\lambda} + b_2 \sum_{n=1}^{\infty} \beta_n u^{n\lambda} + c_2 \left(\sum_{n=1}^{\infty} \alpha_n u^{n\lambda} \right)^2 + \cdots$$

We think of the series on the right as squared, cubed, $\ldots$ and terms of same degree collected. We then compare terms of same degree in u^λ on the left and right. For $n = 1$, that is, for terms in u^λ, we find

$$\alpha_1 \lambda = a_1 \alpha_1 + b_1 \beta_1, \qquad \beta_1 \lambda = a_2 \alpha_1 + b_2 \beta_1,$$

which can be written

$$(a_1 - \lambda)\alpha_1 + b_1 \beta_1 = 0, \qquad a_2 \alpha_1 + (b_2 - \lambda)\beta_1 = 0. \tag{9-65}$$

These have a solution other than $\alpha_1 = 0$, $\beta_1 = 0$, provided

$$\begin{vmatrix} a_1 - \lambda & b_1 \\ a_2 & b_2 - \lambda \end{vmatrix} = 0. \tag{9-66}$$

Equation (9–66), of second degree in λ, is called the *characteristic equation;* it is the characteristic equation of the linear system obtained from Eqs. (9–62) by dropping terms above first degree on the right (Section 6–2). *We assume this equation has distinct nonzero real roots* λ_1, λ_2. The results described can be extended to the case of complex roots.

With λ chosen as λ_1 or λ_2, Eqs. (9–65) can now be solved for α_1, β_1 as in Section 6–2. We find $\alpha_1 = kA_1$, $\beta_1 = kB_1$, where A_1, B_1 are not both 0 and k is arbitrary. We choose $k = 1$, so that $\alpha_1 = A_1$, $\beta_1 = B_1$. (It can be shown that no solutions are lost thereby.)

Comparison of terms of second degree now gives the equations

$$2\alpha_2\lambda = a_1\alpha_2 + b_1\beta_2 + c_1\alpha_1^2 + d_1\alpha_1\beta_1 + e_1\beta_1^2,$$

$$2\beta_2\lambda = a_2\alpha_2 + b_2\beta_2 + c_2\alpha_1^2 + d_2\alpha_1\beta_1 + e_2\beta_2^2,$$

which can be written

$$(a_1 - 2\lambda)\alpha_2 + b_1\beta_2 = -c_1\alpha_1^2 \cdots$$

$$a_2\alpha_2 + (b_2 - 2\lambda)\beta_2 = -c_2\alpha_1^2 \cdots$$

Hence they are linear equations for α_2, β_2 and can be solved for α_2, β_2 in terms of α_1, β_1 unless

$$\begin{vmatrix} a_1 - 2\lambda & b_1 \\ a_2 & b_2 - 2\lambda \end{vmatrix} = 0;$$

that is, provided 2λ is not also a root of the characteristic equation.

For terms of degree n we find similarly that α_n, β_n can be expressed in terms of $\alpha_1, \ldots, \alpha_{n-1}, \beta_1, \ldots, \beta_{n-1}$, provided $n\lambda$ is not a root of the characteristic equation. *We therefore make the assumption* $\lambda_1 \neq n\lambda_2, \lambda_2 \neq n\lambda_1$, $(n = 1, 2, \ldots)$. Then for $\lambda = \lambda_1$ or $\lambda = \lambda_2$ we obtain a series solution (9–64).

Let us now suppose that $\lambda_1 > 0$. Then the solution (9–64), with $\lambda = \lambda_1$, will approach $(0, 0)$ as $u \to 0$. This provides one solution through the singular point (more precisely, one pair of solution paths leading to the singular point). From (9–64) $y/x \to \beta_1/\alpha_1$ as $u \to 0$; hence the solution approaches the origin with a definite limiting slope $\beta_1/\alpha_1 = (\lambda_1 - a_1)/b_1$. A similar remark holds if $\lambda_2 > 0$, the limiting slope being $(\lambda_2 - a_1)/b_1$. Since $\lambda_1 \neq \lambda_2$, the slopes are unequal.

If $\lambda_1 < 0$, then Eqs. (9–64) are series in *negative* powers of u. If we let $u \to \infty$, so that $1/u \to 0$, then again x and $y \to 0$ with the same limiting slope as before. A similar remark applies to the second solution.

If $\lambda_1 < 0$ and $\lambda_2 > 0$, then one solution approaches $(0, 0)$ as $u \to \infty$, the other as $u \to 0$. It can be shown that in this case (as in Fig. 9–2) *these are the only solutions through the singular point.*

When λ_1 and λ_2 have the same sign, we have obtained only two solutions through the singular point, but there are actually infinitely many. To obtain all the solutions through the singular point, we set

$$v_1 = u^{\lambda_1}, \qquad v_2 = u^{\lambda_2} \tag{9–67}$$

and seek solutions

$$x = \alpha_1 v_1 + \alpha_2 v_2 + \alpha_3 v_1^2 + \alpha_4 v_1 v_2 + \cdots ,$$
$$y = \beta_1 v_1 + \beta_2 v_2 + \beta_3 v_1^2 + \beta_4 v_1 v_2 + \cdots \tag{9–68}$$

Substitution in the first differential equation (9–63) leads to the relation

$$\alpha_1 \lambda_1 v_1 + \alpha_2 \lambda_2 v_2 + 2\alpha_3 \lambda_1 v_1^2 + \cdots = a_1(\alpha_1 v_1 + \alpha_2 v_2 + \cdots)$$
$$+ b_1(\beta_1 v_1 + \beta_2 v_2 + \cdots) + c_1(\alpha_1^2 v_1^2 + \cdots) + \cdots$$

We now consider v_1 and v_2 as *independent* variables and choose the coefficients $\alpha_1, \alpha_2, \ldots, \beta_1, \beta_2, \ldots$ so that the equation is an identity. We obtain the conditions

$$(a_1 - \lambda_1)\alpha_1 + b_1\beta_1 = 0,$$
$$(a_1 - \lambda_2)\alpha_2 + b_1\beta_2 = 0,$$
$$(a_1 - 2\lambda_1)\alpha_3 + b_1\beta_3 + c_1\alpha_1^2 + d_1\alpha_1\beta_1 + e_1\beta_1^2 = 0,$$
$$\vdots$$

Similar equations are obtained from the differential equation for y. Because of the choice of λ_1, λ_2, these can be satisfied by choosing $\alpha_1 = k_1 b_1$, $\beta_1 = k_1(\lambda_1 - a_1)$, $\alpha_2 = k_2 b_1$, $\beta_2 = b_2(\lambda_2 - a_1)$, where k_1, k_2 are arbitrary, and then expressing all later constants in terms of α_1, β_1, α_2, β_2, hence in terms of k_1, k_2. Thus we eventually obtain series solutions

$$x = \phi(v_1, v_2, k_1, k_2) = \phi(u^{\lambda_1}, u^{\lambda_2}, k_1, k_2),$$
$$y = \psi(v_1, v_2, k_1, k_2) = \psi(u^{\lambda_1}, u^{\lambda_2}, k_1, k_2),$$

which contain two arbitrary constants k_1, k_2. It can be shown that if $\lambda_1 > 0$, $\lambda_2 > 0$, these provide solutions ending at the singular point $(0, 0)$ as $u \to 0$; if $\lambda_1 < 0$, $\lambda_2 < 0$, the same conclusion holds as $u \to \infty$. Furthermore, in both cases, the solutions found provide all solutions in a neighborhood of the singular point, as in Fig. 9–1.

For a complete discussion of the method and proof of convergence (including systems of higher order) the reader is referred to Chapters 1, 2 of Vol. 3 of Reference 6.

EXAMPLE 1. $y' = (3y + 2xy)/2x$. The parametric equations are

$$u \frac{dx}{du} = 2x, \qquad u \frac{dy}{du} = 3y + 2xy.$$

The characteristic equation is

$$\begin{vmatrix} 2 - \lambda & 0 \\ 0 & 3 - \lambda \end{vmatrix} = 0.$$

The roots are 2, 3; neither is an integral multiple of the other. Since both are positive, we at once seek solutions (9–68), with $v_1 = u^2$, $v_2 = u^3$. Substitution in the differential equation for x gives

$$2\alpha_1 v_1 + 3\alpha_2 v_2 + 4\alpha_3 v_1^2 + \cdots = 2(\alpha_1 v_1 + \alpha_2 v_2 + \alpha_3 v_1^2 + \cdots).$$

From this we conclude at once that α_1 is arbitrary and that α_2, α_3, ... are 0, since

$$3\alpha_2 = 2\alpha_2, \qquad 4\alpha_3 = 2\alpha_3 \ldots$$

Hence x reduces to $\alpha_1 v_1$. The differential equation for y now gives

$$2\beta_1 v_1 + 3\beta_2 v_2 + 4\beta_3 v_1^2 + \cdots = 3(\beta_1 v_1 + \beta_2 v_2 + \beta_3 v_1^2 + \cdots)$$
$$+ 2\alpha_1 v_1 (\beta_1 v_1 + \beta_2 v_2 + \cdots).$$

Hence we find

$$2\beta_1 = 3\beta_1, \qquad 3\beta_2 = 3\beta_2, \qquad 4\beta_3 = 3\beta_3 + 2\alpha_1\beta_1,$$
$$5\beta_4 = 3\beta_4 + 2\alpha_1\beta_2, \qquad 6\beta_5 = 3\beta_5, \qquad \ldots$$

From these equations we conclude that $\beta_1 = 0$, β_2 is arbitrary, $\beta_3 = 0$, $\beta_4 = \alpha_1\beta_2$, $\beta_5 = 0$, ... Hence

$$x = \alpha_1 v_1 = \alpha_1 u^2,$$
$$y = \beta_2 v_2 + \beta_4 v_1 v_2 + \cdots = \beta_2 u^3 + \alpha_1 \beta_2 u^5 + \cdots$$

In this example we can find x, y explicitly:

$$x = c_1 u^2, \qquad y = c_2 u^3 e^{c_1 u^2} = c_2 u^3 (1 + c_1 u^2 + \cdots),$$

where c_1, c_2 are arbitrary. Thus $c_1 = \alpha_1, c_2 = \beta_2$. The solutions are graphed in Fig. 9–4. When y is expressed in terms of x, only one arbitrary constant remains:

$$y = c\,(\pm\,x)^{3/2}e^{x},$$

$$c = c_2\,(\pm\,c_1)^{-3/2}.$$

EXAMPLE 2.

$$y' = (x - x^2)/(y + xy + y^2).$$

There is a singular point at $(0, 0)$. The parametric equations are

$$u\,\frac{dx}{du} = y + xy + y^2,$$

$$u\,\frac{dy}{du} = x - x^2.$$

The characteristic equation is

$$\begin{vmatrix} -\lambda & 1 \\ 1 & -\lambda \end{vmatrix} = 0.$$

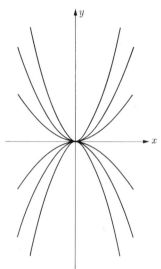

FIG. 9–4. Nonlinear equation with infinitely many solutions leading to singular point.

The roots are ± 1. Hence we seek only two solutions through the singular point. For $\lambda = 1$ we set

$$x = \sum_{n=1}^{\infty} \alpha_n u^n, \qquad y = \sum_{n=1}^{\infty} \beta_n u^n.$$

Upon substituting in the parametric differential equations and comparing terms of same degree, we find

$$\alpha_1 = \beta_1, \qquad \beta_1 = \alpha_1,$$

$$2\alpha_2 - \beta_2 = \alpha_1\beta_1 + \beta_1^2,$$

$$\alpha_2 - 2\beta_2 = \alpha_1^2,$$

$$3\alpha_3 - \beta_3 = \alpha_1\beta_2 + \alpha_2\beta_1 + 2\beta_1\beta_2,$$

$$\alpha_3 - 3\beta_3 = 2\alpha_1\alpha_2, \qquad \ldots$$

Hence we can set $\alpha_1 = k$ and express all coefficients in terms of k:

$$\alpha_1 = k = \beta_1, \qquad \alpha_2 = k^2, \qquad \beta_2 = 0,$$

$$\alpha_3 = \frac{k^3}{8}, \qquad \beta_3 = -\frac{5k^3}{8}, \qquad \ldots$$

Accordingly, the solution sought is

$$x = ku + k^2u^2 + \tfrac{1}{8}k^3u^3 + \cdots,$$

$$y = ku - \tfrac{5}{8}k^3u^3 + \cdots$$

Different choices of k merely change the parameter scale and do not change the path in the xy-plane. With $k = 1$ the path is

$$x = u + u^2 + \tfrac{1}{8}u^3 + \cdots, \qquad y = u - \tfrac{5}{8}u^3 + \cdots \qquad (9\text{--}69)$$

For $\lambda = -1$ we set

$$x = \sum_{n=1}^{\infty} \alpha_n u^{-n}, \qquad y = \sum_{n=1}^{\infty} \beta_n u^{-n}.$$

A similar computation provides the solutions

$$x = u^{-1} - \tfrac{1}{3}u^{-2} + \tfrac{5}{24}u^{-3} + \cdots, \qquad y = -u^{-1} + \tfrac{2}{3}u^{-2} - \tfrac{7}{24}u^{-3} + \cdots$$

Here we wish to let $u^{-1} \to 0$. Accordingly, we set $v = u^{-1}$ and use v as parameter near $v = 0$.

$$x = v - \tfrac{1}{3}v^2 + \tfrac{5}{24}v^3 + \cdots, \qquad y = -v + \tfrac{2}{3}v^2 - \tfrac{7}{24}v^3 + \cdots \qquad (9\text{--}70)$$

The two solutions (9–69) and (9–70) are graphed in Fig. 9–5. The other solutions shown have not been computed; since they cannot end at the singular point, it is easily verified that they must have the appearance shown.

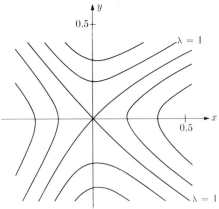

Fig. 9–5. Nonlinear equation with two pairs of solutions leading to singular point.

9-14 The stability problem. In many physical problems (see Sections 6-13, 6-14) we are led to a system of equations

$$\frac{dx_i}{dt} = P_i(x_1, \ldots, x_n) \qquad (i = 1, \ldots, n), \qquad (9\text{-}71)$$

where the P_i are continuous and have continuous first partial derivatives for $|x_1| < a, \ldots, |x_n| < a$, and $P_1 = P_2 = \ldots = P_n = 0$ when $x_1 = 0$, $\ldots, x_n = 0$. Thus the equations have the solution

$$x_1 \equiv 0, \qquad \ldots, \qquad x_n \equiv 0, \qquad (9\text{-}72)$$

which is termed an *equilibrium solution*. For each point $(x_1^0, \ldots, x_n^0)$ such that $|x_i^0| < a$, there is a solution $x_i = f_i(t)$, $(i = 1, \ldots, n)$, such that $f_i(0) = x_i^0$. The equilibrium solution is called *stable* if we can find a number a_1 such that, whenever $|x_i^0| < a_1$, the corresponding solution $x_i = f_i(t)$ is defined for $t > 0$ and $x_i \to 0$ as $t \to +\infty$. (See remarks at the end of Section 5-2.)

The stability problem is closely related to the problem of series solutions at a singular point, as discussed in the preceding sections. When the P_i can be expanded in power series, we can apply the methods described above to obtain series for the x_i in terms of u^λ, where $u = e^t$. If $\lambda < 0$, then the corresponding solution approaches the singular point $(0, \ldots, 0)$ as $u \to +\infty$, hence as $t \to +\infty$. When $n = 2$ and $\lambda_1 < 0$, $\lambda_2 < 0$, we have seen that *all* nearby solutions approach the singular point as $u \to +\infty$, hence as $t \to +\infty$; therefore the equilibrium is stable. A similar result holds for general n. If there are complex characteristic roots λ, the conclusion continues to hold if for every λ, Re $(\lambda) < 0$.

Even when the power series expansions of the $P_i(x_1, \ldots, x_n)$ are not assumed, we can write

$$P_i(x_1, \ldots, x_n) = a_{i1}x_1 + \cdots + a_{in}x_n + g_i(x_1, \ldots, x_n), \qquad (9\text{-}73)$$

where the a_{ij} are constants, namely:

$$a_{ij} = \left.\frac{\partial P_i}{\partial x_j}\right|_{(0, \ldots, 0)}, \qquad (9\text{-}74)$$

and the g_i are of higher than first order. Equation (9-73) is simply the statement that the increment of P_i can be approximated by the differential

$$P_i(x_1, \ldots, x_n) - P_i(0, \ldots, 0) = dP_i + g_i,$$

where $P_i(0, \ldots, 0) = 0$. To study stability, we can ignore the higher order terms g_i and replace Eqs. (9–71) by the *linearized equations*

$$\frac{dx_i}{dt} = \sum_{j=1}^{n} a_{ij} x_j \qquad (i = 1, \ldots, n). \qquad (9\text{--}75)$$

The stability problem for the linear system (9–75) is studied in Chapter 6 (Section 6–11). There it is shown that stability is equivalent to the condition that all roots λ of the characteristic equation

$$\begin{vmatrix} a_{11} - \lambda & \cdots & a_{1n} \\ \vdots & & \vdots \\ a_{n1} & \cdots & a_{nn} - \lambda \end{vmatrix} = 0 \qquad (9\text{--}76)$$

have negative real parts. It can be shown that if all characteristic roots λ do have negative real parts, then the equilibrium solution (9–72) of the given *nonlinear* equations (9–71) is stable; if any one root λ has positive real part, then the equilibrium solution is unstable. If Re $(\lambda) \leq 0$ for every λ, so that some λ may be 0 or pure imaginary, the nonlinear system may be either stable or unstable. For proofs the reader is referred to Chapter 13 of Reference 2 and to Reference 1.

PROBLEMS

1. Replace by equivalent parametric equations $Dx = P(x, y)$, $Dy = Q(x, y)$, where $D = d/dt$, obtain the general solution, and sketch the solutions in the xy-plane:

(a) $y' = \dfrac{5x + 3y}{x - y}$,

(b) $y' = \dfrac{-x + 3y}{x + y}$.

2. For the following equations, verify that there is a singular point at $(0, 0)$, replace by equivalent parametric equations $u\,dx/du = P(x, y)$, $u\,dy/du = Q(x, y)$, show that there are two solutions through the singular point, and obtain series expansions for these solutions:

(a) $y' = \dfrac{\sin x}{y}$,

(b) $y' = \dfrac{\sin (x + 2y)}{\sin (x + 6y)}$.

3. For the following equations, verify that there is a singular point at $(0, 0)$, replace by equivalent parametric equations $u\,dx/du = P(x, y)$, $u\,dy/du = Q(x, y)$, show that all solutions near the origin pass through the origin, and obtain series expansions for these solutions:

(a) $y' = \dfrac{x + 2y + xy}{5x - 2y - xy}$,

(b) $y' = \dfrac{2x + 3y + x^2 - y^2}{4x + y - xy}$.

4. Show that there is an equilibrium point at $x = y = \cdots = 0$ and test for stability:

(a) $\dfrac{dx}{dt} = \log(1 + x + y)$, $\dfrac{dy}{dt} = x - y + x^2$;

(b) $\dfrac{dx}{dt} = e^x \sin y + \sin x + e^z - 1$, $\dfrac{dy}{dt} = \sin(x + y)$,

$\dfrac{dz}{dt} = \tan(x + z)$.

ANSWERS

1. (a) $x = e^{2t}(c_1 \cos 2t + c_2 \sin 2t)$,
   $y = e^{2t}[(-c_1 - 2c_2)\cos t + (2c_1 - c_2)\sin 2t]$;
   (b) $x = c_1 e^{2t} + c_2 t e^{2t}$, $y = c_1 e^{2t} + c_2 e^{2t}(t + 1)$.

2. (a) $x = u - (u^3/48) + \cdots, y = u - (u^3/16) + \cdots$,
   $x = u^{-1} - (u^{-3}/48) + \cdots, y = -u^{-1} + (u^{-3}/16) + \cdots$;
   (b) $x = 2u^4 - (296u^{12}/39) + \cdots, y = u^4 + (12u^{12}/39) + \cdots$;
   $x = -3u^{-1} + (47u^{-3}/28) + \cdots, y = u^{-1} - (31u^{-3}/84) + \cdots$

3. (a) $x = k_1 u^3 + 2k_2 u^4 - k_1^2 u^6 - \frac{7}{4}k_1 k_2 u^7 - \frac{4}{5}k_2^2 u^8 + \cdots$,
   $y = k_1 u^3 + k_2 u^4 + \frac{1}{4}k_1 k_2 u^7 + \frac{1}{5}k_2^2 u^8 + \cdots$;
   (b) $x = k_1 u^2 + \frac{1}{2}k_1^2 u^4 + k_2 u^5 + \frac{1}{2}k_1^3 u^6 + k_1 k_2 u^7 + \cdots$,
   $y = -2k_1 u^2 - 2k_1^2 u^4 + k_2 u^5 - 2k_1^3 u^6 + 2k_1 k_2 u^7 + \cdots$

4. (a) Unstable, (b) unstable.

SUGGESTED REFERENCES

1. BELLMAN, RICHARD, *Stability Theory of Differential Equations*. New York: McGraw-Hill, 1953.

2. CODDINGTON, EARL A., and LEVINSON, NORMAN, *Theory of Ordinary Differential Equations*. New York: McGraw-Hill, 1955.

3. JORDAN, KÁROLY, *Calculus of Finite Differences*. New York: Chelsea, 1947.

4. KAPLAN, WILFRED, *Advanced Calculus*. Reading, Mass.: Addison-Wesley, 1952.

5. KAPLAN, WILFRED, *Stability Theory*. Proceedings of the Symposium on Non-linear Circuit Analysis, Vol. VI, pp. 3–21. New York: Polytechnic Institute of Brooklyn, 1957.

6. PICARD, EMILE, *Traité d'Analyse* (3 vols.), 3rd ed. Paris: Gauthier-Villars, 1922, 1925, 1928.

7. RAINVILLE, EARL D., *Intermediate Differential Equations*. New York: John Wiley and Sons, Inc., 1943.

8. WHITTAKER, E. T., and WATSON, G. N., *A Course of Modern Analysis*, 4th ed. Cambridge, Eng.: Cambridge University Press, 1940.

CHAPTER 10

NUMERICAL METHODS

10–1 General remarks. In particular applications we are often required to tabulate a solution of a differential equation over a given interval. If an explicit formula for the solution has been found, we can use the formula, with tables of sines, cosines, logarithms, etc., to carry out the tabulation. If the solution is known in series form, the series itself becomes an explicit formula from which the solution can be computed to any desired accuracy (Chapter 9).

If no explicit formula has been found, we can consider the differential equation itself as a formula that enables us to evaluate the solution. One way of carrying this out is the method of *step-by-step integration* described in Section 1–5. In this chapter several alternative procedures are described. It will be seen that all the methods are closely related to the method of power series. Even when a series solution is available, the new methods may turn out to provide the desired accuracy with considerably less effort; indeed, they may be easier to use than an explicit formula for the solution.

With the development of digital computers of extreme rapidity, the interest in numerical methods has become very great. We have almost reached a stage where we can consider all solutions of all differential equations as tabulated and available for use. The table is "stored" in the computer. To get the table out for examination, we need only feed the correct program to the computer. From this point of view, finding solutions to differential equations is no more challenging than finding a square root or logarithm. We do of course need to learn how to prepare a program for the computer; that is one of the goals of the present chapter. Furthermore, the tabulated or graphed solution may appear to be very complicated, and a deeper mathematical study of the differential equation may provide better answers to questions about the nature of the solutions.

10–2 Step-by-step integration. Let us review briefly the method of step-by-step integration. For the equation

$$y' = F(x, y) \tag{10–1}$$

we consider the corresponding *difference equation*

$$\Delta y = F(x, y) \, \Delta x. \tag{10–2}$$

If (x_0, y_0) is the given initial point, we then compute successively the values $y_1, y_2, \ldots$, at $x_0 + \Delta x, x_0 + 2\Delta x, \ldots$, by the formulas

$$y_1 = y_0 + F(x_0, y_0)\, \Delta x, \qquad y_2 = y_1 + F(x_1, y_1)\, \Delta x, \qquad \ldots \quad (10\text{–}3)$$

The increment Δx can also be varied from step to step.

The step-by-step procedure corresponds to a series method that uses only terms up to degree 1. For if $y = f(x)$ is the solution sought, then

$$f(x_0) = y_0, \qquad f(x_0 + \Delta x) = f(x_0) + f'(x_0)\, \Delta x + f''(x_0)\, \frac{\overline{\Delta x}^2}{2} + \cdots ,$$

$$\Delta y = f(x_0 + \Delta x) - f(x_0) = f'(x_0)\, \Delta x + \cdots = F(x_0, y_0)\, \Delta x + \cdots$$

Unless the series is very rapidly convergent, the error made by omitting terms of second and higher degree will be large; however, if Δx is very small, the error in turn will be small. Making Δx smaller requires more steps if y is to be found over a given interval. However, this is the only way to ensure accuracy. As $\Delta x \to 0$, the errors all approach 0 over the interval. By an elaboration of the reasoning of Section 9–4 to 9–6, we can determine how small Δx must be in order to ensure that all errors are less than a given amount. In practice we usually settle this question by experiment. We compute with a given value of Δx over a given interval; then we recompute with a smaller value of Δx, say one-half of the first. If the values of y are changed to a negligible extent, we consider the solution to have been found to desired accuracy.

The method of step-by-step integration can be extended to systems of first order equations:

$$\frac{dy_i}{dx} = F_i(x, y_1, \ldots, y_n) \qquad (i = 1, \ldots, n). \qquad (10\text{–}4)$$

The corresponding difference equations are

$$\Delta y_i = F_i(x, y_1, \ldots, y_n)\, \Delta x \qquad (i = 1, \ldots, n). \qquad (10\text{–}5)$$

The method also extends to equations of higher order. For example, the equation

$$y'' = F(x, y, y')$$

is replaced by the system

$$\frac{dy_1}{dx} = y_2, \qquad \frac{dy_2}{dx} = F(x, y_1, y_2),$$

which is then analyzed as a special case of (10–4).

10–3 Series solution. For purposes of comparison we give here the series solution of Eq. (10–1) through terms of degree 4, as computed by the method of Section 9–1. With $h = \Delta x$, we find (Prob. 7 below)

$$y_1 = f(x_0 + h) = y_0 + Fh + (F_x + F_yF)\,\frac{h^2}{2}$$

$$+ \,(F_{xx} + 2FF_{xy} + F^2F_{yy} + F_xF_y + FF_y^2)\,\frac{h^3}{6}$$

$$+ \,[F_{xxx} + 3F_xF_{xy} + F_yF_{xx} + F_xF_y^2 + F(3F_{xxy} + 5F_yF_{xy}$$

$$+ \,3F_xF_{yy} + F_y^3) + F^2(3F_{xyy} + 4F_yF_{yy}) + F_{yyy}F^3]\,\frac{h^4}{24} + \cdots,$$

$$(10\text{–}6)$$

where $F, F_x = \partial F/\partial x , \ldots$ are all evaluated at (x_0, y_0).

10–4 Heun's method. For the equation

$$y' = F(x), \qquad (10\text{–}7)$$

finding the solution $f(x)$ over the interval $x_0 \leqq x \leqq b$, with $y_0 = f(x_0)$ given, is equivalent to evaluating a definite integral:

$$f(x) = y_0 + \int_{x_0}^{x} F(t)\, dt \qquad (x_0 \leqq x \leqq b). \qquad (10\text{–}8)$$

The definite integral can in turn be evaluated by the *trapezoidal rule.* For example, at $x_1 = x_0 + \Delta x$,

$$f(x_1) = y_0 + \Delta x\,\frac{F(x_0) + F(x_1)}{2}$$

$$= y_0 + \Delta x\,\frac{y_0' + y_1'}{2}. \qquad (10\text{–}9)$$

For the more general equation $y' = F(x, y)$ we cannot apply (10–9) because the slope y_1' at (x_1, y_1) is unknown. A reasonable estimate for y_1 is provided by step-by-step integration:

$$y_1 = y_0 + F(x_0, y_0)\,\Delta x.$$

Hence a reasonable estimate for y_1' is

$$y_1' = F[x_1, y_0 + F(x_0, y_0)\,\Delta x].$$

With this expression for y_1', (10–9) gives a new value for y_1:

$$y_1 = f(x_1) = y_0 + \frac{\Delta x}{2} \{F(x_0, y_0) + F[x_1, y_0 + F(x_0, y_0) \Delta x]\}.$$

$$(10\text{–}10)$$

Equation (10–10) is the basis of *Heun's method*. If we write $\Delta x = h$, $x_1 = x_0 + h$, then

$$F[x_1, y_0 + F(x_0, y_0)h] = F[x_0 + h, y_0 + F(x_0, y_0)h]$$

$$= F + h(F_x + FF_y) + \frac{h^2}{2} (F_{xx} + 2F_{xy}F + F_{yy}F^2) + \cdots,$$

$$(10\text{–}11)$$

by a Taylor series expansion (Prob. 8 below), where $F, F_x, \ldots$ are evaluated at (x_0, y_0). Hence Eq. (10–10) gives

$$f(x_0 + h) = f(x_0) + \frac{h}{2} [2F + h(F_x + FF_y) + \cdots]$$

$$= f(x_0) + hF + \frac{h^2}{2} (F_x + FF_y)$$

$$+ \frac{h^3}{4} (F_{xx} + 2F_{xy}F + F_{yy}F^2) + \cdots \qquad (10\text{–}12)$$

A comparison of Eqs. (10–12) and (10–6) shows agreement through the term of *second* degree in h. Accordingly, Heun's method can be considered an alternative to computing a series solution through terms of second degree.

The steps in application of Heun's method are summarized in the following formulas:

$$m_0 = F(x_0, y_0),$$

$$x_1 = x_0 + \Delta x,$$

$$y_1^* = y_0 + m_0 \Delta x,$$

$$m_1 = F(x_1, y_1^*),$$

$$m = \tfrac{1}{2}(m_0 + m_1),$$

$$\Delta y = m \Delta x,$$

$$y_1 = y_0 + \Delta y.$$

$$(10\text{–}13)$$

EXAMPLE. The equation chosen is

$$y' = xy^2 - y,$$

with $y = 1$ for $x = 0$. Trying to obtain an accuracy of two significant figures, we seek the solution for $0 \le x \le 1$.

TABLE 10-1

x_0	y_0	x_1	m_0	y_1^*	m_1	m	Δy
0	1.000	0.5	−1.000	0.500	−0.375	−0.688	−0.344
0.5	0.656	1.0	−0.441	0.435	−0.246	−0.344	−0.172
1.0	0.484						

TABLE 10-2

x_0	y_0	x_1	m_0	y_1^*	m_1	m	Δy
0	1.000	0.25	−1.000	0.75	−0.609	−0.805	−0.201
0.25	0.799	0.50	−0.639	0.639	−0.435	−0.537	−0.134
0.50	0.665	0.75	−0.444	0.554	−0.324	−0.384	−0.096
0.75	0.569	1.00	−0.326	0.487	−0.250	−0.0288	−0.072
1.00	0.497						

In Table 10–1 Δx is chosen as 0.5; in Table 10–2 Δx is reduced to 0.25. Since there is a change in the second significant figure, we should try to reduce Δx further, say to 0.125. If this is done, we find that the results of Table 10–2 are unchanged, up to the first two significant figures; hence Table 10–2 gives satisfactory results.

If we solve the problem by a Taylor series, we find

$$y = 1 - x + x^2 - x^3 + x^4 - \cdots \qquad (10\text{–}14)$$

This suggests that the solution is given exactly by

$$y = \frac{1}{1 + x},$$

and substitution in the differential equation proves the surmise to be correct. Accordingly, the exact values of y at $x = 0, 0.25, \ldots$ are 1, 0.800, 0.6666 ... , 0.5714 ... , 0.500, in very close agreement with Table 10–2. It should be noted that the series converges only for $|x| < 1$, and hence cannot be used for the whole interval. Even for $x = 0.5$, as many as nine terms of the series are needed to give an accuracy as good as that of Table 10–2. Heun's method is equivalent to recomputation of the series, through terms of degree 2, at each successive value of x.

Heun's method can be extended to equations of higher order and to systems (Prob. 4 below).

1. For the differential equation, with initial conditions,

$$y' = x - y \qquad (y = 1 \text{ for } x = 0),$$

carry out the following steps:

(a) obtain the solution for $x = 0, 0.1, 0.2, 0.3, 0.4$ by step-by-step integration;

(b) obtain the solution for $x = 0, 0.2, 0.4$ by Heun's method;

(c) obtain the solution for $x = 0, 0.1, 0.2, 0.3, 0.4$ by Heun's method; discuss the number of significant figures in the result;

(d) obtain the Taylor series solution up to terms of degree 4 and evaluate for $x = 0, 0.1, 0.2, 0.3, 0.4$;

(e) verify that the exact solution is $y = 2e^{-x} + x - 1$ and evaluate at $x = 0, 0.1, 0.2, 0.3, 0.4$.

2. Carry out the program of Prob. 1 for the following differential equation with initial conditions:

$$y' = 1 + (x + x^2)e^{-y} \qquad (y = 0 \text{ for } x = 0).$$

In part (e) show that $y = \log (4e^x - 3 - 3x - x^2)$ is the exact solution.

3. Carry out the program of Prob. 1 for the following differential equation with initial conditions:

$$y' = y^2 e^{-x} + \tfrac{1}{2}y \qquad (y = \tfrac{1}{2} \text{ for } x = 0).$$

In part (e) show that $y = \tfrac{1}{2}e^x$ is the exact solution.

4. Show that the reasoning which leads to Heun's method can be extended to a system

$$\frac{dy}{dx} = F(x, y, z), \qquad \frac{dz}{dx} = G(x, y, z)$$

to yield the following rule. Given (x_0, y_0, z_0) and Δx, we evaluate $\Delta y, \Delta z$ by the rules

$$\Delta y = \tfrac{1}{2} \Delta x[m_0 + F(x_0 + \Delta x, y_0 + m_0 \Delta x, z_0 + n_0 \Delta x)],$$

$$\Delta z = \tfrac{1}{2} \Delta x[n_0 + G(x_0 + \Delta x, y_0 + m_0 \Delta x, z_0 + n_0 \Delta x)],$$

where $m_0 = F(x_0, y_0, z_0)$, $n_0 = G(x_0, y_0, z_0)$.

5. Apply the method of Prob. 4 to evaluate the solution of the problem

$$\frac{dy}{dx} = y - 4z, \quad \frac{dz}{dx} = y - 3z \qquad (y = 2, z = 1 \text{ for } x = 0)$$

for $x = 0, 0.5, 1$. Compare the results with the exact solution $y = 2e^{-x}, z = e^{-x}$.

6. Given the second order equation, with initial conditions,

$$y'' + x^2 y = 0 \qquad (y = 1, y' = 1 \text{ for } x = 0),$$

(a) replace by the system $y' = z$, $z' = -x^2 y$ and obtain the solution as in Prob. 4 for $x = 0, 0.5, 1$.

(b) Obtain the Taylor series solution through terms of degree 5 and compare with the result of part (a).

7. Obtain the series expansion (10–6) for the solution of the equation $y' = F(x, y)$ with given initial values. [*Hint:* As in Prob. 3 following Section 9–1, we have

$$y' = F, \qquad y'' = F_x + F_y y' = F_x + F F_y, \qquad \ldots]$$

8. Obtain the series expansion (10–11). [*Hint:* As in Section 9–4, Eqs. (9–11) and (9–12),

$$F(x_0 + h, y_0 + k) = F + h F_x + k F_y + \frac{1}{2!} (h^2 F_{xx} + 2hk F_{xy} + k^2 F_{yy})$$

$$+ \frac{1}{3!} (h^3 \cdot F_{xxx} + 3h^2 k F_{xxy} + 3hk^2 F_{xyy} + k^3 F_{yyy}) + \cdots,$$

where $F, F_x, F_y, \ldots$ are evaluated at (x_0, y_0).]

ANSWERS

1. (a) 1, 0.9, 0.82, 0.758, 0.712; (b) 1, 0.840, 0.745; (c) 1, 0.910, 0.8381, 0.7825, 0.7417; (d) 1, 0.9097, 0.8375, 0.7817, 0.7408; (e) 1, 0.9097, 0.8375, 0.7817, 0.7406.

2. (a) 0, 0.1, 0.210, 0.329, 0.457; (b) 0, 0.220. 0.474; (c) 0, 0.105, 0.220, 0.343, 0.474; (d) and (e) same as (c).

3. (a) 0.5, 0.55, 0.605, 0.665, 0.731; (b) 0.5, 0.609, 0.742; (c) 0.5, 0.553, 0.611, 0.675, 0.746; (d) and (e) same as (c).

5. $y = 2, 1.25, 0.781, z = 1, 0.625, 0.391$. Exact solution: $y = 2, 1.213, 0.736, z = 1, 0.607, 0.368$.

6. (a) $y = 1, 1.50, 1.91, z = 1, 0.91, 0.32$. (b) $y = 1 + x - (x^4/12) - (x^5/20)$, $y = 1, 1.49, 1.87, z = 1, 0.94, 0.42$.

10–5 Runge's method.

Following the reasoning which led to Heun's method, we can try to attain greater accuracy by using *Simpson's rule*. For the area under the curve $y = F(x)$ from x_0 to $x_0 + h$, the rule gives the expression

$$\frac{h}{6} \left[F(x_0) + 4F\left(x_0 + \frac{h}{2}\right) + F(x_0 + h) \right].$$

Hence for the differential equation

$$y' = F(x, y),$$

we are led to the formula

$$\Delta y = \frac{\Delta x}{6}\left[y_0' + 4y_1' + y_2' \right],$$

where $y_0' = F(x_0, y_0)$ is the slope of the solution at $x = x_0$, y_1' is an estimate for the slope at $x_0 + \frac{1}{2}\Delta x$, and y_2' is an estimate for the slope at $x_0 + \Delta x$.

To obtain y_1', y_2', the Runge method uses the formulas

$$m_0 = F(x_0, y_0),$$
$$m_1 = F(x_0 + \tfrac{1}{2}\Delta x, y_0 + \tfrac{1}{2}m_0\,\Delta x),$$
$$m_2 = F(x_0 + \Delta x, y_0 + m_0\,\Delta x),$$
$$m_3 = F(x_0 + \Delta x, y_0 + m_2\,\Delta x).$$

$$(10\text{–}15)$$

Then $y_0' = m_0$ and y_1' is chosen as m_1, y_2' as m_3 (*not* as m_2). Finally,

$$\Delta y = \frac{\Delta x}{6}(m_0 + 4m_1 + m_3) = m\,\Delta x. \qquad (10\text{–}16)$$

The geometric meaning of the slopes $m_0, \ldots, m_3$ is shown in Fig. 10–1.

Taylor series expansions can be used to show that Runge's method is in agreement with the series solution through terms of degree 3 (Prob. 4 below). Hence it should give greater accuracy than Heun's method.

As an example we choose the equation, with initial condition,

$$y' = x - y \qquad (y = 1 \text{ for } x = 0).$$

$$(10\text{–}17)$$

This is the same as Prob. 1 following Section 10–3, so that we can compare the results obtained there with those given by the Runge method. The numerical work is shown in Table 10–3, with $\Delta x = 0.2$. The value obtained at $x = 0.2$ is 0.83733. Heun's method with $\Delta x = 0.2$ gives 0.840, and with $\Delta x = 0.1$ gives 0.8381. The exact solution gives 0.83746.

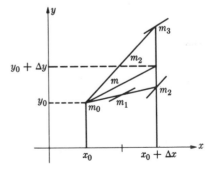

FIG. 10–1. Slopes used in Runge's method.

TABLE 10–3

x_0	0	0.2
y_0	1	0.83733
m_0	-1	-0.63733
$x_0 + \frac{1}{2}\,\Delta x$	0.1	0.3
$y_0 + \frac{1}{2}m_0\,\Delta x$	0.9	etc.
m_1	-0.8	
$x_0 + \Delta x$	0.2	
$y_0 + m_0\,\Delta x$	0.8	
m_2	-0.6	
$y_0 + m_2\,\Delta x$	0.88	
m_3	-0.68	
m	-0.81333	
Δy	-0.16267	

Runge's method can also be extended to systems and to equations of higher order (Prob. 5 below). The method has been modified by Kutta and others to give several procedures which are called *Runge-Kutta methods*. For descriptions of these see, for example, Chapter 5 of Reference 5 and Chapter 6 of Reference 9.

10–6 Milne's method. The methods of Heun and Runge have the defect that certain slopes are first estimated by extrapolation and then recomputed. Since computation of slopes is generally the most time-consuming part of the work, it is desirable to find a method which computes each slope only once. Milne's method has this advantage; instead of extrapolating, we use slopes already found in order to find the best estimate for Δy.

Let the differential equation $y' = F(x, y)$ be given, with initial point (x_0, y_0). Let us suppose that the successive points (x_1, y_1), (x_2, y_2), (x_3, y_3) have already been found, where $x_1 = x_0 + \Delta x$, $x_2 = x_0 + 2\,\Delta x$, $x_3 = x_0 + 3\,\Delta x$. The values y_1, y_2, y_3 can be found, for example, by Runge's method, or by infinite series; Milne's formula cannot be used to find these three values. It does give the succeeding values $y_4, y_5, \ldots$ In each case, four successive values are used to compute the fifth.

Given y_0, y_1, y_2, y_3, we now find y_4 as follows. We write

$$m_1 = F(x_1, y_1), \qquad m_2 = F(x_2, y_2), \qquad m_3 = F(x_3, y_3); \qquad (10\text{–}18)$$

then

$$y_4 = y_0 + \frac{4\,\Delta x}{3}\,(2m_1 - m_2 + 2m_3). \qquad (10\text{–}19)$$

It can be verified that this formula gives a result in agreement with the series solution, up to terms of degree 3. Indeed, if we seek a formula of the type

$$y_4 = y_0 + \Delta x (A m_1 + B m_2 + C m_3),$$

the requirement that the solution agree with the series solution up to terms of degree 3 leads to three conditions on A, B, C, from which we find $A = \frac{8}{3}$, $B = -\frac{4}{3}$, $C = \frac{8}{3}$, so that (10–19) is obtained (Prob. 6 below). This suggests generalizations of Milne's formula to achieve greater accuracy.

EXAMPLE. We consider again the problem of Eq. (10–17). The values of y for $x = 0.1, 0.2, 0.3$ are taken as the exact values (assumed to have been found to four significant figures by Runge's method). The values of y at $x = 0.4$ and 0.5 are computed by the Milne formula. Thus, for $x = 0.4$, $y_0 = 1$ and $m_1 = -0.8097$, $m_2 = -0.6375$, $m_3 = -0.4816$; for $x = 0.5$, $m_1 = -0.6375$, $m_2 = -0.4816$, $m_3 = -0.3407$, and $y_0 = -0.9097$. The results are shown in Table 10–4. The exact values for $x = 0.4$ and 0.5 are 0.740640 and 0.713062, respectively.

TABLE 10–4

x	0	0.1	0.2	0.3	0.4	0.5
y	1	0.9097	0.8375	0.7816	0.7407	0.7131
$F(x, y)$		−0.8097	−0.6375	−0.4816	−0.3407	

10–7 Accuracy of results. The problem of evaluating the accuracy of the tabulated solution is not particularly simple. The simplest test is to reduce the size of the Δx and determine whether the values of y are changed to the number of significant figures we wish to have. There are other check formulas, based on recalculation of the solution by the trapezoidal or Simpson's rule (or some generalization of these). We can of course return to the series solution and use the estimates of Sections 9–4 to 9–6; these give reasonable bounds for methods which agree with the series solution up to terms of kth degree.

One basic difficulty is the accumulation of errors. If the solution is to be computed over a long interval of x, the errors at each stage influence those at the next, so that there is considerably less accuracy at the end than at the beginning. Under some favorable circumstances the computation is stable and errors introduced at each stage have little effect after a few stages have been passed. This can be shown graphically for the case of step-by-step integration (Fig. 10–2). If all solutions approach a given one as x increases, so that there is essentially a unique output, as in Chapter 3, then the step-by-step solution will also remain close to the output, so that errors will not accumulate.

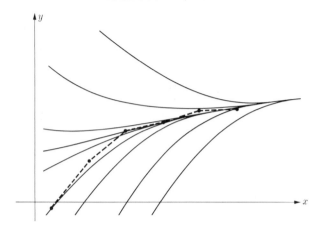

FIG. 10–2. Stability of numerical method.

When accuracy is very important, we must pay considerable attention to errors that arise from rounding off numbers in decimal form. We must compute to at least one and perhaps several decimal places beyond the one at which accuracy is desired. In a long computation, errors due to rounding off can accumulate to a disastrous extent.

10–8 Use of analog computers. The methods described in this chapter are ideally suited to digital computers (Reference 8). For some problems the relevant information can be obtained much more easily by means of differential analyzers or their most modern form, analog computers (see References 3 and 7). These are physical mechanisms which are built to obey given differential equations. Their accuracy is limited by the physical nature of the components used; however, in skilled hands they can give very accurate results. The computers are very flexible, so that we can modify coefficients in a differential equation or change initial conditions with great ease. Families of solution curves can be rapidly produced and qualitative features of the solutions noted visually.

PROBLEMS

1. Given the differential equation $y' = x + y$, with initial condition $y = 1$ for $x = 0$, determine the values of y for $x = 0.1, 0.2, 0.3, 0.4, 0.5$

　(a) by step-by-step integration, $\Delta x = 0.1$;
　(b) by infinite series, using terms up to degree 5;
　(c) by Runge's method;
　(d) from the exact solution.

2. The following values of y on the exact solution of Prob. 1 are found from tables: 1.11034, 1.24281, 1.39972, for $x = 0.1, 0.2, 0.3$. Use Milne's method to obtain the values of y for $x = 0.4, 0.5$ and compare with the results of Prob. 1.

3. Let the differential equation

$$y' = xy + xy^3$$

be given, with $y = 1$ for $x = 0$. Let it be known that the values of y for $x = 0.1, 0.2, 0.3$ are 1.010, 1.043, 1.104.

(a) Find the value of y for $x = 0.4$ by Runge's method.

(b) Find the value of y for $x = 0.4$ by Milne's method.

4. Let $h = \Delta x$. Show from the definitions (10–15) that by series expansions (Prob. 8 following Section 10–4)

$$m_0 = F, \qquad m_1 = F + \frac{a}{2} h + \frac{b}{8} h^2 + \cdots, \qquad m_2 = F + ah + \frac{b}{2} h^2 + \cdots,$$

where $a = F_x + FF_y$, $b = F_{xx} + 2FF_{xy} + F_{yy}F^2$, $c = F_xF_y + FF_y^2$, and $F, F_x, \ldots$ are all evaluated at (x_0, y_0). Show further that

$$m_3 = m_0 + h(F_x + m_2F_y) + \frac{h^2}{2}(F_{xx} + 2m_2F_{xy} + m_2^2F_{yy}) + \cdots$$

$$= F + ah + \frac{b + 2c}{2} h^2 + \cdots$$

Hence conclude that

$$m = F + \frac{a}{2} h + \frac{b + c}{6} h^2 + \cdots,$$

$$y = y_0 + Fh + \frac{a}{2} h^2 + \frac{b + c}{6} h^3 + \cdots$$

and verify that this is in agreement with the series solution (10–6) through terms of degree 3.

5. Runge's method for a system

$$\frac{dy}{dx} = F(x, y, z), \quad \frac{dz}{dx} = G(x, y, z) \qquad (y = y_0, z = z_0 \text{ for } x = x_0)$$

gives the increments Δy, Δz corresponding to the increment $\Delta x = h$ by the following formulas:

$$m_0 = F(x_0, y_0, z_0), \qquad n_0 = G(x_0, y_0, z_0),$$

$$x_1^* = x_0 + \tfrac{1}{2}h, \qquad y_1^* = y_0 + \tfrac{1}{2}m_0h, \qquad z_1^* = z_0 + \tfrac{1}{2}n_0h,$$

$$x_2^* = x_0 + h, \qquad y_2^* = y_0 + m_0h, \qquad z_2^* = z_0 + n_0h,$$

$$m_i = F(x_i^*, y_i^*, z_i^*), \qquad n_i = G(x_i^*, y_i^*, z_i^*) \qquad (i = 1, 2),$$

$$m_3 = F(x_2^*, y_0 + m_2h, z_0 + n_2h), \qquad n_3 = G(x_2^*, y_0 + m_2h, z_0 + n_2h),$$

$$m = \tfrac{1}{6}(m_0 + 4m_1 + m_3), \qquad n = \tfrac{1}{6}(n_0 + 4n_1 + n_3),$$

$$\Delta y = mh, \qquad \Delta z = nh.$$

It can be verified, as in Prob. 4, that these formulas give values that are in agreement with the series solution through terms of degree 3.

Apply Runge's method to obtain the values of y and z for $x = 0.2$ for the system

$$\frac{dy}{dx} = xy^2 + z, \quad \frac{dz}{dx} = xz^2 + y \quad (y = 0, \, z = -1 \text{ for } x = 0).$$

6. To compare Milne's method with a series solution, we consider expansions about (x_2, y_2) and assume that the values y_0, y_1, y_2, y_3 are exact. Let $\Delta x = h$, and

$$a = F_x + F_y F, \quad b = F_{xx} + 2FF_{xy} + F_{yy}F^2, \quad c = F_x F_y + FF_y^2,$$

where $F, F_x, \ldots$ are all evaluated at (x_2, y_2). Show that by Eq. (10–6)

$$y_1 = y_2 - Fh + \frac{a}{2}h^2 - \frac{b+c}{6}h^3 + \cdots,$$

and hence

$$m_1 = F(x_1, y_1) = F(x_2 - h, \, y_2 - Fh + \frac{a}{2}h^2 + \cdots)$$

$$= F - ah + \frac{b+c}{2}h^2 + \cdots$$

Similarly show that $m_2 = F$,

$$m_3 = F + ah + \frac{b+c}{2}h^2 + \cdots,$$

and by Eq. (10–6) that

$$y_0 = y_2 - 2Fh + 2ah^2 - \tfrac{4}{3}(b+c)h^3 + \cdots$$

Let

$$y_4 = y_0 + h(Am_1 + Bm_2 + Cm_3).$$

Show that the requirement that y_4 agree with the series solution of Eq. (10–6):

$$y_4 = y_2 + 2Fh + 2ah^2 + \tfrac{4}{3}(b+c)h^3 + \cdots$$

through the term in h^3 leads to the equations

$$A + B + C = 4, \quad A - C = 0, \quad A + C = \tfrac{16}{3}$$

and that hence $A = C = \tfrac{8}{3}$, $B = -\tfrac{4}{3}$.

7. Milne's method can be extended to the system

$$\frac{dy}{dx} = F(x, y, z), \quad \frac{dz}{dx} = G(x, y, z) \quad (y = y_0, z = z_0 \text{ for } x = x_0)$$

as follows. Let the solution points (x_i, y_i, z_i) be known for $i = 1, 2, 3$, where $x_1 = x_0 + \Delta x$, $x_2 = x_0 + 2\Delta x$, $x_3 = x_0 + 3\Delta x$. Let

$$m_i = F(x_i, y_i, z_i), \quad n_i = G(x_i, y_i, z_i) \qquad (i = 1, 2, 3).$$

Then we choose

$$y_4 = y_0 + \tfrac{4}{3}\Delta x(2m_1 - m_2 + 2m_3),$$

$$z_4 = z_0 + \tfrac{4}{3}\Delta x(2n_1 - n_2 + 2n_3).$$

Apply Milne's method in order to tabulate $\sin x$ as solution of the equation $y'' + y = 0$. To this end we consider the system $y' = z$, $z' = -y$ and assume the values $(0, 0, 1)$, $(0.1, 0.0998, 0.9950)$, $(0.2, 0.1987, 0.9801)$, $(0.3, 0.2955, 0.9553)$ known. We then compute y and z for $x = 0.4, 0.5$, etc.

ANSWERS

1. (a) 1.1, 1.22, 1.362, 1.528, 1.721; (b) 1.11034, 1.24281, 1.39972, 1.58364, 1.79740; (c) 1.11033, 1.24279, 1.39969, 1.58361, 1.79739; (d) exact solution is $2e^x - x - 1$; values are 1.11034, 1.24281, 1.39972, 1.58364, 1.79744.

2. 1.58364, 1.79743.

3. (a) 1.198, (b) 1.192.

5. $y = 0.9165$, $z = 0.9032$.

7. $x = 0.4$, $y = 0.3894$, $z = 0.9211$; $x = 0.5$, $y = 0.4794$, $z = 0.8776$.

SUGGESTED REFERENCES

1. BUCKINGHAM, R. A., *Numerical Methods*. New York: Pitman, 1957.

2. COLLATZ, LOTHAR, *Numerische Behandlung von Differentialgleichungen*, 2nd ed. Berlin: Springer, 1955.

3. JOHNSON, CLARENCE L., *Analog Computer Techniques*. New York: McGraw-Hill, 1956.

4. LEVY, H., and BAGGOTT, E. A., *Numerical Studies in Differential Equations*, Vol. 1. London: Watts and Co., 1934.

5. MILNE, W. E., *Numerical Solution of Differential Equations*. New York: John Wiley and Sons, Inc., 1953.

6. SCARBOROUGH, JAMES B. *Numerical Mathematical Analysis*, 2nd ed. Baltimore: Johns Hopkins Press, 1950.

7. SOROKA, WALTER W., *Analog Methods in Computation and Simulation*. New York: McGraw-Hill, 1954.

8. WILKES, M. V., WHEELER, D. J., and GILL, STANLEY, *The Preparation of Programs for an Electronic Digital Computer*, 2nd ed. Reading, Mass.: Addison-Wesley, 1957.

9. WILLERS, F. A., *Practical Analysis*, transl. by R. T. Beyer. New York: Dover, 1948.

CHAPTER 11

PHASE-PLANE ANALYSIS

11–1 Formulation of program. In this chapter a study is made of the system

$$\frac{dx}{dt} = F(x, y), \qquad \frac{dy}{dt} = G(x, y). \tag{11-1}$$

The goal will not be that of finding solutions of (11–1), but rather that of determining the possible *geometrical configurations* which the solution curves form. The variable t will be treated as a parameter, so that the solutions are curves in the xy-plane.

The equation of second order

$$\frac{d^2x}{dt^2} = \phi\left(x, \frac{dx}{dt}\right) \tag{11-2}$$

is equivalent to a special case of (11–1):

$$\frac{dx}{dt} = y, \qquad \frac{dy}{dt} = \phi(x, y). \tag{11-3}$$

In various physical problems Eqs. (11–3) arise naturally. The pair (x, y) is often termed a *phase* of the physical system, and hence the xy-plane is called the *phase plane*.

A general equation of first order

$$M(x, y)\, dx + N(x, y)\, dy = 0 \tag{11-4}$$

can also be replaced by a system (11–1):

$$\frac{dx}{dt} = N(x, y), \qquad \frac{dy}{dt} = -M(x, y). \tag{11-5}$$

As shown in Section 9–12, introduction of the parameter t may simplify greatly the problem of finding solutions. In particular, the relatively complicated equation

$$(a_1x + b_1y)\, dx + (a_2x + b_2y)\, dy = 0 \tag{11-6}$$

can be replaced by the simpler linear system

$$\frac{dx}{dt} = a_2x + b_2y, \qquad \frac{dy}{dt} = -(a_1x + b_1y). \tag{11-7}$$

It will be assumed that in Eqs. (11–1) the functions F, G have continuous partial derivatives for all (x, y). A point (x, y) at which both F and G are zero will be called an *equilibrium point;* the term *critical point* is also used. Each equilibrium point of Eqs. (11–5) is a *singular point* of Eq. (11–4). It will be assumed that each equilibrium point (x, y) is *isolated;* that is, that the point (x, y) can be enclosed in a circle with (x, y) as center containing no other equilibrium point. (For some purposes exceptional points will be permitted at which the continuity assumptions on F, G are violated; these are *singular points* of the *given* equations 11–1.)

The existence theorem of Section 1–8 implies that for each (x_0, y_0) there is a unique solution $x = f(t)$, $y = g(t)$ of Eqs. (11–1) such that $x = x_0$, $y = y_0$ when $t = 0$. We note that $x = f(t - t_0)$, $y = g(t - t_0)$ is then also a solution of Eqs. (11–1) such that $x = x_0$, $y = y_0$ when $t = t_0$; however, this path is merely a reparametrization of the path $x = f(t)$, $y = g(t)$ (t is replaced by $t - t_0$). Apart from this possible change in parameter, there is a unique solution path through each point (x_0, y_0). If (x_0, y_0) happens to be an equilibrium point, then the solution is given by the equations $x \equiv x_0$, $y \equiv y_0$ for all t.

The appearance of a typical solution family is suggested in Fig. 11–1. The arrows indicate the direction of increasing t. At point P a vector is shown, with x-component F and y-component G. Since $dx/dt = F$, $dy/dt = G$, the vector is tangent to the path and points in the direction of increasing t. We can interpret this vector as the *velocity vector* of the point $[x(t), y(t)]$ as it traces the solution path. We can think of particles moving simultaneously along all solution paths. This is a *fluid motion*, and since the velocity vector does not change with time at each point (x, y), it is a *stationary fluid motion*. The equilibrium points are points of zero velocity;

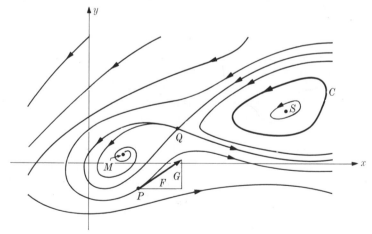

FIG. 11–1.   Solutions in the phase plane.

they are stagnation points of the fluid motion. At the stagnation point Q there are two solution paths approaching Q from opposite directions; by the uniqueness theorem, t cannot have a finite limit as $(x, y) \to Q$, and we conclude that t must approach $+\infty$ as $(x, y) \to Q$. Similarly, there are two paths leaving $Q;$ as (x, y) moves backwards on these paths, $t \to -\infty$.

In the configuration of Fig. 11–1 the most striking features are the equilibrium points, the paths ending at the equilibrium points, and the paths which are *closed* (such as the one labeled C). Indeed, if these essentials are known, there is little choice as to how the pattern can be completed. Accordingly, the greatest effort will be concentrated on determination of these features. The equilibrium points are simply the solutions of the simultaneous equations $F(x, y) = 0$, $G(x, y) = 0$. The paths ending at the equilibrium points can be found by the methods of Section 9–12; we shall study them further here. However, finding the closed paths turns out to be very difficult, but very important for applications. We shall give some information on their location.

Stability of equilibrium point. The isolated equilibrium point (x_0, y_0) is termed *stable* if for each $\epsilon > 0$, a $\delta > 0$ can be found such that every solution passing through a point (x_1, y_1) within distance δ of (x_0, y_0) at time t_1 remains within distance ϵ of (x_0, y_0) for all $t > t_1$ (Fig. 11–2) and approaches (x_0, y_0) as $t \to +\infty$. If the preceding conditions are satisfied, except for the requirement that the solutions approach (x_0, y_0) as $t \to +\infty$, then the equilibrium point is termed *neutrally stable*. If the

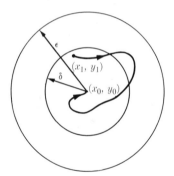

Fig. 11–2. Stability of equilibrium point.

equilibrium point is neither stable nor neutrally stable, then it is called *unstable*. [These definitions are adapted to physical applications, for which it is generally important that deviations from equilibrium die out as $t \to +\infty$; neutral stability guarantees only that initially small deviations will remain small as $t \to +\infty$. In mathematical literature the term *asymptotic stability* corresponds to stability as defined here. (See Chapter 13 of Reference 3, and Reference 12, Vol. 6, pp. 3–21).] In Fig. 11–1 the point M is stable, the point S is neutrally stable and the point Q is unstable.

11–2 The linear case. For the study of equilibrium points a knowledge of the configurations to be expected for the linear system

$$\frac{dx}{dt} = a_1 x + b_1 y, \qquad \frac{dy}{dt} = a_2 x + b_2 y \tag{11–8}$$

turns out to be indispensable. Here we take advantage of the known theory of Eqs. (11–8), as described in Chapter. 6, to classify the configurations which arise.

We introduce the characteristic equation

$$\begin{vmatrix} a_1 - \lambda & b_1 \\ a_2 & b_2 - \lambda \end{vmatrix} = 0, \qquad (11\text{–}9)$$

with roots λ_1, λ_2. This can be written

$$\lambda^2 - p\lambda + q = 0, \qquad (11\text{–}10)$$

$$p = a_1 + b_2 = \lambda_1 + \lambda_2, \qquad (11\text{–}11)$$

$$q = a_1 b_2 - a_2 b_1 = \lambda_1 \lambda_2. \qquad (11\text{–}12)$$

We also introduce the discriminant

$$\begin{aligned} \Delta = p^2 - 4q &= (a_1 + b_2)^2 - 4(a_1 b_2 - a_2 b_1) \\ &= (a_1 - b_2)^2 + 4a_2 b_1. \end{aligned} \qquad (11\text{–}13)$$

The configurations which arise then form three major cases and four borderline cases.

The major cases:

> *Case*   I. $q < 0$, *(Saddle point)*;
>
> *Case*   II. $q > 0, \Delta > 0$, *(Node)*;
>
> *Case* III. $\Delta < 0, p \neq 0$, *(Focus)*.

The borderline cases:

> *Case* IV. $q = 0, p \neq 0$;
>
> *Case* V. $\Delta = 0, p \neq 0$;
>
> *Case* VI. $q > 0, p = 0$, *(Center)*;
>
> *Case* VII. $p = 0, q = 0$.

The curve $\Delta = 0$ is a parabola in the pq-plane. This parabola, the p-axis, and the positive q-axis together divide the pq-plane into regions that correspond to the major cases, as shown in Fig. 11–3.

Stability. The stability of the equilibrium point in the various regions is indicated in Fig. 11–3. There is precisely one stable zone, the second quadrant. On its edges there is neutral stability or, at the origin, possible instability. All other combinations (p, q) yield instability. We can also

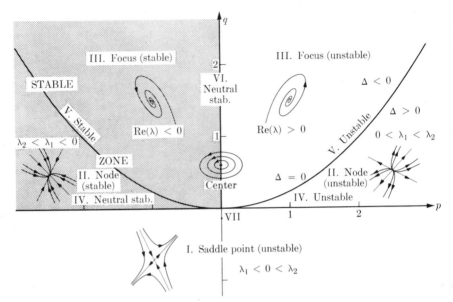

FIG. 11–3. Nature of equilibrium point at origin of $dx/dt = a_1 x + b_1 y$, $dy/dt = a_2 x + b_2 y$ in terms of $p = \lambda_1 + \lambda_2$, $q = \lambda_1 \lambda_2$, $\Delta = p^2 - 4q$; λ_1 and λ_2 are the characteristic roots.

describe the stability in terms of the characteristic roots: *when* $\operatorname{Re}(\lambda) < 0$ *for both characteristic roots* λ, *the equilibrium is stable; if* $\operatorname{Re}(\lambda) > 0$ *for one characteristic root* λ, *the equilibrium is unstable.* If $\operatorname{Re}(\lambda) = 0$ for both characteristic roots λ, or if $\lambda_1 = 0, \lambda_2 < 0$, we have neutral stability, except possibly when $p = q = 0$, so that $\lambda_1 = \lambda_2 = 0$.

11–3 Major cases. *Case* I. Since $q < 0$, the discriminant is positive and the roots λ_1, λ_2 are real and distinct. By Eq. (11–12) the roots have opposite signs; we assume $\lambda_1 < 0 < \lambda_2$. By the theory of Chapter 6 the general solution of Eqs. (11–8) has form

$$x = c_1 \alpha_1 e^{\lambda_1 t} + c_2 \alpha_2 e^{\lambda_2 t}, \qquad y = c_1 \beta_1 e^{\lambda_1 t} + c_2 \beta_2 e^{\lambda_2 t}, \quad (11\text{–}14)$$

where c_1, c_2 are arbitrary constants, and $\alpha_1 \beta_2 - \alpha_2 \beta_1 \neq 0$ (Prob. 4 below, following Section 11–4). For $c_2 = 0, c_1 = \pm 1$ we obtain the solutions

$$x = \pm \alpha_1 e^{\lambda_1 t}, \qquad y = \pm \beta_1 e^{\lambda_1 t}, \qquad (11\text{–}15)$$

which represent two rays approaching $(0, 0)$ from opposite directions; since $\lambda_1 < 0$, x and $y \to 0$ as $t \to +\infty$. For $c_1 = 0, c_2 = \pm 1$ we obtain another pair of rays:

$$x = \pm \alpha_2 e^{\lambda_2 t}, \qquad y = \pm \beta_2 e^{\lambda_2 t}. \qquad (11\text{–}16)$$

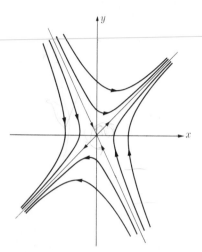

FIG. 11–4. Saddle point.

Since $\lambda_2 > 0$, x and $y \to 0$ as $t \to -\infty$. If neither c_1 nor $c_2 = 0$, the curve represented by (11–14) cannot approach $(0, 0)$ either as $t \to +\infty$ or as $t \to -\infty$ (Prob. 4 below). Furthermore, by the uniqueness theorem, the curve cannot pass through $(0, 0)$ for any finite value of t. Hence there are precisely four solutions leading to $(0, 0)$, namely the four rays (11–15), (11–16). By a graphical analysis of Eq. (11–14) we conclude that the solutions, other than the straight lines, are curves asymptotic to the rays, as suggested in Fig. 11–4 (Prob. 4 below). The configuration resembles that of the level curves of $z = x^2 - y^2$, whose graph in xyz-space is a saddle surface; hence the configuration is termed a *saddle point*. We note that all solutions recede from $(0, 0)$ as $t \to +\infty$, except for one pair of rays. Hence the equilibrium is *unstable*.

Case II. Since $\Delta > 0$, the roots are real and distinct; since $q > 0$, the roots are of same sign. When $p > 0$, both roots are positive; when $p < 0$, both are negative. Let us first assume that both roots are negative. Then the formulas (11–14) again provide the general solution. Furthermore, the formulas (11–15) and (11–16) again provide straight-line solutions; however, since $\lambda_1 < 0, \lambda_2 < 0$, on these solutions $(x, y) \to (0, 0)$ as $t \to +\infty$. For the same reason, along every curve (11–14) $x \to 0$ and $y \to 0$ as $t \to +\infty$. Hence the solutions all approach $(0, 0)$ as $t \to +\infty$ and the equilibrium is stable. A study of the slopes shows that if $\lambda_2 < \lambda_1$, then the solutions (other than 11–16) all approach tangency to the line (11–15) as $t \to +\infty$, as in Fig. 11–5(a) (Prob. 5 below). The solutions are said to form a *node* at $(0, 0)$.

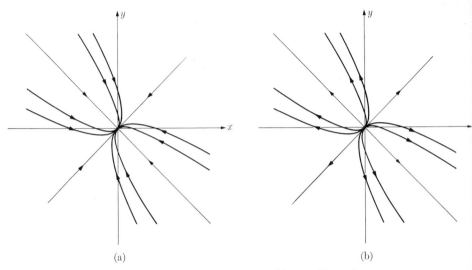

(a)   (b)

FIG. 11–5. (a) Stable node, (b) unstable node.

If $p > 0$, the roots λ_1, λ_2 are positive and we obtain the same conclusions by replacing $t \to +\infty$ by $t \to -\infty$. The configuration is that of Fig. 11–5(b) and the equilibrium is unstable.

Case III. Since $\Delta < 0$, the roots are complex: $\lambda = \rho \pm i\omega$. If $p > 0$, then $\lambda_1 + \lambda_2 = 2\rho > 0$, so that $\rho > 0$; if $p < 0$, then $\rho < 0$. Let us first assume that $p < 0$. The general solution can then be written

$$x = \text{Re}\,[c\alpha e^{(\rho+i\omega)t}], \qquad y = \text{Re}\,[c\beta e^{(\rho+i\omega)t}], \qquad (11\text{–}17)$$

where c is an arbitrary complex constant, and α and β are complex constants, not both zero. If we write

$$c = c_1 + ic_2, \qquad \alpha = \alpha_1 + i\alpha_2, \qquad \beta = \beta_1 + i\beta_2,$$

we find

$$x = e^{\rho t}[c_1(\alpha_1 \cos \omega t - \alpha_2 \sin \omega t) - c_2(\alpha_1 \sin \omega t + \alpha_2 \cos \omega t)],$$

$$y = e^{\rho t}[c_1(\beta_1 \cos \omega t - \beta_2 \sin \omega t) - c_2(\beta_1 \sin \omega t + \beta_2 \cos \omega t)]. \qquad (11\text{–}18)$$

Since $\rho < 0$, x and y both execute damped vibrations (Section 5–4), and hence $x \to 0$, $y \to 0$ as $t \to +\infty$.

Furthermore, each path approaches the origin on a spiral as shown in Fig. 11–6(a) or (b). To prove this, we introduce the polar coordinate θ and compute $d\theta/dt$:

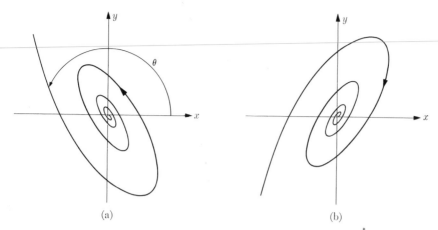

FIG. 11–6. Stable foci: (a) $d\theta/dt > 0$, (b) $d\theta/dt < 0$.

$$\theta = \tan^{-1}\frac{y}{x},$$

$$\frac{d\theta}{dt} = \frac{x\,(dy/dt) - y\,(dx/dt)}{x^2 + y^2},$$

$$\frac{d\theta}{dt} = \frac{a_2 x^2 + xy(b_2 - a_1) - b_1 y^2}{x^2 + y^2}, \qquad (11\text{–}19)$$

by Eqs. (11–8). We assume $x^2 + y^2 \neq 0$, as we are interested in solutions other than the equilibrium solution. Now by assumption $\Delta < 0$. From Eq. (11–13) we find

$$\Delta = (a_1 - b_2)^2 + 4a_2 b_1.$$

Hence $a_2 b_1 < 0$; that is, a_2 and b_1 have opposite signs. Let us suppose $a_2 > 0$, so that $b_1 < 0$. Then for $y = 0$ Eq. (11–19) gives $d\theta/dt = a_2 > 0$; for $x = 0$ Eq. (11–19) gives $d\theta/dt = -b_1 > 0$. If neither x nor y is 0, $d\theta/dt$ cannot be 0; for if it were, Eq. (11–19) would give

$$a_2 x^2 + xy(b_2 - a_1) - b_1 y^2 = 0,$$

$$a_2 \left(\frac{x}{y}\right)^2 + (b_2 - a_1)\frac{x}{y} - b_1 = 0.$$

This is a quadratic equation for x/y, which we are assuming to be satisfied for some real value of x/y. Hence the discriminant of the equation must be positive, but this discriminant is

$$(b_2 - a_1)^2 + 4a_2 b_1 = \Delta < 0,$$

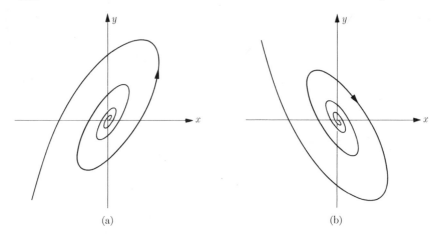

Fig. 11–7. Unstable foci: (a) $d\theta/dt > 0$, (b) $d\theta/dt < 0$.

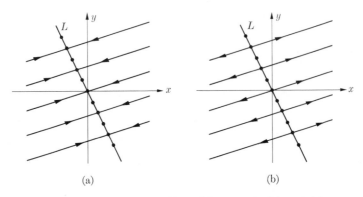

Fig. 11–8. Case IV. (a) Neutrally stable, $p < 0$; (b) unstable, $p > 0$.

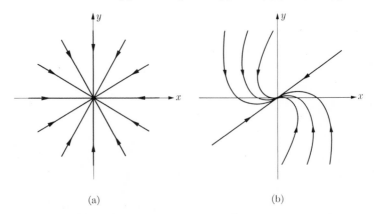

Fig. 11–9. Case V. Curves are drawn for $\lambda < 0$. When $\lambda > 0$, directions are reversed. (a) $a_2 = b_1 = 0$, (b) $a_2 b_1 \neq 0$.

which is a contradiction. Hence $d\theta/dt$ is never 0. Since $d\theta/dt$ is continuous for $x^2 + y^2 \neq 0$, it always has the same sign, which must be a plus sign. Thus θ increases steadily on the solution. Since x and y change sign infinitely often, the path must spiral into the origin as in Fig. 11–6(a). If $a_2 < 0$, then $b_1 > 0$, $d\theta/dt < 0$ and the path spirals into the origin as in Fig. 11–6(b).

If $p > 0$, then $\rho > 0$, and we obtain similar conclusions for $t \to -\infty$. When $a_2 > 0$, again $d\theta/dt > 0$. As $t \to -\infty$, θ now decreases steadily, so that the curve approaches the origin as in Fig. 11–7(a). When $a_2 < 0$, θ increases as $t \to -\infty$, as in Fig. 11–7(b).

11–4 Borderline cases. For applications these cases are of somewhat less interest, since the slightest change in one of the constants will change the configuration to that of one of the major cases. We summarize briefly the appearances in the four cases and leave the discussion to the problems.

Case IV. There is a line L of equilibrium points. The other solutions are parallel rays, each ending at an equilibrium point on L (Fig. 11–8a and b). In this case $\lambda_1 = 0$, $\lambda_2 \neq 0$. When $\lambda_2 < 0$, the origin has neutral stability; when $\lambda_2 > 0$, the origin is an unstable equilibrium point.

Case V. Here $\lambda_1 = \lambda_2 \neq 0$. There is stability for $\lambda_1 < 0$ and instability for $\lambda_1 > 0$. The solutions are either rays ending at the origin, as in Fig. 11–9(a), or else as in Fig. 11–9(b), a family that consists of two oppositely directed rays plus curves approaching the origin at a direction tangent to one of them.

Case VI. The characteristic roots are pure imaginary. The solutions are ellipses with center at the origin, as in Fig. 11–10. The equilibrium point is called a *center*. Although the nearby solutions do not approach the equilibrium point as $t \to +\infty$, they do not recede indefinitely, and the equilibrium has neutral stability.

Case VII. Either every point is an equilibrium point or else there is a line L of equilibrium points through $(0, 0)$ and the other solutions are lines parallel to L (Fig. 11–11).

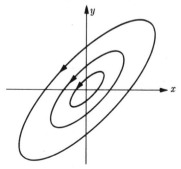

FIG. 11–10. Case VI. Center.

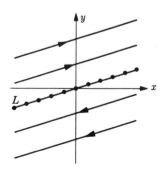

FIG. 11–11. Case VII.

PROBLEMS

1. Determine which of the seven cases of Section 11–2 is applicable to the following systems, find the solutions and graph:

(a) $dx/dt = 3x - 2y, dy/dt = 2x + 3y$;

(b) $dx/dt = 2x + y, dy/dt = x + 2y$.

2. Determine which of the seven cases of Section 11–2 is applicable to the following systems and whether the equilibrium is stable:

(a) $dx/dt = 2x + 5y, dy/dt = x + 2y$;

(b) $dx/dt = -3x + 6y, dy/dt = -2x$;

(c) $dx/dt = x - y, dy/dt = 2x - y$;

(d) $dx/dt = 2x + y, dy/dt = 4x + 2y$;

(e) $dx/dt = -2x - y, dy/dt = 2x + y$;

(f) $dx/dt = 0, dy/dt = 2x$.

3. Consider the equation for the damped *vibration of a spring* (Section 5–10):

$$m \frac{d^2 x}{dt^2} + 2h \frac{dx}{dt} + k^2 x = 0 \qquad (m > 0, h \geqq 0, k > 0).$$

Replace the second order equation by the two first order equations

$$\frac{dx}{dt} = y, \qquad \frac{dy}{dt} = -\frac{k^2}{m} x - \frac{2h}{m} y,$$

which are a special case of Eqs. (11–8).

(a) Find p, q, Δ

(b) Which case corresponds to absence of friction ($h = 0$)?

(c) Which case corresponds to undercritical damping? to overcritical damping? to critical damping?

4. Consider Eqs. (11–8) in Case I: $\lambda_1 < 0 < \lambda_2$.

(a) Obtain the general solution in form (11–14) and show that $\alpha_1 \beta_2 - \alpha_2 \beta_1 \neq 0$.

(b) Show that if $c_1 \neq 0, c_2 \neq 0$, then on the solution (11–14) $|x|$ or $|y|$ or both approach $+\infty$ as $t \to +\infty$ and as $t \to -\infty$.

(c) Under the assumptions of part (b) show that $y/x \to \beta_2/\alpha_2$ as $t \to +\infty$ and $y/x \to \beta_1/\alpha_1$ as $t \to -\infty$.

(d) Show that the isoclines $(a_2 x + b_2 y)/(a_1 x + b_1 y) = m$ are straight lines through the origin and hence that in Case I, except for the rays (11–15), (11–16), no solution is tangent to an isocline. Show that this implies that for the solutions discussed in part (b) the polar angle θ must be steadily increasing or steadily decreasing as t increases. Interpret in terms of Fig. 11–4.

5. In Eqs. (11–14) let $\lambda_2 < \lambda_1 < 0, c_1 \neq 0, c_2 \neq 0, \alpha_1^2 + \beta_1^2 \neq 0, \alpha_2^2 + \beta_2^2 \neq 0$. Show that as $t \to +\infty, y/x \to \beta_1/\alpha_1$ and, as $t \to -\infty, y/x \to \beta_2/\alpha_2$. Interpret in terms of Fig. 11–5(a).

6. Let Case IV hold, so that $q = 0$, $p \neq 0$, $\lambda_1 = 0$, $\lambda_2 \neq 0$. Show that Eqs. (11–14) are applicable, with $\lambda_1 = 0$, $\lambda_2 = p$. Show that the solutions with $c_2 = 0$ form a line of equilibrium points and that the solutions with $c_2 \neq 0$ form parallel rays which approach the equilibrium points as $t \to +\infty$, (for $p < 0$), or as $t \to -\infty$, (for $p > 0$). (See Fig. 11–8.)

7. Let Case V hold, so that $\Delta = 0$, $p \neq 0$, $\lambda_1 = \lambda_2 \neq 0$. By the theory of Chapter 6 the general solution has the form

$$x = e^{\lambda_1 t}[c_1\alpha_1 + c_2(\alpha_2 + \alpha_3 t)],$$

$$y = e^{\lambda_1 t}[c_1\beta_1 + c_2(\beta_2 + \beta_3 t)].$$

(a) Show that if $\lambda_1 < 0$, then all solutions approach $(0, 0)$ as $t \to +\infty$; if $\lambda_1 > 0$, all solutions approach $(0, 0)$ as $t \to -\infty$.

(b) Show that the isoclines $(a_2 x + b_2 y)/(a_1 x + b_1 y) = m$ are rays from the origin. Show that if $a_2 = b_1 = 0$, then $a_1 = b_2$ and all the isoclines are solutions. (Fig. 11–9a.)

(c) Show that if $a_2 \neq 0$ or $b_1 \neq 0$, only two isoclines are solutions, namely those of slope $(b_2 - a_1)/2b_1 = 2a_2/(a_1 - b_2)$. To study the other solutions, show that if $a_2 \neq 0$, then Eq. (11–19) can be written in the form

$$\frac{d\theta}{dt} = \frac{[2a_2 \cos\theta + (b_2 - a_1)\sin\theta]^2}{4a_2} = \frac{R^2}{4a_2}\cos^2(\theta - \alpha),$$

where $2a_2 = R\cos\alpha$, $b_2 - a_1 = R\sin\alpha$. Show that $\tan(\theta - \alpha) = (R^2/4a_2)t + c$, and conclude that each solution approaches the origin in one of the two directions $\alpha \pm \frac{1}{2}\pi$, which are the directions of the two solution rays. (See Fig. 11–9b. In this figure a_2 is >0, so that $d\theta/dt > 0$. If $a_2 < 0$, then $d\theta/dt < 0$ and the figure should be reflected in the line of the solution rays.)

8. Let Case VI hold, so that $q > 0$, $p = 0$, $\Delta < 0$. Show that the characteristic roots are pure imaginary and that Eqs. (11–18) give the general solution, with $\rho = 0$; hence conclude that x and y execute sinusoidal oscillations. Show that since $p = 0$, elimination of t leads to an exact equation

$$(a_2 x + b_2 y)\, dx - (a_1 x + b_1 y)\, dy = 0.$$

Obtain the general solution and show that since $\Delta < 0$, the solutions are similar ellipses.

9. Let Case VII hold, so that $p = 0$, $q = 0$, $\lambda_1 = \lambda_2 = 0$. Show that either $a_1 = a_2 = b_1 = b_2 = 0$ or else one of the two equations $a_1 x + b_1 y = 0$, $a_2 x + b_2 y = 0$ defines a line L each point of which is an equilibrium point, and that the lines parallel to L are solution paths (Fig. 11–11). Show that when $a_1 = a_2 = b_1 = b_2 = 0$, every point (x, y) is an equilibrium point.

1. (a) III, unstable focus, $x = e^{3t}(c_1 \cos 2t + c_2 \sin 2t)$, $y = e^{3t}(c_1 \sin 2t - c_2 \cos 2t)$; (b) II, unstable node, $x = c_1 e^{3t} + c_2 e^t$, $y = c_1 e^{3t} - c_2 e^t$.

2. (a) I, unstable saddle point; (b) III, stable focus; (c) VI, neutrally stable center; (d) IV, unstable; (e) IV, neutrally stable; (f) VII, unstable.

3. (a) $p = -2h/m$, $q = k^2/m$, $\Delta = 4(h^2 - k^2 m)/m^2$; (b) VI, neutrally stable center; (c) III, stable focus; II, stable node; V, stable.

8. Solutions: $a_2 x^2 + 2b_2 xy - b_1 y^2 = $ const, where $b_2^2 + a_2 b_1 < 0$.

11–5 Isolated equilibrium points of nonlinear systems.

We return to the general problem of Section 11–1:

$$\frac{dx}{dt} = F(x, y), \qquad \frac{dy}{dt} = G(x, y). \tag{11–20}$$

We assume that each equilibrium point (x_0, y_0) is *isolated;* that is, that we can find a circle with (x_0, y_0) as center and containing no equilibrium point other than (x_0, y_0). We now ask what configurations the solutions can form near such an isolated equilibrium point.

In the linear case of Sections 11–2 to 11–4 a considerable variety of configurations was found. All these occur for nonlinear systems and are indeed the most common types. The nonlinearity does change the appearance somewhat. For example, at a saddle point (Fig. 11–4) the straight-line solutions ending at the equilibrium point are replaced by smooth curves with the same property; at a center (Fig. 11–10) the ellipses are replaced by other ovals. However, in most cases the configuration is indistinguishable from a linear one if we remain sufficiently close to the equilibrium point. The reason for this will be made clear later in this section.

In addition to the linear configurations, a great variety of others can arise. Some are shown in Fig. 11–12. Despite the involved patterns of these examples, they can arise in differential equations of very simple appearance. For example, the equations

$$\frac{dx}{dt} = 2xy, \qquad \frac{dy}{dt} = y^2 - x^2 \tag{11–21}$$

have solutions of the form of Fig. 11–12(b).

Isolated singular points. By a singular point of Eqs. (11–20) is meant a point (x_0, y_0) at which F or G (or both) fail to have continuous first partial derivatives. For example, the equations

$$\frac{dx}{dt} = \frac{x + x^3 - xy}{x^2 + y^2}, \qquad \frac{dy}{dt} = \frac{2x - y + x^2}{x^2 + y^2} \tag{11–22}$$

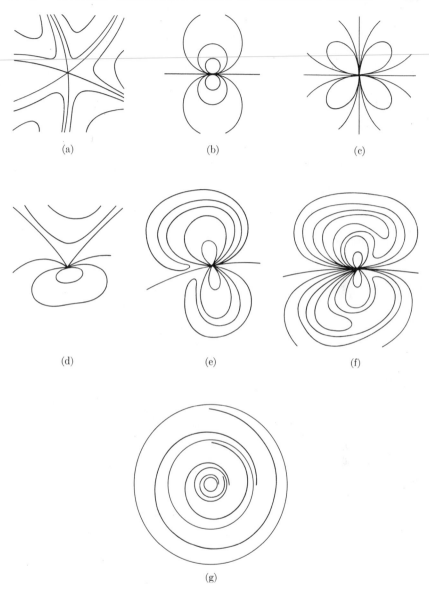

(a)   (b)   (c)

(d)   (e)   (f)

(g)

Fig. 11–12. Isolated equilibrium points of nonlinear systems.

have a singular point at the origin $(0, 0)$. The singular point is said to be *isolated* if there is no other singular point within some circle with center at (x_0, y_0). The configurations formed by the solutions near an isolated singular point resemble those near an isolated equilibrium point, as in Fig. 11–12. In fact, an appropriate change of parameter replaces the singular point by an equilibrium point; for Eqs. (11–22) we can write

$$\frac{dx}{du} = x + x^3 - xy, \qquad \frac{dy}{du} = 2x - y + x^2, \qquad (11\text{–}23)$$

$$du = \frac{dt}{x^2 + y^2}. \qquad (11\text{–}24)$$

Here we think of x, y expressed as functions of t along each solution of Eqs. (11–22), and u is obtained from t by integration along such a curve; that is,

$$u = \int \frac{dt}{x^2(t) + y^2(t)} + \text{const},$$

for some choice of the indefinite integral. We can also consider Eqs. (11–22) and (11–23) as parametrizations of the equations

$$\frac{2y - x + x^2}{x^2 + y^2} dx - \frac{x + x^3 - xy}{x^2 + y^2} dy = 0,$$

$$(2y - x + x^2) dx - (x + x^3 - xy) dy = 0,$$

which are equivalent, except at $(0, 0)$. The replacement of the parametrization (11–22) by the new parametrization (11–23) has the effect of changing the length of the tangent vector without changing its direction; the length of the tangent vector defined by Eqs. (11–22) can be shown to approach ∞ as $(x, y) \to (0, 0)$, but the length of the tangent vector for the new equations (11–23) approaches *zero* as $(x, y) \to (0, 0)$.

For the general equations (11–20) multiplication of the tangent vectors by a scale factor $\mu(x, y)$, $[\mu > 0$ except at $(x_0, y_0)]$, leads to the new equations

$$\frac{dx}{du} = \mu(x, y)F(x, y), \qquad \frac{dy}{du} = \mu(x, y)G(x, y), \qquad (11\text{–}25)$$

$$du = \frac{dt}{\mu(x, y)}; \qquad (11\text{–}26)$$

for appropriate choice of μ we can force the new equations to have an equilibrium point at (x_0, y_0). In the example of the preceding paragraph, $\mu = x^2 + y^2$. The method is illustrated in Probs. 6, 7 below, following Section 11–6.

Linearization at an isolated equilibrium point. Let the system (11–20) have an isolated equilibrium point at (x_0, y_0); for simplicity, we assume $x_0 = 0, y_0 = 0$. To study the nearby solutions, we can (if possible) expand F and G in power series (see Section 9–4):

$$\frac{dx}{dt} = a_1x + b_1y + c_1x^2 + d_1xy + e_1y^2 + \cdots ,$$

$$\frac{dy}{dt} = a_2x + b_2y + c_2x^2 + d_2xy + c_2y^2 + \cdots \tag{11-27}$$

(The constant terms are missing, since $F(0, 0) = G(0, 0) = 0$.) If $|x|$ and $|y|$ are very small, so that (x, y) is very close to $(0, 0)$, the terms of second and higher degree in Eqs. (11–27) are very small compared to those of first degree. Hence it is reasonable to approximate Eqs. (11–22) by the linear system

$$\frac{dx}{dt} = a_1x + b_1y, \qquad \frac{dy}{dt} = a_2x + b_2y. \tag{11-28}$$

It is also natural to conjecture that the configuration of the solutions of (11–27) near the equilibrium point $(0, 0)$ will be the same as that of the solutions of the linear system (11–28). This is indeed so, except in the borderline cases of Section 11–4. *The configuration of a nonlinear system at an equilibrium point is in general the same as that of the approximating linear system.*

In the borderline cases the linear terms fail to determine the pattern. We can then consider the terms through the second degree. Except in certain borderline cases, these will determine the configuration. In the exceptional cases we must consider terms through degree 3, and so on. Equations (11–21) are an example of a system in which the linear terms are all 0 and the second degree terms determine the configuration.

The series expansions of Eqs. (11–27) are very valuable in studying all the possibilities. In particular, as in Section 9–13, they can be used to find the solution curves ending at the equilibrium point. However, for the present analysis it is sufficient to replace Eqs. (11–27) by the simpler statement that the right-hand members consist of terms of first order and terms of higher order. More precisely, we assume that

$$\frac{dx}{dt} = a_1x + b_1y + g_1(x, y),$$

$$\frac{dy}{dt} = a_2x + b_2y + g_2(x, y), \tag{11-29}$$

where $g_1(x, y)$, $g_2(x, y)$ have continuous first partial derivatives for all (x, y) and

$$g_1(x, y) = (x^2 + y^2)^{1/2}h_1(x, y), \qquad g_2(x, y) = (x^2 + y^2)^{1/2}h_2(x, y), \tag{11-30}$$

where $h_1(x, y)$, $h_2(x, y)$ are continuous at $(0, 0)$, and

$$h_1(0, 0) = 0, \qquad h_2(0, 0) = 0. \tag{11-31}$$

For example, if

$$\frac{dx}{dt} = x - y + x^2 + 2y^2,$$

then $g_1 = x^2 + 2y^2 = r[(x^2 + 2y^2)/r]$, where $r = (x^2 + y^2)^{1/2}$. Thus $h_1 = (x^2 + 2y^2)/r = r \cos^2 \theta + 2r \sin^2 \theta$ and h_1 does have value 0 at the origin. Similarly, Eqs. (11–29), with the conditions described, will be valid whenever we have series expansions (11–27). They also hold whenever $F(x, y)$ and $G(x, y)$ have continuous first partial derivatives, and necessarily $a_1 = F_x(0, 0)$, $b_1 = F_y(0, 0)$, $a_2 = G_x(0, 0)$, $b_2 = G_y(0, 0)$. (See pp. 82–84 of Reference 7.)

At an equilibrium point (x_0, y_0) other than the origin, a translation $u = x - x_0$, $v = y - y_0$ leads to similar equations

$$\frac{du}{dt} = a_1 u + b_1 v + g_1(u, v),$$

$$\frac{dv}{dt} = a_2 u + b_2 v + g_2(u, v), \tag{11-29a}$$

where

$$a_1 = F_x(x_0, y_0), \qquad b_1 = F_y(x_0, y_0),$$

$$a_2 = G_x(x_0, y_0), \qquad b_2 = G_y(x_0, y_0).$$

11–6 Stability theorems. We now formulate basic relationships between the nonlinear system and the approximating linear one.

THEOREM 1. *Let the system* (11–20) *be given, satisfying the hypotheses of Section* 11–5, *so that there is an isolated equilibrium point at* $(0, 0)$ *and Eqs.* (11–29), (11–30), (11–31) *hold. Let the characteristic roots of the linear system* (11–28) *have negative real parts, so that the linear system has a stable equilibrium point at* $(0, 0)$. *Then the given system* (11–20) *has a stable equilibrium point at* $(0, 0)$.

Proof. Since the linear system is stable, the corresponding point (p, q) is in the second quadrant of Fig. 11–3, so that

$$p = a_1 + b_2 < 0,$$

$$q = a_1 b_2 - a_2 b_1 > 0. \tag{11-32}$$

(Thus we have Case II, III, or V for the linear system.)

We introduce the functions

$$Q(x, y) = (a_1 x + b_1 y)^2 + (a_2 x + b_2 y)^2, \qquad (11\text{–}33)$$

$$V(x, y) = Q(x, y) + q(x^2 + y^2). \qquad (11\text{–}34)$$

[For the linear system $V = v_x^2 + v_y^2 + q(x^2 + y^2)$, so that $\frac{1}{2}V$ has the appearance of kinetic plus potential energy (Section 7–6). The stability of the system is related to the fact that the total energy is being dissipated by friction or some similar effect; hence along each solution the energy decreases steadily and approaches 0. This is precisely what we shall show.]

We remark that $Q(x, y) = 0$ only for $x = 0$, $y = 0$. For the equation $Q(x, y) = 0$ is equivalent to the two simultaneous equations

$$a_1 x + b_1 y = 0, \qquad a_2 x + b_2 y = 0.$$

The determinant of coefficients of these equations is q, which is positive by assumption; hence the only solution is $x = 0$, $y = 0$. In polar coordinates r, θ we have

$$Q = r^2[(a_1 \cos \theta + b_1 \sin \theta)^2 + (a_2 \cos \theta + b_2 \sin \theta)^2].$$

The function in brackets is a continuous function of θ for $0 \leq \theta \leq 2\pi$; since $Q \neq 0$ for $r \neq 0$, this function is never equal to 0. The function is always positive, and has therefore a maximum $\beta > 0$ and a minimum $\alpha > 0$. Therefore

$$\beta r^2 \geq Q \geq \alpha r^2 \qquad (\beta > \alpha > 0), \qquad (11\text{–}35)$$

for all r, θ.

Now on each solution of Eqs. (11–29)

$$\frac{1}{2}\frac{dV}{dt} = (a_1 x + b_1 y)\left(a_1 \frac{dx}{dt} + b_1 \frac{dy}{dt}\right) + (a_2 x + b_2 y)\left(a_2 \frac{dx}{dt} + b_2 \frac{dy}{dt}\right)$$

$$+ q\left(x \frac{dx}{dt} + y \frac{dy}{dt}\right),$$

where dx/dt and dy/dt are given by Eqs. (11–29). After substituting for dx/dt, dy/dt and expanding the right-hand side, we find (Prob. 4 below) that

$$\frac{1}{2}\frac{dV}{dt} = pQ(x, y) + (Ax + By)g_1(x, y) + (Bx + Cy)g_2(x, y), \quad (11\text{–}36)$$

where

$$A = a_1^2 + a_2^2 + q, \qquad B = a_1 b_1 + a_2 b_2, \qquad C = b_1^2 + b_2^2 + q.$$

$$(11\text{–}37)$$

Accordingly, by Eqs. (11–30),

$$\frac{1}{2}\frac{dV}{dt} = pr^2\left[\frac{Q}{r^2} + h_1\left(A\frac{x}{r} + B\frac{y}{r}\right) + h_2\left(B\frac{x}{r} + C\frac{y}{r}\right)\right]$$

$$= pr^2\left[\frac{Q}{r^2} + h_1(A\cos\theta + B\sin\theta) + h_2(B\cos\theta + C\sin\theta)\right].$$

By assumption, h_1 and $h_2 \to 0$ as $r \to 0$, so that the second and third terms in the brackets are as small as desired for r sufficiently small; by inequalities (11–35) the first term is at least equal to α. Hence for appropriate choice of r_1 the whole bracket exceeds $\frac{1}{2}\alpha$ for $r < r_1$. Since $p < 0$, we have

$$\frac{1}{2}\frac{dV}{dt} < pr^2\frac{\alpha}{2} < 0 \qquad (0 < r < r_1). \tag{11–38}$$

Therefore V is steadily decreasing as t increases, as long as $r < r_1$.

Now from Eq. (11–34) the curves $V = \text{const}$ are conic sections; since $q > 0$, both terms in V are positive or 0. Hence the curves $V = \text{const} = k > 0$ must be ellipses with center at the origin (Prob. 5 below). As k decreases, the ellipse shrinks and for $k = 0$ the ellipse reduces to the point $(0, 0)$ (Fig. 11–13). By the preceding paragraph we can choose k_1 so small that for each solution that starts inside the ellipse $V = k_1$ we have $dV/dt < 0$, so that V decreases and the solution must move into successively smaller ellipses as t increases. Each such solution must, in fact, approach the origin, for by inequalities (11–35) and (11–38)

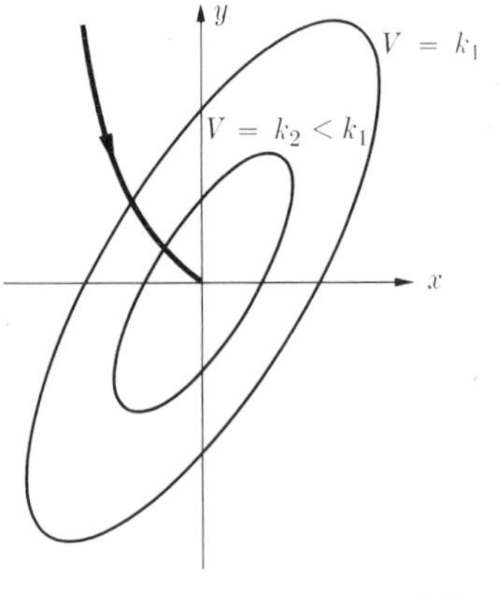

$$V = Q + qr^2 \leqq (\beta + q)r^2,$$

$$\frac{dV}{dt} < par^2 \leqq \frac{p\alpha V}{\beta + q}.$$

Therefore we can write

$$\frac{dV}{dt} + \gamma V = \phi(t) \qquad (\phi(t) < 0),$$

FIG. 11–13. Proof of stability theorem.

where $\gamma = -p\alpha/(\beta + q) > 0$. This is a first order linear differential equation. If $V = V_0$ for $t = 0$ we find

$$V = V_0 e^{-\gamma t} + e^{-\gamma t} \int_0^t e^{\gamma u} \phi(u) \, du < V_0 e^{-\gamma t} \qquad (t > 0).$$

Therefore $V \to 0$ as $t \to \infty$ and every solution (inside the ellipse $V = k_1$) approaches $(0, 0)$ as $t \to +\infty$. Accordingly, the equilibrium is stable.

Remark 1. When $\Delta = a_1 b_2 - a_2 b_1 < 0$ and $p < 0$, the linear system (11–28) falls under Case III and has a stable *focus* at $(0, 0)$. It can be shown that the nonlinear system (11–29) has the same configuration. To prove that the curves spiral into $(0, 0)$, we can compute $d\theta/dt$ as in Section 11–3 and verify that because of the smallness of g_1, g_2, θ is either steadily increasing to $+\infty$ or else steadily decreasing to $-\infty$ on each path. When $\Delta > 0$ and $p < 0$, $q > 0$, the linear system has a stable *node* at $(0, 0)$; it can be shown that the nonlinear system has a similar configuration (pp. 48–59 of Reference 5). When $\Delta = 0$, $p < 0$, $q > 0$, the linear system falls under the borderline case V. Theorem 1 shows that the nonlinear system has a stable equilibrium point at $(0, 0)$, so that all nearby paths approach $(0, 0)$ as $t \to +\infty$; however, the paths may spiral or behave otherwise and no simple statement can be made about the behavior of θ.

Remark 2. The method of proof of Theorem 1 is based on the general procedure developed by Liapounoff and known as *Liapounoff's second method* (Reference 10 and Reference 12, Vol. 6, pp. 3–21). Whenever a continuous function V can be found which is greater than 0 except at the equilibrium point and which is steadily decreasing on each path near the equilibrium point, we can conclude that there is stability.

THEOREM 2. *Let the system* (11–20) *be given, satisfying the hypotheses of Section* 11–5, *so that there is an isolated equilibrium point at* $(0, 0)$ *and Eqs.* (11–29), (11–30), (11–31) *hold. Let at least one characteristic root of the linear system* (11–28) *have positive real part, so that the linear system has an unstable equilibrium point at* $(0, 0)$. *Then the given system* (11–20) *has an unstable equilibrium point at* $(0, 0)$.

Proof. We first assume that $p > 0$, $q > 0$, so that Re $(\lambda_1) > 0$, Re $(\lambda_2) > 0$. For the linear system each solution approaches $(0, 0)$ as $t \to -\infty$. Indeed, if we replace t by $-t$, we obtain a linear system with a stable equilibrium point at $(0, 0)$; hence if t is replaced by $-t$ in (11–20), Theorem 1 shows that the resulting system has a stable equilibrium point at $(0, 0)$. This implies that for the original system (11–20) all solutions near $(0, 0)$ approach $(0, 0)$ as $t \to -\infty$, which, in turn, implies instability of the equilibrium point.

We next assume that $q < 0$, so that the linear system (11–28) has a saddle point at $(0, 0)$. Let the characteristic roots be λ_1, λ_2, with $\lambda_1 < 0 < \lambda_2$. Corresponding to the root λ_2, there is a solution $x = \alpha_2 e^{\lambda_2 t}$,

$y = \beta_2 e^{\lambda_2 t}$, with α_2, β_2 not both 0. Let us assume, for example, that $\alpha_2 > 0$, so that $x > 0$. Then the solution in question is a ray $y = (\beta_2/\alpha_2)x$; as t increases, (x, y) recedes from $(0, 0)$ on the ray (Fig. 11–14).

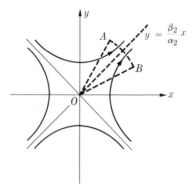

It is natural to conjecture that for the nonlinear system (11–29), the solutions near the ray are all receding from $(0, 0)$. To prove this, we begin by considering the behavior of dr/dt and $d\theta/dt$ along a ray

FIGURE 11–14

$$y = \left(\frac{\beta_2}{\alpha_2} + k\right) x \qquad (-\epsilon \leqq k \leqq \epsilon, x > 0). \qquad (11\text{–}39)$$

The value of ϵ (> 0) will be assigned below.

Now α_2, β_2 form a solution of the equations

$$(a_1 - \lambda_2)\alpha_2 + b_1\beta_2 = 0, \qquad a_2\alpha_2 + (b_2 - \lambda_2)\beta_2 = 0. \qquad (11\text{–}40)$$

Hence on the ray (11–39), Eqs. (11–29) give

$$\frac{dx}{dt} = \left(a_1 + \frac{b_1\beta_2}{\alpha_2} + b_1k\right) x + g_1 = \lambda_2 x + b_1 k x + g_1,$$

$$\frac{dy}{dt} = \left(a_2 + \frac{b_2\beta_2}{\alpha_2} + b_2k\right) x + g_2 = \lambda_2 y + (b_2 - \lambda_2)kx + g_2.$$

$$(11\text{–}41)$$

Therefore along the ray (11–39),

$$r\frac{dr}{dt} = x\frac{dx}{dt} + y\frac{dy}{dt}$$

$$= \lambda_2(x^2 + y^2) + b_1kx^2 + (b_2 - \lambda_2)kx\left(\frac{\beta_2}{\alpha_2} + k\right) x + xg_1 + yg_2$$

$$= x^2\left[\lambda_2 + b_1k + \frac{(b_2 - \lambda_2)\beta_2}{\alpha_2}k + (b_2 - \lambda_2)k^2\right] + \lambda_2 y^2$$

$$+ xg_1 + yg_2.$$

If $k = 0$, the coefficient of x^2 is $\lambda_2 > 0$, and hence for $|k| \leqq \epsilon$ and ϵ sufficiently small the coefficient of x^2 is positive. Since g_1, g_2 are of second or higher order, an argument similar to the proof of Theorem 1 shows that the terms xg_1, yg_2 do not affect the sign of $r(dr/dt)$, for $x^2 + y^2$ sufficiently small. Hence $dr/dt > 0$ in a sufficiently small sector, as shown in Fig. 11–14.

We also consider the behavior of $d\theta/dt$ in such a sector. Along each ray (11–39), Eqs. (11–41) give

$$r^2 \frac{d\theta}{dt} = x \frac{dy}{dt} - y \frac{dx}{dt} = (b_2 - \lambda_2)kx^2 - b_1kx^2 \left(\frac{\beta_2}{\alpha_2} + k \right) + xg_2 - yg_1$$

$$= kx^2 \left(b_2 - \lambda_2 - \frac{b_1\beta_2}{\alpha_2} - b_1k \right) + xg_2 - yg_1.$$

By Eqs. (11–40), $-b_1\beta_2 = \alpha_2(a_1 - \lambda_2)$. Hence

$$r^2 \frac{d\theta}{dt} = kx^2(b_2 + a_1 - 2\lambda_2 - b_1k^2) + xg_2 - yg_1.$$

But $\lambda_1 + \lambda_2 = p = b_2 + a_1$. Hence

$$r^2 \frac{d\theta}{dt} = kx^2(\lambda_1 - \lambda_2 - b_1k^2) + xg_2 - yg_1.$$

Here $\lambda_1 - \lambda_2 < 0$ so that the coefficient of kx^2 is negative, if $|k| = \epsilon > 0$ and ϵ is sufficiently small. Again the higher order terms $xg_2 - yg_1$, have no effect if r^2 is sufficiently small. In particular, $d\theta/dt$ is positive for $k = -\epsilon$ and negative for $k = +\epsilon$; that is, along the rays OA and OB the paths enter the sector with increasing t. Once a path has entered, r will increase steadily, and hence the path must leave the sector by the arc AB. Accordingly, solutions exist that start as close as we wish to $(0, 0)$, at $t = 0$, and which recede from 0 to AB as t increases. This clearly implies that $(0, 0)$ is an unstable equilibrium point.

In the borderline case $q = 0$, $p > 0$ we have roots $\lambda_1 = 0 < \lambda_2$, and the reasoning just given can be repeated.

Remark. A more detailed analysis in the saddle-point case, $(q < 0)$, shows that the nonlinear system has a saddle point also. For a proof the reader is referred to pp. 48–59 of Reference 5.

PROBLEMS

1. For each of the following systems verify that there is an equilibrium point at $(0, 0)$, verify that the solutions are level curves of the function $u(x, y)$ given (i.e., that $du/dt = 0$ on each solution), and graph the solutions:

(a) $\dfrac{dx}{dt} = e^x - 1$, $\dfrac{dy}{dt} = ye^x$, $u = \dfrac{e^x - 1}{y}$;

$$\left[Hint: \frac{du}{dt} = \frac{e^x}{y} \frac{dx}{dt} - \frac{e^x - 1}{y^2} \frac{dy}{dt}. \right]$$

(b) $\dfrac{dx}{dt} = \sin y$, $\dfrac{dy}{dt} = x$, $u = x^2 + 2 \cos y$;

(c) $\dfrac{dx}{dt} = 2xy$, $\dfrac{dy}{dt} = x^2 - y^2$, $u = x^3 - 3xy^2$;

(d) $\dfrac{dx}{dt} = y^3 - 3x^2 y$, $\dfrac{dy}{dt} = x^3 - 3xy^2$, $u = \dfrac{x^2 - y^2}{(x^2 + y^2)^2}$.

2. Describe the nature of the equilibrium point at $(0, 0)$:
(a) $dx/dt = x + 2y + x^3 - y^3$, $dy/dt = 5x + y - x^3 + y^3$;
(b) $dx/dt = 3x - \dot{y} + x^2 - y^4$, $dy/dt = x + xy - xy^3$;
(c) $dx/dt = \sin x - 3 \sin y$, $dy/dt = 2x - 2y$;
(d) $dx/dt = e^{-x-2y} - 1$, $dy/dt = -x(1 - y)^2$.

3. Find all equilibrium points and determine the stability of each:
(a) $dx/dt = x - 2y$, $dy/dt = x^2 + 2y^2 - 6$,
(b) $dx/dt = x^2 - y^2 - 7$, $dy/dt = x^2 + y^2 - 25$.

4. Prove that along each solution of Eqs. (11–29) the function V defined by Eq. (11–34) has a derivative dV/dt satisfying Eq. (11–36).

5. Let $V(x, y)$ be given by Eqs. (11–33), (11–34). Show that if $q > 0$ and $k > 0$, then each ray $x = at$, $y = bt$, $(0 \leq t < \infty)$, meets each curve $V(x, y) = k$ in exactly one point. Hence conclude that the curve $V = k$, which is a conic section, must be an ellipse. [*Hint:* Show that as t goes from 0 to $+\infty$, $V(at, bt)$ increases from 0 to $+\infty$.]

6. Let the differential equations $dx/dt = F(x, y)$, $dy/dt = G(x, y)$ have an isolated singular point at (x_0, y_0) and let $F^2 + G^2 \neq 0$ in some circle with center at (x_0, y_0). Show that the equations

$$\frac{dx}{du} = r \frac{F}{(F^2 + G^2)^{1/2}}, \qquad \frac{dy}{du} = r \frac{G}{(F^2 + G^2)^{1/2}},$$

where $r = [(x - x_0)^2 + (y - y_0)^2]^{1/2}$, have an isolated equilibrium point at (x_0, y_0).

7. For each of the following systems verify that there is an isolated singular point at $(0, 0)$ and introduce a new parameter in terms of which the origin is an isolated equilibrium point:

(a) $\dfrac{dx}{dt} = \dfrac{x + y - x^2}{(x^2 + y^2)^{1/3}}$, $\qquad \dfrac{dy}{dt} = \dfrac{2xy + y^3}{(x^2 + y^2)^{1/3}}$;

(b) $\dfrac{dx}{dt} = \dfrac{1 + x^2}{x^2 + xy + y^2}$, $\qquad \dfrac{dy}{dt} = \dfrac{1 + y^2}{x^2 + xy + y^2}$,

(c) $\dfrac{dx}{dt} = \log(x^4 + y^4)$, $\qquad \dfrac{dy}{dt} = \csc(x^4 + y^4)$. [*Hint:* See Prob. 6.]

2. (a) Saddle point, unstable; (b) unstable node; (c) stable focus; (d) saddle point, unstable.

3. (a) $(2, 1)$ and $(-2, -1)$, both unstable; (b) $(4, 3)$, $(4, -3)$, $(-4, 3)$ unstable and $(-4, -3)$ stable.

7. Possible solutions:

(a) $dx/du = x + y - x^2$, $dy/du = 2xy + x^3$, $du = dt/(x^2 + y^2)^{1/3}$;

(b) $dx/du = (x^2 + xy + y^2)(1 + x^2)$, $dy/du = (x^2 + xy + y^2)(1 + y^2)$, $du = dt/(x^2 + xy + y^2)^2$;

(c) as in Prob. 6, with $r = (x^2 + y^2)^{1/2}$, $F = \log(x^4 + y^4)$, $G = \csc(x^4 + y^4)$.

11–7 Behavior of solutions at infinity. It is convenient to introduce a *point at infinity* as an ideal point adjoined to the xy-plane. We denote this point by Ω. By definition the point (x, y) approaches Ω if

$$r = (x^2 + y^2)^{1/2} \to +\infty.$$

We can picture Ω by means of a device familiar in the theory of complex variables, *stereographic projection*. We consider the xy-plane as part of a three-dimensional xyu-coordinate system (Fig. 11–15). The sphere S, $x^2 + y^2 + (u - \frac{1}{2})^2 = \frac{1}{4}$, is then tangent to the xy-plane at the origin. The point $(0, 0, 1)$, the "north pole" of the sphere, we denote by N. To each point P in the xy-plane we then assign the point P' on the sphere, where P' is the point at which the line NP meets the sphere. This describes the stereographic projection of the xy-plane on the sphere S. (Conversely, we can consider the assignment of P to P' as a projection of sphere on plane; this is commonly done in map-making.) The point N itself can then be regarded as Ω', the projection of Ω on the sphere. It is clear that as

$$r = (x^2 + y^2)^{1/2} \to +\infty,$$

the point P' does approach $N = \Omega'$.

To study the behavior of the solutions of a system

$$\frac{dx}{dt} = F(x, y), \qquad \frac{dy}{dt} = G(x, y) \tag{11–42}$$

for large r (that is, near Ω), we can project the family of solutions on the sphere S. This can be very revealing. For example, the equations

$$\frac{dx}{dt} = x, \qquad \frac{dy}{dt} = -y \tag{11–43}$$

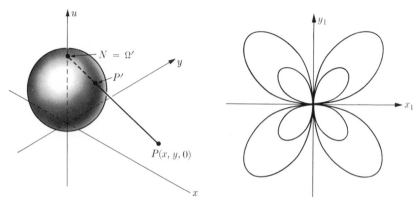

FIG. 11–15. Stereographic projection.   FIG. 11–16. Singular point at infinity.

have as solutions the hyperbolas $xy = $ const. Thus there is a saddle point at $(0, 0)$. If we project the solutions on the sphere, we obtain the configuration of Fig. 11–16, which shows only a neighborhood of Ω. The configuration is clearly quite different.

The behavior of the solutions near the point at infinity can also be studied by means of the substitution

$$x_1 = \frac{x}{r^2}, \qquad y_1 = \frac{-y}{r^2}. \tag{11–44}$$

(This corresponds to the substitution $z_1 = 1/z$, which is much used in theory of functions of complex variables $z = x + iy$, $z_1 = x_1 + iy_1$.) Equations (11–44) assign a unique point (x_1, y_1) in an x_1y_1-plane. Furthermore, Eqs. (11–44) can be solved uniquely for x, y:

$$x = \frac{x_1}{r_1^2}, \qquad y = \frac{-y_1}{r_1^2}, \qquad r_1 = (x_1^2 + y_1^2)^{1/2}, \tag{11–45}$$

which shows that the correspondence is one-to-one. Now from (11–44)

$$r_1^2 = x_1^2 + y_1^2 = \frac{x^2 + y^2}{r^4} = \frac{1}{r^2}. \tag{11–46}$$

Hence, as $r \to +\infty$, $r_1 \to 0$. Thus *the point at infinity, Ω, of the xy-plane, corresponds to the point $(0, 0)$ in the x_1y_1-plane.* Accordingly, we can study the solutions of Eqs. (11–42) near Ω by studying the solutions of the corresponding equations in (x_1, y_1) near the point $(0, 0)$ of the x_1y_1-plane. From Eqs. (11–44), (11–45) we find the new differential equations to be as follows (Prob. 2 below, following Section 11-9):

$$\frac{dx_1}{dt} = (y_1^2 - x_1^2)F + 2x_1y_1G = F_1(x_1, y_1),$$

$$\frac{dy_1}{dt} = (y_1^2 - x_1^2)G - 2x_1y_1F = G_1(x_1, y_1),$$

(11–47)

where F, G are expressed in terms of x_1, y_1 by Eqs. (11–45).

EXAMPLE. The system (11–43) becomes

$$\frac{dx_1}{dt} = \frac{3x_1y_1^2 - x_1^3}{r_1^2}, \qquad \frac{dy_1}{dt} = \frac{y_1^3 - 3x_1^2y_1}{r_1^2}.$$

(11–43a)

Unfortunately, the new system has a singular point when $x_1 = 0$, $y_1 = 0$, but we can remedy the situation by introducing a new parameter u:

$$du = \frac{dt}{r_1^2}.$$

(11–48)

The differential equations become

$$\frac{dx_1}{du} = 3x_1y_1^2 - x_1^3, \qquad \frac{dy_1}{du} = y_1^3 - 3x_1^2y_1.$$

(11–43b)

These have an isolated equilibrium point at $(0, 0)$, and hence stability can be discussed.

The solutions of Eqs. (11–43b) are the curves

$$r_1^4 + cx_1y_1 = 0.$$

When graphed, they have the appearance of the curves of Fig. 11–16. We can indeed consider x_1, y_1 as coordinates on the sphere in the neighborhood of Ω, as suggested in that figure.

11–8 Index of an isolated equilibrium point or singular point. Now let a system (11–42) be given, with F and G having continuous first partial derivatives for all (x, y), and with all equilibrium points isolated. To each equilibrium point (x_0, y_0) we assign an integer I, called the *index* of the equilibrium point. The index is defined as follows. We choose a circle C about (x_0, y_0) such that no other equilibrium point lies on or within C. At each point (x, y) on C there is a vector with components F and G, $F = F(x, y)$, $G = G(x, y)$. Let ϕ be the angle from the x-direction

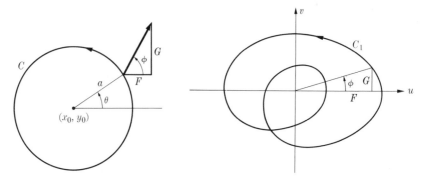

FIG. 11–17. Index of equilibrium point.

FIG. 11–18. Index as a winding number.

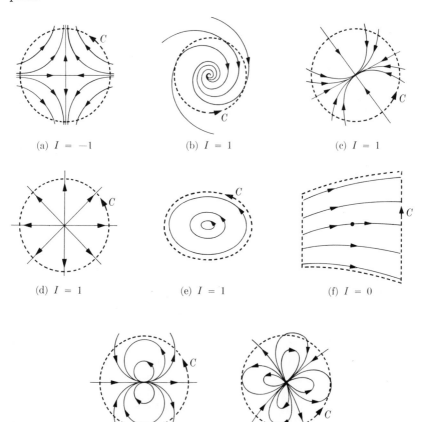

FIG. 11–19. Indices of equilibrium points.

to this vector, so that in particular $\tan \phi = G/F$ (see Fig. 11–17). If we start at a particular point (x_1, y_1) on C and move around C in the counterclockwise direction, then ϕ will vary continuously and, after one circuit of C, will return to its initial value *possibly increased or decreased by a multiple of* 2π. We now define:

$$I = \frac{1}{2\pi} \quad \text{(increase in } \phi \text{ in one circuit)}, \qquad (11\text{–}49)$$

where I is negative if ϕ has decreased.

We can compute I as follows. Let θ be a polar angle relative to (x_0, y_0), as in Fig. 11–17. Then

$$\frac{d\phi}{d\theta} = \frac{d}{d\theta} \tan^{-1} \frac{G}{F} = \frac{F \,(dG/d\theta) - G \,(dF/d\theta)}{F^2 + G^2},$$

$$I = \frac{1}{2\pi} \int_0^{2\pi} \frac{d\phi}{d\theta} \, d\theta = \frac{1}{2\pi} \int_0^{2\pi} \frac{F \,(dG/d\theta) - G \,(dF/d\theta)}{F^2 + G^2} \, d\theta, \quad (11\text{–}50)$$

where $F = F(x_0 + a \cos \theta, y_0 + a \sin \theta)$, $G = G(x_0 + a \cos \theta, y_0 + a \sin \theta)$, ($a$ is the radius of C). We can also let $u = F(x, y)$, $v = G(x, y)$. Then as (x, y) moves about C, (u, v) moves along a path C_1 in a uv-plane, and ϕ is the polar angle of (u, v). The total change of ϕ is simply 2π times the *winding number* of C_1 about the origin. (The winding number is the number of times that C_1 effectively encircles the origin.) This is illustrated in Fig. 11–18, in which $I = 2$. We can also interpret the integral (11–50) as a *line integral* (Section 2–6) on C_1:

$$I = \frac{1}{2\pi} \int_{C_1} \frac{u \, dv - v \, du}{u^2 + v^2} = \text{winding number of } C_1. \qquad (11\text{–}51)$$

The definition of I has apparently been with reference to a particular choice of C. However, I does not depend on the choice of C, for from (11–50) it can be seen that I depends continuously on the radius a of C. But I is always an integer; hence I must be identically constant. By extension of this argument we can show that I can be computed as the total change of ϕ in one circuit of any simple closed path C which encircles (x_0, y_0) once in the counterclockwise direction, provided always that there is no other equilibrium point on C or interior to C. (A simple closed path is a path which starts and ends at a point P, without passing through any other point more than once.)

EXAMPLES. Figure 11–19 shows eight equilibrium points and suggested simple closed paths C. The resulting indices are indicated on the figure. The verification is left as a problem (Prob. 3 below, following Section 11–9).

Index of a singular point. The words "equilibrium point" can be replaced by "singular point" in the above definition and discussion. Thus we can refer to the index of an isolated singular point of a system (11–42). As remarked in Section 11–5, we can convert each singular point into an equilibrium point by changing the length of the vector (F, G) appropriately. This does not affect the angle ϕ, which depends only on the *direction* of the vector. Hence the index of the singular point is the same as the index of the corresponding equilibrium point. (The singular point must be isolated both with respect to the singular points and the equilibrium points; that is, inside a sufficiently small circle about the point there must be no equilibrium point or other singular point.)

Index of point at infinity. The point at infinity Ω can be considered as a singular point of the system (11–42). We can define the index of Ω geometrically by reference to a stereographic projection. Equivalently, we can define the index of Ω relative to (11–42) to be the index of $(0, 0)$ relative to the equations (11–47) in (x_1, y_1).

Index at a nonsingular point not an equilibrium point. If (x_0, y_0) is a point at which F, G have continuous first partial derivatives and at which $F^2 + G^2 \neq 0$, then a unique solution curve of Eqs. (11–42) passes through (x_0, y_0). A similar statement will apply to points (x, y) sufficiently close to (x_0, y_0). If we compute the index of (x_0, y_0) as above, then we must find $I = 0$; for since $F^2 + G^2 \neq 0$ at (x_0, y_0), $\phi = \tan^{-1}(G/F)$ can be chosen as a continuous function in a neighborhood of (x_0, y_0). On a sufficiently small circle about (x_0, y_0), ϕ can have a total change that is as small as we wish. This implies that I must be as close to 0 as we wish. But we know that I is a constant; therefore $I = 0$.

11–9 Index theorems. In this section we present two theorems relative to sums of indices of equilibrium points or singular points. By means of these theorems we can deduce information about the presence of such points in specified regions; we can also make statements about possible configurations at the points.

THEOREM 3. *Let the system*

$$\frac{dx}{dt} = F(x, y), \qquad \frac{dy}{dt} = G(x, y) \qquad (11\text{–}52)$$

be given, where F and G have continuous first partial derivatives and $F^2 + G^2 \neq 0$ except at a finite number of points (x_j, y_j), $(j = 1, \ldots, N)$. Let C be a simple closed path in the xy-plane, and let none of the points (x_j, y_j) be on C. Let α be the total change in $\phi = \tan^{-1}(G/F)$ in one circuit of C in the counterclockwise direction. Then

$$\frac{\alpha}{2\pi} = \text{sum of indices of all points } (x_j, y_j) \text{ inside } C. \qquad (11\text{–}53)$$

Proof. Under the hypotheses stated, each point (x_j, y_j) is either a singular point or an equilibrium point, so that its index is well defined. Let us suppose, for example, that (x_1, y_1) and (x_2, y_2) lie within C, as in Fig. 11–20. Then we choose P_1, P_3 on C and join them by a simple path $P_1P_4P_3$ which separates (x_1, y_1) from (x_2, y_2) within C, as in the figure. Let I_j be the index of (x_j, y_j). Then

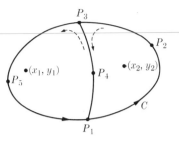

Fig. 11–20. Index theorem.

$$2\pi I_1 = \text{change in } \phi \text{ on } P_1P_4P_3 + \text{change in } \phi \text{ on } P_3P_5P_1,$$

$$2\pi I_2 = \text{change in } \phi \text{ on } P_1P_2P_3 + \text{change in } \phi \text{ on } P_3P_4P_1,$$

$$2\pi(I_1 + I_2) = \text{change in } \phi \text{ on } P_1P_2P_3 + \text{change in } \phi \text{ on } P_3P_5P_1,$$

since the changes on $P_1P_4P_3$ and on $P_3P_4P_1$ cancel. Hence $2\pi(I_1 + I_2) =$ change in ϕ on $C = \alpha$ and the assertion is proved. The argument is easily extended to any number of exceptional points.

COROLLARY. *If the path C of Theorem 3 is itself a solution curve of the differential equations, then the sum of the indices of all (x_j, y_j) inside C is 1, so that there is at least one exceptional point inside C.*

Proof. When C is a solution curve, the vector (F, G) is tangent to the path at each point. In one circuit about C the direction of the tangent changes by 2π. (This geometrically evident theorem can be proved as a limiting case of the theorem that the sum of the exterior angles of a polygon is 2π. For another proof see pp. 399–400 of Reference 3.) Hence $\alpha = 2\pi$ and by Eq. (11–53) there must be at least one exceptional point inside C.

THEOREM 4. *Under the hypotheses of Theorem 3 let C be chosen to include all the exceptional points in its interior. Then the index of Ω, the point at infinity, is*

$$I_\Omega = 2 - \frac{\alpha}{2\pi}. \tag{11–54}$$

Proof. By definition (Section 11–8) the index I_Ω is the index of $(0, 0)$ relative to the equations (11–47). By Theorem 3 we get the same value of α for all choices of C, as long as all exceptional points are inside C. We can therefore choose C to be a circle $x^2 + y^2 = R^2$. By the substitution (11–44) C corresponds to a circle C_1: $x_1^2 + y_1^2 = R^{-2}$ in the x_1y_1-plane. We also verify that as (x, y) describes the circle C in the xy-plane in the counterclockwise direction, (x_1, y_1) describes the circle C_1 in the *clockwise* direction (Fig. 11–21).

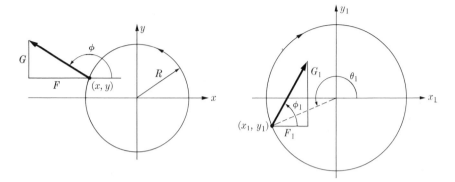

FIG. 11–21. Theorem on index of point at infinity.

Now if we write

$$F + iG = \rho e^{i\phi}, \qquad F_1 + iG_1 = \rho_1 e^{i\phi_1}, \qquad x_1 + iy_1 = r_1 e^{i\theta_1},$$

then by Eqs. (11–47)

$$
\begin{aligned}
(x_1 + iy_1)^2 (F + iG) &= r_1^2 e^{2i\theta_1} \rho e^{i\phi} \\
&= [(x_1^2 - y_1^2) + 2ix_1 y_1][F + iG] \\
&= F(x_1^2 - y_1^2) - 2x_1 y_1 G \\
&\quad + i[2x_1 y_1 F + (x_1^2 - y_1^2)G] \\
&= -F_1 - iG_1 = -\rho_1 e^{i\phi_1} \\
&= \rho_1 e^{i(\phi_1 + \pi)}.
\end{aligned}
$$

Hence

$$2\theta_1 + \phi = \phi_1 + \pi \qquad \text{(up to multiples of } 2\pi\text{).} \qquad (11\text{–}55)$$

If (x, y) now traces C in the counterclockwise direction, then (x_1, y_1) traces C_1 in the clockwise direction; therefore as ϕ increases by α, θ_1 decreases by 2π, and ϕ_1 decreases by $2\pi I_\Omega$. Hence from (11–55)

$$-4\pi + \alpha = -2\pi I_\Omega,$$

from which (11–54) follows at once.

COROLLARY. *Under the hypotheses of Theorem 3 the sum of the indices of all exceptional points, including the point at infinity, is 2.*

Proof. By (11–54) and Theorem 3, $2 = (\alpha/2\pi) + I_\Omega =$ sum of indices of all (x_j, y_j) plus I_Ω.

From this corollary we can deduce various consequences. For example, if it is known that there are two finite exceptional points (x_1, y_1), (x_2, y_2)

and that there is a saddle point at (x_1, y_1), and a focus at Ω, then by Fig. 11–19 $I_1 = -1$, $I_\Omega = 1$. Hence I_2 must be 2, so that there can be no saddle point, focus, node, or center at (x_2, y_2); there may be a configuration such as that of Fig. 11–19(g).

Relation to stability. We first remark that replacing t by $-t$ in the differential equations (11–52) has the same effect as replacing F, G by $-F$, $-G$, hence the same effect as rotating the vector (F, G) through 180°. The total change in the angle ϕ is the same as the change in the angle $\phi + 180°$. Therefore, replacing t by $-t$ has no effect on the index. This shows that the index will not permit us to distinguish, for example, between a stable focus and an unstable focus. However, if an equilibrium point is stable, then all nearby paths must be approaching the point as $t \to +\infty$. From this we can show that the index at a stable equilibrium point must be $+1$. If as above we have two equilibrium points (x_1, y_1), (x_2, y_2), and I_1 and I_Ω can be found, then (x_2, y_2) is surely unstable if $I_1 + I_\Omega \neq 1$. If $I_1 + I_\Omega = 1$, then $I_2 = 1$ and (x_2, y_2) may be stable or unstable.

PROBLEMS

1. For the following linear systems describe the type of equilibrium point at $(0, 0)$, determine the configuration at the point at infinity by stereographic projection, and obtain the differential equations (11–47) in terms of (x_1, y_1):

(a) $dx/dt = y$, $dy/dt = -x$;   (b) $dx/dt = y$, $dy/dt = -3x - 2y$;

(c) $dx/dt = 2x$, $dy/dt = 2y$.

2. (a) Show that after the substitution (11–44) the equations (11–42) become Eqs. (11–47).

(b) Let $z = x + iy$, $z_1 = x_1 + iy_1$. Show that the systems (11–42), (11–47) can be written

$$\frac{dz}{dt} = F + iG, \qquad \frac{dz_1}{dt} = -z_1^2(F + iG).$$

Show that by Eqs. (11–44) $z_1 = 1/z$ and $dz_1/dt = -z^{-2}\, dz/dt$ and hence obtain Eqs. (11–47) from Eqs. (11–42).

3. Verify the assignment of indices in Fig. 11–19(a), . . . , (h).

4. Given the linear system

$$\frac{dx}{dt} = a_1 x + b_1 y, \qquad \frac{dy}{dt} = a_2 x + b_2 y,$$

obtain the index of the equilibrium point at $(0, 0)$ in terms of $p = a_1 + b_2$, $q = a_1 b_2 - a_2 b_1$. Point out the values of p, q for which the index is undefined.

5. For the system of Prob. 4 obtain the index of the singular point at infinity in terms of p, q.

6. A system $dx/dt = F(x, y)$, $dy/dt = G(x, y)$ is given with equilibrium points only at $P_1:(1, 0)$, $P_2:(2, 0)$ and a singular point only at $P_3 = \Omega$.

(a) If there is a saddle point at P_1 and a focus at P_3, can there be a center at P_2?

(b) If there is a saddle point at Ω and P_1 is stable, can P_2 be stable?

(c) If P_1 is stable and there is either a saddle point or a center at Ω, can P_2 be stable?

7. A system $dx/dt = F(x, y)$, $dy/dt = G(x, y)$ is given with a singular point only at Ω and equilibrium points only at $P_1:(1, 0)$, $P_2:(2, 0)$, $P_3:(3, 0)$. At Ω there is a focus and at each of the points P_j there is either a saddle point or a focus. List all possible combinations of saddle and focus and draw a sketch suggesting a realization of each.

8. Let C_1, C_2, C_3 be closed solution curves of a system $dx/dt = F(x, y)$ $dy/dt = G(x, y)$ as in Theorem 3, and let C_1, C_2, C_3 together form the boundary' of a domain D. Show that there must be at least one singular point or equilibrium point in D. What is the sum of the indices of these points?

ANSWERS

1. (a) Center at $(0, 0)$ and at Ω, $dx_1/dt = -y_1$, $dy_1/dt = x_1$; (b) stable focus at $(0, 0)$, unstable focus at Ω, $dx_1/dt = r_1^{-2}(4x_1y_1^2 - 5x_1^2y_1 - y_1^3)$, $dy_1/dt = r_1^{-2}(3x_1^3 - 2x_1^2y_1 - x_1y_1^2 + 2y_1^3)$; (c) unstable node at $(0, 0)$, stable node at Ω, $dx_1/dt = -2x_1$, $dy_1/dt = -2y_1$.

4. $I = 1$ for $q > 0$, $I = -1$ for $q < 0$, I undefined for $q = 0$.

5. $I = 1$ for $q > 0$, $I = 3$ for $q < 0$, I undefined for $q = 0$.

6. (a) No, (b) no, (c) no.

7. Two must be foci, one a saddle point.

8. -1.

11–10 Periodic solutions. Limit cycles. Let C be a closed solution curve of the system (11–52). Let $x = f(t)$, $y = g(t)$ be the corresponding parametric equations, and let $x = x_0$, $y = y_0$ for $t = 0$. Since the path is closed, (x, y) must return to the position (x_0, y_0) at some positive value T of t. Thus

$$f(T) = f(0) = x_0, \qquad g(T) = g(0) = y_0.$$

We assert that for all t

$$f(t + T) = f(t), \qquad g(t + T) = g(t); \tag{11–56}$$

that is, f and g have period T, so that $x = f(t)$, $y = g(t)$ is a *periodic solution* of Eqs. (11–52). Indeed, the equations $x = f(t + T)$, $y = g(t + T)$

also define a solution of the system (11–52) and at $t = 0$, $x = f(T) = x_0$, $y = g(T) = y_0$. Hence, by the uniqueness property, this solution must coincide with the given solution for all t. Conversely, every periodic solution of Eqs. (11–52) (other than an equilibrium solution) is represented by a closed solution curve. Thus the study of periodic solutions of Eqs. (11–52) (other than the equilibrium solutions) is the same as the study of the closed solution curves.

It should be noted that there may be a solution curve C which approaches an equilibrium point (x_0, y_0) as $t \to +\infty$ and as $t \to -\infty$ (curve C_1 of Fig. 11–22). This we do not consider as a closed solution curve, even though its graph (except for one point) is the same as that of a closed solution curve. Curve C_2 in Fig. 11–22 is a closed solution curve.

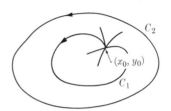

FIG. 11–22. Limit cycle versus closed solution through equilibrium point.

Closed solution curves are also called *limit cycles* (following Poincaré) and we shall use this term also.

A given system (11–52) may have no limit cycles, as the simplest examples show. However, under certain conditions we can be sure that there is at least one limit cycle in a given region.

THEOREM 5. (*Poincaré-Bendixson*) *Let D be a domain in the xy-plane bounded by a finite number of simple closed curves $\gamma_1, \ldots, \gamma_n$ and let D lie inside γ_1. Let R denote D plus its boundary. Let the system*

$$\frac{dx}{dt} = F(x, y), \qquad \frac{dy}{dt} = G(x, y) \qquad (11\text{–}57)$$

satisfy the hypotheses of Theorem 3, but let no singular point or equilibrium point lie in R. If there exists a solution $x = f(t)$, $y = g(t)$ of Eqs. (11–57) for which (x, y) remains in R for $t \geqq t_1$, for some t_1, then there is a limit cycle of Eqs. (11–57) in R.

Before considering the proof of the theorem, we discuss the example shown in Fig. 11–23. The region R is bounded by two simple closed curves γ_1, γ_2 (not solution curves). The vectors (F, G) at the various boundary points of R all point *into* R. Hence each solution C through such a boundary point (at $t = t_1$) must enter R and remain in R for $t > t_1$. Theorem 5 permits us to conclude that if there are no singular points or equilibrium points of Eqs. (11–57) in R, then there must be at least one limit cycle C_0 in R. In particular, each curve entering R at a boundary point must spiral towards a limit cycle as $t \to +\infty$.

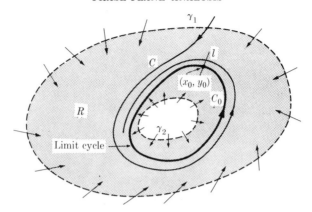

FIG. 11-23. Poincaré-Bendixson theorem.

Proof of Theorem 5. A complete proof requires somewhat advanced tools. Here we present the main ideas and refer to Chapter 11 of Reference 3 for the missing details.

In the statement of the theorem a solution C: $x = f(t), y = g(t)$ is given, $(t \geqq t_1)$. If C is itself a limit cycle, then C must lie in R and the theorem is proved. We therefore assume that C is not a limit cycle. Let $x_n = f(t_1 + n), y_n = g(t_1 + n),$ $(n = 1, 2, \ldots)$. Thus (x_n, y_n) is a sequence of points on C and in R. Since R is a bounded closed region, the sequence (x_n, y_n) has at least one limit point (x_0, y_0) in R; that is, there is at least one point (x_0, y_0) such that for every $\epsilon > 0$ and every integer $N > 0$, there is at least one integer $n > N$ such that $|x_n - x_0| < \epsilon$, $|y_n - y_0| < \epsilon$ (p. 306, p. 571 of Reference 7). Let C_0 be the solution curve of (11-57) passing through (x_0, y_0) at $t = 0$. Since C lies in R for $t \geqq t_1$, we can show that C_0 lies in R for $t > 0$. At (x_0, y_0) we draw a line segment l with midpoint at (x_0, y_0) and normal to C_0 at (x_0, y_0). If l is sufficiently short, continuity ensures that all solution curves cross l in the same direction, as suggested in Fig. 11-24. Since there are points (x_n, y_n) arbitrarily close to $(x_0, y_0),$ we conclude that C must cross l infinitely often as t increases. The path C_0 must also cross l infinitely often as t

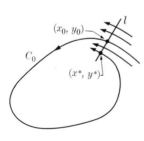

FIGURE 11-24

increases. For if C_0 does not meet l for $t > t_2$, a continuity argument shows that each curve passing sufficiently close to (x_0, y_0) does not meet l for t sufficiently large; but C itself would then have to fail to meet l for t sufficiently large, contrary to assumption. Let (x^*, y^*) be the first point at which C_0 meets l after (x_0, y_0). We assert that (x^*, y^*) must coincide with (x_0, y_0). For otherwise, if we let l^* be the portion of l between (x_0, y_0) and (x^*, y^*), then C_0 crosses l^* at (x^*, y^*) but can never cross l^* again as t increases (a subsequent crossing would be in the wrong direction). By continuity, all curves passing sufficiently close to (x_0, y_0) will also cross l again near (x^*, y^*) and can never meet l^* again. This applies in particular to the curve C. Hence C cannot have (x_0, y_0) as limit point. Therefore there is a contradiction and (x^*, y^*) must coincide with (x_0, y_0), so that C_0 is a limit cycle. We can show that C must spiral towards C_0 as shown in Fig. 11–23.

In seeking limit cycles for a particular system (11–57), we can take advantage of the indices of equilibrium points (or singular points). By the corollary to Theorem 3 the sum of the indices of the exceptional points inside a limit cycle C must be $+1$. If, for example, there are four equilibrium points with indices 4, 2, -2, -2, then there can be no limit cycle; for no combination of the indices adds up to $+1$. The following theorem also provides information.

THEOREM 6. (*Bendixson*) *Let D be a region enclosed by a simple closed curve in the xy-plane. Let $F(x, y)$ and $G(x, y)$ have continuous first partial derivatives in D and let $F_x + G_y$ be different from 0 in D. Then Eqs. (11–57) have no limit cycle in D.*

Proof. By continuity, $F_x + G_y$ must have the same sign (for example, plus) in D. If C is a limit cycle in D, then by Green's theorem (p. 239 of Reference 7),

$$\oint_C G \, dx - F \, dy = \iint_R (- F_x - G_y) \, dx \, dy < 0,$$

where R is the region bounded by C. But

$$\oint_C G \, dx - F \, dy = \int_0^T \left(G \frac{dx}{dt} - F \frac{dy}{dt} \right) dt = \int_0^T (GF - FG) \, dt = 0,$$

where T is the period of C. Hence there is a contradiction and there can be no limit cycle.

11–11 Stability of periodic solutions.   Let $C: x = f(t), y = g(t)$ be a periodic solution (limit cycle) of Eqs. (11–57). We can then consider the possible configurations of the solutions near C. Some typical formations are indicated in Fig. 11–25. By an elaboration of the proof of the Poincaré-Bendixson theorem (Theorem 5) we can deduce all the possible configurations. The inside and outside of C must be considered separately. On the inside there may be curves spiraling against C, either for increasing or for decreasing t, or else there is a sequence of limit cycles approaching C. All

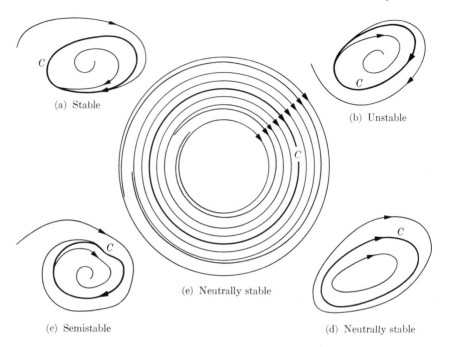

(a) Stable

(b) Unstable

(e) Neutrally stable

(c) Semistable   (d) Neutrally stable

FIG. 11–25. Stability question for periodic solutions.

the curves sufficiently close to C may be limit cycles, or there may be spirals between limit cycles, as in Fig. 11–25(e). A similar description applies to the outside.

It is natural to call only the configuration 11–25(a) *stable*. Here all curves sufficiently close to C, on both inside and outside, approach C in the direction of increasing t. We term C *orbitally stable*. Note that orbital stability does not imply that for each nearby solution C_1: $x = f_1(t)$, $y = g_1(t)$ we have

$$f(t) - f_1(t) \to 0, \quad g(t) - g_1(t) \to 0 \quad \text{as } t \to +\infty. \quad (11\text{–}58)$$

Indeed, for each constant k, $x = f_1(t + k)$, $y = g_1(t + k)$ is also a solution whose graph is the curve C_1; even if for some k we have

$$f(t) - f_1(t + k) \to 0, \quad g(t) - g_1(t + k) \to 0 \quad \text{as } t \to +\infty, \quad (11\text{–}59)$$

this will not hold for all k. We can give examples to show that even though C_1 spirals towards C, (11–59) need not hold for any choice of k. (This phenomenon is easily justified on physical grounds. For example, we can think of the earth's orbit as orbitally stable; the same path is continually retraced. However, perturbations lead to slight differences in the length of a year, and there is no mechanism which will compensate for these differences; they can accumulate over centuries.)

By symmetry we call the configuration of Fig. 11–25(b) unstable and call C *orbitally unstable*. The configuration of Fig. 11–25(c) provides stability on one side, instability on the other; C is called *orbitally semistable*. In Figs. 11–25(d), (e), the nearby paths do not approach C, but their maximum deviation from C can be made as small as desired; we call C *neutrally stable* in these cases.

In many cases we can reason geometrically to establish orbital stability. For example, in Fig. 11–23 if we know that there is precisely one limit cycle in R, then it must be orbitally stable. Indeed, for an arbitrary limit cycle C we can construct a thin annular region R about C. A study of the pattern formed by the vectors (F, G) in R will often reveal stability or instability.

11–12 Characteristic exponents. The determination of stability of an equilibrium point (Section 11–6) is based on expanding the functions F, G in power series and retaining only linear terms. A similar procedure can be applied to analysis of stability of a periodic solution.

Let $x = f(t)$, $y = g(t)$ be a periodic solution, of period T, of Eqs. (11–57). We then introduce new variables u, v:

$$u = x - f(t), \qquad v = y - g(t); \tag{11–60}$$

for each solution $x = x(t)$, $y = y(t)$ of Eqs. (11–57), u, v will depend on t and will measure the deviation of the solution $x(t)$, $y(t)$ from the periodic solution. Since x, y satisfy (11–57), we find that u, v also satisfy differential equations

$$\frac{du}{dt} = \frac{dx}{dt} - f'(t) = F(x, y) - F[f(t), g(t)]$$

$$= F[u + f(t), v + g(t)] - F[f(t), g(t)],$$

$$\frac{dv}{dt} = G[u + f(t), v + g(t)] - G[f(t), g(t)].$$

For each t we assume a Taylor's series expansion (or, more generally, as in Section 11–5, a decomposition into linear terms plus terms of higher order):

$$F[f(t) + u, g(t) + v] = F[f(t), g(t)] + F_x u + F_y v + F_{xx} \frac{u^2}{2} + \cdots ,$$

where

$$F_x = F_x[f(t), g(t)] = A_1(t),$$
$$F_y = F_y[f(t), g(t)] = B_1(t). \tag{11–61}$$

A similar expansion is assumed for G. If we retain only the linear terms, we obtain the equations

$$\frac{du}{dt} = A_1(t)u + B_1(t)v,$$

$$\frac{dv}{dt} = A_2(t)u + B_2(t)v,$$

(11–62)

where $A_1(t)$, $B_1(t)$, $A_2(t)$, $B_2(t)$ are F_x, F_y, G_x, G_y respectively, evaluated for $x = f(t)$, $y = g(t)$. Equations (11–62) are known as the *variational equations* associated with the periodic solution $x = f(t)$, $y = g(t)$ of Eqs. (11–57). One particular solution of Eqs. (11–62) is $u = 0$, $v = 0$; this corresponds to the periodic solution itself. We wish to know whether all other solutions approach this in some sense.

Equations (11–62) form a system *of homogeneous linear equations* for u, v. Furthermore the coefficients $A_1(t), \ldots, B_2(t)$ are *periodic functions* of t, with period T. For $f(t)$, $g(t)$ are periodic. Hence

$$A_1(t + T) = F_x[f(t + T), g(t + T)]$$

$$= F_x[f(t), g(t)] = A_1(t),$$

and a similar argument applies to the other coefficients. Concerning such linear differential equations with periodic coefficients, we have a general theorem.

THEOREM 7. (*Floquet-Poincaré*) *Let the functions* $A_1(t)$, $B_1(t)$, $A_2(t)$, $B_2(t)$ *be continuous functions of* t, *for all* t, *with period* T. *There is then a pair of numbers* ρ_1, ρ_2, *called characteristic exponents, such that the general solution of Eqs. (11–62) has the form (for* $\rho_1 \neq \rho_2$)

$$u = c_1 e^{\rho_1 t}P_1(t) + c_2 e^{\rho_2 t}P_2(t),$$

$$v = c_1 e^{\rho_1 t}Q_1(t) + c_2 e^{\rho_2 t}Q_2(t),$$

(11–63)

or if $\rho_1 = \rho_2$, *the form*

$$u = e^{\rho_1 t}[c_1 P_1(t) + c_2\{tP_1(t) + P_2(t)\}],$$

$$v = e^{\rho_1 t}[c_1 Q_1(t) + c_2\{tQ_1(t) + Q_2(t)\}],$$

(11–64)

where $P_1(t)$, $P_2(t)$, $Q_1(t)$, $Q_2(t)$ *are periodic, of period* T.

For a proof of this theorem the reader is referred to pp. 78–81 of Reference 3.

The characteristic exponents ρ_1, ρ_2 may be real or complex. They are uniquely determined up to addition or subtraction of multiples of $i\omega$, where $\omega = 2\pi/T$. If n is an integer, replacement of $e^{\rho_1 t}$ by $e^{(\rho_1 + n i \omega)t}$ is equivalent to multiplication of $e^{\rho_1 t}$ by

$$e^{n i \omega t} = \cos n\omega t + i \sin n\omega t,$$

which is a periodic function of period $2\pi/\omega = T$. Hence this replacement does not affect the representations (11–63) or (11–64).

When ρ_1, ρ_2 are real, the remaining quantities in Eqs. (11–63), (11–64) can be chosen to be real and the equations then give all the real solutions. When ρ_1, ρ_2 are complex, we can allow c_1, c_2 to be arbitrary complex constants and obtain all real solutions by taking the real parts of the complex solutions, as in Section 4–8.

It is clear that the characteristic exponents are the counterparts of the characteristic roots at an equilibrium point. Indeed, in the special case when $A_1(t)$, ... are all *constants*, $P_1(t)$, P_2, ... reduce to constants and ρ_1, ρ_2 to the characteristic roots λ_1, λ_2. If ρ_1, ρ_2 have negative real parts, then $e^{\rho_1 t} \to 0$ and $e^{\rho_2 t} \to 0$ as $t \to +\infty$, and multiplication by $P(t)$ or $tP(t)$, where P is periodic, cannot affect these relations. Hence *when the characteristic exponents have negative real parts, for every solution of (11–62) we have*

$$u(t) \to 0, \quad v(t) \to 0 \qquad \text{as } t \to +\infty.$$

This implies that the origin is a stable equilibrium point of Eqs. (11–62). If one characteristic exponent has positive real part, we similarly conclude instability.

Application of characteristic exponents to orbital stability. We now return to Eqs. (11–62) as variational equations for the periodic solution $x = f(t)$, $y = g(t)$ of Eqs. (11–57). We might hope that orbital stability would follow from the property that the characteristic exponents have negative real parts. However, for these variational equations *one characteristic exponent must be zero.* This complication turns out not to be disastrous, for we have the following theorem.

THEOREM 8. *Let C: $x = f(t)$, $y = g(t)$ be a periodic solution, of period T, of the differential equations $dx/dt = F(x, y)$, $dy/dt = G(x, y)$ and let Eqs. (11–62) be the corresponding variational equations. Then one characteristic exponent, ρ_1, of Eqs. (11–62), is zero and the second characteristic exponent, ρ_2, can be chosen to be real. If ρ_2 is negative, then the periodic solution C is orbitally stable and, in fact, for every solution $x = f_1(t)$, $y = g_1(t)$ spiraling to C we have for some constant k*

$$f_1(t + k) - f(t) \to 0, \qquad g_1(t + k) - g(t) \to 0,$$

as $t \to +\infty$. If ρ_2 is positive, C is orbitally unstable. If ρ_2 is 0, there is no conclusion as to stability.

For a proof the reader is referred to pp. 323–327 of Reference 3.

11–13 Evaluation of characteristic exponents. In order to apply the results of Section 11–12, we would have to form the variational equations (11–62) and somehow obtain the characteristic exponents. Unfortunately, the characteristic exponents cannot be found merely by solving an algebraic equation and, indeed, their determination for general linear systems with periodic coefficients is exceedingly difficult.

For the two-dimensional problem considered here the problem can be solved with ease. First we remark that the general solution of the variational equations can be obtained by elementary methods. One solution of (11–62) is in fact known; it is given by the equations,

$$u = f'(t), \qquad v = g'(t). \tag{11–65}$$

To verify this, we note that

$$f'(t) = F[f(t), g(t)], \qquad g'(t) = G[f(t), g(t)],$$

so that on differentiation with respect to t,

$$f''(t) = F_x f' + F_y g', \qquad g'' = G_x f' + G_y g',$$

where $F_x, \ldots$ are evaluated at $[f(t), g(t)]$; that is,

$$f''(t) = A_1(t)f'(t) + B_1(t)g'(t),$$
$$g''(t) = A_2(t)f'(t) + B_2(t)g'(t).$$

Hence (11–65) defines a solution of Eqs. (11–62). (From the form of this solution we see that one characteristic exponent must be zero, as stated in Theorem 8.)

If we now eliminate v from Eqs. (11–62), we obtain a homogeneous second order linear equation for u. One solution of this equation is $u = f'(t)$. Hence the general solution can be found as in Section 7–4. Once u is known, v can be found.

We illustrate the procedure by an example. The system chosen is

$$\frac{dx}{dt} = x - y - x^3 - xy^2, \qquad \frac{dy}{dt} = x + y - x^2 y - y^3. \tag{11–66}$$

It is clear that the circular path C: $x = \cos t$, $y = \sin t$ is a solution of Eqs. (11–66). By a graphical analysis we find that the solutions are as shown in Fig. 11–26, so that C appears to be stable. Since $F_x = 1 - 3x^2 - y^2, \ldots$, the variational equations are

$$\frac{du}{dt} = -2u \cos^2 t - (1 + 2 \sin t \cos t)v,$$
$$\tag{11–67}$$
$$\frac{dv}{dt} = (1 - 2 \sin t \cos t)u - 2v \sin^2 t.$$

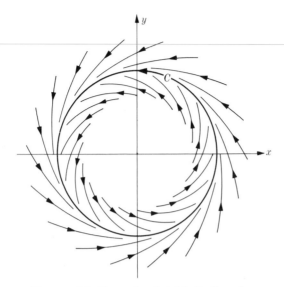

FIG. 11–26. Example of stable limit cycle.

We verify that $u = -\sin t$, $v = \cos t$ is a solution of Eqs. (11–67). To find the general solution, we eliminate v. Upon differentiating the first of Eqs. (11–67) and then eliminating v, we obtain the equation for u (Prob. 5 below)

$$(1 + 2 \sin t \cos t)u'' + (4 \sin^2 t + 4 \sin t \cos t)\, u'$$

$$+ (1 - 2 \sin t \cos t - 4 \cos^2 t)\, u = 0. \qquad (11\text{–}68)$$

Since $u = -\sin t$ is a solution, so also is $u = \sin t$. We set

$$u = w \sin t, \qquad u' = w' \sin t + w \cos t,$$

$$u'' = w'' \sin t + 2\, w' \cos t - w \sin t$$

and obtain the equation for w

$$(\sin t + 2 \sin^2 t \cos t)w'' + (2 \cos t + 4 \sin t \cos^2 t$$

$$+ 4 \sin^3 t + 4 \sin^2 t \cos t)w' = 0. \qquad (11\text{–}69)$$

If we set $w' = z$, we have a linear equation for z, whose general solution we find to be

$$z = w' = \frac{c_1 e^{-2t}(1 + \sin 2t)}{\sin^2 t}.$$

Accordingly,

$$w = c_1 \int e^{-2t}(\csc^2 t + 2 \cot t)\, dt + c_2$$
$$= -c_1 e^{-2t} \cot t + c_2,$$
$$u = w \sin t = -c_1 e^{-2t} \cos t + c_2 \sin t.$$

From the first of Eqs. (11–67) we obtain v and hence finally

$$u = -c_1 e^{-2t} \cos t + c_2 \sin t, \qquad v = -c_1 e^{-2t} \sin t - c_2 \cos t \quad (11\text{–}70)$$

as the general solution of the variational equations. Thus the characteristic exponents are 0, -2 and the periodic solution is stable.

In general, it is not necessary to solve the variational equations, for we have the following theorem.

THEOREM 9. *In the notations of Theorem 8 the characteristic exponent* ρ_2 *is equal to*

$$\frac{1}{T} \int_0^T [A_1(t) + B_2(t)]\, dt.$$

For example, in Eqs. (11–67) $A_1 = -2 \cos^2 t$, $B_2 = -2 \sin^2 t$, $T = 2\pi$,

$$\rho_2 = \frac{1}{2\pi} \int_0^{2\pi} (-2 \cos^2 t - 2 \sin^2 t)\, dt = -2,$$

in agreement with the results found by computation.

For a proof see Prob. 7 below.

PROBLEMS

1. Find all limit cycles for each of the following systems:
(a) $dx/dt = 2x - 3y$, $dy/dt = 3x - 2y$;
(b) $dx/dt = -y^3$, $dy/dt = x^3$; [*Hint:* Eliminate t.]
(c) $dx/dt = 2y$, $dy/dt = -\sin x$.

2. Apply the Poincaré-Bendixson theorem (Theorem 5) to show that there is a limit cycle between the circles $x^2 + y^2 = 1$ and $x^2 + y^2 = 16$:

(a) $\dfrac{dx}{dt} = 4x - 4y - x(x^2 + y^2)$, $\quad \dfrac{dy}{dt} = 4x + 4y - y(x^2 + y^2)$;

(b) $\dfrac{dx}{dt} = x \sin \dfrac{x^2 + y^2}{4} - y$, $\quad \dfrac{dy}{dt} = x + y \sin \dfrac{x^2 + y^2}{4}$.

3. A system $dx/dt = F(x, y)$, $dy/dt = G(x, y)$ is known to have no singular points and only one equilibrium point, at $(0, 0)$, which is stable. If there are

three limit cycles in all and none is semistable, describe the position and stability of the limit cycles.

4. A system $dx/dt = F(x, y)$, $dy/dt = G(x, y)$ has no singular point or equilibrium point on a certain closed curve C, which has a continuously turning tangent vector. If the vector (F, G) is never tangent to C, show that there must be a singular point or equilibrium point inside C.

5. For the example of Eqs. (11–66) obtain the variational equations (11–67) and fill in all missing details to obtain the solutions (11–70).

6. For each of the following systems verify that the path given is a limit cycle, obtain the variational equations, and apply Theorem 9 to determine ρ_2 and stability of the limit cycle.

(a) Eqs. of Prob. 2(a), $x = 2 \cos 4t$, $y = 2 \sin 4t$.
(b) Eqs. of Prob. 2(b), $x = 2\sqrt{\pi} \cos t$, $y = 2\sqrt{\pi} \sin t$.

7. *Proof of Theorem* 9. Let Eqs. (11–63) define the general solution of Eqs. (11–62) and let the characteristic exponents ρ_1, ρ_2 be real. Let

$$u_1(t) = e^{\rho_1 t} P_1(t), \qquad u_2(t) = e^{\rho_2 t} P_2(t),$$

$$v_1(t) = e^{\rho_1 t} Q_1(t), \qquad v_2(t) = e^{\rho_2 t} Q_2(t),$$

$$W(t) = \begin{vmatrix} u_1(t) & u_2(t) \\ v_1(t) & v_2(t) \end{vmatrix}, \qquad q(t) = A_1(t) + B_2(t).$$

Since the pairs $u_1(t)$, $v_1(t)$ and $u_2(t)$, $v_2(t)$ are necessarily linearly independent solutions of Eqs. (11–62), we can conclude (see Prob. 6 following Section 6–1) that $W(t) \neq 0$ for all t.

(a) Show that $dW/dt = q(t)W$. [*Hint:* Expand the determinant, differentiate, and use the relations $Du_j = A_1 u_j + B_1 v_j$, $Dv_j = A_2 u_j + B_2 v_j$, $(j = 1, 2)$, where $D = d/dt$.]

(b) From the result of part (a) conclude that

$$W(t) = W(0)e^{\int_0^t q(u)\, du}.$$

(c) From the periodicity of $P_j(t)$, $Q_j(t)$, $(j = 1, 2)$, conclude that

$$W(T) = e^{(\rho_1 + \rho_2)T} W(0).$$

(d) Combine the results of parts (b), (c) to conclude that

$$e^{(\rho_1 + \rho_2)T} = e^{\int_0^T q(t)\, dt},$$

$$\rho_1 + \rho_2 = \frac{1}{T} \int_0^T q(t)\, dt$$

and deduce Theorem 9. (If ρ_1, ρ_2 are not necessarily real, the same reasoning shows that

$$\rho_1 + \rho_2 = \frac{1}{T} \int_0^T q(t)\, dt + n\omega i \qquad \left(\omega = \frac{2\pi}{T} \right),$$

for some integer n.)

ANSWERS

1. (a) $x = 3c \cos \omega t$, $y = 2c \cos \omega t + c\omega \sin \omega t$, $\omega = \sqrt{5}$, $c > 0$;
   (b) $x^4 + y^4 = c^4$, $c > 0$ (t obtainable by integration:
   $t = \pm \int [c^4 - x^4]^{-3/4}\, dx)$;
   (c) $y^2 = \cos x + c$, $|x - 2n\pi| \leq \cos^{-1}(-c)$, $n = 0, \pm 1, \ldots$,
   $-1 < c < 1$, (t obtainable by integration).

3. They must be concentric, with $(0, 0)$ within all. The smallest must be unstable, the next stable, the last unstable.

6. (a) $du/dt = -8u \cos^2 4t - 4v(1 + 2 \sin 4t \cos 4t)$,
   $dv/dt = 4u(1 - 2 \sin 4t \cos 4t) - 8v \sin^2 4t$, $\rho_2 = -8$, stable;
   (b) $du/dt = -2\pi u \cos^2 t - v(1 + 2\pi \sin t \cos t)$,
   $dv/dt = u(1 - 2\pi \sin t \cos t) - 2\pi v \sin^2 t$, $\rho_2 = -2\pi$, stable.

11–14 The van der Pol equation. This is the differential equation

$$\frac{d^2x}{dt^2} + \mu(x^2 - 1) \frac{dx}{dt} + x = 0, \tag{11–71}$$

where μ is a positive constant. Certain electric circuits containing vacuum tubes are governed by such an equation. A special case of this equation, ($\mu = 1$), is studied in Chapter 1 (Fig. 1–6 and Prob. 7 following Section 1–4). A graphical analysis in the phase plane reveals a single stable limit cycle. A similar result is found for arbitrary μ, though the shape of the limit cycle varies considerably (Fig. 11–27). With $v = dx/dt$, the differential equations in the phase plane are

$$\frac{dx}{dt} = v, \qquad \frac{dv}{dt} = \mu(1 - x^2)v - x. \tag{11–72}$$

From these we see that there is one equilibrium point, at $x = 0$, $v = 0$. The corresponding linear equations are

$$\frac{dx}{dt} = v, \qquad \frac{dv}{dt} = \mu v - x. \tag{11–73}$$

Since $p = \mu$, $q = 1$, $\Delta = \mu^2 - 4$, we conclude that the origin is unstable (an unstable node for $\mu > 2$, an unstable focus for $\mu < 2$). From a study of the isoclines we can find two simple closed curves C_1, C_2 such

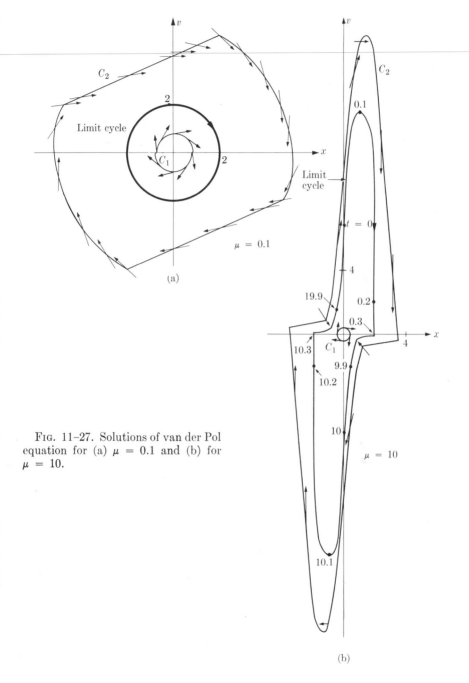

Fig. 11–27. Solutions of van der Pol equation for (a) $\mu = 0.1$ and (b) for $\mu = 10$.

that every solution meeting C_1 or C_2 enters the region R bounded by C_1, C_2 and never leaves, while R contains no equilibrium point. Such curves are sketched in Fig. 11–27. (A proof that such curves exist can be obtained from the stability theorem stated below.) From the Poincaré-Bendixson theorem we conclude that there must be at least one limit cycle. It is less simple to establish that there is only one.

THEOREM 10. (*Liénard*) *For every positive value of μ the van der Pol equation* (11–71) *has exactly one stable limit cycle in the phase plane.*

Proof. Instead of the variables x, v, we use the variables x and y, where

$$y = v + g(x), \qquad g(x) = \mu\left(\frac{x^3}{3} - x\right).$$

There is a one-to-one correspondence (continuous both ways) between the pairs (x, y) and the pairs (x, v), so that the configuration in the xy-plane must be similar to that in the xv-plane. We can interpret (x, y) as curvilinear coordinates in the xv-plane, as shown in Fig. 11–28. The differential equations in the xy-plane are obtained as follows:

$$\frac{dx}{dt} = v = y - g(x),$$

$$\frac{dy}{dt} = \frac{dv}{dt} + g'(x)\,\frac{dx}{dt} = \mu(1 - x^2)v - x + \mu(x^2 - 1)v.$$

Accordingly, the new differential equations are

$$\frac{dx}{dt} = y - g(x), \qquad \frac{dy}{dt} = -x. \qquad (11\text{–}74)$$

If we replace x by $-x$, y by $-y$ in Eqs. (11–74), the equations are unchanged; hence reflection of a solution curve in the origin of the xy-plane

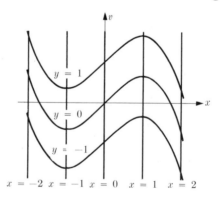

FIGURE 11–28

gives a solution curve. Thus the solutions have a certain *symmetry*. If we know the solutions in the right half-plane, $(x > 0)$, those in the left half-plane, $(x < 0)$, can be obtained at once by reflection. We therefore concentrate attention on the right half-plane and note that $dy/dt < 0$ for $x > 0$.

We shall denote by C_0 the curve

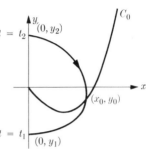

$$y = g(x) = \mu\left(\frac{x^3}{3} - x\right) \qquad (x \geqq 0).$$
$$(11\text{–}75)$$

Along C_0, dx/dt is 0, so that dy/dx is infinite. Above C_0, $dx/dt > 0$; below C_0, $dx/dt < 0$. We note that on C_0, y has a minimum at $x = 1$, equal to $-2\mu/3$. (See Fig. 11–29.)

FIGURE 11–29

Let (x_0, y_0) be a point on C_0 $(x_0 > 0)$. We assert that on the solution passing through (x_0, y_0) at $t = 0$, we have the following properties: *as t increases through positive values, y decreases and x decreases until the y-axis is reached at a point $(0, y_1)$; as t decreases through negative values, y increases and x decreases until the y-axis is reached at a point $(0, y_2)$.* To prove these assertions, we choose a point $(0, b)$ on the y-axis and let

$$r^2 = x^2 + (y - b)^2.$$

Then on the solution in question

$$r\frac{dr}{dt} = x\frac{dx}{dt} + y\frac{dy}{dt} - b\frac{dy}{dt} = x[y - g(x)] - x(y - b)$$

$$= x[b - g(x)] = [g(x) - b]\frac{dy}{dt}. \qquad (11\text{–}76)$$

If b is now less than the minimum of $g(x)$, then $g(x) - b$ is always positive; as t increases, y decreases, so that r decreases. The path must move into the region below C_0 and, since $dx/dt < 0$, it moves steadily to the left; since $dr/dt < 0$, it must eventually reach the y-axis. Next we choose b positive and greater than y_0. As t *decreases* from 0, y must increase, so that the path moves into the region above C_0, where $dx/dt > 0$, so that the path has to move to the left. Hence $g(x)$ remains less than b and by (11–76) $dr/dt > 0$; thus r decreases as t decreases and again we conclude that the path reaches the y-axis. (Without the restriction on dr/dt, the path might have a vertical asymptote $x = x_1 > 0$.) Thus the assertions are proved.

The values y_1, y_2 depend on the choice of x_0. We now assert that *there is a unique value* x^* *such that for* $x_0 < x^*$, $y_2 < -y_1$; *for* $x_0 = x^*$, $y_2 = -y_1$; *for* $x_0 > x^*$, $y_2 > -y_1$. It will be seen that the solution through (x^*, y^*), $y^* = g(x^*)$, is the limit cycle. To prove the assertion, we choose $b = 0$ and let $r^2 = x^2 + y^2$. Then Eq. (11–76) gives

$$r \frac{dr}{dt} = g(x) \frac{dy}{dt} \tag{11–77}$$

on each solution C. We now introduce the line integral (Section 2–6)

$$\phi = \int_{(0,y_1)}^{(0,y_2)} g(x)\, dy, \tag{11–78}$$

where the path is the solution curve, followed in the direction of decreasing t. The value of ϕ depends on the choice of x_0, so that $\phi = \phi(x_0)$. By (11–77)

$$\phi(x_0) = \int_{t_1}^{t_2} g(x) \frac{dy}{dt}\, dt = \int_{t_1}^{t_2} r \frac{dr}{dt}\, dt = \frac{1}{2}\left(r_2^2 - r_1^2\right),$$

where t_1, t_2 are the corresponding t-values and $r_1 = -y_1$, $r_2 = y_2$. Hence our assertions will be proved if we show that for some x^*, $\phi(x_0)$ is negative for $x_0 < x^*$, positive for $x_0 > x^*$.

Now $\phi(x_0)$ is surely negative for $x_0 < \bar{x} = \sqrt{3}$. Indeed, $\bar{x}$ is the positive root of $g(x)$, hence on the solution path $g(x) < 0$; since $dy > 0$, Eq. (11–78) shows that $\phi(x_0) < 0$. For $x_0 > \bar{x}$ we can show that $\phi(x_0)$ steadily increases and approaches $+\infty$ as $x_0 \to +\infty$. Indeed, with reference to Fig. 11–30,

$$\phi(x_0) = \int_{A_1}^{B_1} g(x)\, dy + \int_{B_1}^{B_2} g(x)\, dy + \int_{B_2}^{A_2} g(x)\, dy. \tag{11–79}$$

Now

$$\int_{A_1}^{B_1} g(x)\, dy < \int_{A_1'}^{B_1'} g(x)\, dy, \qquad \int_{B_2}^{A_2} g(x)\, dy < \int_{B_2'}^{A_2'} g(x)\, dy.$$

$$\tag{11–80}$$

For on $A_1 B_1$

$$\int_{A_1}^{B_1} g(x)\, dy = \int_0^{\bar{x}} g(x) \frac{dy}{dx}\, dx = \int_0^{\bar{x}} g(x) \frac{x}{g(x) - y}\, dx;$$

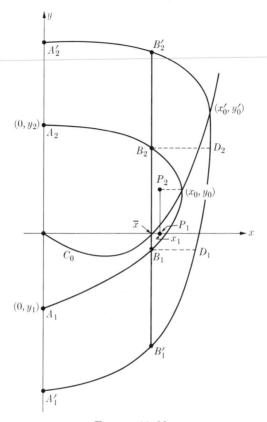

FIGURE 11–30

a similar expression is obtained for the integral from A_1' to B_1', with $g(x) - y$ replaced by a larger value. Since $g(x)$ itself is negative, the integral on $A_1'B_1'$ will be greater than that on A_1B_1. A similar argument gives the second inequality (11–80). Finally,

$$\int_{B_1}^{B_2} g(x)\, dy < \int_{D_1}^{D_2} g(x)\, dy < \int_{B_1'}^{B_2'} g(x)\, dy;$$

for in replacing B_1B_2 by D_1D_2, we are increasing the values of $g(x)$ and along $B_1'D_1$ and D_2B_2', $g(x)$ is positive. Hence all three terms of which $\phi(x_0)$ is composed in (11–79) increase if x_0 is replaced by x_0', so that $\phi(x_0)$ is steadily increasing as x_0 increases beyond $\bar{x}$. The second term on the right of (11–79) is at least equal to the integral on the line segment from P_1 to P_2, that is, to $g(x_1)\cdot y_0$. If x_1 is kept fixed and $x_0 \to +\infty$, then $y_0 \to +\infty$; hence this second term approaches $+\infty$ as $x_0 \to +\infty$, and we conclude that $\phi(x_0) \to +\infty$ as $x_0 \to +\infty$.

From the properties of ϕ it follows that $\phi(x_0)$ is positive for x_0 sufficiently large. But $\phi(x_0)$ is negative for $x < \bar{x}$. We can verify that $\phi(x_0)$ is continuous (Prob. 3 below). Hence there is a unique value x^* such that $\phi(x^*) = 0$, and x^* has all the properties asserted above. For the solution through (x^*, y^*) we have $-y_1 = y_2$. The symmetry property implies that the reflection of this solution in the origin is also a solution; the two curves thus obtained clearly fit together to form a limit cycle C^*. From the fact that $-y_1 > y_2$ for $x_0 < x^*$ and the symmetry, we conclude that the solutions inside C^* spiral out to C^*; from the fact that $-y_1 < y_2$ for $x_0 > x^*$, we conclude that the solutions outside C^* spiral in to C^*. Thus C^* is a stable limit cycle. Any other limit cycle would have to enclose the equilibrium point at $(0, 0)$, and hence would pass through a point (x_0, y_0) on C_0; except for C^*, the solution curves meeting C_0 are spirals, and hence there are no other limit cycles.

Remark 1. If we study the above proof, we see that the explicit formula for $g(x)$ is not actually used, but merely the fact that $g(x) = \int_0^x f(u)\, du$, where

(a) $f(x)$ is continuous for all x,

(b) $f(-x) = f(x)$,

(c) $g(x) < 0$ for $0 < x < \bar{x}, f(x) > 0$ and $g(x) > 0$ for $x > \bar{x}$,

(d) $g(x) \to +\infty$ as $x \to +\infty$.

Accordingly, the differential equation

$$\frac{d^2x}{dt^2} + f(x)\frac{dx}{dt} + x = 0 \qquad (11\text{–}81)$$

has a unique stable limit cycle in the xv-plane, $(v = dx/dt)$, whenever these conditions hold. By similar methods we can extend the conclusion to the differential equation

$$\frac{d^2x}{dt^2} + f(x)\frac{dx}{dt} + p(x) = 0, \qquad (11\text{–}82)$$

where $f(x)$ is as above, $p(x)$ has a continuous derivative for all x, $p(-x) = -p(x)$, $p(x) > 0$ for $x > 0$, and $\int_0^x p(u)\, du \to +\infty$ as $x \to +\infty$. For a proof see pp. 402–403 of Reference 3.

Remark 2. Various attempts have been made to obtain an explicit formula for the periodic solution of the van der Pol equation. Some approximate methods have had striking success, notably those of Kryloff and Bogoliuboff and Ritz-Galerkin. (Reference 8 and Reference 12, Vol. 2, pp. 234–257.)

PROBLEMS

1. Show that the differential equation

$$a \frac{d^2x}{dt^2} + b(x^2 - 1) \frac{dx}{dt} + cx = 0 \qquad (a > 0, b > 0, c > 0),$$

where a, b, c are constants, can be reduced to the van der Pol equation (11–71) by an appropriate change of time scale.

2. Apply numerical integration to obtain the t-values shown on the limit cycle of Fig. 11–27(b). [*Hint:* Use the equations

$$dt = dx/v = dv/\{10(1 - x^2)v - x\}.]$$

Show that the graphs of x and v against t have the appearance of Fig. 11–31.

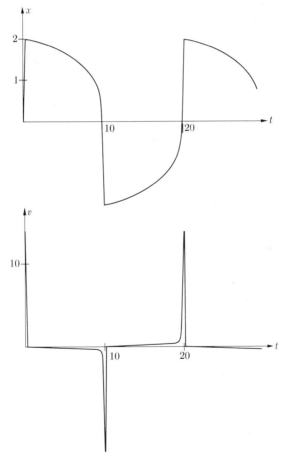

FIG. 11–31. Periodic solution of van der Pol equation ($v = dx/dt$).

3. Give a geometric argument to show that in Fig. 11–29 y_2 and y_1 are continuous functions of x_0 and conclude that $\phi(x_0)$ is continuous for $x_0 > 0$.

4. Given the differential equation

$$\frac{d^2x}{dt^2} = f(x),$$

let $f(x)$ be continuous for all x and let $F(x) = \int_0^x f(u)\, du$.

(a) Show that the solutions in the xv-plane, $(v = dx/dt)$, are the curves

$$v^2 = 2[F(x) - c].$$

(b) Show that if $F(x_1) = F(x_2) = c$, $F(x) > c$ for $x_1 < x < x_2$, then there is a limit cycle through $(x_1, 0)$, $(x_2, 0)$.

(c) Show that if $f(x) \neq 0$ for $0 < |x - x_0| < a$, then there is a center at $(x_0, 0)$ if $F(x)$ has a relative maximum at x_0, and a saddle point at $(x_0, 0)$ if $F(x)$ has a relative minimum at $(x_0, 0)$.

(d) Show that all equilibrium points are on the x-axis. Show that all limit cycles meet the x-axis and are symmetric with respect to the x-axis.

5. Analyze the following equations graphically in the phase plane, on the basis of the results of Prob. 4 (where applicable):

(a) $d^2x/dt^2 = 32$ (falling body),   (b) $d^2x/dt^2 = -x$ (spring),
(c) $d^2x/dt^2 = x$,   (d) $d^2x/dt^2 = -1/x^2$ (gravitation),
(e) $d^2x/dt^2 = -\sin x$ (pendulum),   (f) $d^2x/dt^2 = x - 2x^3$.

6. *Pulling into step of a synchronous motor.* The problem is described (Reference 4) by the differential equation

$$\frac{d^2\theta}{d\lambda^2} + k(1 - b\cos 2\theta)\frac{d\theta}{d\lambda} + \frac{P_r}{P_m}\sin 2\theta + \sin \theta = \frac{P_1}{P_m},$$

where

θ = angle between axes of armature and rotor fields,

λ = a variable proportional to time,

k = relative damping coefficient, a constant,

b = an empirical constant,

P_r = a constant, $P_r \sin 2\theta$ = reluctance torque,

P_m = a constant, $P_m \sin \theta$ = synchronizing torque,

P_1 = load torque, a constant.

We assign numerical values as follows:

$$b = 0.25, \qquad \frac{P_r}{P_m} = 0.25, \qquad \frac{P_1}{P_m} = 0.5.$$

We let $x = \theta$, $t = \lambda$ and consider the equations

$$\frac{dx}{dt} = y,$$

$$\frac{dy}{dt} = 0.5 - k(1 - 0.25 \cos 2x)y - 0.25 \sin 2x - \sin x$$

in the xy-plane.

(a) Show that for $k = 0$ the analysis of Prob. 4 is applicable and classify the equilibrium points.

(b) Show that for $k > 0$ the equilibrium points are located as in part (a) and classify them.

(c) Show that the solutions ending at the saddle points divide the xy-plane into regions from each of which at most one stable equilibrium point can be approached.

ANSWERS

5. (a) Solutions $v^2 = 64x - 2c$, no limit cycles, no equilibrium points;

(b) solutions $v^2 = -x^2 + c^2$, all are limit cycles, except for equilibrium point at $(0,0)$, which is a center;

(c) solutions $v^2 = x^2 - c$, no limit cycles, equilibrium point at $(0,0)$ is a saddle point;

(d) solutions $v^2 = (2/x) - c$, no limit cycles, no equilibrium points;

(e) solutions $v^2 = 2 \cos x - c$, formed of limit cycles for $-2 < c < 2$, equilibrium points at $(n \cdot 2\pi, 0)$ are centers, at $[(2n + 1)\pi, 0]$ are saddle points, $(n = 0, \pm 1, \ldots)$;

(f) solutions $v^2 = x^2 - x^4 - c$ are limit cycles for $c < 0$ and for $0 < c < \frac{1}{4}$, equilibrium point at $(0, 0)$ is a saddle point, equilibrium points at $(\pm 1/\sqrt{2}, 0)$ are centers.

6. (a) Centers on x-axis at $x = 0.35 + n \cdot 2\pi$, saddle points at $x = 2.3 + n \cdot 2\pi$, $(n = 0, \pm 1, \ldots)$; (b) same as (a) except that centers become stable foci for $0 < k < 2.8$, stable nodes for $k > 2.8$.

11–15 Analysis in phase space of more than two dimensions. Many of the concepts and theorems of this chapter can be generalized to systems of higher order, which we consider in the form

$$\frac{dx_i}{dt} = F_i(x_1, \ldots, x_n) \qquad (i = 1, \ldots, n). \qquad (11\text{–}83)$$

As shown in Section 1–7, quite general differential equations and systems of equations can be reduced to the form

$$\frac{dx_i}{dt} = F_i(x_1, \ldots, x_n, t) \qquad (i = 1, \ldots, n). \qquad (11\text{–}84)$$

These equations can be considered as a special case of (11–83). We write $x_{n+1} = t$ and replace Eqs. (11–84) by the system of form (11–83):

$$\frac{dx_i}{dt} = F_i(x_1, \ldots, x_n, x_{n+1}) \qquad (i = 1, \ldots, n), \qquad \frac{dx_{n+1}}{dt} = 1.$$

$$(11\text{–}85)$$

Thus Eqs. (11–83) are, in a sense, the general form.

We consider Eqs. (11–83) as describing a flow in the n-dimensional space of $(x_1, \ldots, x_n)$, each point of which is a *phase*. The velocity vector at each point has components $F_1, \ldots, F_n$. The velocity does not change with time; hence we have a stationary fluid motion in the phase space. The points where $F_1 = F_2 = \ldots = F_n = 0$ are *equilibrium points*. The periodic solutions of (11–83) are represented by closed curves in the phase space; they are the *limit cycles*.

The analysis of stability of equilibrium points follows the pattern for two dimensions. If $(\bar{x}_1, \ldots, \bar{x}_n)$ is an equilibrium point, then we expand all F_i in Taylor series about this point and retain only the linear terms. The resulting linear differential equations have constant coefficients and can be solved as in Chapter 6. When all characteristic roots have negative real parts, the equilibrium point is stable; if one characteristic root has positive real part, the equilibrium point is unstable. For proofs the reader is referred to Chapter 13 of Reference 3. For $n > 2$ the geometrical classification of the configurations at equilibrium points becomes far more complicated and difficult to visualize.

The location of limit cycles for $n > 2$ is in general very difficult. The analog of the Poincaré-Bendixson theorem (Section 11–10) does not hold for $n > 2$. In particular, there exist bounded solutions which are *almost periodic;* these are illustrated by a spherical pendulum that follows the patterns of *Lissajous figures*. The solution paths return infinitely often to the immediate vicinity of an initial point, but never return exactly.

For the topic of existence of limit cycles and of more complicated solutions, the reader is referred to Reference 2.

If a limit cycle has been found, its stability can be analyzed with the aid of variational equations and characteristic exponents. This is discussed in Chapter 13 of Reference 3.

PROBLEMS

1. Find all equilibrium points and test for stability:
   (a) $dx/dt = y - z$, $dy/dt = x - z$, $dz/dt = x^2 - y$;
   (b) $dx/dt = (1 + x^2)y + z^2$, $dy/dt = e^z - 1$,
       $dz/dt = -x^3 - \sin(x + y) - 3z$.

2. *Restricted problem of three bodies.* Let two bodies, each of mass $\frac{1}{2}$, rotate at the same constant angular velocity about the origin of the $\xi\eta$-plane, with

their center of mass always at the origin. Let a third body, of negligible mass, move in the plane, subject to the gravitational attraction of the first two bodies. We then have a special case of the restricted problem of three bodies, an important problem in celestial mechanics. If we choose axes x, y rotating with the large bodies, so that the bodies always lie on the x-axis, then (for appropriate choice of units) the differential equations take the form (pp. 351–352 of Reference 15)

$$\frac{d^2x}{dt^2} - 2\frac{dy}{dt} = x - \frac{1}{2}\left(\frac{x + \frac{1}{2}}{r_1^3} + \frac{x - \frac{1}{2}}{r_2^3}\right),$$

$$\frac{d^2y}{dt^2} + 2\frac{dx}{dt} = y - \frac{1}{2}\left(\frac{y}{r_1^3} + \frac{y}{r_2^3}\right),$$

where (x, y) are the coordinates of the third body, the large bodies are at $x = \pm\frac{1}{2}$, $y = 0$, and

$$r_1 = [(x + \tfrac{1}{2})^2 + y^2]^{1/2}, \qquad r_2 = [(x - \tfrac{1}{2})^2 + y^2]^{1/2}.$$

(a) Show that for the motion in the phase space ($x_1 = x$, $x_2 = dx/dt$, $x_3 = y$, $x_4 = dy/dt$), there is an equilibrium point for $x_1 = 0$, $x_2 = 0$, $x_3 = \frac{1}{2}\sqrt{3}$, $x_4 = 0$, at which the three bodies form an equilateral triangle.

(b) Show that the equilibrium point of part (a) is unstable.

3. Consider the system

$$\frac{d^2x}{dt^2} + (x^2 - 1)\frac{dx}{dt} + x = 0, \qquad 2\frac{d^2y}{dt^2} + \sqrt{2}(y^2 - 1)\frac{dy}{dt} + y = 0,$$

for which x is governed by a van der Pol equation and y by a van der Pol equation with modified time scale (see Prob. 1 following Section 11–14).

(a) Show that there is a particular solution $x = f(t)$, $y = g(t)$, where f and g are periodic with periods τ, $\sqrt{2}\tau$ respectively.

(b) Show that in the phase space $x_1 = x$, $x_2 = dx/dt$, $x_3 = y$, $x_4 = dy/dt$ the solution of part (a) is not periodic.

(An interesting interpretation of the solution is obtained by introducing a phase angle on the van der Pol limit cycle, running from 0 to 2π. Let θ be the phase angle for x, and ϕ the phase angle for y, so that $\theta = \theta(t)$, $\phi = \phi(t)$ where $\theta(t)$ increases by 2π when t increases by τ, and $\phi(t)$ increases by 2π when t increases by $\sqrt{2}\tau$. The pair (θ, ϕ) can be considered as angular coordinates on a torus. The path $\theta = \theta(t)$, $\phi = \phi(t)$ then winds about the torus without ever crossing itself.)

ANSWERS

1. (a) $(0, 0, 0)$ unstable and $(1, 0, 0)$ unstable, (b) $(0, 0, 0)$ stable.

SUGGESTED REFERENCES

1. ANDRONOFF, A., and CHAIKIN, C. E., *Theory of Oscillations.* Princeton, N. J.: Princeton University Press, 1949.

2. BIRKHOFF, GEORGE D., *Dynamical Systems.* American Mathematical Society Colloquium Publications, Vol. 9. New York: American Mathematical Society, 1927.

3. CODDINGTON, EARL A., and LEVINSON, NORMAN, *Theory of Ordinary Differential Equations.* New York: McGraw-Hill, 1955.

4. EDGERTON, H. E., and FOURMARIER, P., "The Pulling into Step of a Salient-pole Synchronous Motor." *Transactions of the American Institute of Electrical Engineers*, Vol. 50 (1931), pp. 769–781.

5. HOHEISEL, G., *Gewöhnliche Differentialgleichungen*, 2nd ed. (*Sammlung Göschen*, No. 920). Berlin: Walter de Gruyter and Co., 1930.

6. INCE, E. L., *Ordinary Differential Equations.* New York: Dover, 1956.

7. KAPLAN, WILFRED, *Advanced Calculus.* Reading, Mass.: Addison-Wesley, 1952.

8. KRYLOFF, N., and BOGOLIUBOFF, N., *Introduction to Non-Linear Mechanics*, transl. by S. Lefschetz (*Annals of Mathematics Studies*, No. 11). Princeton: Princeton University Press, 1943.

9. LAWDEN, DEREK F., *Mathematics of Engineering Systems.* New York: John Wiley and Sons, Inc., 1954.

10. LIAPOUNOFF, A., *Problème général de la stabilité du mouvement* (*Annals of Mathematics Studies*, No. 17). Princeton: Princeton University Press, 1947.

11. MINORSKY, N., *Introduction to Non-Linear Mechanics.* Ann Arbor, Mich.: J. W. Edwards, 1947.

12. *Proceedings of the Symposium on Nonlinear Circuit Analysis*, Vols. 2, 6. Brooklyn, N. Y.: Polytechnic Institute of Brooklyn, 1953, 1957.

13. LEFSCHETZ, S., *Differential Equations: Geometric Theory.* New York: Interscience, 1957.

14. TRUXAL, JOHN H., *Automatic Feedback Control System Synthesis*, New York: McGraw-Hill, 1955.

15. WHITTAKER, E. T., *Analytical Dynamics*, 4th ed. Cambridge, Eng.: Cambridge University Press, 1937.

CHAPTER 12

FUNDAMENTAL THEORY

12–1 Introduction. Throughout this book there has been continual application of the basic existence theorem of Section 1–8: *under appropriate hypotheses of continuity a differential equation has a unique solution satisfying given initial conditions.* In this chapter proofs are given of this theorem and related theorems.

The postponement of these proofs until now is motivated by several considerations. First, the proofs are fairly difficult, though not excessively so. Second, if we studied the proofs before any acquaintance with particular differential equations, we might fail to see the importance and wide applicability of the theorems. Finally, by studying the theory last, we can approach the subject with a rich background of examples, which indicate that the hypotheses and conclusions are reasonable.

12–2 Uniform convergence. The concept of uniform convergence is a basic tool in the proof of the existence theorems. Here we consider this topic briefly. For a detailed discussion and proofs of the theorems the reader is referred to pp. 338–349 of Reference 6.

Let $f_n(x)$ be a sequence of functions of x, $(n = 1, 2, \ldots)$, all defined in an interval $a \leq x \leq b$. The sequence is said to *converge to* $f(x)$ in this interval if, for each x of the interval,

$$\lim_{n \to \infty} f_n(x) = f(x). \tag{12–1}$$

When (12–1) holds, we can make $|f_n(x) - f(x)|$ as small as we wish (for fixed x) by choosing n sufficiently large. For example, we can make $|f_n(x) - f(x)| < 0.0001$ by choosing $n > N$, for appropriate choice of N. However, the choice of N may vary as we vary the choice of x; for one value of x, N may have to be 100, for another, N may have to be 1000. The condition of uniform convergence is simply the requirement that such an irregularity does not arise:

Definition. The sequence $f_n(x)$ *converges uniformly* to $f(x)$ for $a \leq x \leq b$ if for every $\epsilon > 0$ an N can be found (depending on ϵ but not on x) such that

$$|f_n(x) - f(x)| < \epsilon \qquad (n > N) \tag{12–2}$$

for all x, $(a \leq x \leq b)$.

EXAMPLE 1.   The sequence $1 + (x/n)$ converges uniformly to 1 for $0 \leq x \leq 1$. To make $|f_n(x) - f(x)| < \epsilon$, we must have $|x/n| < \epsilon$ for all x, $(0 \leq x \leq 1)$. The condition is satisfied if $n > N$, provided $N > 1/\epsilon$, for then

$$\left| \frac{x}{n} \right| < \left| \frac{x}{N} \right| < \epsilon |x| \leq \epsilon \qquad (0 \leq x \leq 1).$$

EXAMPLE 2.   The sequence x^n converges for $\frac{1}{2} \leq x \leq 1$. However, the limit function $f(x)$ is *discontinuous:* $f(x) = 0$ for $\frac{1}{2} \leq x < 1, f(1) = 1$. The convergence is not uniform. The condition: $|f_n(x) - f(x)| < \epsilon$ for $\frac{1}{2} \leq x \leq 1$ becomes $x^n < \epsilon$ or

$$n > \frac{\log \epsilon}{\log x}.$$

By choosing x sufficiently close to 1, we can make $\log x$ as close to 0 as desired; this forces n to be larger than any preassigned value. Thus no N can be chosen which will suffice for all x of the interval.

THEOREM A.   *If the sequence $f_n(x)$ converges uniformly to $f(x)$ for $a \leq x \leq b$ and, for each n, $f_n(x)$ is continuous for $a \leq x \leq b$, then $f(x)$ is continuous for $a \leq x \leq b$.*

Thus the limit of a uniformly convergent sequence of continuous functions is continuous. This theorem by itself implies that the sequence x^n, considered above, cannot be uniformly convergent for $\frac{1}{2} \leq x \leq 1$; for the limit function is discontinuous.

THEOREM B.   *Under the hypotheses of Theorem A, for $a \leq a_1 < b_1 \leq b$ we have*

$$\lim_{n\to\infty} \int_{a_1}^{b_1} f_n(x)\, dx = \int_{a_1}^{b_1} f(x)\, dx. \qquad (12\text{--}3)$$

Thus we can proceed to the limit under the integral sign, provided the convergence is uniform.

Definition.   Let the functions $u_n(x)$ be defined for $a \leq x \leq b$. Then the *series*

$$\sum_{n=1}^{\infty} u_n(x)$$

converges uniformly to $f(x)$ for $a \leq x \leq b$ if the sequence of partial sums

$$f_n(x) = u_1(x) + \cdots + u_n(x)$$

converges uniformly to $f(x)$ for $a \leq x \leq b$.

THEOREM C. (*M-test*). *If a sequence of constants M_n can be found such that the series $\sum_{n=1}^{\infty} M_n$ converges and such that.*

$$|u_n(x)| \leq M_n \qquad (a \leq x \leq b),$$

then the series $\sum_{n=1}^{\infty} u_n(x)$ converges uniformly for $a \leq x \leq b$.

EXAMPLE. The series $\Sigma x^n/n!$ converges uniformly for $-1 \leq x \leq 1$, for

$$\left| \frac{x^n}{n!} \right| \leq \frac{1}{n!} = M_n$$

and

$$\sum_{n=0}^{\infty} M_n = \sum_{n=0}^{\infty} \frac{1}{n!} = e.$$

(The fact that the series is numbered starting with 0 is of no import.)

In obtaining constants M_n for which $|u_n(x)| \leq M_n$ holds, we shall apply the principle that a function continuous on an interval $a \leq x \leq b$ has a finite maximum. An analogous principle holds for functions of two or more variables $x, y_1, \ldots$; the interval is replaced by a "rectangle" $a \leq x \leq b$, $a_1 \leq y_1 \leq b_1$, $a_2 \leq y_2 \leq b_2$, ... or by a *bounded closed region* (pp. 72–74 of Reference 6). On occasion we also apply the principle that the absolute value of a sum is at most equal to the sum of the absolute values of the summands and the limiting form of this principle:

$$\left| \int_a^b f(x) \, dx \right| \leq \int_a^b |f(x)| \, dx \leq M(b - a),$$

if $|f(x)| \leq M$ for $a \leq x \leq b$ (pp. 172–174 of Reference 6).

12–3 Existence theorem for $y' = F(x, y)$.

THEOREM 1. *Let $F(x, y)$ be continuous in a domain D of the xy-plane. Let $F_y(x, y)$ be continuous in D and let (x_0, y_0) be a point of D. Then a solution $y = f(x)$, $(|x - x_0| < h)$, of the differential equation*

$$y' = F(x, y) \tag{12–4}$$

exists such that $f(x_0) = y_0$ and, for each x, $[x, f(x)]$ is in D.

Proof. We first choose h_1, k_1 so small that the rectangle

$$R_1: \quad |x - x_0| \leq h_1, \quad |y - y_0| \leq k_1$$

is in D (Fig. 12–1). Now if (x, y_1), (x, y_2) are in R_1, by the law of the mean we have

$$F(x, y_2) - F(x, y_1) = (y_2 - y_1) F_y(x, y^*),$$

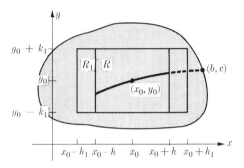

FIG. 12–1. Proof of existence theorem.

where y^* is between y_1 and y_2. Since F_y is continuous in R_1, $|F_y| \leqq K$ for (x, y) in R_1, for some constant K (see end of Section 12–2). Hence

$$|F(x, y_2) - F(x, y_1)| \leqq K|y_2 - y_1|. \qquad (12\text{–}5)$$

Since $F(x, y)$ is continuous in R_1, we have for some constant M

$$|F(x, y)| \leqq M \qquad [(x, y) \text{ in } R_1]. \qquad (12\text{–}6)$$

We now choose $h > 0$ such that

$$h \leqq h_1, \qquad h \leqq \frac{k_1}{M}. \qquad (12\text{–}7)$$

Henceforth, we restrict (x, y) to the rectangle

$$R: \quad |x - x_0| \leqq h, \quad |y - y_0| \leqq k_1,$$

which is contained in R_1.

We now define a sequence of continuous functions $f_n(x)$, $(|x - x_0| \leqq h)$, whose graphs lie in R and which, as we shall see, converge to the desired solution $f(x)$. We choose

$$f_0(x) \equiv y_0,$$

$$f_1(x) = \int_{x_0}^{x} F[u, f_0(u)]\, du + y_0 = \int_{x_0}^{x} F(u, y_0)\, du + y_0,$$

$$f_2(x) = \int_{x_0}^{x} F[u, f_1(u)]\, du + y_0,$$

and, inductively,

$$f_n(x) = \int_{x_0}^{x} F[u, f_{n-1}(u)]\, du + y_0 \qquad (n = 1, 2, \ldots). \qquad (12\text{–}8)$$

We first verify that the $f_n(x)$ all have graphs in R, so that all are well defined. This is clearly true for $f_0(x)$. If it is true for $f_{n-1}(x)$, then $F[u, f_{n-1}(u)]$ is defined for $|u - x_0| \leq h$ and, by (12–8), (12–6), (12–7)

$$|f_n(x) - y_0| \leq \left| \int_{x_0}^{x} F[u, f_{n-1}(u)]\, du \right| \leq M|x - x_0| \leq Mh \leq k_1.$$

Hence $f_n(x)$ has a graph in R and, by induction, this holds for all n.

Next we study the convergence of the sequence $f_n(x)$. We have by (12–5), (12–6)

$$|f_1(x) - f_0(x)| = \left| \int_{x_0}^{x} F(u, y_0)\, du \right| \leq M|\, x - x_0|,$$

$$|f_2(x) - f_1(x)| = \left| \int_{x_0}^{x} \{F[u, f_1(u)] - F[u, f_0(u)]\}\, du \right|$$

$$\leq K \left| \int_{x_0}^{x} |f_1(u) - f_0(u)|\, du \right|$$

$$\leq KM \left| \int_{x_0}^{x} |x - x_0|\, dx \right|$$

$$\leq KM \frac{|x - x_0|^2}{2}.$$

Similarly, we find that

$$|f_3(x) - f_2(x)| \leq MK^2 \frac{|x - x_0|^3}{3!}$$

and, in general (by induction),

$$|f_{n+1}(x) - f_n(x)| \leq MK^n \frac{|x - x_0|^{n+1}}{(n+1)!}.$$

Accordingly, for $|x - x_0| \leq h$

$$|f_{n+1}(x) - f_n(x)| \leq MK^n \frac{h^{n+1}}{(n+1)!} = M_n.$$

Therefore the series

$$f_0(x) + [f_1(x) - f_0(x)] + \cdots + [f_{n+1}(x) - f_n(x)] + \cdots$$

is uniformly convergent to a function $f(x)$ for $|x - x_0| \leq h$; for ΣM_n

converges by the ratio test and Theorem C is applicable. The nth partial sum of the series is

$$f_0(x) + \cdots + [f_n(x) - f_{n-1}(x)] = f_n(x).$$

Hence $f_n(x)$ converges uniformly to $f(x)$ for $|x - x_0| \leq h$. By Theorem A, $f(x)$ is continuous. Since $f_n(x) \to f(x)$, the graph of $f(x)$ is in R.

Now the sequence

$$g_n(x) = F[x, f_n(x)]$$

is also uniformly convergent, for $|x - x_0| \leq h$, to $F[x, f(x)]$. This follows from the uniform convergence of $f_n(x)$, for by (12–5)

$$|F[x, f(x)] - F[x, f_n(x)]| \leq K|f(x) - f_n(x)|.$$

Therefore by Theorem B

$$f(x) = \lim_{n \to \infty} f_{n+1}(x) = \lim_{n \to \infty} \left\{ \int_{x_0}^x F[u, f_n(u)]\, dx + y_0 \right\}$$

$$= \int_{x_0}^x F[u, f(u)]\, du + y_0.$$

Accordingly,

$$f'(x) = \frac{d}{dx} \int_{x_0}^x F[u, f(u)]\, du = F[x, f(x)];$$

thus $y = f(x)$ is a solution of the differential equation (12–4). Also

$$f(x_0) = \int_{x_0}^{x_0} F[u, f(u)]\, du + y_0 = y_0.$$

Hence the initial condition is satisfied and the theorem is completely proved.

Remark 1. In the proof the continuity of the derivative F_y was used only in order to obtain the inequality (12–5). If for each rectangle R_1 in D a constant K can be chosen so that (12–5) holds for (x, y_1), (x, y_2) in R_1, then $F(x, y)$ is said to satisfy a *local Lipschitz condition* in D. Whenever $F(x, y)$ is continuous and satisfies such a condition, the existence theorem is applicable. If, for example, $F(x, y) = x|y|$, then F is continuous for all (x, y) and satisfies a local Lipschitz condition (Prob. 2 below), but F_y does not exist for $y = 0$.

Remark 2. The solution $y = f(x)$ is obtained as a limit of a sequence $f_n(x)$. The functions $f_n(x)$ are called *successive approximations* of $f(x)$ and the method of proof is called *Picard's method of successive approximations.* The crucial formula is (12–8), which defines each approximation in terms of the preceding one. The choice of initial function, $f_0(x)$, can be varied, so long as its graph is within a rectangle such as R, for which convergence can be established. We can use the Picard method as an alternative to the numerical methods of Chapter 10; however, in most cases the successive integrations make the method unwieldy.

EXAMPLE. $y' = xy + 2x - x^3$, $y = 0$ for $x = 0$.   With $f_0(x) \equiv 0$, we find

$$f_1(x) = \int_0^x (2u - u^3)\, du = x^2 - \frac{x^4}{4},$$

$$f_2(x) = \int_0^x \left(2u - \frac{u^5}{4}\right) du = x^2 - \frac{x^6}{24},$$

$$\vdots$$

$$f_n(x) = x^2 - \frac{x^{2n+2}}{4 \cdot 6 \cdots (2n+2)}.$$

The exact solution is $y = x^2$, and the convergence appears to be rapid for small $|x|$. (In fact, $f(x) - f_n(x)$ is the general term of the power series

$$\sum_{n=1}^{\infty} \frac{x^{2n+2}}{4 \cdot 6 \cdots (2n+2)}.$$

Since this series converges for all x (by the ratio test), the nth term approaches 0 and $f_n(x) \to f(x)$ for all x.)

12–4 Complete solutions. The solution $y = f(x)$ obtained in Theorem 1 is defined only in a sufficiently small interval $|x - x_0| < h$. We may, however, be able to prolong the solution. Let us suppose, for example, that $f(x)$ has a limit as $x \to x_0 + h$; we define the value $f(x_0 + h)$ as this limit, so that $f(x)$ is continuous (to the left) at $x_0 + h$. Let us suppose the point $[x_0 + h, f(x_0 + h)]$ to be in D. Then in the equation

$$f(x) = \int_{x_0}^x F[u, f(u)]\, du + y_0 \qquad (|x - x_0| < h)$$

we can let $x \to x_0 + h$, so that the equation remains valid at this point. If we differentiate, we find

$$f'(x) = F[x, f(x)],$$

even at $x = x_0 + h$, where the derivative is a derivative *to the left*. Now let $x_1 = x_0 + h$, $y_1 = f(x_1)$. Then Theorem 1 is applicable again, with (x_1, y_1) replacing (x_0, y_0); hence we obtain a solution $g(x)$, defined in some interval $|x - x_1| < h'$. It will be seen below that $g(x)$ must coincide with $f(x)$, where the intervals of definition overlap. Disregarding this fact, we ignore the values of $g(x)$ for $x < x_1$ and consider $g(x)$ to be defined only for $x_1 \leqq x < x_1 + h' = x_2$. The function $g_1(x)$ equal to $f(x)$ for $x_0 - h < x < x_0 + h$ and equal to $g(x)$ for $x_1 \leqq x < x_2$ is then continuous and satisfies the differential equation for $x_0 - h < x < x_2$ (see Fig. 12–2). For $f(x)$ is continuous to the left at x_1, $g(x)$ is continuous to the right at x_1, and $f(x_1) = g(x_1) = g_1(x_1)$; hence $g_1(x)$ is continuous at x_1; the continuity at the other points follows from that of f and g. A similar argument shows that $g_1'(x_1)$ exists and equals $F(x_1, y_1)$, where $y_1 = f(x_1) = g_1(x_1)$, and for other x, $g_1'(x) = F[x, g_1(x)]$ since f and g are solutions. Thus $g_1(x)$ is a solution of the differential equation for $x_0 - h < x < x_2$.

Continuing thus, we may be able to obtain a solution over longer and longer intervals $x_0 - h < x < x_2$, $x_0 - h < x < x_3$, ..., $x_0 - h < x < x_n$, ... A similar process can be carried out for values of x less than $x_0 - h$, that is, for $x_{-2} < x \leqq x_{-1} = x_0 - h$, $x_{-3} < x \leqq x_{-1}, \ldots$ If $x_n \to +\infty$ and $x_{-n} \to -\infty$ as $n \to \infty$, then we have obtained a solution valid for all x. However, it can happen that the x_n (which form a monotone increasing sequence) have a finite limit b and the x_{-n} a finite limit a. We then obtain a solution valid in a finite interval $a < x < b$. If the solution cannot be extended to a larger interval, we call it a *complete solution;* otherwise we call the solution *incomplete.*

Definition. A solution $y = \phi(x)$, $(a < x < b)$, of the differential equation $y' = F(x, y)$ is a *complete* solution if there does not exist a solution $\phi_1(x)$ defined in a larger interval and agreeing with $\phi(x)$ for $a < x < b$; if $\phi(x)$ is not a complete solution, then we call $\phi(x)$ an *incomplete* solution. (In the definition we allow a to be $-\infty$ and b to be $+\infty$; a solution valid for $-\infty < x < \infty$ is necessarily complete.)

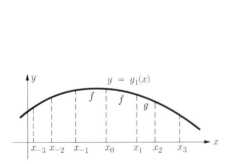

FIG. 12–2. Complete solution.

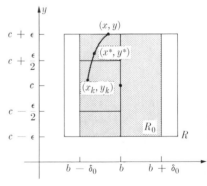

FIGURE 12–3

It will be proved in Section 12–5 that there is a unique complete solution through every point (x_0, y_0) of D. Each such complete solution must tend to the boundary of D as $x \to b-$ or $x \to a+$, as the following theorem shows.

THEOREM 2. *Let* $y = \phi(x)$, $(a < x < b)$, *be a complete solution of the differential equation* $y' = F(x, y)$ *considered in Theorem 1. If* $b \neq +\infty$ *and* x_n *is a sequence such that* $x_n \to b$, $(x_n < b)$, *and the sequence* $y_n = \phi(x_n)$ *has a limit* c, *then the point* (b, c) *is a boundary point of* D.

Remark. As suggested in Fig. 12–1, a point (b, c) is a boundary point of D if there are points of D arbitrarily close to (b, c) and points not in D arbitrarily close to (b, c). Since D is a domain (open region, see Section 2–1), the boundary points of D do not belong to D.

Proof of Theorem 2. Since the points (x_n, y_n) are in D, the limit point (b, c) must be in D or on the boundary of D. We suppose (b, c) is in D and show that this leads to a contradiction. Let $\phi_1(x)$ be defined to equal $\phi(x)$ for $a < x < b$ and to equal c for $x = b$. We then assert that $\phi_1(x)$ *is continuous at* $x = b$. To prove this, we choose δ, ϵ so small that the rectangle R: $|x - b| \leq \delta$, $|y - c| \leq \epsilon$ is in D (Fig. 12–3). Let M be chosen so that $|F(x, y)| \leq M$ in R. Let δ_0 be chosen less than δ and less than $\epsilon/(2M)$, and let R_0 be the rectangle $|x - b| < \delta_0$, $|y - c| < \epsilon$. Now since $(x_n, y_n) \to (b, c)$, we can find a point (x_k, y_k) such that $b - \delta_0 < x_k < b$, $c - (\epsilon/2) < y_k < c + (\epsilon/2)$. The graph of $y = \phi(x)$ must remain in R_0 for $x_k \leq x < b$. For if the graph leaves R_0, it must cross the edge $y = c + \epsilon$ or the edge $y = c - \epsilon$. If (x, y) is the first crossing point, then by the law of the mean

$$\frac{y - y_k}{x - x_k} = \phi'(x^*) = F[x^*, \phi(x^*)] \qquad (x_k < x^* < x).$$

Since $[x^*, \phi(x^*)]$ is in R_0, $|F| \leq M$ at this point, and

$$\left| \frac{y - y_k}{x - x_k} \right| \leq M, \qquad |y - y_k| \leq M|x - x_k| \leq M \delta_0 < \frac{\epsilon}{2}.$$

This is impossible, since $y = c \pm \epsilon$, $|y_k - c| < \frac{1}{2}\epsilon$. Hence the graph of $\phi(x)$ must remain in R_0 for $x_k \leq x < b$; that is, $|\phi(x) - c| < \epsilon$ for $x_k \leq x < b$. Since ϵ is arbitrary, we conclude that

$$\lim_{x \to b-} \phi(x) = c = \phi_1(b)$$

and $\phi_1(x)$ is continuous for $a < x \leq b$. Since (b, c) is in D, we can now prolong the solution to a larger interval $a < x < b + k$, $(k > 0)$, as done for $f(x)$ at the beginning of this section. Thus $\phi(x)$ is not a complete solution, contrary to assumption. Accordingly, (b, c) must be on the boundary of D.

Remark 1. A similar theorem can be stated for sequences (x_n, y_n) for which $x_n \to a$ and $a \neq -\infty$.

Remark 2. If $F(x, y)$ and F_y are continuous for all (x, y), then D has no boundary points. It does not follow that all complete solutions are defined for $-\infty < x < \infty$. For example,

$$y' = 1 + y^2$$

has the complete solution $y = \tan x$, $(-\tfrac{1}{2}\pi < x < \tfrac{1}{2}\pi)$. However, as the example illustrates, a solution can terminate for finite x only if $y \to +\infty$ or $y \to -\infty$. For if $y = \phi(x)$ is defined for $a < x < b$, $(b \neq +\infty)$, and y does not approach $+\infty$ or $-\infty$ as $x \to b$, then we can find a sequence (x_n, y_n), $x_n \to b$, $y_n \to c$, where c is finite; this contradicts Theorem 2. By a similar argument we can show that if $F(x, y)$ and F_y are continuous except at one point (b, c) and $y = \phi(x)$ is a complete solution defined for $a < x < b$, then as $x \to b$, one and only one of the following alternatives occurs: $y \to +\infty$; $y \to -\infty$; $y \to c$.

Under certain conditions we can be sure that a complete solution does not become infinite for finite x.

THEOREM 3. *Let $F(x, y)$ and $F_y(x, y)$ be continuous for $A < x < B$, $-\infty < y < \infty$ and let*

$$|F_y(x, y)| \leqq K$$

for $A < x < B$, $-\infty < y < \infty$. If $y = \phi(x)$ is a co plete solution of the differential equation (12–4), then $y = \phi(x)$ is defined for $A < x < B$. In particular, if $A = -\infty$, $B = +\infty$, $y = \phi(x)$ is defined for

$$-\infty < x < \infty.$$

Proof. We first establish a lemma which will find other applications below.

LEMMA. *Let $v(x)$, $v'(x)$, and $\psi(x)$ be defined and continuous for $x_0 \leqq x \leqq x_1$; let K be a constant. If $v(x_0) = 0$ and*

$$v'(x) \leqq Kv(x) + \psi(x) \qquad (x_0 \leqq x \leqq x_1),$$

then

$$v(x) \leqq e^{Kx} \int_{x_0}^{x} \psi(u) e^{-Ku}\, du \qquad (x_0 \leqq x \leqq x_1).$$

Proof. Under the assumptions made,

$$v'(x) - Kv(x) = \psi(x) + \chi(x),$$

where $\chi(x)$ is continuous and $\chi(x) \leqq 0$. This is a linear differential equation for v. Since $v(x_0) = 0$, on solving we find

$$v(x) = e^{Kx} \int_{x_0}^{x} e^{-Ku}[\psi(u) + \chi(u)] \, du$$

$$\leqq e^{Kx} \int_{x_0}^{x} e^{-Ku}\psi(u) \, du.$$

We now prove Theorem 3. Under the hypotheses made we have

$$|F(x, y_1) - F(x, y_2)| \leqq K|y_1 - y_2|$$

for arbitrary y_1, y_2 and $A < x < B$. Now since $\phi(x)$ is a solution of the differential equation, we can write

$$\phi(x) = \int_{x_0}^{x} F[u, \phi(u)] \, du + y_0 \qquad (a < x < b),$$

where (x_0, y_0) is a point on the graph of $\phi(x)$. Hence for $x > x_0$

$$|\phi(x) - y_0| = \left| \int_{x_0}^{x} F[u, \phi(u)] \, du \right|$$

$$= \left| \int_{x_0}^{x} \{F[u, \phi(u)] - F(u, y_0)\} \, du + \int_{x_0}^{x} F(u, y_0) \, du \right|$$

$$\leqq \int_{x_0}^{x} K|\phi(u) - y_0| \, du + \left| \int_{x_0}^{x} F(u, y_0) \, du \right|.$$

If we write

$$v(x) = \int_{x_0}^{x} |\phi(u) - y_0| \, du, \qquad \psi(x) = \left| \int_{x_0}^{x} F(u, y_0) \, du \right|,$$

then $v(x_0) = 0$, $v(x)$ is continuous, with continuous derivative

$$v'(x) = |\phi(x) - y_0| \qquad (a < x < b),$$

and $\psi(x)$ is continuous for $A < x < B$. The inequality obtained above becomes

$$v'(x) \leqq Kv(x) + \psi(x) \qquad (x_0 < x < b). \tag{12–9}$$

Hence, by the Lemma,

$$v(x) \leqq e^{Kx} \int_{x_0}^{x} \psi(u)e^{-Ku} \, du, \qquad x_0 \leqq x < b. \tag{12–10}$$

The right-hand side is continuous for $A < x < B$. If b were less than B, then $\phi(x)$ would have to approach $\pm\infty$ as $x \to b-$, for there are no boundary points on the line $x = b$, $b < B$. Hence $v'(x) \to +\infty$ as $x \to b$; by (12–9) this implies that $v(x) \to +\infty$ as $x \to b$; this contradicts (12–10). Hence $b = B$. Similarly $a = A$.

12–5 Uniqueness theorem.

THEOREM 4. *Under the hypotheses of Theorem 1 let $f(x)$ and $g(x)$ both be solutions of the differential equation $y' = F(x, y)$ for $|x - x_0| < h$, with $f(x_0) = g(x_0) = y_0$. Then $f(x) \equiv g(x)$ for $|x - x_0| < h$.*

Proof: We first prove $f(x) \equiv g(x)$ for $|x - x_0| < h_1$, for h_1 sufficiently small. We choose h_1 so small that there is a rectangle

$$R: \quad |x - x_0| \leq h_1, \quad |y - y_0| \leq k_1$$

in D containing the graphs of $f(x)$, $g(x)$ for $|x - x_0| \leq h_1$. As in the proof of Theorem 1, we choose K so that

$$|F(x, y_1) - F(x, y_2)| \leq K|y_1 - y_2|$$

for (x, y_1), (x, y_2) in R. Now for $x \geq x_0$

$$f(x) = \int_{x_0}^{x} F[u, f(u)] \, du + y_0,$$

$$g(x) = \int_{x_0}^{x} F[u, g(u)] \, du + y_0,$$

$$f(x) - g(x) = \int_{x_0}^{x} \{F[u, f(u)] - F[u, g(u)]\} \, du,$$

$$|f(x) - g(x)| \leq \int_{x_0}^{x} |F[u, f(u)] - F[u, g(u)]| \, du$$

$$\leq K \int_{x_0}^{x} |f(u) - g(u)| \, du.$$

As in the proof of Theorem 3, we set

$$v(x) = \int_{x_0}^{x} |f(u) - g(u)| \, du, \quad v'(x) = |f(x) - g(x)|, \quad v(x_0) = 0.$$

Then $v'(x) \leqq Kv(x)$, and by the Lemma of Section 12–4 we conclude that $v(x) \leqq 0$. But $v(x) \geqq 0$ for $x \geqq x_0$. Hence $v(x) \equiv 0$; therefore $v'(x) \equiv 0$ and $f(x) \equiv g(x)$. A similar argument holds for $x \leqq x_0$; therefore $f(x) \equiv g(x)$ for $|x - x_0| \leqq h_1$.

If $f(x) \not\equiv g(x)$ for $x_0 \leqq x < x_0 + h$, then there must be a value x_1 such that $f(x) \equiv g(x)$ for $x_0 \leqq x < x_1$ and such that $f(x) \not\equiv g(x)$ for $x_0 \leqq x < x_2$ for $x_2 > x_1$. Since f and g are continuous, we conclude that $f(x_1) = g(x_1)$. The argument given above now establishes that $f(x) \equiv g(x)$ in some interval $|x - x_1| \leqq h_2$. This contradicts the choice of x_1. Hence $f(x) \equiv g(x)$ for $x_0 \leqq x < x_0 + h$ and similarly $f(x) \equiv g(x)$ for $x_0 - h < x \leqq x_0$. The theorem is proved.

THEOREM 5. *Let $F(x, y)$ satisfy the hypotheses of Theorem 1 and let (x_0, y_0) be a point of D. Then a unique complete solution, $y = \phi(x)$, $(a < x < b)$, passes through (x_0, y_0).*

Proof. For each x there may or may not be a solution of the differential equation, passing through (x_0, y_0) and defined at x. If there is one, let $\phi(x)$ be the value of y at x. This assigns only one value to each x, by virtue of the uniqueness theorem. Furthermore, $y = \phi(x)$ is defined in some interval $a < x < b$, where a may be $-\infty$ and b may be $+\infty$. If $\phi(x)$ is not defined for all $x > x_0$, then there must be a value x_1 such that $\phi(x)$ is defined for $x_0 \leqq x < x_1$ and not defined for $x > x_1$; $\phi(x)$ cannot be defined at $x = x_1$, for if it were, there would be a solution through $[x_1, \phi(x_1)]$; this solution must exist in an interval about x_1, so that $\phi(x)$ would be defined for $x_0 \leqq x < x_2$, with $x_2 > x_1$. This contradicts the choice of x_1. Hence we can choose x_1 as b. Similarly, if $\phi(x)$ is not defined for all $x < x_0$, we can choose a as the value such that $\phi(x)$ is defined for $a < x < x_0$ but undefined for $x \leqq a$.

We can denote the complete solution through (x_0, y_0) by $\phi(x; x_0, y_0)$. This function of three variables then describes all solutions of the differential equation in the given domain D; it is truly the general solution in D. The differential equation is of first order, but ϕ depends on *two* arbitrary constants, because each solution is repeated many times. If $y = f(x)$ is a solution, then $f(x) \equiv \phi(x; x_0, y_0)$ for each choice of (x_0, y_0) on the graph of $f(x)$. In many cases we can describe all solutions by keeping x_0 fixed, e.g., at 0. If all solutions meet the line $x_0 = 0$ (the y-axis) as in Fig. 12–4(a), then $\phi(x; 0, y_0)$ does indeed describe all solutions in terms of one arbitrary constant, y_0. In other cases this may be impossible, as for the tangent curves of Fig. 12–4(b); in this example, each curve meets the line $y = 0$ just once and $\phi(x; x_0, 0)$ gives all solutions: $y = \tan(x - x_0)$. In the example of Fig. 12–4(c), we cannot obtain all solutions by fixing either x_0 or y_0. As suggested in the figure, we can nevertheless represent the solutions as level curves of a function $u(x, y)$, that is, as the curves $u = c$.

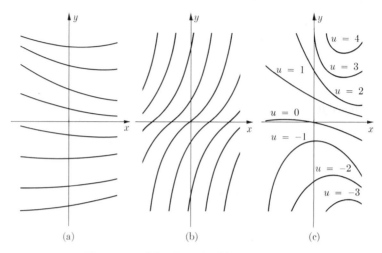

FIG. 12–4. Meaning of arbitrary constant.

These examples show that the statement that the general solution depends on one arbitrary constant requires elaboration (see Section 12–11).

PROBLEMS

1. For each of the following differential equations with initial conditions, show that the solution exists at least in the interval given and apply the method of successive approximations to obtain $f_0(x) = y_0, f_1(x), f_2(x), f_3(x)$.

(a) $y' = y$, $x_0 = 0$, $y_0 = 1$, $|x| < 0.5$. [*Hint:* The maximum of $|F|$ in a rectangle $|x| \leqq h$, $|y - 1| \leqq k$ is $M = 1 + k$. As in the proof of Theorem 1, we must choose h so that $h < k/M = k(1 + k)$. Show that for appropriate k this permits h to be 0.5.]

(b) $y' = -x/y$, $x_0 = 0$, $y_0 = 1$, $|x| < 0.4$. [*Hint:* Reason as in part (a). Here $M = h/(1 - k)$ for $k < 1$.]

2. (a) Show that $F(x, y) = x|y|$ is continuous for all (x, y), that F_y exists and is continuous except when $y = 0$, and that F satisfies a local Lipschitz condition for all (x, y).

(b) Find all solutions of the differential equation $y' = x|y|$ and show that a unique solution passes through each point.

3. (a) Show that $F(x, y) = y^{4/3}$ is continuous and has a continuous derivative F_y for all (x, y).

(b) Find all solutions of the differential equation $y' = y^{4/3}$ and show that there is a unique solution through each point.

4. (a) Show that $F(x, y) = y^{1/3}$ is continuous for all (x, y), but that F_y does not exist for $y = 0$ and that $F(x, y)$ does not satisfy a local Lipschitz condition for all (x, y).

(b) Obtain all solutions of the differential equation $y' = y^{1/3}$ and verify that there is more than one solution through each point on the x-axis.

5. Let $p(x)$, $q(x)$ be continuous for $A < x < B$. Show that every complete solution of the differential equation $y' + p(x)y = q(x)$ exists for $A < x < B$. [*Hint:* Choose a, b so that $A < a < b < B$. Let K be the maximum of $|p(x)|$ for $a \leqq x \leqq b$. Apply Theorem 3 to show existence of solutions for $a \leqq x \leqq b$. Since a, b are arbitrary, conclude that all complete solutions exist for $A < x < B$.]

6. Show that the conclusion of Theorem 3 remains valid if we replace the condition $|F_y| \leqq K$ by the condition $|F_y| \leqq |p(x)|$, where $p(x)$ is a continuous function of x, $(A < x < B)$. [*Hint:* Reason as in Prob. 5.]

7. Show that every complete solution of the differential equation

$$y' = x^2 \arctan y + e^x$$

is defined for $-\infty < x < \infty$. (See Prob. 6.)

8. For each of the following differential equations find $\phi(x; x_0, y_0)$, the complete solution through (x_0, y_0), and determine whether $\phi(x; 0, y_0)$ gives all solutions:

(a) $y' - 2y = e^x$,

(b) $y' = x \sin y$,

(c) $y' = e^x y^2$,

(d) $y' + \dfrac{2xy}{x^2 + y^2} = 0$ $(y > 0)$,

(e) $y' + \dfrac{2xy}{x^2 + y^2} = 0$, all (x, y) except $(0, 0)$.

ANSWERS

1. (a) $f_0 = 1$, $f_1 = 1 + x$, $f_2 = 1 + x + (x^2/2)$, $f_3 = 1 + x + (x^2/2) + (x^3/6)$;

(b) $f_0 = 1$, $f_1 = 1 - \frac{1}{2}x^2$, $f_2 = 1 + \log(1 - \frac{1}{2}x^2)$, $f_3 = 1 - \int_0^x [u/f_2(u)]\, du$.

2. (b) $y = ce^{x^2/2}$, $c \geqq 0$, $y = ce^{-x^2/2}$, $c < 0$.

3. (b) $y = (c - \frac{1}{3}x)^{-3}$ and $y = 0$.

4. (b) $y = \pm[c + (2x/3)]^{3/2}$ and $y = 0$, and combinations of the two forms of solution.

8. (a) $(y_0 + e^{x_0})e^{2(x - x_0)} - e^x$; (b) $y = 2 \arctan [e^{(x^2 - x_0^2)/2} \tan \frac{1}{2}y_0]$, branch of arctan chosen to agree with value $y = y_0$ for $x = x_0$; if $y_0 = n\pi$, then $y \equiv y_0$; (c) $y = y_0/[1 + y_0(e^{x_0} - e^x)]$; (d) and (e) ϕ defined implicitly by equation $y^3 + 3x^2 y - y_0^3 - 3x_0^2 y_0 = 0$. The function $\phi(x; 0, y_0)$ gives all solutions in parts (a), (b), (d).

12–6 Existence and uniqueness theorems for systems of first order equations. The results established above for the differential equation $y' = F(x, y)$ can be extended to arbitrary systems of first order equations

$$\frac{dy_i}{dx} = F_i(x, y_1, \ldots, y_n) \qquad (i = 1, \ldots, n). \qquad (12\text{–}11)$$

It will be assumed that the functions F_i are defined in a *domain* D of the $(n + 1)$-dimensional space of the variables $x, y_1, \ldots, y_n$. By a domain D (see Section 2–1) is meant a portion D of the space (i.e., set of points) with the following two properties:

(a) if $(\bar{x}, \bar{y}_1, \ldots, \bar{y}_n)$ is in D, then for h sufficiently small all points $(x, y_1, \ldots, y_n)$ for which

$$|x - \bar{x}| < h, \qquad |y_1 - \bar{y}_1| < h, \qquad \ldots, \qquad |y_n - \bar{y}_n| < h$$

are in D;

(b) any two points of D can be joined by a continuous path

$$x = \phi_0(t), \qquad y_1 = \phi_1(t), \qquad \ldots, \qquad y_n = \phi_n(t)$$

in D.

Of these two properties, only the first actually concerns us. The second merely guarantees that D does not consist of several separate pieces; when D is thus split up, we prefer to concentrate on one piece at a time, for which both properties (a) and (b) will then hold.

THEOREM 6. *Let the functions* $F_i(x, y_1, \ldots, y_n)$ *be defined and continuous in a domain* D *and let the partial derivatives* $\partial F_i/\partial y_j$, $(j = 1, \ldots, n)$, *be continuous in* D *for* $i = 1, \ldots, n$. *Then for each point* $(\bar{x}, \bar{y}_1, \ldots, \bar{y}_n)$ *in* D *there exists a solution*

$$y_1 = \phi_1(x), \qquad \ldots, \qquad y_n = \phi_n(x) \qquad (|x - \bar{x}| < h), \qquad (12\text{–}12)$$

of the system (12–11) *for which*

$$\phi_i(\bar{x}) = \bar{y}_i \qquad (i = 1, \ldots, n).$$

Proof. We choose an $(n + 1)$-dimensional rectangle

$$R_1: \quad |x - \bar{x}| \leqq h_1, \quad |y - \bar{y}_1| \leqq k_1, \quad \ldots, \quad |y - \bar{y}_n| \leqq k_n$$

in D. Let $(x, y_1', \ldots, y_n')$, $(x, y_1'', \ldots, y_n'')$ be two points in R_1 (the primes do not refer to derivatives). By the law of the mean for functions of several variables (see p. 371 of Reference 6),

$$F_i(x, y'_1, \ldots, y'_n) - F_i(x, y''_1, \ldots, y''_n)$$

$$= (y'_1 - y''_1) \frac{\partial F_i}{\partial y_1}(x, y_1^*, \ldots, y_n^*) + \cdots$$

$$+ (y'_n - y''_n) \frac{\partial F_i}{\partial y_n}(x, y_1^*, \ldots, y_n^*), \qquad (12\text{–}13)$$

where $(x, y_1^*, \ldots, y_n^*)$ lies on the line segment joining $(x, y'_1, \ldots, y'_n)$ to $(x, y''_1, \ldots, y''_n)$. Since all the first partial derivatives are continuous in R_1, we can choose K so large that

$$\left| \frac{\partial F_i}{\partial y_j} \right| \leqq K \text{ in } R_1.$$

Hence from (12–13)

$$|F_i(x, y'_1, \ldots, y'_n) - F_i(x, y''_1, \ldots, y''_n)| \leqq K(|y'_1 - y''_n| + \cdots$$
$$+ |y'_n - y''_n|). \qquad (12\text{–}14)$$

Since for each rectangular region in D an inequality such as (12–14) holds, we say that the F_i satisfy *local Lipschitz conditions* in D. It is these conditions which are used in the proof, rather than the hypothesis that the $\partial F_i / \partial y_j$ are continuous.

Since the F_i are continuous, we can also choose M so that

$$|F_i(x, y_1, \ldots, y_n)| \leqq M \text{ in } R_1. \qquad (12\text{–}15)$$

We now choose h so that

$$0 < h \leqq h_1, \quad h \leqq \frac{k_1}{M}, \quad \cdots, \quad h \leqq \frac{k_n}{M}. \qquad (12\text{–}16)$$

Henceforth we restrict attention to the $(n + 1)$-dimensional rectangle

$$R: \quad |x - \bar{x}| \leqq h, \quad |y_1 - \bar{y}_1| \leqq k_1, \quad \ldots, \quad |y_n - \bar{y}_n| \leqq k_n.$$

Our solution is now defined as the limit of a sequence of successive approximations (method of Picard):

$$\phi_i^{(0)}(x) \equiv \bar{y}_i \qquad (i = 1, \ldots, n),$$

$$\phi_i^{(1)}(x) = \int_{\bar{x}}^{x} F_i(u, \bar{y}_1, \ldots, \bar{y}_n)\, du + \bar{y}_i,$$

$$\phi_i^{(2)}(x) = \int_{\bar{x}}^{x} F_i[u, \phi_1^{(1)}(u), \ldots, \phi_n^{(1)}(u)]\, du + \bar{y}_i, \qquad \cdots$$

and inductively

$$\phi_i^{(N)}(x) = \int_{\bar{x}}^{x} F_i[u, \phi_1^{(N-1)}(u), \ldots, \phi_n^{(N-1)}(u)] \, du + \bar{y}_i, \quad (12\text{-}17)$$

for $N = 1, 2, \ldots$ Throughout, $|x - \bar{x}| \leq h$. We must verify that the $\phi_i^{(N)}(x)$ are well defined. This we do by showing inductively that at each stage, the graph of $y_i = \phi_i^{(N)}(x)$, $(i = 1, \ldots, n)$, in the $(n + 1)$-dimensional space lies in R. This is clearly true for $N = 0$. If it is true for $N = k$, then Eq. (12-17) is meaningful for $N = k + 1$ and defines the functions $\phi_i^{(k+1)}(x)$. Furthermore

$$|\phi_i^{(k+1)}(x) - \bar{y}_i| \leq M|x - \bar{x}| \leq Mh \leq k_i$$

by (12-15) and (12-16). Hence the graph lies in R for $N = k + 1$ and, by induction, the $\phi_i^{(N)}(x)$ are well defined for all N.

By their definition the $\phi_i^{(N)}(x)$ are continuous functions of x. We now show that they converge uniformly for $|x - \bar{x}| \leq h$ to the desired functions $\phi_i(x)$, $(i = 1, \ldots, n)$. We have

$$|\phi_i^{(0)}(x) - \bar{y}_i| \leq M \left| \int_{\bar{x}}^{x} du \right| = M|x - \bar{x}|,$$

$$|\phi_i^{(2)}(x) - \phi_i^{(1)}(x)| = \left| \int_{\bar{x}}^{x} \{F_i[u, \phi_1^{(1)}(u), \ldots] - F_i(u, \bar{y}_1, \ldots, \bar{y}_n)\} \, du \right|$$

$$\leq K \left| \int_{\bar{x}}^{x} [|\phi_1^{(1)}(u) - \bar{y}_1| + \cdots + |\phi_n^{(1)}(u) - \bar{y}_n|] \, du \right|$$

$$\leq KnM \left| \int_{\bar{x}}^{x} |u - \bar{x}| \, du \right| = KnM \frac{|x - \bar{x}|^2}{2},$$

by (12-14). Proceeding inductively, we find that

$$|\phi_i^{N+1}(x) - \phi_i^{(N)}(x)| \leq K^N n^N M \frac{|x - \bar{x}|^{N+1}}{(N+1)!} = M_N.$$

We then conclude, as in the proof of Theorem 1, that for each i, $\phi_i^N(x)$ converges uniformly to $\phi_i(x)$:

$$\phi_i(x) = \phi_i^{(0)}(x) + \sum_{N=0}^{\infty} [\phi_i^{(N+1)}(x) - \phi_i^{(N)}(x)],$$

and that

$$\phi_i(x) = \int_{\bar{x}}^{x} F_i[u, \phi_1(u), \ldots, \phi_n(u)] \, du + \bar{y}_i,$$

so that $\phi_i(\bar{x}) = \bar{y}_i$ and

$$\frac{d\phi_i}{dx} = F_i[x, \phi_1(x), \ldots, \phi_n(x)] \qquad (|x - \bar{x}| < h).$$

Thus the theorem is proved.

Complete solutions and their properties can now be discussed as in Section 12–4. A solution defined for $a < x < b$ is complete if there is no solution defined in a larger interval and agreeing with the given solution for $a < x < b$. If $b < +\infty$ and the given solution can be defined at $x = b$ so that it is continuous (to the left) and remains in D, then we can prolong the solution beyond $x = b$, so that the solution is not complete. If $b < +\infty$ and the given solution is complete, then as $x \to b-$, the solution must approach the boundary of D; the formulation and proof of a precise result are left as a problem (Prob. 4 below, following Section 12–7). If the F_i satisfy the condition: $|\partial F_i/\partial y_j| \leqq K$ for some constant K, for $A < x < B$ and all $y_1, \ldots, y_n$, then each complete solution in the domain $A < x < B$ must be defined for $A < x < B$; see Prob. 5 below.

Now the uniqueness of solutions will be proved in detail.

THEOREM 7.   *Under the hypotheses of Theorem* 6 *let* $y_i = \phi_i(x)$ *and* $y_i = \psi_i(x)$, $(i = 1, \ldots, n)$, *both be solutions of the system* (12–11) *for* $|x - \bar{x}| < h$. *If* $\phi_i(\bar{x}) = \psi_i(\bar{x}) = \bar{y}_i$, $(i = 1, \ldots, n)$, *then*

$$\phi_i(x) \equiv \psi_i(x).$$

Proof. We first choose $h_1 < h$ and a rectangular region

$$R_1: \quad |x - \bar{x}| \leqq h_1, \quad |y_i - \bar{y}_i| \leqq k_i \qquad (i = 1, \ldots, n)$$

in D which contains the graphs of the two solutions for $|x - \bar{x}| \leqq h_1$. As in the proof of Theorem 6, we can choose K so that (12–14) holds in R_1. From the fact that $\phi_i(x)$ and $\psi_i(x)$, $(i = 1, \ldots, n)$, are solutions of (12–11), we deduce that for $\bar{x} \leqq x \leqq \bar{x} + h_1$

$$\phi_i(x) = \int_{\bar{x}}^{x} F_i[u, \phi_1(u), \ldots, \phi_n(u)]\, du + \bar{y}_i,$$

$$\psi_i(x) = \int_{\bar{x}}^{x} F_i[u, \psi_1(u), \ldots, \psi_n(u)]\, du + \bar{y}_i,$$

$$\phi_i(x) - \psi_i(x) = \int_{\bar{x}}^{x} \{F_i[u, \phi_1(u), \ldots] - F_i[u, \psi_1(u), \ldots]\}\, du,$$

$$|\phi_i(x) - \psi_i(x)| \leqq \int_{\overline{x}}^{x} |F_i[u, \phi_1(u), \ldots] - F_i[u, \psi_1(u), \ldots]| \, du$$

$$\leqq K \int_{\overline{x}}^{x} \sum_{i=1}^{n} |\phi_i(u) - \psi_i(u)| \, du.$$

If we add these inequalities for $i = 1, \ldots, n$, we find

$$\sum_{i=1}^{n} |\phi_i(x) - \psi_i(x)| \leqq nK \int_{\overline{x}}^{x} \sum_{i=1}^{n} |\phi_i(u) - \psi_i(u)| \, du.$$

Hence if

$$v(x) = \int_{\overline{x}}^{x} \sum_{i=1}^{n} |\phi_i(u) - \psi_i(u)| \, du \qquad (\overline{x} \leqq x \leqq \overline{x} + h_1),$$

we have $v(x) \geqq 0$, $v(\overline{x}) = 0$, and

$$v'(x) = \sum_{i=1}^{n} |\phi_i(x) - \psi_i(x)| \, dx \leqq nKv(x).$$

Thus, by the Lemma of Section 12–4, $v(x) \leqq 0$. Since $v(x) \geqq 0$, we conclude that $v(x) \equiv 0$ for $\overline{x} \leqq x \leqq \overline{x} + h_1$. Therefore $v'(x) \equiv 0$; accordingly, each term $|\phi_i(x) - \psi_i(x)|$ must be 0, so that

$$\phi_i(x) \equiv \psi_i(x) \text{ for } \overline{x} \leqq x \leqq \overline{x} + h_1 \qquad (i = 1, \ldots, n).$$

Similar reasoning proves the identity for $\overline{x} - h_1 \leqq x \leqq \overline{x}$. As in the proof of Theorem 4, we now conclude that the solutions coincide over the whole of the given interval $|x - \overline{x}| \leqq h$.

Now we can prove a theorem analogous to Theorem 5: there is a unique complete solution through each point of D. With very slight changes, the proof of Theorem 5 can be repeated. The complete solution becomes a set of n functions

$$\phi_i(x; \overline{x}, \overline{y}_1, \ldots, \overline{y}_n)$$

of $n + 2$ variables, defined for $(\overline{x}, \overline{y}_1, \ldots, \overline{y}_n)$ in D and $a < x < b$, where a, b depend on $(\overline{x}, \overline{y}_1, \ldots, \overline{y}_n)$. This set of functions can again be considered to be the general solution of the system of differential equations; it depends on $n + 1$ arbitrary constants. As in Section 12–5, we may be able to obtain all the solutions by restricting $(\overline{x}, \overline{y}_1, \ldots, \overline{y}_n)$; for example, by restricting $\overline{x}$ to one fixed value.

12-7 The equation of order n. As an immediate application of the theorem on systems of first order equations, we can establish the basic theorem on the equation of order n:

$$y^{(n)} = F(x, y, y', \ldots, y^{(n-1)}). \tag{12-18}$$

THEOREM 8. *Let F be defined and continuous in a domain D of the $(n + 1)$-dimensional space of the variables x, y, y', $\ldots$, $y^{(n-1)}$ and let the partial derivatives*

$$\frac{\partial F}{\partial y}, \quad \frac{\partial F}{\partial y'}, \quad \ldots, \quad \frac{\partial F}{\partial y^{(n-1)}}$$

be continuous in D. Let $(x_0, y_0, y_0', \ldots, y_0^{(n-1)})$ be a point of D. Then there is a solution $y = f(x)$ of Eq. (12-18), $|x - x_0| < h$, such that

$$f(x_0) = y_0, \quad f'(x_0) = y_0', \quad \ldots, \quad f^{(n-1)}(x_0) = y_0^{(n-1)}. \tag{12-19}$$

If $y = g(x)$ is a second solution of Eq. (12-18), $|x - x_0| < h$, such that $g(x_0) = y_0$, $g'(x_0) = y_0', \ldots$, then $f(x) \equiv g(x)$.

Proof. We let $y_1 = y$, $y_2 = y'$, $\ldots$, $y_n = y^{(n-1)}$. Then the statement that $y = f(x)$ is a solution of Eq. (12-18) is equivalent to the statement that $y_i = \phi_i(x)$, $(i = 1, \ldots, n)$, is a solution of the system

$$\frac{dy_1}{dx} = y_2, \quad \ldots, \quad \frac{dy_{n-1}}{dx} = y_n, \quad \frac{dy_n}{dx} = F(x, y_1, \ldots, y_n) \tag{12-20}$$

and that $\phi_1(x) = f(x)$. Under the assumptions on F all the hypotheses for Theorem 6 are satisfied and we conclude that a solution $y_i = \phi_i(x)$ of (12-20) exists for $|x - x_0| < h$. If we set $f(x) = \phi_1(x)$, we obtain the desired solution of (12-18). The uniqueness of the solution follows from Theorem 7.

Again we can define complete solutions and conclude that (12-18) has a unique complete solution $y = f(x)$, $(a < x < b)$, that satisfies (12-19).

Remark. Systems of form more general than that of (12-20) can be replaced by a single equation of order n. However, there are distinct advantages in considering the equations in system form, as the following example illustrates:

$$\frac{dy}{dx} = yg(x) + z, \quad \frac{dz}{dx} = yh(x) + z,$$

where g and h are continuous for all x. The existence theorem for systems guarantees existence of a solution for arbitrary initial conditions. If we try to eliminate y or z to obtain a single equation of order n, we meet difficulties. For example, elimination of z proceeds as follows:

$$\frac{d^2y}{dx} = yg'(x) + g(x)\frac{dy}{dx} + \frac{dz}{dx}$$

$$= yg'(x) + g(x)\,[yg(x) + z] + yh(x) + z$$

$$= y[g'(x) + g^2(x) + h(x)] + z[g(x) + 1]$$

$$= y[g'(x) + g^2(x) + h(x)] + \left[\frac{dy}{dx} - yg(x)\right][g(x) + 1]$$

$$= [g(x) + 1]\frac{dy}{dx} + y[g'(x) + h(x) - g(x)].$$

But $g'(x)$ may fail to exist for many values of x, even though $g(x)$ is continuous. Hence the second order equation is not in a form to which the existence theorem applies.

PROBLEMS

1. Let the following system, with initial conditions, be given:

$$\frac{dy}{dx} = xy - z, \qquad \frac{dz}{dx} = xy^2 - z, \qquad y = 1 \text{ and } z = 1 \text{ for } x = 0.$$

(a) Show that the solution can be obtained for $|x| < 0.3$ by successive approximations.

(b) Obtain the first three successive approximations for y and z.

(c) Obtain the power series solution through terms in x^3 and compare with the result of part (b).

2. Replace the differential equation $y'' + y = 0$ by a system of first order equations and obtain the first four successive approximations to the solution for which $y = 0$, $y' = 1$ when $x = 0$. Compare with the exact solution.

3. Obtain the complete solution satisfying the given initial conditions:

(a) $y'' = e^y y'$, $y = 0$ and $y' = 1$ for $x = 0$;

(b) $dy/dx = z \cos x + x + y$, $dz/dx = 1 + z^2$, $y = 0$, $z = 0$ for $x = 0$;

(c) $dy/dx = 4y - 5z$, $dz/dx = y - 2z$, $y = y_0$, $z = z_0$ for $x = x_0$.

4. Prove: if $y_i = \phi_i(x)$, $(i = 1, \ldots, n)$, $a < x < b$ is a complete solution of the system (12–11) in the domain D of Theorem 6, then there can exist no convergent sequence x_m such that $x_m < b$, $x_m \to b$ as $m \to \infty$, and the sequences $\phi_i(x_m)$ converge to values $c_1, \ldots, c_n$ such that $(b, c_1, \ldots, c_n)$ is in D. [*Hint:* Show, as in the proof of Theorem 2, that if we define $\phi_i(b)$ to be c_i for $i = 1, \ldots, n$, then existence of such a sequence x_m implies that $\phi_i(x)$ is continuous to the left at $x = b$ and hence that the solution can be prolonged to a larger interval.]

5. (a) Prove: under the hypotheses of Theorem 6 if $|\partial F_i/\partial y_j| < K$, $(i = 1, \ldots, n, j = 1, \ldots, n)$, for $A < x < B$, where K is a constant and D is the domain $A < x < B$, $(y_i$ unrestricted), then every complete solution of system (12–11) exists for $A < x < B$. [*Hint:* Proceed as in the proofs of Theorems 3 and 7 to show that

$$v'(x) \leqq nKv(x) + \psi(x) \qquad (x \geqq \bar{x}),$$

where

$$v(x) = \int_{\bar{x}}^{x} \sum_{i=1}^{n} |\phi_i(u) - \bar{y}_i|\, du$$

and $\psi(x)$ is continuous for $A < x < B$. Then complete the proof as for Theorem 3.]

(b) Show that the conclusion of part (a) remains valid if we replace the condition $|\partial F_i/\partial y_j| < K$ by the condition: $|\partial F_i/\partial y_j| \leqq Q(x)$, where $Q(x)$ is continuous for $A < x < B$. [*Hint:* See Probs. 5, 6 following Section 12–5.]

6. Consider a system

$$\frac{dx_i}{dt} = F_i(x_1, \ldots, x_n) \qquad (i = 1, \ldots, n)$$

which describes a stationary flow in the n-dimensional phase space D of $(x_1, \ldots, x_n)$ (see Chapter 11). Let the F_i be continuous and have continuous first partial derivatives in D.

(a) Let $(\bar{x}_1, \ldots, \bar{x}_n)$ be an equilibrium point; that is, a point of D at which all F_i are 0. Let $x_i = \phi_i(t)$, $(i = 1, \ldots, n)$, be a solution, not the equilibrium solution $x_i \equiv \bar{x}_i$, for which $\lim \phi_i(t) = \bar{x}_i$ as $t \to \bar{t}$. Show that $\bar{t} = +\infty$ or $-\infty$.

(b) Let $\phi_i(t; \bar{t}, \bar{x}_1, \ldots, \bar{x}_n)$, $(i = 1, \ldots, n)$, be the complete solution passing through $(\bar{x}_1, \ldots, \bar{x}_n)$ when $t = \bar{t}$. Show that

$$\phi_i(t; \bar{t}, \bar{x}_1, \ldots, \bar{x}_n) \equiv \phi_i(t + k; \bar{t} + k, \bar{x}_1, \ldots, \bar{x}_n)$$

$$\equiv \phi_i(t - \bar{t}; 0, \bar{x}_1, \ldots, \bar{x}_n).$$

ANSWERS

1. (b) $y_0 = 1$, $z_0 = 1$; $y_1 = 1 - x + \frac{1}{2}x^2$, $z_1 = 1 - x + \frac{1}{2}x^2$, $y_2 = 1 - x + x^2 - \frac{1}{2}x^3 + \frac{1}{8}x^4$, $z_2 = 1 - x + x^2 - x^3 + (x^4/2) - (x^5/5) + (x^6/24)$;

(c) $y = 1 - x + x^2 - \frac{2}{3}x^3 + \cdots$, $z = 1 - x + x^2 - x^3 + \cdots$

2. $y_0 = 0$, $y_1 = y_2 = x$, $y_3 = y_4 = x - \frac{1}{6}x^3$. Exact solution is $\sin x$.

3. (a) $y = \log [1/(1 - x)]$, $-\infty < x < 1$;

(b) $y = \frac{3}{2}e^x - \frac{1}{2}(\cos x + \sin x) - x - 1$, $z = \tan x$, $-\frac{1}{2}\pi < x < \frac{1}{2}\pi$;

(c) $y = \frac{1}{4}[5(y_0 - z_0)e^{3(x-x_0)} + (5z_0 - y_0)e^{-(x-x_0)}]$,

$z = \frac{1}{4}[(y_0 - z_0)e^{3(x-x_0)} + (5z_0 - y_0)e^{-(x-x_0)}]$.

12–8 Linear differential equations. As in Chapter 6, the independent variable will be denoted by t. We consider a system

$$\frac{dx_i}{dt} = \sum_{j=1}^{n} a_{ij}(t)x_j + F_i(t) \qquad (i = 1, \ldots, n). \qquad (12\text{–}21)$$

THEOREM 9. *Let the functions $a_{ij}(t)$ and $F_i(t)$,*

$$(i = 1, \ldots, n, j = 1, \ldots, n),$$

be defined and continuous for $A < t < B$, and let t_0 be a number between A and B. Every complete solution of the system (12–21) exists for $A < t < B$. For each set of values $x_1^0, \ldots, x_n^0$, there is a unique complete solution $x_i = \phi_i(t)$, $(i = 1, \ldots, n)$, such that

$$\phi_1(t_0) = x_1^0, \quad \ldots, \quad \phi_n(t_0) = x_n^0. \qquad (12\text{–}22)$$

When $F_i(t) \equiv 0$, there exist n linearly independent solutions of (12–21) for $A < t < B$. If $x_i = \phi_{ki}(t)$ is a set of n linearly independent solutions, $(k = 1, \ldots, n)$, for $A < t < B$, then the general solution for $A < t < B$ is

$$x_i = \phi_i(t) = \sum_{k=1}^{n} c_k \phi_{ki}(t) \qquad (i = 1, \ldots, n), \qquad (12\text{–}23)$$

where $c_1, \ldots, c_n$ are arbitrary constants.

For arbitrary $F_i(t)$ the general solution for $A < t < B$ is

$$x_i = \phi_i(t) = \phi_i^*(t) + \sum_{k=1}^{n} c_k \phi_{ki}(t) \qquad (i = 1, \ldots, n), \qquad (12\text{–}24)$$

where $\phi_i^*(t)$, $(i = 1, \ldots, n)$, is one solution and $\Sigma c_k \phi_{ki}$ is, as in (12–23), the general solution of the related homogeneous system.

Remark. Throughout this section the indices k and i run from 1 to n. The first subscript, usually k, gives the number of the solution; the second gives the x-coordinate referred to. Thus $\phi_{ki}(t)$ is the function $x_i(t)$ in the kth solution. For each fixed k the n functions $\phi_{ki}(t)$, $(i = 1, \ldots, n)$, provide an n-tuple of functions which may be a solution of the system of differential equations.

Proof of Theorem 9. The quantities $\partial F_i/\partial y_j$ of Section 12–6 become the partial derivatives of the right-hand members of (12–21) with respect to the dependent variables x_j; hence they are simply the functions $a_{ij}(t)$. Since the $a_{ij}(t)$ and $F_i(t)$ are continuous, we conclude that the hypotheses of Theorems 6 and 7 are satisfied and that there is a unique solution of

(12–21) that satisfies given initial conditions (12–22). Since the a_{ij} are continuous, we conclude that each complete solution exists for $A < t < B$ (Prob. 5 following Section 12–7).

To obtain n linearly independent solutions of the related homogeneous system ($F_i \equiv 0$), we construct what is called a *fundamental set of solutions*. The first solution is $\phi_{1i}(t)$, which is chosen to satisfy the initial conditions

$$\phi_{11}(t_0) = 1, \qquad \phi_{12}(t_0) = 0, \qquad \ldots, \qquad \phi_{1n}(t_0) = 0.$$

The second, $\phi_{2i}(t)$, is chosen to have $x_2 = 1$ at $t = t_0$ and all other initial values 0, and so on. In general, the kth solution is $\phi_{ki}(t)$, where

$$\phi_{ki}(t_0) = \delta_{ki} = \begin{cases} 1, k = i, \\ 0, k \neq i. \end{cases} \tag{12–25}$$

(The symbol δ_{ki} is known as the *Kronecker delta*). The n solutions thus defined are linearly independent for $A < t < B$. For if for some constants $c_1, \ldots, c_n$ we have

$$c_1\phi_{1i}(t) + \cdots + c_n\phi_{ni}(t) \equiv 0 \qquad (A < t < B),$$

for $i = 1, \ldots, n$, then for $t = t_0$ by (12–25)

$$c_1\delta_{1i} + \cdots + c_i\delta_{ii} + \cdots + c_n\delta_{ni} = 0;$$

that is, $c_i = 0$ for $i = 1, \ldots, n$.

Remark. We can also prove that if the $\phi_{ki}(t)$ are linearly independent solutions and

$$c_1\phi_{1i}(t) + \cdots + c_n\phi_{ni}(t) = 0 \qquad (i = 1, \ldots, n)$$

for at least one value of t, ($A < t < B$), then $c_1 = 0, \ldots, c_n = 0$. (See Prob. 1 below, following Section 12–9.)

Now let $c_1, \ldots, c_n$ be arbitrary constants and form the functions

$$\phi_i(t) = \sum_{k=1}^{n} c_k\phi_{ki}(t) \qquad (i = 1, \ldots, n). \tag{12–26}$$

Since the differential equations (12–21) are linear and the $\phi_{ki}(t)$ are solutions of the related homogeneous system, the functions

$$\phi_i(t), \qquad (i = 1, \ldots, n),$$

define a solution of the homogeneous system for each choice of the c's. By (12–25), when $t = t_0$,

$$\phi_i(t_0) = c_1\delta_{1i} + \cdots + c_i\delta_{ii} + \cdots + c_n\delta_{ni} = c_i.$$

Hence the c's are simply the initial values of the $\phi_i(t)$. This shows that Eq. (12–26) is the general solution of the homogeneous system; for to obtain the solution with prescribed initial conditions (12–22), we need only choose $c_1 = x_1^0, \ldots, c_n = x_n^0$.

More generally, let $\phi_{ki}(t)$ be any set of n linearly independent solutions of the homogeneous system. Then Eq. (12–26) again represents a solution of the homogeneous system for every choice of the c's and is also the general solution. For to satisfy arbitrary initial conditions (12–22), we must choose the c's so that

$$\sum_{k=1}^{n} c_k \phi_{ki}(t_0) = x_i^0 \qquad (i = 1, \ldots, n). \tag{12–27}$$

These equations can always be solved, unless the determinant (Wronskian determinant)

$$W = \begin{vmatrix} \phi_{11}(t_0) & \phi_{21}(t_0) & \cdots & \phi_{n1}(t_0) \\ \vdots & & & \vdots \\ \phi_{1n}(t_0) & \phi_{2n}(t_0) & \cdots & \phi_{nn}(t_0) \end{vmatrix} \tag{12–28}$$

is 0. If W were 0, then we could choose constants $C_1, \ldots, C_n$, not all 0, so that

$$\sum_{k=1}^{n} C_k \phi_{ki}(t_0) = 0 \qquad (i = 1, \ldots, n).$$

By the remark above (see Prob. 1 following Section 12–9), this contradicts the linear independence of the solutions $\phi_{ki}(t)$. Hence W cannot be 0 and Eqs. (12–27) can be solved uniquely for $c_1, \ldots, c_n$. Thus Eqs. (12–26) do define the general solution.

Now let us consider the nonhomogeneous system. We already know that solutions exist for $A < t < B$. If $\phi_i^*(t)$, $(i = 1, \ldots, n)$, is one solution, then by linearity Eqs. (12–24) define a solution, for every choice of the c's. Furthermore all solutions are so obtained. For if $\phi_i(t)$, $(i = 1, \ldots, n)$, form a solution of the nonhomogeneous system, then by linearity the functions $\phi_i(t) - \phi_i^*(t)$ form a solution of the homogeneous system; hence

$$\phi_i(t) - \phi_i^*(t) = \sum_{k=1}^{n} c_k \phi_{ki}(t) \qquad (i = 1, \ldots, n),$$

for appropriate choice of the c's. The proof of Theorem 9 is now complete.

Remark 1. If ϕ_{ki}, $(i = 1, \ldots, n)$, $(k = 1, \ldots, n)$, form n solutions of the homogeneous system, then we can form the Wronskian determinant

$$W(t) = \begin{vmatrix} \phi_{11}(t) & \cdots & \phi_{n1}(t) \\ \vdots & & \vdots \\ \phi_{1n}(t) & \cdots & \phi_{nn}(t) \end{vmatrix} \tag{12–29}$$

If the solutions are linearly independent then, as above, $W(t)$ is never 0 for $A < t < B$. Conversely, if $W(t)$ is never 0 for $A < t < B$, then the functions must be linearly independent. For otherwise we could choose constants $C_1, \ldots, C_n$, not all 0, so that

$$\sum_{k=1}^{n} C_k \phi_{ki}(t) \equiv 0, \qquad A < t < B \qquad (i = 1, \ldots, n);$$

this in turn implies that the determinant of the coefficients, which is W, is 0 for every t.

Remark 2. Let $\phi_{ki}(t)$ and $\psi_{ki}(t)$ be two sets of linearly independent solutions of the homogeneous system. Then we can choose constants c_{kj} such that

$$\phi_{ki}(t) = \sum_{j=1}^{n} c_{kj} \psi_{ji}(t) \qquad (i = 1, \ldots n, k = 1, \ldots, n). \qquad (12\text{–}30)$$

If k is considered as fixed, we are merely stating that the solution $\phi_{ki}(t)$ is expressible as a linear combination of the solutions $\psi_{ji}(t)$. A similar relation must hold, expressing the ψ_{ki} in terms of the ϕ_{ji}; that is, for each i Eqs. (12–30) can be solved uniquely for the $\psi_{ji}(t)$ in terms of the $\phi_{ki}(t)$. This implies that the determinant

$$C = \begin{vmatrix} c_{11} & \cdots & c_{1n} \\ \vdots & & \vdots \\ c_{n1} & \cdots & c_{nn} \end{vmatrix} \qquad (12\text{–}31)$$

is not 0 (see Section 6–18). Conversely, if $\psi_{ki}(t)$ is a set of linearly independent solutions and the determinant (12–31) is not 0, then Eqs. (12–30) define a new set of solutions $\phi_{ki}(t)$, which are also linearly independent. Indeed, from the rules for multiplication of determinants we can verify that

$$W_\phi(t) = C W_\psi(t), \qquad (12\text{–}32)$$

where W_ϕ and W_ψ are the Wronskian determinants of the two sets of solutions. Since $W_\psi \neq 0$, $W_\phi \neq 0$ and the ϕ_{ki} are linearly independent.

As an application of Theorem 9 we obtain a theorem on the linear equation of order n.

THEOREM 10. *Let $a_0(t), a_1(t), \ldots, a_n(t), Q(t)$ be defined and continuous for $A < t < B$, and let $a_0(t) \neq 0$ for $A < t < B$. Let t_0 be a number between A and B. Then the linear differential equation*

$$a_0(t) \frac{d^n x}{dt^n} + \cdots + a_n(t)x = Q(t) \qquad (12\text{–}33)$$

has a unique complete solution $x = \phi(t)$ *that satisfies given initial conditions*

$$\phi(t_0) = x_0, \quad \phi'(t_0) = x_0', \quad \ldots, \quad \phi^{(n-1)}(t_0) = x_0^{(n-1)}.$$
$$(12\text{--}34)$$

Every complete solution is defined for $A < t < B$.

When $Q(t) \equiv 0$, *there exist* n *linearly independent solutions of Eq. (12–33) for* $A < t < B$. *If* $\phi_1(t), \ldots, \phi_n(t)$ *is a set of* n *linearly independent solutions for* $A < t < B$, *then the general solution for* $A < t < B$ *is*

$$x = \phi(t) = \sum_{k=1}^{n} c_k \phi_k(t), \quad (12\text{--}35)$$

where $c_1, \ldots, c_n$ *are arbitrary constants.*

When $Q(t) \not\equiv 0$, *the general solution for* $A < t < B$ *is*

$$x = \phi(t) = \phi^*(t) + \sum_{k=1}^{n} c_k \phi_k(t), \quad (12\text{--}36)$$

where $x = \phi^*(t)$ *is one solution and* $\Sigma c_k \phi_k(t)$ *is the complementary function; that is, the general solution of the related homogeneous equation.*

Proof. The theorem follows from Theorem 9 if we set $x = x_1$,

$$\frac{dx_1}{dt} = x_2, \quad \frac{dx_2}{dt} = x_3, \quad \ldots, \quad \frac{dx_{n-1}}{dt} = x_n,$$
$$\frac{dx_n}{dt} = \frac{1}{a_0(t)} [Q(t) - a_1(t)x_n - \cdots - a_n(t)x_1]. \quad (12\text{--}37)$$

Thus, as in Section 12–7, the equation of order n is replaced by an equivalent system of first order equations. Here the system is linear and we can apply Theorem 9. Hence there is a unique complete solution $x_i = \phi_i(t)$ with given initial values at $t = t_0$; if we require

$$\phi_1(t) = x_0, \quad \phi_2(t_0) = x_0', \quad \ldots, \quad \phi_n(t_0) = x_0^{(n-1)},$$

then by Eqs. (12–37) the function $x = x_1 = \phi_1(t)$ satisfies Eqs. (12–34) and is a solution of Eq. (12–33). By Theorem 9 the solution exists for $A < t < B$.

When $Q(t) \equiv 0$, the system (12–37) is homogeneous. Let $\phi_{ki}(t)$ be, as above, a set of n linearly independent solutions for $A < t < B$. Then the n functions $\phi_{11}(t), \phi_{21}(t), \ldots, \phi_{n1}(t)$ satisfy (12–33), with $Q(t) \equiv 0$, and are linearly independent for $A < t < B$. For let a relation

$$c_1\phi_{11}(t) + \cdots + c_n\phi_{n1}(t) \equiv 0 \quad (12\text{--}38)$$

hold for $A < t < B$, with $c_1, \ldots, c_n$ constant and not all 0. We differentiate this relation $(n - 1)$ times. Since, for fixed k, ϕ_{ki} $(i = 1, \ldots, n)$, is a solution of (12–37), we have

$$\frac{d\phi_{k1}}{dt} = \phi_{k2}, \qquad \frac{d\phi_{k2}}{dt} = \phi_{k3}, \qquad \cdots, \qquad \frac{d\phi_{k,n-1}}{dt} = \phi_{kn}. \quad (12\text{–}39)$$

Hence the repeated differentiation of (12–38) yields the identities

$$
\begin{aligned}
c_1\phi_{11}(t) + \cdots + c_n\phi_{n1}(t) &\equiv 0, \\
c_1\phi_{12}(t) + \cdots + c_n\phi_{n2}(t) &\equiv 0, \\
&\ \vdots \\
c_1\phi_{1n}(t) + \cdots + c_n\phi_{nn}(t) &\equiv 0.
\end{aligned}
\qquad (12\text{–}40)
$$

Accordingly, the n solutions $\phi_{ki}(t)$ would be linearly *dependent* for $A < t < B$, contrary to assumption. Therefore a relation (12–38) cannot hold and the functions

$$\phi_1(t) = \phi_{11}(t), \qquad \cdots, \qquad \phi_n(t) = \phi_{n1}(t) \qquad (12\text{–}41)$$

form n linearly independent solutions of (12–33), with $Q(t) \equiv 0$. Conversely, if the n-functions (12–41) are linearly independent solutions of (12–33), with $Q(t) \equiv 0$, then Eqs. (12–39), (12–41) define functions $\phi_{ki}(t)$ for $k = 1, \ldots, n$ which are solutions of (12–37) and are linearly independent for $A < t < B$. For if they were linearly dependent, then relations (12–40) would hold; the first of these would imply the linear dependence of $\phi_1(t), \ldots, \phi_n(t)$. The general solution of (12–37), with $Q \equiv 0$, is now given by

$$x_i = \sum_{k=1}^{n} C_k\phi_{ki}(t) \qquad (i = 1, \ldots, n),$$

and we conclude that the general solution of (12–33), with $Q \equiv 0$, is given by

$$x = x_1 = \sum_{k=1}^{n} C_k\phi_{k1}(t) = \sum_{k=1}^{n} C_k\phi_k(t).$$

The discussion of the nonhomogeneous equation (12–33) is the same as for the system. Accordingly, Theorem 10 is established.

12–9 Complex linear equations. The results of Section 12–8 remain valid if we permit the dependent variables $x, x_1, \ldots, x_n$ to be complex functions of t, $(A < t < B)$, and permit the functions $a_{ij}(t)$, $F_i(t)$, $a_i(t)$,

$Q(t)$ to be complex functions of t. As in Section 4–8, each complex function is of form

$$F(t) = f(t) + ig(t) \qquad (i = \sqrt{-1}),$$

where f and g are real. Addition, subtraction, and multiplication of such functions follow the rules for such operations on complex numbers. Differentiation is defined by the equation

$$F'(t) = f'(t) + ig'(t).$$

Accordingly, complex solutions of linear differential equations with complex coefficients can be defined just as for real equations.

To formulate the counterpart of Theorem 9 for the complex case, we rewrite the differential equations (12–21) as follows:

$$\frac{dz_j}{dt} = \sum_{h=1}^{n} A_{jh}(t)z_h + F_j(t) \qquad (j = 1, \ldots, n). \tag{12–42}$$

The z's are complex variables and we write

$$z_1 = x_1 + ix_{n+1}, \qquad z_2 = x_2 + ix_{n+2}, \qquad \ldots, \qquad z_n = x_n + ix_{2n}; \tag{12–43}$$

thus the real and imaginary parts of the z's form $2n$ real variables $x_1, \ldots, x_{2n}$. Similarly, we write

$$A_{jh}(t) = a_{jh}(t) + ib_{jh}(t), \tag{12–44}$$

$$F_j(t) = f_j(t) + ig_j(t). \tag{12–45}$$

We assume the functions $A_{jh}(t)$, $F_j(t)$ to be continuous for $A < t < B$; that is, that the real functions $a_{jh}(t)$, $b_{jh}(t)$, $f_j(t)$, $g_j(t)$ are continuous in this interval. *Equations (12–42) are now equivalent to a system of $2n$ real differential equations for the variables $x_1, \ldots, x_{2n}$.* For (12–42) can be written

$$\frac{dz_j}{dt} = \frac{dx_j}{dt} + i\frac{dx_{n+j}}{dt} = \sum_{h=1}^{n} (a_{jh} + ib_{jh})(x_h + ix_{n+h}) + f_j + ig_j.$$

The equality holds if and only if real and imaginary parts are equal:

$$\frac{dx_j}{dt} = \sum_{h=1}^{n} [a_{jh}(t)x_h - b_{jh}(t)x_{n+h}] + f_j(t),$$

$$\frac{dx_{n+j}}{dt} = \sum_{h=1}^{n} [b_{jh}(t)x_h + a_{jh}(t)x_{n+h}] + g_j(t), \tag{12–46}$$

where $j = 1, \ldots, n$. This is the equivalent real system of order $2n$.

We now apply Theorem 9 to the system (12–46) and deduce the desired information concerning the complex system (12–42). For example, Eqs. (12–42) have a unique solution with given initial values (*complex*) $z_1^0, \ldots, z_n^0$ at t_0, $(A < t_0 < B)$. This follows from existence of a solution of (12–46) with given initial values $x_1^0, \ldots, x_{2n}^0$, where

$$z_1^0 = x_1^0 + ix_{n+1}^0, \qquad \ldots, \qquad z_n^0 = x_n^0 + ix_{2n}^0.$$

If the $F_j(t)$ are all 0, we can obtain n solutions of Eqs. (12–42):

$$z_j = \Phi_{kj}(t) = \phi_{kj}(t) + i\phi_{k,n+j}(t), \qquad (k = 1, \ldots, n)$$

which are linearly independent with respect to *complex* coefficients. For example, we can choose a *fundamental set*, defined by initial conditions $\Phi_{kj}(t_0) = \delta_{kj}$, where δ_{kj} is the Kronecker delta (Section 12–8). The proof of linear independence, with complex coefficients, is the same as in Section 12–8. In particular, the solution with initial values $z_1^0, \ldots, z_n^0$ is given by

$$z_j = \sum_{k=1}^{n} C_k \Phi_{kj}(t) \qquad (j = 1, \ldots, n),$$

where $C_1 = z_1^0, \ldots, C_n = z_n^0$.

A similar discussion can be carried out for the complex equation of order n

$$A_0(t) \frac{d^n z}{dt^n} + \cdots + A_n(t)z = F(t). \tag{12–47}$$

As in Section 12–8, this can be replaced by a system of first order equations.

PROBLEMS

1. Prove: if ϕ_{ki}, $(k = 1, \ldots, n; \; i = 1, \ldots, n)$, is a set of n solutions of Eqs. (12–21), with $F_i(t) \equiv 0$ for $i = 1, \ldots, n$, and $c_1\phi_{1i}(t_0) + \cdots + c_n\phi_{ni}(t_0) = 0$, $(i = 1, \ldots, n)$, for some constants $c_1, \ldots, c_n$ not all 0, then $c_1\phi_{1i}(t) + \cdots + c_n\phi_{ni}(t) \equiv 0$, so that the solutions are linearly dependent. [*Hint:* Apply the uniqueness theorem, Theorem 7.]

2. (a) Consider a homogeneous system (12–21), with $n = 3$. Let ϕ_{ki}, $(k = 1, 2, 3)$, be three solutions, so that $D\phi_{ki} = a_{i1}\phi_{k1} + a_{i2}\phi_{k2} + a_{i3}\phi_{k3}$ for $i = 1, 2, 3$, $D = d/dt$. Let W be the corresponding Wronskian determinant

$$W = \begin{vmatrix} \phi_{11} & \phi_{21} & \phi_{31} \\ \phi_{12} & \phi_{22} & \phi_{32} \\ \phi_{13} & \phi_{23} & \phi_{33} \end{vmatrix}$$

By the rule for differentiation of products, it follows that DW is the sum of three determinants, obtained by differentiating the three rows of W in turn:

$$DW = \begin{vmatrix} D\phi_{11} & D\phi_{21} & D\phi_{31} \\ \phi_{12} & \phi_{22} & \phi_{32} \\ \phi_{13} & \phi_{23} & \phi_{33} \end{vmatrix} + \cdots$$

Show that

$$DW = (a_{11} + a_{22} + a_{33})W,$$

and hence that

$$W = W(t_0) \exp \int_{t_0}^{t} (a_{11} + a_{22} + a_{33})\, dt.$$

(b) From the result of part (a) prove that if $W(t)$ is 0 at one point of the interval $A < t < B$, then $W \equiv 0$.

(c) Extend the results of parts (a) and (b) to a system of order n.

(d) Prove the result of part (b) by considering the linear independence of the three solutions (see Prob. 1).

3. The Wronskian determinant for n solutions $\phi_1(t), \ldots, \phi_n(t)$ of an equation of order n

$$(a_0 D^n + \cdots + a_{n-1}D + a_n)x = 0$$

is defined (Section 4–3) as the determinant

$$W = \begin{vmatrix} \phi_1(t) & \cdots \phi_n(t) \\ \phi_1'(t) & \phi_n'(t) \\ \vdots & \vdots \\ \phi_1^{(n-1)}(t) & \cdots \phi_n^{(n-1)}(t) \end{vmatrix}$$

Replace the equation of order n by a system of n first order equations, as in Section 12–8, and show that W can be interpreted as the Wronskian determinant of n solutions of the system.

4. Given the differential equation

$$(D^3 - 5D^2 + 8D - 4)x = 0,$$

obtain an equivalent system of first order equations, as in Section 12–8, and obtain the general solution of the system.

5. Let $x_i = \phi_i(t; t_0, x_1^0, \ldots, x_n^0)$ be the general solution of a linear system (12–21), $(A < t < B)$. Show that the ϕ_i are linear functions of $x_1^0, \ldots, x_n^0$:

$$\phi_i(t; t_0, x_1^0, \ldots, x_n^0) = x_1^0 \psi_{1i} + x_2^0 \psi_{2i} + \cdots + x_n^0 \psi_{ni} + \psi_{n+1,i}$$

where the ψ's are independent of the initial values $x_1^0, \ldots, x_n^0$.

6. Obtain a fundamental set of solutions, with $t_0 = 0$, of the complex system

$$Dz_1 = 2z_1 + iz_2, \qquad Dz_2 = -2z_1 + 2z_2.$$

7. (a) Prove: if $F(t) = f(t) + ig(t)$, where f and g have continuous derivatives for $A < t < B$, then

$$\frac{d}{dt} e^{F(t)} = e^{F(t)} F'(t).$$

(b) Obtain the general solution of the complex linear equation of first order

$$\frac{dz}{dt} + p(t)z = q(t),$$

where $p(t)$, $q(t)$ are continuous complex functions for $A < t < B$.

(c) Let $p(t) = p_1(t) + ip_2(t)$, $q(t) = q_1(t) + iq_2(t)$, $z = x + iy$ in the equation of part (a) and obtain differential equations for x and y.

(d) Obtain the general solution of the system

$$Dx + 2tx - 2ty = 3t, \qquad Dy + 2tx + 2ty = t.$$

[*Hint:* Use the result of part (b) to replace the system by one complex equation as in part (a).]

8. Show that the proof of Theorem 1 (Section 12–3) can be applied to the complex equation of Prob. 7(b), so that a solution can be obtained by successive approximations:

$$z = \lim \phi_n(t), \quad \phi_0(t) = z_0, \quad \phi_1(t) = z_0 + \int_{t_0}^{t} [q(u) - p(u)\phi_0(u)] \, du, \quad \dots$$

9. Let $z_j = \Phi_{kj}(t)$ be n complex solutions of Eqs. (12–42), with $F_j(t) \equiv 0$, $(k = 1, \dots, n, j = 1, \dots, n)$. Let $\phi_{kj} = \text{Re} [\Phi_{kj}]$, $\phi_{k,n+j} = \text{Im} [\Phi_{kj}]$, so that $x_j = \phi_{kj}(t)$, $x_{n+j} = \phi_{k,n+j}(t)$ defines n real solutions of (12–46). Show that linear independence of the complex solutions with respect to complex coefficients does not imply, but is implied by linear independence of the real solutions.

ANSWERS

4. $x_1 = x = c_1 e^t + e^{2t}(c_2 + c_3 t)$, $x_2 = Dx = c_1 e^t + e^{2t}(c_3 + 2c_2 + 2c_3 t)$, $x_3 = D^2 x = c_1 e^t + e^{2t}(4c_3 + 4c_2 + 4c_3 t)$.

6. $z_1 = C_1 e^{(1+i)t} + C_2 e^{(3-i)t}$, $z_2 = C_1(1 + i)e^{(1+i)t} + C_2(-1 - i)e^{(3-i)t}$.

7. (b) $z = v^{-1}(\int vq(t) \, dt + C)$, $v = \exp \int p \, dt$; (c) $(D + p_1)x - p_2 y = q_1$, $(D + p_1)y + p_2 x = q_2$; (d) $x = 1 + e^{-t^2}(c_1 \cos t^2 + c_2 \sin t^2)$, $y = -\frac{1}{2} + e^{-t^2}(c_2 \cos t^2 - c_1 \sin t^2)$.

12–10 Dependence of solutions on initial conditions and on parameters.

For a general system

$$\frac{dy_i}{dx} = F_i(x, y_1, \ldots, y_n) \qquad (i = 1, \ldots, n) \qquad (12\text{–}48)$$

we have obtained (Section 12–6) a complete solution passing through an arbitrary initial point $(x_0, y_1^0, \ldots, y_n^0)$. This solution we denote by

$$y_i = \phi_i(x; x_0, y_1^0, \ldots, y_n^0) \qquad (i = 1, \ldots, n). \qquad (12\text{–}49)$$

For the applications it is important to know that if the initial values $x_0, y_1^0, \ldots, y_n^0$ are changed slightly, then the solution (12–49) is modified only slightly. In mathematical terms, we would like to be sure that the functions $\phi_i(x; x_0, y_1^0, \ldots, y_n^0)$ depend continuously on $x_0, y_1^0, \ldots, y_n^0$.

A similar question arises concerning the effect on the solutions (12–49) of slightly changing the differential equations themselves. For example, we can allow the F_i to depend on a parameter μ; we consider the equations

$$\frac{dy_i}{dx} = F_i(x, y_1, \ldots, y_n, \mu) \qquad (i = 1, \ldots, n). \qquad (12\text{–}50)$$

For each fixed μ, Eqs. (12–50) are differential equations for the y_i. If we consider only fixed initial conditions, the solution for each μ is a set of functions

$$y_i = \phi_i(x, \mu). \qquad (12\text{–}41)$$

The question arises whether the ϕ_i depend continuously on μ.

In this section we prove the continuous dependence of solutions on initial conditions and parameters. For the sake of simplicity, detailed proof is given only for the single equation of first order.

THEOREM 11. *Let $F(x, y)$ be defined in a domain D of the xy-plane. Let $F(x, y)$ and $\partial F/\partial y$ be continuous in D. Let $\phi(x; x_0, y_0)$ be the complete solution of the differential equation*

$$\frac{dy}{dx} = F(x, y) \qquad (12\text{–}52)$$

passing through the point (x_0, y_0) of D. Then ϕ and $\partial\phi/\partial x$ are continuous functions of all three variables.

Proof. We wish to show that $\phi(x; x_0, y_0)$ is continuous wherever it is defined. Let ϕ be defined for $x = \bar{x}$, $x_0 = \bar{x}_0$, $y_0 = \bar{y}_0$, where $\bar{x}_0 \leqq \bar{x}$; let $\bar{y} = \phi(\bar{x}; \bar{x}_0, \bar{y}_0)$. Then there is a solution curve defined for $\bar{x}_0 \leqq x \leqq \bar{x}$ and passing through $(\bar{x}_0, \bar{y}_0)$ (Fig. 12–5). Since $(\bar{x}_0, \bar{y}_0)$ and $(\bar{x}, \bar{y})$ are in D, we can continue the solution to a slightly larger interval, $\bar{x}_0 - \alpha \leqq x \leqq$

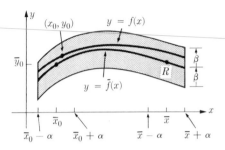

Fig. 12–5. Continuous dependence of solution on initial conditions.

$\bar{x} + \alpha$, where $\alpha > 0$. We denote the solution thus obtained by $\bar{f}(x)$. We next choose β so small and positive that the strip region

$$R: \quad \bar{x}_0 - \alpha \leqq x \leqq \bar{x} + \alpha, \quad \bar{f}(x) - \beta \leqq y \leqq \bar{f}(x) + \beta \quad (12\text{--}53)$$

is within D. (If β could not be so chosen, we would have boundary points of D arbitrarily close to the curve $y = \bar{f}(x)$; a limit point of such boundary points would then lie on the curve $y = \bar{f}(x)$ and would be a boundary point of D; this is impossible since the curve $y = \bar{f}(x)$ lies inside D.) Since $\partial F/\partial y$ is continuous in R, we can choose a constant K so that

$$\left| \frac{\partial F}{\partial y} \right| \leqq K \text{ in } R. \quad (12\text{--}54)$$

We now consider a solution

$$f(x) = \phi(x; x_0, y_0), \quad (12\text{--}55)$$

where $|x_0 - \bar{x}_0| < \delta_1$, $|y_0 - \bar{y}_0| < \delta_2$. The value of δ_2 will be specified below. Once δ_2 has been chosen, δ_1 will be chosen less than α and so that $|\bar{y}_0 - \bar{f}(x_0)| < \delta_2$ for $|x_0 - \bar{x}_0| < \delta_1$; such a choice is possible since $\bar{f}(x)$ is continuous at $\bar{x}_0$. We assert that for proper choice of δ_1, δ_2 the solution $f(x)$ is defined at least for $\bar{x}_0 - \alpha \leqq x \leqq \bar{x} + \alpha$ and satisfies the inequality $|f(x) - \bar{f}(x)| < \beta$. Indeed, if we write

$$v(x) = \int_{x_0}^{x} |f(x) - \bar{f}(x)| \, dx \quad (x \geqq x_0), \quad (12\text{--}56)$$

then we have, as in Section 12–5,

$$f(x) = \int_{x_0}^{x} F[x, f(x)] \, dx + y_0,$$

$$\bar{f}(x) = \int_{x_0}^{x} F[x, \bar{f}(x)] \, dx + \bar{f}(x_0),$$

$$|f(x) - \bar{f}(x)| = \left| \int_{x_0}^{x} \{F[x, f(x)] - F[x, \bar{f}(x)]\} \, dx + y_0 - \bar{f}(x_0) \right|$$

$$\leq \int_{x_0}^{x} |F[x, f(x)] - F[x, \bar{f}(x)]| \, dx + |y_0 - \bar{y}_0| + |\bar{y}_0 - \bar{f}(x_0)|$$

$$\leq K \int_{x_0}^{x} |f(x) - \bar{f}(x)| \, dx + 2\delta_2,$$

so long as the curve $y = f(x)$ remains in R. Hence

$$v'(x) \leq Kv(x) + 2\delta_2,$$

$$v(x) \leq 2\delta_2 \frac{e^{K(x-x_0)} - 1}{K},$$

$$v'(x) = |f(x) - \bar{f}(x)| \leq 2\delta_2 e^{K(x-x_0)}$$

by the Lemma of Section 12–4. Hence if

$$0 < \delta_2 < \tfrac{1}{2}\beta e^{-K(\bar{x}-\bar{x}_0+2\alpha)}, \tag{12–57}$$

we have

$$|f(x) - \bar{f}(x)| < \beta e^{-K[\bar{x}+\alpha-x+x_0-(\bar{x}_0-\alpha)]} \leq \beta$$

for $x_0 \leq x \leq \bar{x} + \alpha$. A similar argument applies for $x \leq x_0$ and we conclude that $|f(x) - \bar{f}(x)| < \beta$ for $\bar{x}_0 - \alpha \leq x \leq \bar{x} + \alpha$. Therefore the solution $y = f(x)$ cannot approach the boundary of R within the interval $\bar{x}_0 - \alpha \leq x \leq \bar{x} + \alpha$. Hence by Theorem 2 the solution is defined throughout the whole interval and remains in R.

Now let ϵ be given, $(\epsilon > 0)$. We repeat the proof of the preceding paragraph, replacing β by $\epsilon/2$ and conclude that δ_1 and δ_2 can be chosen so that $\delta_1 > 0$, $\delta_2 > 0$ and that for $|x_0 - \bar{x}_0| < \delta_1$, $|y_0 - \bar{y}_0| < \delta_2$ we have not only $|f(x) - \bar{f}(x)| < \beta$, but also

$$|f(x) - \bar{f}(x)| < \frac{\epsilon}{2}, \qquad \bar{x}_0 - \alpha \leq x \leq \bar{x} + \alpha.$$

Since $\bar{f}(x)$ is continuous at $\bar{x}$, we can choose $\delta_3 > 0$ so that for $|x - \bar{x}| < \delta_3$ we have

$$|\bar{f}(x) - \bar{f}(\bar{x})| < \frac{\epsilon}{2}.$$

Accordingly, for $|x_0 - \bar{x}_0| < \delta_1$, $|y_0 - \bar{y}_0| < \delta_2$, $|x - \bar{x}| < \delta_3$,

$$|\phi(x; x_0, y_0) - \phi(\bar{x}; \bar{x}_0, \bar{y}_0)| \equiv |f(x) - \bar{f}(\bar{x})|$$

$$\leq |f(x) - \bar{f}(x)| + |\bar{f}(x) - \bar{f}(\bar{x})|$$

$$< \frac{\epsilon}{2} + \frac{\epsilon}{2} = \epsilon.$$

Therefore $\phi(x; x_0, y_0)$ is continuous in all three variables. Since

$$\frac{\partial}{\partial x} \phi(x; x_0, y_0) = F[x, \phi(x; x_0, y_0)], \tag{12–58}$$

we conclude that $\partial\phi/\partial x$ is also continuous in all three variables.

If we differentiate Eq. (12–58) formally with respect to y_0, we find

$$\frac{\partial}{\partial x} \frac{\partial\phi}{\partial y_0} = F_y \frac{\partial\phi}{\partial y_0}. \tag{12–59}$$

For fixed (x_0, y_0) this is a linear differential equation for $\partial\phi/\partial y_0$, from which we conclude that

$$\frac{\partial\phi}{\partial y_0} = c e^{\int_{x_0}^{x} F_y \, dx},$$

where c is a constant. Since $\phi(x; x_0, y_0) \equiv y_0$ when $x = x_0$, we conclude that $\partial\phi/\partial y_0 = 1$ when $x = x_0$. Accordingly, $c = 1$ and

$$\frac{\partial\phi}{\partial y_0} = e^{\int_{x_0}^{x} F_y \, dx}, \tag{12–60}$$

where $F_y = F_y[x; \phi(x; x_0, y_0)]$. From Eq. (12–60) we deduce the inequality

$$\frac{\partial\phi}{\partial y_0} > 0. \tag{12–61}$$

Accordingly, *for fixed x_0 the value of ϕ at each x increases as y_0 increases.* As shown in Fig. 12–6, the conclusion has a simple geometric meaning. If ϕ were first to increase and then to decrease, two solutions would have to cross, in contradiction to the uniqueness property. The manipulations leading to (12–60) and (12–61) were purely formal, but they can be justified (pp. 25–27 of Reference 3).

Systems of higher order. Theorem 11 can be generalized and proved in the same manner for a system of order n

$$\frac{dy_i}{dx} = F_i(x, y_1, \ldots, y_n) \quad (i = 1, \ldots, n). \tag{12–62}$$

The complete solution

$$y_i = \phi_i(x; x_0, y_1^0, \ldots, y_n^0)$$
$$(i = 1, \ldots, n) \tag{12–63}$$

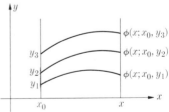

FIGURE 12–6

is formed of functions ϕ_i which depend continuously on all $n + 2$ variables, and the derivatives $\partial\phi_i/\partial x$ are continuous in all $n + 2$ variables. Substitution in Eq. (12–62) and differentiation leads to the system of n^2 linear differential equations for the functions $\partial\phi_i/\partial y_j^0$:

$$\frac{\partial}{\partial x}\left(\frac{\partial \phi_i}{\partial y_j^0}\right) = \sum_{k=1}^{n} \frac{\partial F_i}{\partial y_k}\frac{\partial \phi_k}{\partial y_j^0} \qquad (i, j = 1, \ldots, n). \qquad (12\text{–}64)$$

Furthermore, from Eqs. (12–64) we deduce (Prob. 3 below) that the Jacobian determinant

$$J = \begin{vmatrix} \dfrac{\partial \phi_1}{\partial y_1^0} & \cdots & \dfrac{\partial \phi_1}{\partial y_n^0} \\ \vdots & & \vdots \\ \dfrac{\partial \phi_n}{\partial y_1^0} & \cdots & \dfrac{\partial \phi_n}{\partial y_n^0} \end{vmatrix} = \frac{\partial(\phi_1, \ldots, \phi_n)}{\partial(y_1^0, \ldots, y_n^0)} \qquad (12\text{–}65)$$

satisfies the linear differential equation

$$\frac{\partial J}{\partial x} = J \sum_{i=1}^{n} \frac{\partial F_i}{\partial y_i} \qquad (12\text{–}66)$$

along each solution (12–63). Furthermore, $\phi_i(x_0; x_0, y_1^0, \ldots, y_n^0) \equiv y_i^0$, so that

$$\frac{\partial \phi_i}{\partial y_j^0} = \delta_{ij} \text{ for } x = x_0, \qquad (12\text{–}67)$$

where δ_{ij} is the Kronecker delta (Section 12–8). Therefore, $J = 1$ for $x = x_0$. Accordingly, from (12–66), $J > 0$ for all values of the variables (cf. the proof of relation 12–61). This inequality has a simple physical interpretation in terms of the "flow" in n-dimensional space described by the differential equations. (See pp. 299–300 of Reference 6.)

Systems depending on a parameter. The analysis of systems depending on a parameter can be reduced to the question of dependence on initial conditions.

THEOREM 12. *Let the functions $F_i(x, y_1, \ldots, y_n, \mu)$, $(i = 1, \ldots, n)$, be defined in a domain D of the space of the $n + 2$ variables $x, y_1, \ldots, y_n, \mu$. Let the F_i and their first partial derivatives with respect to $y_1, \ldots, y_n, \mu$ be continuous in D. Then for each fixed μ and fixed $x_0, y_1^0, \ldots, y_n^0$ such that $(x_0, y_1^0, \ldots, y_n^0, \mu)$ is in D, the differential equations*

$$\frac{dy_i}{dx} = F_i(x, y_1, \ldots, y_n, \mu) \qquad (i = 1, \ldots, n) \qquad (12\text{–}68)$$

have a unique complete solution

$$y_i = \phi_i(x; x_0, y_1^0, \ldots, y_n^0, \mu) \tag{12–69}$$

passing through $(x_0, y_1^0, \ldots, y_n^0)$. *The functions* ϕ_i *and* $\partial\phi_i/\partial x$ *are continuous in all* $n + 3$ *variables* $x, x_0, y_1^0, \ldots, y_n^0, \mu$.

Proof. We write $y_{n+1} = \mu$ and consider the system

$$\frac{dy_i}{dx} = F_i(x, y_1, \ldots, y_n, y_{n+1}) \qquad (i = 1, \ldots, n),$$

$$\frac{dy_{n+1}}{dx} = 0. \tag{12–70}$$

The existence and uniqueness theorems (Theorems 6, 7) are applicable to this system and we conclude that there is a unique complete solution

$$y_i = \phi_i(x; x_0, y_1^0, \ldots, y_n^0, y_{n+1}^0) \tag{12–71}$$

through each point $(x_0, y_1^0, \ldots, y_n^0, y_{n+1}^0)$ in D. By Theorem 11 (as generalized to the system of order n) the functions ϕ_i and $\partial\phi_i/\partial x$ are continuous in all $n + 3$ variables. Since $dy_{n+1}/dt = 0$, $y_{n+1} = \text{const} = \mu$ on each solution. Hence the functions (12–71) are precisely the solutions (12–69) sought and have the properties asserted.

Since μ can be interpreted as an initial value y_{n+1}^0, the equation analogous to Eq. (12–64) is valid:

$$\frac{\partial}{\partial x}\left(\frac{\partial\phi_i}{\partial\mu}\right) = \sum_{k=1}^{n} \frac{\partial F_i}{\partial y_k}\frac{\partial\phi_k}{\partial\mu} + \frac{\partial F_i}{\partial\mu}, \tag{12–72}$$

where $\partial F_i/\partial y_k$ and $\partial F_i/\partial\mu$ are evaluated along a fixed solution (12–69). This is a system of linear equations for $\partial\phi_i/\partial\mu$.

The results described can be extended at once to equations depending on several parameters:

$$\frac{dy_i}{dx} = F_i(x, y_1, \ldots, y_n, \mu_1, \ldots, \mu_k) \qquad (i = 1, \ldots, n).$$

PROBLEMS

1. For each of the following differential equations, containing a parameter μ, find the solution through $(0, 0)$, discuss its continuity, and find the linear differential equation satisfied by $v = \partial y/\partial\mu$ along the solution:

(a) $y' + \mu y = 1$,   (b) $y' = (1 + y)\sqrt{1 - \mu^2 x^2}$.

2. For each of the following differential equations find the solution $\phi(x; x_0, y_0)$, find $\partial\phi/\partial y_0$, and verify that $\partial\phi/\partial y_0 = 1$ when $x = x_0$, $\partial\phi/\partial y_0 > 0$ in general:

(a) $y' - 3y = e^x$,

(b) $y' = 3x^2 e^y$.

3. Prove the relation (12–66). [*Hint:* Compute $\partial J/\partial x$ as a sum of determinants, as for Prob. 2 following Section 12–9. Use the relations (12–64) and simplify.]

4. Prove: for linear equations $y'_j = \sum a_{ij}(x)y_j$ the Jacobian J of Eq. (12–65) reduces to the Wronskian determinant of a fundamental set of solutions.

ANSWERS

1. (a) $y = (1 - e^{-\mu x})/\mu$, $\mu \neq 0$, $y = x$ for $\mu = 0$; y continuous for all x, μ; $(dv/dx) + \mu v + \mu^{-1}(1 - e^{-\mu x}) = 0$, $\mu \neq 0$, $(dv/dx) + x = 0$ for $\mu = 0$.
(b) $y = -1 + \exp \frac{1}{2}[x(1 - \mu^2 x^2)^{1/2} + \mu^{-1}\sin^{-1}(x\mu)]$, $\mu \neq 0$; $y = e^x - 1$ for $\mu = 0$; continuous for $|x\mu| < 1$, $(dv/dx) - v(1 - \mu^2 x^2)^{1/2} = -\mu x^2(1 - \mu^2 x^2)^{-1/2} \exp \frac{1}{2}[x(1 - \mu^2 x^2)^{1/2} + \mu^{-1}\sin^{-1}(x\mu)]$, $\mu \neq 0$; $dv/dx - v = 0$ for $\mu = 0$.

2. (a) $\phi = -\frac{1}{2}e^x + (y_0 + \frac{1}{2}e^{x_0})e^{3(x-x_0)}$, $\partial\phi/\partial y_0 = e^{-3(x-x_0)}$;
(b) $\phi = y_0 - \log[1 + e^{y_0}(x_0^3 - x^3)]$, $x < (x_0^3 + e^{-y_0})^{1/3}$, $\partial\phi/\partial y_0 = [1 + e^{y_0}(x_0^3 - x^3)]^{-1}$.

12–11 Existence of integrals. Integrals of differential equations are discussed in Chapter 7. Here we state the basic definitions again. Let a system

$$\frac{dy_i}{dx} = F_i(x, y_1, \ldots, y_n) \qquad (i = 1, \ldots, n) \qquad (12–73)$$

be given in domain D as in Theorem 6 (Section 12–6). A function $G(x, y_1, \ldots, y_n)$ is an integral of the system (12–73) if G remains constant on each solution. We consider only functions G which are continuous and have continuous first partial derivatives in D, and to rule out trivial integrals, we require also that G not be identically constant in any smaller domain contained in D.

The fact that G is an integral is expressed by the relation

$$\frac{\partial G}{\partial x} + F_1\frac{\partial G}{\partial y_1} + \cdots + F_n\frac{\partial G}{\partial y_n} \equiv 0. \qquad (12–74)$$

For the left-hand side of Eq. (12–74) is simply dG/dx along a solution of (12–73); since G is constant, $dG/dx \equiv 0$.

When $G(x, y_1, \ldots, y_n)$ is an integral of (12–73) we can use the equation

$$G(x, y_1, \ldots, y_n) = c_1 = \text{const}$$

to eliminate one of the y's from (12–73) and hence lower the order by 1. In general, we seek n integrals $G_i(x, y_1, \ldots, y_n)$ and use the equations

$$G_1(x, y_1, \ldots, y_n) = c_1, \qquad \ldots, \qquad G_n(x, y_1, \ldots, y_n) = c_n \qquad (12\text{–}75)$$

to solve for $y_1, \ldots, y_n$ in terms of x and n arbitrary constants $c_1, \ldots, c_n$. The equations thus describe the solution curves in implicit form. In order to be sure that Eqs. (12–75) define actual curves, it is sufficient that the G_i be *functionally independent;* that is, that the determinant

$$\frac{\partial(G_1, \ldots, G_n)}{\partial(y_1, \ldots, y_n)} = \begin{vmatrix} \dfrac{\partial G_1}{\partial y_1} & \cdots & \dfrac{\partial G_1}{\partial y_n} \\ \vdots & & \vdots \\ \dfrac{\partial G_n}{\partial y_1} & \cdots & \dfrac{\partial G_n}{\partial y_n} \end{vmatrix} \qquad (12\text{–}76)$$

be different from 0 in D. (See pp. 132–136 of Reference 6.)

We can now prove existence of such functionally independent integrals, at least in a restricted portion of D. For the single equation

$$y' = F(x, y) \qquad (12\text{–}77)$$

the single desired integral is obtained as follows. Let (a, b) be a point of D and let L be the largest interval of the line $x = a$ in D and containing the point (a, b) (Fig. 12–7). Let $\phi(x; x_0, y_0)$ be the complete solution of Eq. (12–77) through a point (x_0, y_0) of D. We now assume that all solution curves meet L or, what is the same thing, we restrict attention to that part of D which is swept out by the solutions that meet L. To each such solution curve we can assign a number, namely the value y^* of y at which the curve meets L. To each point (x, y) on the solution curve we assign the same value y^*. Hence y^* is defined throughout D and is constant on each solution curve. We can compute y^* as a function of (x, y)

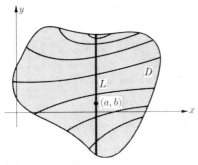

Fig. 12–7. Existence of an integral of $y' = F(x, y)$.

as follows. We consider (x, y) as *initial point* and seek the value at $x = a$ of the solution through (x, y); this is clearly equal to y^*. But it is also equal to $\phi(a; x, y)$. Hence

$$y^* = \phi(a; x, y) = G(x, y) \qquad (12\text{--}78)$$

is the integral sought. We already know that (if F and $\partial F/\partial y$ are continuous) ϕ has continuous partial derivatives. The inequality (12–64) now tells us that $\partial G/\partial y > 0$. (Thus the determinant (12–76), for $n = 1$, is different from 0.)

The same reasoning carries over to systems of arbitrary order. We state the general result.

THEOREM 13. *Let the differential equations*

$$\frac{dy_i}{dx} = F_i(x, y_1, \ldots, y_n) \qquad (i = 1, \ldots, n) \qquad (12\text{--}79)$$

be given, where the F_i and $\partial F_i/\partial y_j$ are continuous in a domain D. Let

$$y_i = \phi_i(x; x_0, y_1^0, \ldots, y_n^0) \qquad (12\text{--}80)$$

be the complete solution of (12–79) through the point $(x_0, y_1^0, \ldots, y_n^0)$ of D. Let L be the portion of the domain D in the hyperplane $x = a$, for some fixed a. Let L have the property that every solution curve (12–80) meets L. Then Eqs. (12–79) have n functionally independent integrals $G_i(x, y_1, \ldots, y_n)$, namely, the functions

$$G_i = \phi_i(a; x, y_1, \ldots, y_n) \qquad (i = 1, \ldots, n), \qquad (12\text{--}81)$$

and the equations

$$G_i(x, y_1, \ldots, y_n) = c_i = \text{const} \qquad (i = 1, \ldots, n) \qquad (12\text{--}82)$$

define the solutions (12–80) implicitly.

Here each G_i gives the value of y_i at the point at which the solution crosses L; it is clear that each G_i is constant on each solution curve and that Eqs. (12–82) define the solutions implicitly. The G_i have the required continuity and differentiability, as pointed out in Section 12–10. Finally, we note that the determinant (12–76) is simply the Jacobian J of Section 12–10 (Eq. 12–71) evaluated at $(x, y_1, \ldots, y_n)$. As shown at the end of Section 12–10, $J > 0$ everywhere. Hence the G_i have all desired properties.

Remark. Theorem 13 provides an existence theorem for the *partial differential equation of first order* (12–74). (See Chapter 3 of Reference 2.)

EXAMPLE 1. For the equation $y' + 2y = 2x - 1$ all solutions meet the line $L\colon x = 0, -\infty < y < \infty$. The function $\phi(x; x_0, y_0)$ is

$$e^{-2(x-x_0)}(y_0 + 1 - x_0) + x - 1.$$

In accordance with (12–78) the integral G is

$$G = \phi(0; x, y) = e^{2x}(y + 1 - x) - 1,$$

and the solutions are given by the curves $G = c = \text{const}, -\infty < c < \infty$, where c is the value of y for $x = 0$.

EXAMPLE 2. The system

$$\frac{dx}{dt} = 4x - 3y, \qquad \frac{dy}{dt} = 5x - 4y$$

(Example 2 of Section 6–2) has the general solution

$$x = c_1 e^t + 3c_2 e^{-t}, \qquad y = c_1 e^t + 5c_2 e^{-t}.$$

The solution through (x_0, y_0) at $t = t_0$ is found to be

$$x = \tfrac{1}{2}[(5x_0 - 3y_0)e^{t-t_0} + 3(y_0 - x_0)e^{-(t-t_0)}] = \phi_1(t; t_0, x_0, y_0),$$

$$y = \tfrac{1}{2}[(5x_0 - 3y_0)e^{t-t_0} + 5(y_0 - x_0)e^{-(t-t_0)}] = \phi_2(t; t_0, x_0, y_0).$$

Hence two functionally independent integrals are given by

$$G_1 = \phi_1(0; t, x, y) = \tfrac{1}{2}[(5x - 3y)e^{-t} + 3(y - x)e^t],$$

$$G_2 = \phi_2(0; t, x, y) = \tfrac{1}{2}[(5x - 3y)e^{-t} + 5(y - x)e^t].$$

PROBLEMS

1. Let the following differential equation be given:

$$y' = -\cot x + 2xe^{-y}\csc x \qquad (0 < x < \pi).$$

(a) Show that $u = e^y \sin x - x^2$ is an integral.

(b) Find the complete solution $\phi(x; x_0, y_0)$ and obtain an integral $G = \phi(\tfrac{1}{2}\pi; x, y)$.

(c) Show that the integrals of parts (a) and (b) are functionally dependent, that is, that G is a function of u.

2. Let the following differential equation be given:

$$y' = -xy^3 \qquad \text{all } (x, y).$$

(a) Find the complete solution $\phi(x; x_0, y_0)$ and the integral $\phi(0; x, y)$.

(b) Show that the domain filled by the curves $\phi(0; x, y) = \text{const}$ is not the whole xy-plane.

3. Given the system

$$\frac{dx}{dt} = 2x + 2y - 1 - 4t, \qquad \frac{dy}{dt} = -x - y + 2 + 2t,$$

find two functionally independent integrals.

4. *Integrals in the phase plane.* For a system

$$\frac{dx}{dt} = f(x, y), \qquad \frac{dy}{dt} = g(x, y)$$

there may exist a *time-invariant* integral, that is, an integral $G(x, y)$ independent of t. Show that such an integral is given by the function G stated for each of the following systems:

(a) $dx/dt = y, dy/dt = -x, G = x^2 + y^2$;

(b) $dx/dt = x, dy/dt = -y, G = xy$;

(c) $\dfrac{dx}{dt} = x, \dfrac{dy}{dt} = y, G = \dfrac{y}{\sqrt{x^2 + y^2}}$ except at $(0, 0)$.

5. Show that the phase-plane equations of the van der Pol equation (Section 11–14)

$$\frac{dx}{dt} = y, \qquad \frac{dy}{dt} = \mu(1 - x^2)y - x$$

cannot have a time-invariant integral valid in a domain including the limit cycle. [*Hint:* Show that since the solutions spiral towards the limit cycle, a function $G(x, y)$, constant on each solution and continuous, must be identically constant.]

6. (a) Show that each time-invariant integral u of the system of Prob. 4 satisfies the partial differential equation

$$f(x, y) \frac{\partial u}{\partial x} + g(x, y) \frac{\partial u}{\partial y} = 0.$$

(b) Find a solution u (not identically 0) of the partial differential equation

$$y \frac{\partial u}{\partial x} - x \frac{\partial u}{\partial y} = 0.$$

(See Prob. 4a.)

ANSWERS

1. (b) $\phi = \log [(x^2 - x_0^2 + e^{y_0} \sin x_0) \csc x]$,
    $G = \log [(\pi^2/4) - x^2 + e^y \sin x] = \log [u + (\pi^2/4)]$.

2. (a) $y = y_0[1 + y_0^2(x^2 - x_0^2)]^{-1/2}, \phi(0; x, y) = y(1 - x^2y^2)^{-1/2}$;
    (b) solutions fill domain $|xy| < 1$.

3. $G_1 = 2 + 3t - x - 2y + 2e^{-t}(-1 - 2t + x + y)$,
    $G_2 = -1 - 3t + x + 2y - e^{-t}(-1 - 2t + x + y)$.

6. (b) $u = x^2 + y^2$.

12–12 Convergence of power series solutions. In Sections 9–4 through 9–6 several theorems concerning the convergence of power series solutions of differential equations are stated. We now provide the proofs of these theorems. It should be remarked that this topic can be discussed much more easily with the aid of analytic functions of a *complex variable*. We shall, however, use real variables as far as possible. The rules for manipulating real power series will be assumed known (Chapter 6 of Reference 6).

We consider first the differential equation

$$y' = F(x, y) \qquad (12\text{--}83)$$

where, as in Section 9–4, $F(x, y)$ is assumed to be analytic at $x = 0$, $y = 0$; that is, F has a series representation

$$F(x, y) = c_{0,0} + (c_{1,0}x + c_{0,1}y) + \cdots \qquad (12\text{--}84)$$

valid for $|x| \leqq a$, $|y| \leqq b$, $a > 0$, $b > 0$. It is further assumed that the series

$$|c_{0,0}| + |c_{1,0}|a + |c_{0,1}|b + \cdots$$
$$+ |c_{i,j}|a^i b^j + \cdots \qquad (12\text{--}85)$$

converges and has a sum at most equal to M. We now wish to show that the formal Taylor series solution of (12–83) through $(0, 0)$ converges in some interval $|x| < \rho$, $\rho > 0$. Once convergence has been established, it will not be difficult to show that the function $f(x)$ represented by the series is an actual solution.

We first recall the procedure by which the series solution is constructed. The value of y at $x = 0$, namely, 0 is given. The value of y' at $x = 0$ is $F(0, 0)$. From (12–83) repeated differentiation gives the higher derivatives

$$y'' = F_x + F_y y', \qquad y''' = F_{xx} + 2F_{xy}y' + F_{yy}y'^2 + F_y y'', \qquad \cdots \qquad (12\text{--}86)$$

It is clear that only positive coefficients appear in these formulas. Hence if F and all successive derivatives are $\geqq 0$ at $(0, 0)$, all derivatives of y will be $\geqq 0$ at $x = 0$. Substitution of $x = 0$, $y = 0$, $y' = F(0, 0)$ in (12–86) gives each higher derivative in turn at $x = 0$. The final series is

$$y_0'x + y_0''\frac{x^2}{2!} + \cdots + y_0^{(n)}\frac{x^n}{n!} + \cdots \qquad (12\text{--}87)$$

Now from the fact that the series (12–85) has sum at most M, we conclude that each term is less than or equal to M:

$$|c_{i,j}|a^i b^j \leqq M;$$

that is,

$$|c_{i,j}| \leqq \frac{M}{a^i b^j}. \tag{12-88}$$

Now the series

$$\sum \frac{M x^i y^j}{a^i b^j} = M\left[1 + \left(\frac{x}{a} + \frac{y}{b}\right) + \left(\frac{x^2}{a^2} + \frac{xy}{ab} + \frac{y^2}{b^2}\right) + \cdots\right] \tag{12-89}$$

converges for $|x| < a$, $|y| < b$. Indeed this series is simply the product

$$M \frac{1}{1 - (x/a)} \cdot \frac{1}{1 - (y/b)} = M(1 + \frac{x}{a} + \cdots + \frac{x^i}{a^i} + \cdots)(1 + \frac{y}{b} + \cdots). \tag{12-90}$$

Since the two geometric series converge absolutely for $|x| < a$, $|y| < b$, we can multiply and arrange terms in any order desired. Let us denote by $G(x, y)$ the function represented by (12–89) or (12–90), so that

$$G(x, y) = \frac{M}{[1 - (x/a)][1 - (y/b)]}. \tag{12-91}$$

Then from (12–89) we see that

$$G(0, 0) = M, \qquad G_x(0, 0) = \frac{M}{a},$$
$$G_y(0, 0) = \frac{M}{b}, \qquad G_{xx}(0, 0) = \frac{2M}{a^2}, \qquad \cdots \tag{12-92}$$

Similarly, from (12–84)

$$F(0, 0) = c_{0,0}, \qquad F_x(0, 0) = c_{1,0}, \qquad F_y(0, 0) = c_{0,1}, \qquad \cdots \tag{12-93}$$

From (12–86), (12–92), (12–93) we now conclude that the derivatives of $F(x, y)$ at $(0, 0)$ are, in absolute value, less than or equal to those of $G(x, y)$ at $(0, 0)$:

$$\left|\frac{\partial^{i+j} F}{\partial x^i \partial y^j}\right|_{(0,0)} \leqq \left.\frac{\partial^{i+j} G}{\partial x^i \partial y^j}\right|_{(0,0)}. \tag{12-94}$$

Accordingly, if we form the power series solution of the differential equation

$$y' = G(x, y) \tag{12-95}$$

through $(0, 0)$, we will obtain *positive* coefficients y_0', y_0'', ... all in absolute value greater than or equal to those obtained for Eq. (12–83). Hence if the series solution for (12–95) converges absolutely in some interval, the comparison test shows that the series solution for (12–83) converges absolutely in the same interval.

The equation (12–95) can be solved explicitly, for it is the equation

$$y' = \frac{M}{[1 - (x/a)][1 - (y/b)]} .$$

Separation of variables and application of the initial condition lead to the solution

$$y = g(x) = b - b\left[1 + \frac{2Ma}{b} \log\left(1 - \frac{x}{a}\right)\right]^{1/2} \qquad (|x| < a). \qquad (12\text{–}96)$$

Since $\log[1 - (x/a)]$ can be represented by a power series for $|x| < a$ and $(1 + u)^{1/2}$ can be represented by a power series in u, for $|u| < 1$, we conclude from the rule for substitution of one power series in another (pp. 361–362 of Reference 6) that $g(x)$ can be represented by a power series for $|x|$ sufficiently small. A more detailed analysis (Prob. 6 below, following Section 12–14) shows that the series for $g(x)$ converges for

$$\frac{2Ma}{b}\left|\log\left(1 - \frac{|x|}{a}\right)\right| < 1. \qquad (12\text{–}97)$$

This inequality is satisfied (Prob. 1 below) when

$$|x| < \rho, \qquad \rho = a(1 - e^{-1/c}), \qquad c = \frac{2Ma}{b} . \qquad (12\text{–}98)$$

Hence we conclude that *the power series solution of Eq. (12–83) converges for $|x| < \rho$.*

Now let $f(x)$ be the function represented by the series solution (12–87). Since the coefficients in the series for $f(x)$ are in absolute value less than the (positive) coefficients in the series for $g(x)$, we conclude that $|f(x)| \leq g(x)$ for $|x| < \rho$. From Eq. (12–96) we see that $|g(x)| \leq b$ for $|x| < \rho$. Therefore $|f(x)| < b$ for $|x| < \rho$. From (12–98) we deduce that $\rho < a$. Therefore, when $|x| < \rho$, we can be sure that $|x| < a$ and $|y| = |f(x)| < b$. Accordingly, $F[x, f(x)]$ is well defined for $|x| < \rho$ and can be represented by the series (12–84). By substitution of the series for $f(x)$, $F[x, f(x)] = H(x)$ becomes a power series in x, converging for $|x| < \rho$. Now

$$H(0) = F[0, f(0)] = F(0, 0),$$

$$H'(x) = F_x + F_y f'(x), \qquad H'(0) = F_x(0, 0) + F_y(0, 0)f'(0).$$

On the other hand,

$$f'(0) = F(0, 0),$$

$$f''(0) = F_x(0, 0) + F_y(0, 0)f'(0),$$

by the way in which the coefficients y_0', y_0'', ... are defined. Hence $f'(0) = H(0)$, $f''(0) = H'(0)$ and a similar reasoning shows that $f^{(n)}(0) = H^{(n-1)}(0)$ for $n = 2, 3, \ldots$ Since $f'(x)$ and $H(x)$ have equal derivatives of all orders at $x = 0$ and both are represented by power series for $|x| < \rho$, we conclude that $f'(x) \equiv H(x)$, $(|x| < \rho)$; that is

$$f'(x) \equiv F[x, f(x)].$$

Accordingly, $f(x)$ is a solution of the differential equation.

Systems of equations. The reasoning given above extends easily to systems of equations

$$y_i' = F_i(x, y_1, \ldots, y_N) \qquad (i = 1, \ldots, N) \qquad (12\text{--}99)$$

as discussed in Section 9–5. It is assumed that the F_i can be expanded in series of powers of x, y_1, ..., y_N for $|x| < a$, $|y_i| < b$, $(i = 1, \ldots, N)$, $a > 0$, $b > 0$, that these series converge when $x = a$, $y_i = b$ and each coefficient is replaced by its absolute value, and that the N sums thus obtained are all less than a fixed number M. Under these hypotheses we reason that the coefficients in the power series for each F_i are in absolute value at most equal to the corresponding coefficients in the power series for

$$G(x, y_1, \ldots, y_N) = \frac{M}{[1 - (x/a)][1 - (y_1/b)] \ldots [1 - (y_N/b)]} \qquad (12\text{--}100)$$

about $(0, 0, \ldots, 0)$. The solution of the system

$$\frac{dy_i}{dx} = G(x, y_1, \ldots, y_N) \qquad (i = 1, \ldots, N) \qquad (12\text{--}101)$$

through $(0, 0, \ldots, 0)$ can be found explicitly. By symmetry, we see that $y_1 = y_2 = \cdots = y_N$ for this solution. Hence, in particular, y_1 satisfies the equation

$$\frac{dy_1}{dx} = \frac{M}{[1 - (x/a)][1 - (y_1/b)]^N}. \qquad (12\text{--}102)$$

Upon separating variables and integrating, we find (for the given initial conditions)

$$y_1 = g(x) = b\left\{1 - \left[1 + \frac{M(N+1)a}{b}\log\left(1 - \frac{x}{a}\right)\right]^{1/(N+1)}\right\} \qquad (12\text{--}103)$$

and $y_2 = y_3 = \cdots = y_N = g(x)$. The power series for $g(x)$ about $x = 0$ has positive coefficients and converges for

$$|x| < \rho, \qquad \rho = a(1 - e^{-1/c}), \qquad c = \frac{M(N+1)a}{b}. \qquad (12\text{--}104)$$

The method of series solution of (12–99) leads to series for $y_1, \ldots, y_N$ whose coefficients are in absolute value at most equal to the corresponding coefficients in the series for $g(x)$. Accordingly, by the comparison test we conclude that the series for $y_1, \ldots, y_N$ converge absolutely for $|x| < \rho$ to functions $f_1(x), \ldots, f_N(x)$. We finally verify that these functions do determine a solution of the given system (12–99).

12–13 Estimates for remainder. In Sections 9–4 and 9–5 estimates are given for the remainder, after n terms, of the series solutions. In this section we explain how the estimates are reached. We consider only the case of the system (12–99); the single equation (12–83) is the special case $N = 1$. Since each term of the series $f_i(x)$ is in absolute value less than or equal to the corresponding term of the series for $g(|x|)$, the remainder R_n for each $f_i(x)$ is at most equal to the corresponding remainder for $g(|x|)$. Accordingly, we could expand $g(x)$, as defined by (12-103), in a power series about $x = 0$ and for a fixed value of n find precisely the difference between $g(|x|)$ and the sum of the first n terms of the series. This leads to a complicated expression, unless n is very small. For example, we find

$$g(x) = Mx + \frac{Mb + M^2 N}{ab} \frac{x^2}{2!} + \cdots$$

Hence

$$|R_1| \leqq g(|x|) - M|x|,$$

$$|R_2| \leqq g(|x|) - M|x| - \frac{Mb + M^2 N}{ab} \frac{x^2}{2!},$$

where $g(|x|)$ is to be evaluated by (12–103).

To obtain simpler formulas, we have recourse to the theory of analytic functions of a complex variable (Chapter 9 of Reference 6). We replace x by z in (12–103) and consider the function

$$g(z) = b\left\{1 - \left[1 + c \log\left(1 - \frac{z}{a}\right)\right]^{1/(N+1)}\right\}, \qquad (12\text{–}105)$$

where c is given by (12–104). The function $g(z)$ is *multiple-valued*, because of the $(N + 1)$-st root. However, if we restrict attention to the circular region $|z| < \rho$ in the complex plane, then a single-valued analytic branch of $g(z)$ can be defined there, reducing to 0 when $z = 0$. Equation (12–105) will be understood to refer only to this branch.

Since $g(z)$ is analytic for $|z| < \rho$, it can be expanded in a power series about $z = 0$:

$$g(z) = c_1 z + c_2 z^2 + \cdots + c_n z^n + \cdots$$

We wish to estimate the remainder after n terms:

$$R_n^*(z) = g(z) - [c_1 z + \cdots + c_n z^n].$$

It is known (see p. 542 of Reference 6), that

$$R_n^*(z) = \frac{z^{n+1}}{2\pi i} \int_{C_r} \frac{g(\zeta)}{\zeta^{n+1}(\zeta - z)}\, d\zeta,$$

where C_r is a circle: $|\zeta| = r$, $r < \rho$, and $|z| < r$. If we denote by $m(r)$ the maximum of $|g(z)|$ on C_r, then

$$|R_n^*(z)| \leqq \frac{|z|^{n+1}}{2\pi} \cdot \frac{m(r)}{r^{n+1}} \cdot \frac{1}{r - |z|} \cdot 2\pi r,$$

since $|\zeta - z| \geqq r - |z|$ for ζ on C_r. Accordingly,

$$|R_n^*(z)| \leqq m(r)\, \frac{|z|^{n+1}}{r^n(r - |z|)}.$$

Now it can be shown (Probs. 6, 7 below) that

$$m(r) \leqq \phi(r) = b\left[2 + \left\{\left|c\right| \log\left(1 - \frac{r}{a}\right)\right\}^{1/(N+1)}\right]. \qquad (12\text{--}106)$$

Therefore

$$|R_n^*(z)| \leqq \phi(r)\, \frac{|z|^{n+1}}{r^n(r - |z|)}.$$

If we replace z by x, we obtain an estimate valid when z reduces to the real number x. For each function $f_i(x)$ we know that the remainder $R_n(x)$ is, in absolute value, at most equal to that for $g(|x|)$; hence

$$|R_n(x)| \leqq \phi(r)\, \frac{|x|^{n+1}}{r^n(r - |x|)} \qquad (|x| < r < \rho), \qquad (12\text{--}107)$$

which is the formula given in Section 9–5 (Theorem 2).

12–14 Series solutions of linear differential equations. In Section 9–6 a convergence theorem (Theorem 4) is formulated for linear differential equations. The essential content of the theorem is that for linear differential equations, convergence of series solutions is guaranteed as far as the coefficients themselves permit. This conclusion can be reached in the same manner as that of the preceding sections.

We consider first a homogeneous system

$$\frac{dy_i}{dx} = \sum_{j=1}^{N} a_{ij}(x)y_j \qquad (i = 1, \ldots, N). \tag{12–108}$$

We assume the coefficients $a_{ij}(x)$ can be represented by convergent power series for $|x| < r$, $r > 0$. Hence the series are absolutely convergent for $|x| \leq a$ for each $a < r$. For each such a (> 0), we can then, as in Section 12–12, choose M so that each term of each series is in absolute value less than or equal to the corresponding term of the series

$$M\left(1 + \frac{|x|}{a} + \frac{|x|^2}{a^2} + \cdots\right) = \frac{M}{1 - (|x|/a)}. \tag{12–109}$$

We now consider a solution of (12–108) for which $y_1 = 1$, $y_2 = 0$, $\ldots$, $y_N = 0$ for $x = 0$. (The solution with all initial values 0 is, of course, identically 0.) The power series method leads us to series for $y_0(x)$, $\ldots$, $y_N(x)$, which are (by the reasoning of Section 12–12) termwise less than or equal, in absolute value, to the terms of the series solution of the differential equations

$$\frac{dy_i}{dx} = \frac{M}{1 - (x/a)} (y_1 + \cdots + y_N) \qquad (i = 1, \ldots, N) \tag{12–110}$$

with the same initial values. The solution of Eqs. (12–110) is found explicitly (Prob. 3 below):

$$y_1 = 1 + \frac{1}{N}\left[\left(1 - \frac{x}{a}\right)^{-aMN} - 1\right],$$

$$y_2 = y_3 = \cdots = y_N = \frac{1}{N}\left[\left(1 - \frac{x}{a}\right)^{-aMN} - 1\right]. \tag{12–111}$$

These functions can be represented by their power series in powers of x for $|x| < a$. Hence the series solution of (12–108) also converges for $|x| < a$ and satisfies (12–108) in this interval. Since a is an arbitrary number less than r, we conclude that the series solution is valid for $|x| < r$. The remainder, after n terms, can also be estimated by considering the remainder, after n terms, of the series for (12–111):

$$y_1 = 1 + \frac{1}{N}\left[MNx + \frac{(MN)(MNa + 1)}{a}\frac{x^2}{2!} + \cdots\right],$$

$$y_2 = \frac{1}{N}\left[MNx + \frac{(MN)(MNa + 1)}{a}\frac{x^2}{2!} + \cdots\right], \cdots$$

These remainders can be analyzed as in Section 12–13 or as in pp. 326–332 of Reference 6.

We have thus far considered only a solution with special initial values. To extend the result to arbitrary initial values $y_1^0, \ldots, y_N^0$, we form a fundamental set of solutions, as in Section 12–8:

$$y_i = \phi_{ki}(x) \qquad (k = 1, \ldots, N, i = 1, \ldots, N).$$

For each k the solution is required to have initial values

$$\phi_{ki}(0) = \delta_{ki}.$$

The particular solution considered above is given by $\phi_{1i}(x)$, $(i = 1, \ldots, n)$. The same analysis applies to all $\phi_{ki}(x)$ and we conclude that all $\phi_{ki}(x)$ are represented by power series in x for $|x| < r$. The solution with initial values $y_1^0, \ldots, y_N^0$ is given by

$$y_i = \sum_{k=1}^{N} y_k^0 \, \phi_{ki}(x) \qquad (i = 1, \ldots, N). \tag{12–112}$$

Hence every solution in powers of x is valid for $|x| < r$ and is represented by convergent series in this interval. The remainders can be estimated by combining the results for each ϕ_{ki} separately. Accordingly, each term in each series is less than or equal, in absolute value, to the corresponding term of the power series for

$$g_i(x) = (|y_1^0| + \cdots + |y_N^0|) \frac{1}{N} \left[\left(1 - \frac{x}{a} \right)^{-MNa} - 1 \right] + |y_i^0|. \tag{12–113}$$

A *nonhomogeneous system* can be discussed in similar fashion. Let the equations be

$$\frac{dy_i}{dx} = \sum_{j=1}^{N} a_{ij}(x)y_j + Q_i(x) \qquad (i = 1, \ldots, N). \tag{12–114}$$

We assume that the $a_{ij}(x)$ satisfy the same conditions as before and that the $Q_i(x)$ satisfy the same conditions as the $a_{ij}(x)$. The general solution of (12–114) for $|x| < r$ is given by

$$y_i = \sum_{k=1}^{N} y_k^0 \, \phi_{ki}(x) + y_i^*(x) \qquad (i = 1, \ldots, N), \tag{12–115}$$

where the functions $\Sigma y_k^0 \phi_{ki}(x)$ are, as in (12–112), the general solution of the related homogeneous system and $y_i = y_i^*(x)$, $(i = 1, \ldots, N)$, is a particular solution of (12–114). We choose $y_i^*(x)$ as the solution with all initial values 0 at $x = 0$. Then (12–115) gives the solution of (12–114) with arbitrary initial values $y_1^0, \ldots, y_N^0$ at $x = 0$.

The convergence properties of the functions (12–112) have already been considered. For the particular solution y_i^* we reason as above. We choose a so that $0 < a < r$ and then choose M so that the series for all $a_{ij}(x)$ and for the $Q_i(x)$ are termwise in absolute value at most equal to the corresponding terms of the series for $M[1 - (x/a)]^{-1}$ in powers of x. A corresponding comparison can then be made between the series for $y_i(x)$ and the series for the solution of the differential equations

$$\frac{dy_i}{dx} = \frac{M}{1 - (x/a)}(y_1 + \cdots + y_N + 1) \qquad (i = 1, \ldots, N) \qquad (12\text{–}116)$$

with initial values 0 at $x = 0$. This solution is given explicitly (Prob. 4 below) by

$$y_i = \frac{1}{N}\left[\left(1 - \frac{x}{a}\right)^{-MNa} - 1\right] \qquad (i = 1, \ldots, N). \qquad (12\text{–}117)$$

Again we conclude that the solution series for $y_i(x)$ converge for $|x| < a$. Since a is arbitrary, $a < r$, we conclude that the series converge for $|x| < r$ and represent the solution in this interval. The series for the function (12–117) can be used to estimate the error after n terms.

A much deeper study of the power series solution of linear equations can be made with the aid of analytic functions of a complex variable. For discussion of this topic see Chapters 3–5 of Reference 3.

PROBLEMS

1. Prove that for $a > 0$, $b > 0$, $M > 0$ the inequality (12–97) is satisfied precisely when (12–98) holds.

2. Separate variables to solve Eq. (12–102) and obtain the particular solution (12–103).

3. Obtain the general solution of Eqs. (12–110) and show that the solution with initial values $1, 0, \ldots, 0$ at $x = 0$ is given by (12–111). [*Hint:* Let $u = y_1 + \cdots + y_N$ and use $u, y_2, \ldots, y_N$ as dependent variables.]

4. Obtain the general solution of Eqs. (12–116) and show that the particular solution with all initial values 0 at $x = 0$ is given by (12–117). [*Hint:* The general solution of the related homogeneous system is found in Prob. 3. Verify that $y_i = -1/N$ defines a particular solution of (12–116).]

5. Let $p(x)$ and $q(x)$ be represented by convergent power series for all x:

$$p(x) = \sum_{n=0}^{\infty} b_n x^n, \qquad q(x) = \sum_{n=0}^{\infty} c_n x^n.$$

For fixed $a > 0$ let

$$\sum_{n=0}^{\infty} |b_n| a^n < M, \qquad \sum_{n=0}^{\infty} |c_n| a^n < M.$$

Show that the series solution of the differential equation with initial condition

$$y' = p(x)y^2 + q(x), \qquad y(0) = 0$$

converges for

$$|x| < a(1 - e^{-\pi/(2aM)}).$$

[*Hint:* Compare with the differential equation

$$y' = \frac{M}{1 - (x/a)} (1 + y^2).]$$

6. If $Z = f(z)$ is an analytic function of the complex variable z for $|z| < \rho$ and $w = \phi(Z)$ is an analytic function of Z for $|Z| < r$, and $|f(z)| < r$ for $|z| < \rho$, then $\phi[f(z)] = \psi(z)$ is analytic for $|z| < \rho$ and can be represented by a power series for $|z| < \rho$ (p. 513 of Reference 6). Hence the function

$$\psi(z) = \left[1 + c \log \left(1 - \frac{z}{a} \right) \right]^{1/2} \qquad [\psi(0) = 1, \log 1 = 0]$$

is analytic for $|z| < \rho$, where ρ is chosen so that

$$\left| c \log \left(1 - \frac{z}{a} \right) \right| < 1, \qquad (|z| < \rho).$$

Show from the power series for $\log [1 - (z/a)]$ that

$$\left| \log \left(1 - \frac{z}{a} \right) \right| \leq \left| \log \left(1 - \frac{|z|}{a} \right) \right| \qquad (|z| < a) \tag{*}$$

and hence that $\psi(z)$ is analytic for

$$|c| \left| \log \left(1 - \frac{|z|}{a} \right) \right| < 1.$$

7. (a) Prove: for any two complex numbers z_1, z_2 and for $N = 0, 1, 2, \ldots$,

$$|z_1 + z_2|^{1/(N+1)} \leq |z_1|^{1/(N+1)} + |z_2|^{1/(N+1)}.$$

[*Hint:* Consider the $(N + 1)$-st power of both sides.]

(b) Apply the inequality (*) of Prob. 6 and the result of Prob. 7 (a) to prove that for $c > 0$, $a > 0$, $|z| < a$, $N = 0, 1, 2, \ldots$,

$$\left| 1 - \left[1 + c \log \left(1 - \frac{z}{a} \right) \right]^{1/(N+1)} \right| \leq 2 + \left\{ c \left| \log \left(1 - \frac{|z|}{a} \right) \right| \right\}^{1/(N+1)}.$$

ANSWERS

3. $y_1 = (c_1/N)[1 - (x/a)]^{-NMa} - c_2 - \cdots - c_N,$
   $y_j = (c_1/N)[1 - (x/a)]^{-NMa} + c_j \text{ for } j = 2, \ldots, N.$

4. $y_1 = (c_1/N)[1 - (x/a)]^{-NMa} - c_2 - \cdots - c_N - 1/N,$
   $y_j = (c_1/N)[1 - (x/a)]^{-NMa} + c_j - 1/N \text{ for } j = 2, \ldots, N.$

SUGGESTED REFERENCES

1. BIRKHOFF, GEORGE D., *Dynamical Systems*. American Mathematical Society Colloquium Publications, Vol. 9. New York: American Mathematical Society, 1927.

2. CARATHÉODORY, CONSTANTIN, *Variationsrechnung und Partielle Differentialgleichungen Erster Ordnung*. Leipzig: Teubner, 1925.

3. CODDINGTON, EARL A., and LEVINSON, NORMAN, *Theory of Ordinary Differential Equations*. New York: McGraw-Hill, 1955.

4. INCE, E. L., *Ordinary Differential Equations*. New York: Dover, 1956.

5. KAMKE, E., *Differentialgleichungen reeller Funktionen*. Leipzig: Akademische Verlagsgesellschaft, 1933.

6. KAPLAN, WILFRED, *Advanced Calculus*. Reading, Mass.: Addison-Wesley, 1952.

7. PICARD, EMILE, *Traité d'Analyse* (3 vols.), 3rd ed. Paris: Gauthier-Villars, 1922, 1925, 1928.

INDEX

INDEX